T0189791

Lecture Notes in Computer Science 973

Edited by G. Goos, J. Hartmanis and J. van Leeuwen

Advisory Board: W. Brauer D. Gries J. Stoer

Springer

Berlin
Heidelberg
New York
Barcelona
Budapest
Hong Kong
London
Milan
Paris
Tokyo

Heimo H. Adelsberger Jiří Lažanský
Vladimír Mařík (Eds.)

Information Management in Computer Integrated Manufacturing

A Comprehensive Guide to
State-of-the-Art CIM Solutions

Springer

Series Editors

Gerhard Goos, Karlsruhe University, Germany

Juris Hartmanis, Cornell University, NY, USA

Jan van Leeuwen, Utrecht University, The Netherlands

Volume Editors

Heimo H. Adelsberger
Universität GH Essen
Universitätsstraße 2, D-45177 Essen, Germany

Jiří Lažanský
Faculty of Electrical Engineering, Czech Technical University
Technická 2, CZ-166 27 Prague 6, Czech Republic

Vladimír Mařík
Faculty of Electrical Engineering, Czech Technical University
Technická 2, CZ-166 27 Prague 6, Czech Republic and
Allen-Bradley Research Center Prague, Ltd.
Americká 22, CZ-120 00 Prague 2, Czech Republic

Cataloging-in-Publication data applied for

Die Deutsche Bibliothek - CIP-Einheitsaufnahme

Information management in computer integrated manufacturing
: a comprehensive guide to state of the art CIM solutions /
Heimo H. Adelsberger ... (ed.). - Berlin ; Heidelberg ; New
York ; Barcelona ; Budapest ; Hong Kong ; London ; Milan ;
Paris ; Tokyo : Springer, 1995
  (Lecture notes in computer science ; Vol. 973)
  ISBN 3-540-60286-0
NE: Adelsberger, Heimo H. [Hrsg.]; GT

CR Subject Classification (1991): J.6, H.2, I.2.1, I.2.5, I.2.8-9, J.1

ISBN 3-540-60286-0 Springer-Verlag Berlin Heidelberg New York

Typesetting: Camera-ready by author
SPIN 10485498   06/3142 – 5 4 3 2 1 0   Printed on acid-free paper

Preface

This book can be considered as one of the main results of the Joint European Project (JEP) No. 2609 **Education in CIM**, sponsored by the Commission of the European Communities (Task Force Human Resources, Education, Training and Youth) in the Trans-European Mobility Scheme for University Studies (TEMPUS). The first part of this project took three years and was completed in 1994. In January 1995 - after evaluation of the project - the EC authorities decided to support it for another two years with the stress on dissemination of results achieved.

The project was contracted by the Danish Technical University, Lyngby Copenhagen (project coordinator: Prof. Dr. Heimo H. Adelsberger), and the following institutions took part: IFW TU Berlin (D), Universität Gesamthochschule Essen (D), FAW J. Kepler University Linz (A), Czech Technical University Prague (CZ), West-Bohemian University Pilsen (CZ), and SPEL Kolín Ltd. (CZ).

The goal of this project was to design the curriculum and to establish the regular education in the specialized area of Information Management in Computer Integrated Manufacturing. The curriculum is aimed at students of control engineering and applied computer science in their last semesters of study.

This book covers the majority of subjects and courses that have been developed and 'verified in real life' within the frame of the project. Some parts of the book (and of the education) are partially devoted to managerial perspectives of the problems. The reason is that these views of computer supported manufacturing activities had been missing in the traditional technical education. This way, the topics covered in this book are highly interdisciplinary. That is why some contributions issue from results of other, tightly cooperating projects, namely from the TEMPUS JEP 1191 "Teaching AI" and TEMPUS JEP 0886 "Higher Education in Control Engineering".

The results of JEP 2609 have brought a quite new view on CIM. This view is concentrated on such questions as reliable and efficient software design, heterogeneous databases, operation research aspects, and artificial intelligence approaches. Comparatively low attention is given to such problems as control theory and automation, although these subjects form an indispensable part of the education process. This is due to the fact that these topics are covered by separate publications and are traditional from a certain point of view.

The contributions included in the book are diverse in nature. Some contributions have been written in standard textbook style and include many introductory concepts without assuming mich background knowledge from the readers. Other articles appearing in the book stress the newest scientific achievements in particular areas. The latter fashion has been preferred especially in the cases of well established, "traditional" areas covered by good, already existing textbooks. In all cases, special attention has been given to the style of writing: readability and comprehensibility.

That is why the book can serve as a good teaching aid for students and young scientists engaged in the CIM area, treating the subject from different perspectives.

The book's contents are logically organized into an introduction and seven parts, namely:

- Databases
- Planning and scheduling
- Modelling and control
- Information technologies in CIM
- Computer supported enterprise activities
- Artificial intelligence
- Examples and case studies.

This splitting reflects the decisive role of particular disciplines of applied informatics in CIM.

The project participants have been invited to organize a specialized workshop **Information Management in Manufacturing** as an organic part of the 6th International Conference DEXA'95 held in London, September 4-8, 1995. We hope this book will serve as introductory background material for discussion within the workshop.

Essen, Prague, July 1995

Heimo Adelsberger
Jiří Lažanský
Vladimír Mařík

Table of Contents

INFORMATION TECHNOLOGIES IN CIM

COMPUTER SUPPORTED ENTERPRISE ACTIVITIES

ARTIFICIAL INTELLIGENCE

EXAMPLES & CASE STUDIES

Introductory Overview

Günter Spur, Karl Mertins

with contributions by: R. Albrecht, F. Duttenhofer, H. Edeler, R. Jochem,
H. Kopp-Jung, U. Wegener, M. Rabe, and B. Schallock

IPK Berlin, Germany

Abstract: This introductory contribution gives an extended survey of the concept of Computer Integrated Manufacturing (CIM) and its contents. It concentrates mainly on the information aspects of CIM and stresses the relationships and connections between the technological and managerial activities in the manufacturing enterprise. Analytical and modelling points of view are considered with the aim of methodological support of CIM systems design. The article contains also many examples of existing CIM (sub-)systems.

1 Approach for Systematic Planning and Factory Management

Computer Integrated Manufacturing (CIM) was developed in the 1980's when it became obvious that the handling of information was hindered by a lack of compatibility. An integration of different applications in design (CAD), process planning (CAPP), production (CAM), and order related business processes like production planning and control (PPC) had to be realized. Apparently, a smooth data exchange could only contribute to an improvement in business results to a limited extent.

Rethinking factory operations and the organizational structure is the most important task in order to meet the requirements of the market. These requirements had originally been fulfilled by low prices; nowadays however, the market requires customer orientation and fast delivery of high quality products. Therefore, the focus of attention, which had previously been on a high utilization rate, shifted to fast reaction and therefore to efficient information handling.

This book's chapters on planning, scheduling, enterprise activities, and case studies, and the articles on decision support represent the importance of the business process aspect of CIM design. The technological aspects of databases, communication, systems engineering, tools and new algorithms represent the other necessary view on CIM.

The following introductory chapters cover the essence of the CIM's aims, approaches and tools for systematic planning and factory management. They reveal that enterprise modeling eases the design of effective information systems, that order controls are in need of new integrated concepts, and that developing human resources requires appropriate organizational structures as well as technology and participative design methods.

Therefore, technology, i.e. databases, networks and application software, follows the organizational structure, but also represents a prerequisite in order to realize flat, selfcontrolling structures.

A method integration is considered as the most urgent task in manufacturing research in these days.

During the process of restructuring an enterprise, different technical, organizational, and social views on the same object "factory" will be employed. It helps if these views were all based on a common enterprise model.

2 Enterprise Modeling

2.1 Object-Oriented Modeling and Analysis of Business Processes

Enterprises face increasing competition. They are confronted with shorter product life cycles combined with customer needs requiring additional product differentiation. They must accommodate increasing levels of technology; they must cope with the shortening of order throughput times and of product development times; they face shorter delivery times while trying to increase product quality. In order to be able to react to these growing demands enterprises must attain a higher flexibility in all areas of activity.

A higher transparency of the processes and the company's organization is therefore required. Flexible manufacturing presupposes a clear structure of the organization, of information and material flows within the enterprise. For the consideration of the complex relationships, models or modeling methods have to be applied to support, to ease, and to systematize the planning and the integration of processes.

Models are useful abstractions of reality to clarify and to manage the highly complex tasks of modern enterprises. They filter-out irrelevant details and represent only information essential to the task. In order to impel enterprise integration the modeling of functions, services and processes of the enterprise is required to facilitate their integration and to improve performance of enterprise functions. The specific modeling technique depends on the purpose of modeling and on the approach to integration. The basic principle describes enterprise models as a way to clarify the interactions of enterprise functions so that these interactions can be planned, rationalized, improved, and optimized.

The planning of integral business processes is characterized by a wide range of subjects. Material and information flows, product development and marketing concepts have to be taken into consideration [2,3]. Commercial, as well as organizational and technical matters are to be dealt with. The common understanding of all participants about the objectives and the core processes of the enterprise, as well as about the project goals is essential for the efficiency of both the planning and the development of business process organization.

In order to provide a complete and comprehensive model of business processes, the IEM method uses the object-oriented modeling approach. The structure of the objects within the enterprise use its main features [1,8,9]:

- treatment of functions and data as an intrinsic whole (encapsulation),
- inheritance, and
- the class concept.

Objects are described by data (static description) and by functions describing the changes of the objects (behavioral description). The object-oriented modeling approach allows the integration of different views of an enterprise in the same consistent enterprise model and it allows the easy adaptation of the model to changes within the enterprise.

Therefore, the following characteristics for the description of a real system can be deduced:

1. Representation of real systems and their properties by objects and related object attributes.
2. The object attributes can be distinguished as descriptive data, functions, and relations between objects.
3. Definition of object classes and inheritance of attributes to subclasses.
4. Assignment of values to attributes at instanciation (creation of objects).

Modeling objects can be created according to real world entities. The properties of real world entities are represented by appropriate attributes.

An object class is characterized by a determined set of attributes and functions (methods) operating on these attributes, which are valid for all objects (instances) of this class. Object classes can be developed in a hierarchical way. This means that subclasses inherit the attributes of the parent class. At instanciation (creation of objects) concrete values will be assigned to the attributes.

Derivation of generic classes of objects

The method of Integrated Enterprise Modeling (IEM) incorporates any view on the model from a standardized model kernel. Further model features can be tied to this kernel if necessary. The object classes "Product", "Resource" and "Order" are the basis for developing the model as a description of an individual enterprise [4,7,11]. Products and resources can consist of material, information or energy. Orders consist only of information. The three kinds of operands lead to the main classes of objects in a manufacturing enterprise from the user's point of view. Each class has a specific generic structure which means that it's possible to predefine a frame for describing the structure and behavior of objects of this class. Within the modeling process of a real enterprise real objects have to be related to one of these three classes. Furtheron, various additional subclasses, reflecting different purposes, have to be derived from the three main classes.

Required enterprise data and business processes, i.e. the tasks referring to objects, are structured according to the object classes during the process of modeling. Furthermore, the relations between objects are determined. The result is a complete description of tasks, business processes, enterprise data, production equipment and information systems of the enterprise at any level of detail [5,6].

The model kernel comprises two main views. The tasks, which are to be executed on objects, and the business processes are the focal point of the view "business process model", whereas the view "information model" primarily regards the object describing data (Figure 1).

Thus, the kernel of the enterprise model consists of the data and process representations of classes of objects. The views are interlinked by using the same objects and activities even though they represent them in different ways, levels of detail and context. The extension towards other views is part of the method. The way of defining views is shown below.

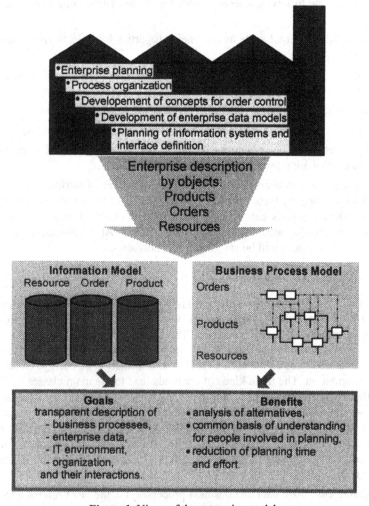

Figure 1. Views of the enterprise model

The three generic object classes and the generic activity model provide the basic constructs for the generation of particular manufacturing enterprise models.

Business Processes as Interactions of Objects

Everything that happens in a manufacturing enterprise as part of the manufacturing process can be described by activities. In general, activities process and modify objects which were classified above as Products, Orders and Resources. The execution of any activity requires direct or indirect planning and scheduling; it is executed by Resources owing the needed capability. The IEM method suggests three levels of describing the essentials of an activity.

- The *Action* is an object-independent description of some work or business, a description of some task, process step or procedure.

- The *Function* describes the processing of objects, as a transformation from one determined (beginning) state to another determined (ending) state.

- The *Activity* specifies the Order controlling the execution of the Function and the Resource(s) being in charge of executing the Function.

Figure 2 graphically represents the Generic Activity Model by connecting the beginning and ending states with the action rectangle by arrows from left to right. The controlling of the activity is represented by an Order state description and a dashed vertical arrow from top; the required or actually assigned capability for executing the function is represented by a Resource state description and a dashed vertical arrow from bottom.

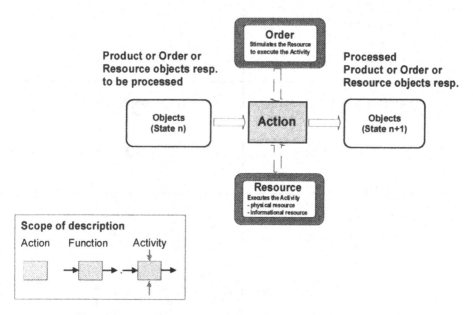

Figure 2. Interactions of Objects described by Generic Activity Model

The Generic Activity Model represents the processing of objects of Product or Order or Resource classes respectively indicating the object interactions at processing. The related organizational structure is described by specific resource classes along with their interrelations.

The description of enterprise processes starts with the analysis of the actual situation, normally applying a top-down procedure. Business processes along with the respective classes of objects to be processed are the starting point when modeling a certain part of the enterprise.

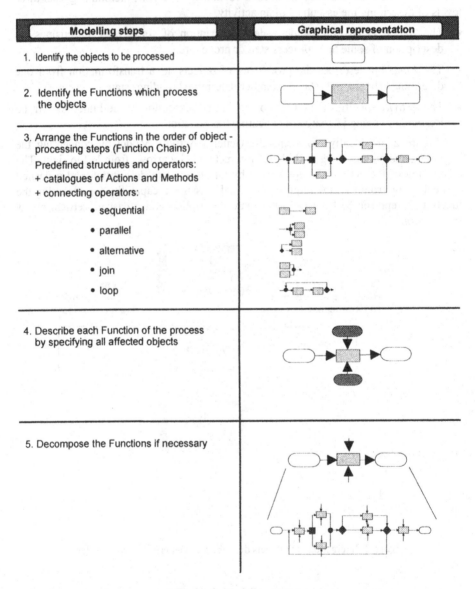

Modelling steps	Graphical representation
1. Identify the objects to be processed	
2. Identify the Functions which process the objects	
3. Arrange the Functions in the order of object - processing steps (Function Chains) Predefined structures and operators: + catalogues of Actions and Methods + connecting operators: • sequential • parallel • alternative • join • loop	
4. Describe each Function of the process by specifying all affected objects	
5. Decompose the Functions if necessary	

Figure 3. Modeling process

This part is delimited in a first step concerning e.g. Product and respective "ordering" with regard to the main task, required resources and the interfaces to the environment as well. The main task and objects of the application area are described by that [8,9,10].

The objects, which are to be changed, must be defined in relation to the functions and processes. The products of an enterprise have to be identified as instances of the IEM object class Product and the business processes have to be modeled independently of organizational structures according to the products, with regard to available rules for applying the IEM constructs (Figure 3).

Resources and controlling orders have to be identified in the next step for each function of the defined enterprise processes.

The development of particular business process models has to be extended by the order and resource flow, the analysis of concurrence of business processes and their mutual influence. For this purpose simulation and other methods have to be used.

Information Modeling

The collection and structuring of the data of all objects which were identified within the modeling process leads to a particular enterprise information model. For this purpose, a structuring frame for representing the relevant data of manufacturing enterprises is required.

The differentiation of the three generic classes of objects and their internal structure yields to a pre-defined structure of the enterprise information model. Three interconnected submodels, the Product, Control (Order) and Resource models have been defined.

Besides the aid for data structuring the IEM data modeling provides the advantage of close relation to the application area by considering real world objects with regard to the generic classes. The independence of data of a specific information processing system secures the extendibility and interchangeability of data between several systems. A particular enterprise information model will provide the preconditions for a general use of data bases and support the recognition of priorities within data exchange [8].

Integration of Organizational Structures and Information System Support by Additional Views

Further modeling aspects related to special modeling purposes can be integrated as additional views into the model. Examples of such views are special representations of control mechanisms, organizational units and costs. The relevant properties of the additional views can be represented by the development of specific subclasses of the generic object classes

• Determination of class specific attributes and

• determination of attribute values.

An example for the integration of additional views related to a special purpose is shown by Figure 4.

The kernel of the model of manufacturing enterprises is the basis for the development of application-oriented modeling constructs, views and partial models. Further existing application-oriented classes, constructs, views and models can be traced back to these main views of an enterprise model. Therefore, business process changes or organizational changes and their impact on information system support (data storage, communication and application systems) can be evaluated.

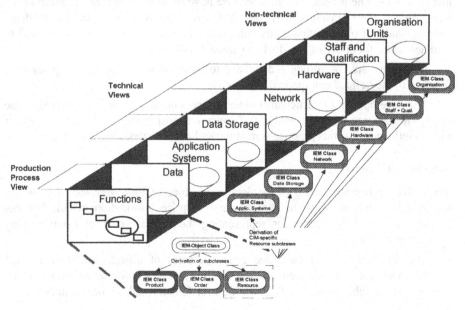

Figure 4. Views of Information System Support and Organization

This example of a specific application within the enterprise modeling area shows how the kernel of the manufacturing enterprise model can be used and extended by additional represented subjects and views. In the same principle way, several other models can be integrated into an entire enterprise model [9].

Figure 4 shows the consistent formalized representation of an enterprise-specific situation of computer integrated manufacturing for the tasks of planning and introduction of CIM systems. By different views on the model the planning information can be modeled separately. By determined relations between the views the model consistency is ensured. A usable model structure and a modeling guideline is derived of the IEM concept.

The scheme comprises eight layers, which represent the interlinked fields of design within the enterprise

The layers "business processes" and "data" represent the manufacturing process itself and the required information support independently of technical solutions. They are equivalent to the kernel of the enterprise model presented above.

The layers *application systems*, *data storage*, *network* and *hardware* are the technical fields of development of information system planning. The description of these subjects is also closely related to the first two layers, i.e. to the kernel of the enterprise model. E.g. the "data storage" view includes the relation of the data storage systems to the data as well as rules for the rights of access or the rules to modify the data and the responsibilities for retaining.

The layers *organizational units, staff and qualification* are non technical fields of design, which have to be worked-out simultaneously with the other fields. These layers contain additional items and subjects about the addressed aspects. All descriptions are related to the business processes and information described at the first two layers, i.e. to the kernel of the enterprise model. The tube across the layers illustrates their interrelations. The respective classes and objects incorporate these interrelations by their functions and data.

The requirements of the planning process caused by the mutual influences of the different views can be handled in an efficient way by using the interlinked layers [9].

For the additional views, such as organization, specific subclasses of the IEM object class 'Resource' are developed. The organization is represented by specific resource classes with additional attributes reflecting organizational aspects. For each view, some new subclasses might have to be developed [10].

The right choice of the level of detail is important for the modeling effort and benefit. For the task of information system planning, an overall, not too detailed modeling of a number of enterprise areas should be preferred. Therefore, alternatives within each view have to be worked out and evaluated against their potential benefit [9].

The described method of Integrated Enterprise Modeling is suitable for various planning and structuring measures in enterprises. It covers the aspects of material and information flows as well. In addition to the systematization of planning in a certain project, the benefits of applying the method are obtained from the reusability of the enterprise model for further projects with different tasks [5]. Examples for possible application of the are:

- Ascertainment of kernel processes in distributed enterprise structures.
- Analysis and ascertainment of organization of production preparation and execution.
- Presentation of potentials for saving time and costs by improved information systems at order execution (Figure 5).
- Development of a CIM architecture and an order control.
- Concept of interfaces between head office and decentralized production plants in distributed company structures.

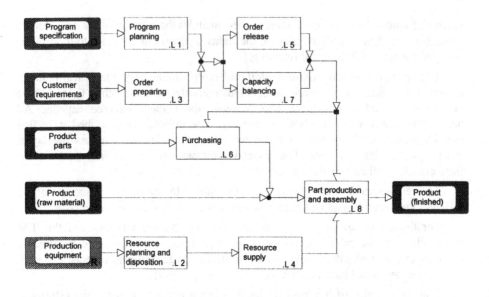

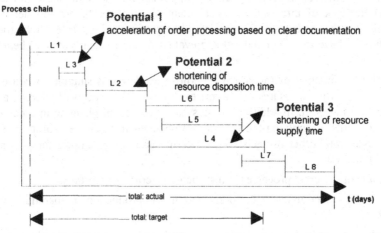

Figure 5. Derivation of potentials to save time at order processing

2.2 Material Process

Exact and reliable planning requires computer-aided methods. The planning and introduction of integrated information and manufacturing systems requires support by a reference model, specified exclusively for one company, for the development of an information flow concept, and supported by a material flow model for the development of a logistic concept. As information and material flow will influence each other very intensively, these models should be based on a common kernel, which is described using the same modeling approach. An efficient integration of information flow and logistic concepts can only be assured on this common basis.

Production System and Material Process

The material process of an enterprise is determined by its production system. Figure 1 shows the typical structure of a production system. The material flow system is the main part of the system. The material process is the result of operating the production system. Thus the material process can be described as a sequence of events and activities combining parts and products with the facilities and resources of the production system for a specified time in order to succeed in reaching a desired change of state.

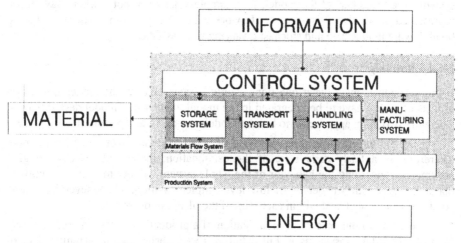

Figure 6. Structure of a production system [10]

Events are timed information quantities. They are the heartbeat of the material process. Each occurrence of an event demands a decision about the next activity to be performed in order to reach the production target.

In the field of production it is one oft the main tasks of CIM management and design to group the events in appropriate control loops.

Modeling the Material Process

Model based experimentation is an iterative process with the phases model definition, model design and model experimentation.

Model Definition

A model is a simplified representation of a real or imagined system. A real production system is a highly complex system, thus the model can only reflect a part of the aspects of the real system. Therefore it is necessary to define the important system aspects according to the target of the model building process.

To determine the most important attributes of the system from the viewpoint of a given planning or management task, it is a good starting point to define an explicit hierarchy of goals. Beginning with the overall goal, subgoals have to be defined down to the level of real system performance indicators. For example, the high level goal *reducing throughput time* can have *increasing transport frequency* as a subgoal, for which one of the indicators is the *number of workpiece carriers*.

Such hierarchies of goals make problem-related model design a much easier task. They support the process or they result in comparing to a great extent. As the goals of the study are broken down to the level of indicators, the result values of the related model elements can be compared directly in terms of how well they support the high level goals.

When the goal definition is finished it can be decided which aspects of the real system have to be part of the model, for example whether or not workers have to be represented or whether it is necessary to work with a complex time pattern, including breakdown, maintenance and stochastic varying process times.

Model Building

If the decision is made what questions should be answered by the model and which aspects of reality have to be integrated in the model the phase of model design begins. There are two main approaches to build a model: top-down and bottom-up.

Using the top-down method one level of detail after the other will be described. Therefore, in a fairly early state of the examination, the planner is able to give statements concerning the behavior of the whole system without making assumptions about the yet unknown details of low level implementation. Thus, several possible solutions can be analyzed within each step of the planning phase.

According to the process of modification of a production system the planner may be able to model subsystems in full detail at a very early stage in planning. Out of these subsystems the planner builds-up the upper level parts of the system by synthesis, until the whole system is modeled. The bottom-up method allows detailed analysis of system parts in short time, but limits the field of solutions at an early stage.

This method won't be used in a strict way but in a combined way in the case of different user-established habits, different requirements, and result ranges of applications. The top-down method should be preferred in most cases because of its integrating and creativity improving character.

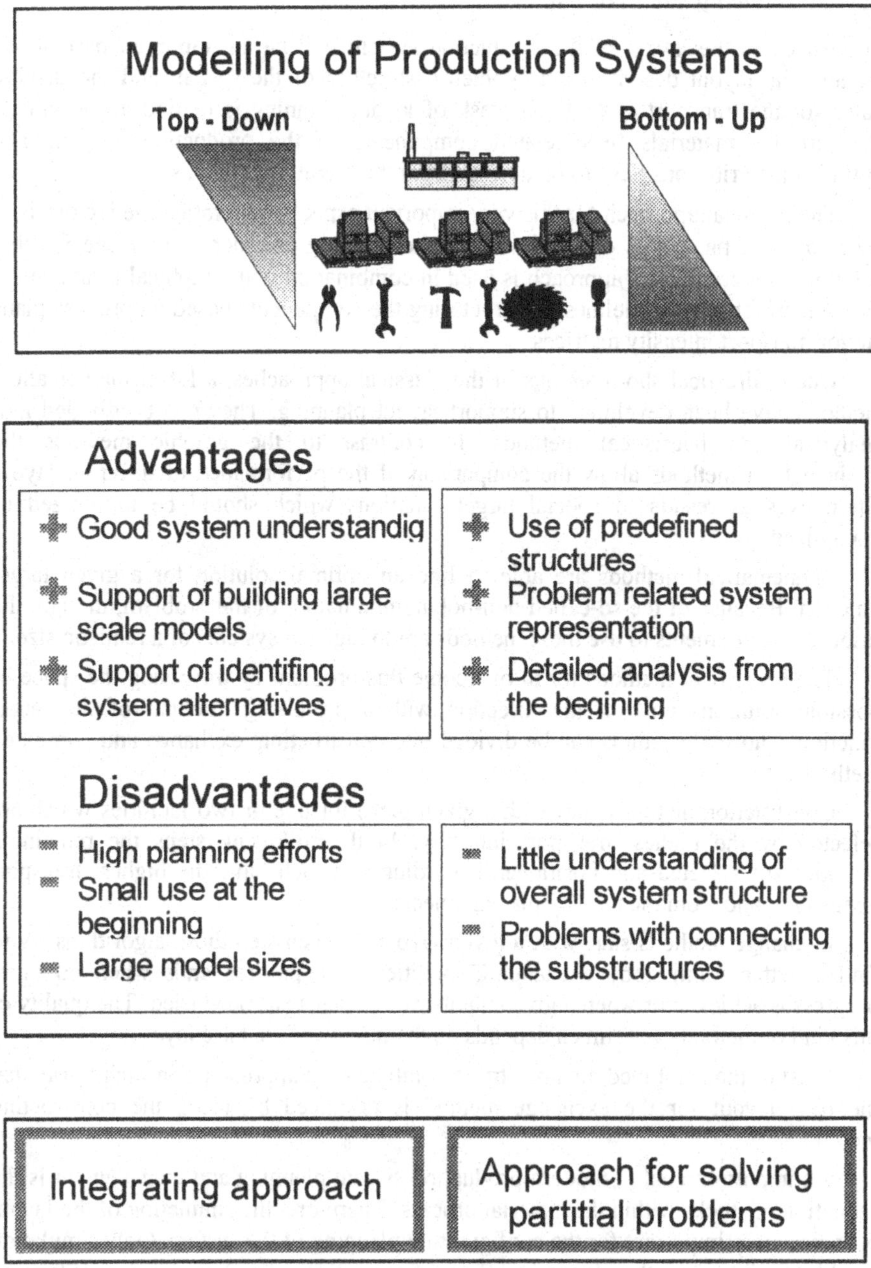

Modelling of Production Systems

Top - Down

Bottom - Up

Advantages

- Good system understandig
- Support of building large scale models
- Support of identifing system alternatives

- Use of predefined structures
- Problem related system representation
- Detailed analysis from the begining

Disadvantages

- High planning efforts
- Small use at the beginning
- Large model sizes

- Little understanding of overall system structure
- Problems with connecting the substructures

Integrating approach

Approach for solving partitial problems

Figure 7. Top-Down and Bottom-Up modeling

Layout Modeling and Planning

In most cases the layout of the production system will be an important part of the model. The layout describes the geometric structure of the system and the driving paths for the transport system. The task of layout planning is to find an optimized structure for materials flow related components of the production system. The optimization criterions have to be derived from the hierarchy of goals.

The classical approach for this very important aspect of operating the factory is to try to arrange paper jigs on the layout of the building in order to find one feasible solution. Sometimes this approach is used in combination with graphical arrangement methods which give guidelines for structuring the components based on process plans and/or transport intensity matrices.

Due to drastical shortcomings of the classical approaches, a lot of mathematical methods have been developed to support layout planning. They can be divided into analytical and heuristical methods. In contrast to the graphic methods the mathematical methods allow the comparison of the performance of different layout alternatives as results of special target functions which should be minimized or maximized.

Mathematical methods are able to find an optimal solution for a given target function. Because of the so-called non-polynomial nature of the problem, though, the resource requirements to use these methods are to high for systems of a realistic size.

The heuristical methods try to overcome this problem by traversing the space of possible solutions only in the direction with a promising gradient of the target function. Those algorithms can be divided into construction, exchange and combined methods.

Construction methods start with a given place for one or two facilities which are selected by the highest transport intensity. In the following steps the remaining facilities are selected and positioned according to which have the highest transport intensity to and from the already placed objects.

Exchange methods start with a given layout. In each step those algorithms try to find a better solution by exchanging facilities. The process ends when no more progress is achieved or when a given number of changes has been tried. The quality of this kind of methods very much depends on the quality of the base layout.

Most of the combined methods try to combine both approaches in such a way that the base layout for the exchange method is produced by using the constructing method.

An important goal in modern production system planning and management is the reduction of stocks within the material process. Therefore, the simulation of the layout becomes more important for the configuration planning of the system. Only simulation is able to integrate the dynamic effect of operating a system into the planning procedure. This is extremely important if a flexible transport system with a net structure is to be used for the examined system.

Different approaches

To describe the operation of a production system within a model, different approaches have been developed. There are some requirements that apply to each description system:

- it has to be complete and in an adequate degree of detail,
- it may not contain contradictions,
- it has to be unmistakable and easy to understand and
- it has to be appropriate for the application of mathematical or numerical methods.

To evaluate the description systems from the production system point of view, it is necessary to make some systematic considerations.

Description systems can be divided into graphical and textual based methods. It is an advantage of textual descriptions that they can be easily fed into and administrated on a computer. Textual descriptions range from natural language over structured languages, like programming languages, to decision tables. Particularly object-oriented programming languages and simulation languages are very flexible and able to describe almost every aspect of a given system. But the task of building a computer program for a complex production system can only be done by expert programmers. On the other hand, programming languages trying to overcome that problem by preparing a set of high level language constructs specialized for simulation lose a lot of the flexibility of that description system, especially because they force the user to build-up the model on a predefined degree of detail.

Graphical description systems are based on a set of symbols and the model is constructed by combining and connecting the symbols in a specific way. They range from drawings and flow charts to petri-nets and systems of functional building blocks. Only the last-named can be dynamically studied. Figure 8 shows the degree of abstraction of the different methods.

Especially for the planning of highly concurrent material flows, which are characteristic for complex production systems, simulation has proved to be an adequate method. Simulation turns out to be a standard tool of the 1990's to analyze material flow systems.

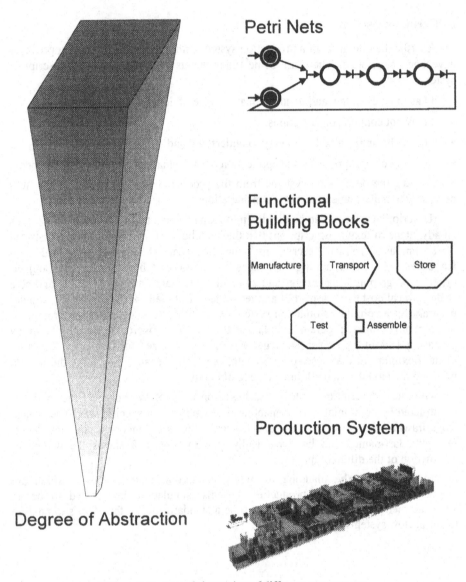

Figure 8. Degree of abstraction of different system types

Planning System MOSYS

MOSYS has been developed as a typical factory planning system. It supports the planning process of production systems and is not specialized on certain industries. With MOSYS it is possible to model the functional structure, the topology, and the control rules of production systems. To make the dynamic relations of the system evident, the integrated model can be animated. For the evaluation of different solutions a special support is given by MOSYS.

The modeling method of MOSYS is function oriented. Five basic functions are applied, Manufacture, Assemble, Store, Transport and Test. These building blocks can be combined freely with each other and thus allow the modeling of every imaginable sequence. Figure 9 gives a survey on the description system of MOSYS. By the use of parameters, the building blocks are customized to the specific applications. The modeling of a system can be carried out stepwise in a hierarchical way. The description system is well suited for the modeling of manufacturing systems as well as for assembly systems. There are sophisticated features available for the modeling of the material flow systems and the implementation of user-specified control strategies. These control rules are specified with an adapted and simplified petri net.

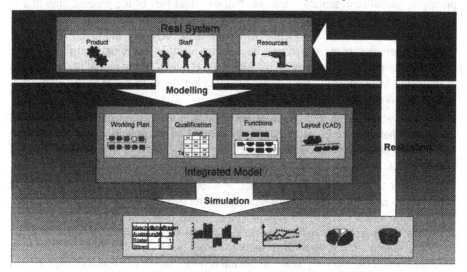

Figure 9. Planning system MOSYS

Functional Building Blocks

The functional behavior of the system is described using the five building blocks Manufacture, Transport, Store, Assemble and Test. These building blocks can be combined freely with each other thus allowing the modeling of every imaginable sequence. Each building block has a distinguished behavior to make it easy for the user to decide which block to use. By way of using parameters they are customized to the specific applications.

Partial Representation Concept

Functional models are well suited for hierarchical structured simulation models. The summary of a group of functions can be looked at as a complex function which can be easily integrated into a higher level model. This approach leads to a concept of partial representation-models. A partial representation is a functional structure, which includes all relevant model information and which can be simulated independently. In order to use a partial representation as part of another model it is merely required to create an according building block for hierarchical elements and refer it to the files which describe the partial representation.

Process Plans

For every product modeled, a process plan is required. This process plan describes all possible paths including alternatives which can lead the product through the system. The process plan includes process times, input information, priorities and branching information which is used to select a branch in a given situation during simulation.

The process plan is part of the partial representation. According to the black-box principle, partial representation files contain only that part of the process plan that is part of the partial representation. If one product can occur more than once in the same partial representation, it is possible to define a theoretically unlimited number of alternative process plans which can be selected using a combination of the object name and the process plan name in the higher level model. If the partial representation is simulated independently, the process plan that has the same name as the object that it describes is selected.

In that way it is possible to work not only with an hierarchical function model. The hierarchical approach is consequently applied on all aspects of the model, and make even large scale models of great complexity easy to handle and free of redundancies.

In order to use the black-box principle throughout all levels of the model, it is necessary to attach the input information for each object to the partial representation. When the complete hierarchy is simulated the information to generate the model input is always taken from the highest level. If a partial representation should always be used as a submodel, it is possible to mark the input behavior of its objects as implicit. This means that objects only enter the partial representation as a result of higher level materials flow and no input information has to be explicitly generated.

Topological Description

The topological description consists of nodes, which have co-ordinates, and tracks connecting these nodes. The building blocks are located at the nodes. Several building blocks may be located at the same node, some building blocks, as pallet buffers, may be distributed to different nodes and other nodes serve only as a base for the tracks. Nodes and tracks may be accessed using numbers. A CAD System supports the input of the topological description. The factory layout can be edited within the CAD system in order to define the particular nodes and tracks for the specific representation.

For the simulation model, several topological descriptions may be combined. There is no limit on the number of nodes, paths or any other elements forming complex descriptions .

Economic efficiency of simulation depends on the effort for simulation studies. Analyses show that the experimental effort (45%) is much bigger than the modeling effort (35%) (Schlüter 1993). Missing methods and aids for the execution of simulation experiments were given as a reason.

Because of economic restrictions it is not possible to examine all possible combinations of system parameters within the simulation of complex factory plants or production systems. The number of necessary experiments is exponentially increasing with the number of parameters and their adjustment possibilities. Furthermore, simulation models include stochastic factors, for example faults of the tools. It should be determined how strong the stochastic factors influence the simulation results. It is problematic to find a good goal-related solution which includes a financially justifiable number of experiments and also considers all relevant effects of the system parameters. In current planning practice the number of necessary experiments depends on the knowledge of the user. Normally, the number of simulation experiments is too high. Even an expert has to use the trial and error method to determine the characteristics of e.g. a production system.

The method of statistical experiment planning is very suitable for the exact planning of simulation experiments. The method guarantees the fulfillment of the statistical requests for number, quality, and expressiveness of the data even if not all possible parameter constellations could be tested.

3 Product Model

One of the most challenging tasks in the engineering profession is to develop new products that have the shortest lead - time, the highest quality, and the lowest cost with an optimal life-cycle consideration. The issue of product modeling is at the center of various new product development paradigms designed to meet this challenge, and has therefore received major attention from application and research communities. Due to the fast development of computer and information technologies and the increasing demands of competitiveness and productivity, the scopes and approaches of product modeling have evolved rapidly in recent years.

Although each of the above-mentioned strategies is somewhat different in its focus and its approach, they all share one fundamental requirement: that is the need for advanced information technologies to integrate and coordinate various life-cycle considerations during product development activities. A central issue among these information technologies is product modeling which generates an information reservoir of complete product data to support various activities at different product development phases. Therefore, product modeling is the key factor in determining the success of various product development strategies and of future industrial competitiveness.

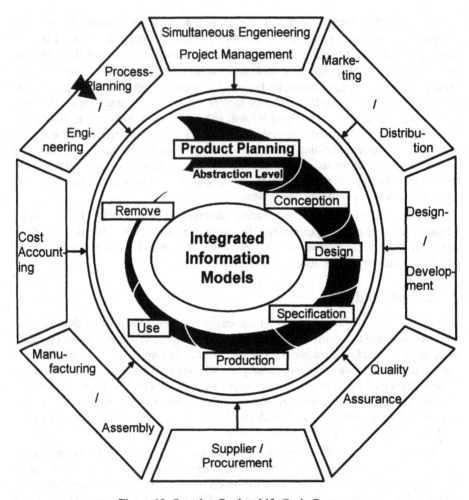

Figure 10. Complete Product Life-Cycle Concerns

Product modeling technologies must not only support those new product development strategies described above, but also deal with other manufacturing-related models, such as factory models which include resource models, equipment models, machine models, and tool models. This indicates that product modeling, while being a unique subject in itself, is a very broad topic which is closely related to many other challenging issues in modern manufacturing engineering and complete product life-cycle concerns as indicated in Figure 10 [14]. It also shows that the issue of information processing for product modeling is very complex in engineering practice. To take full advantage of the available information technology and to minimize capital investments, it is desirable to adapt to an open system architecture while implementing product modeling technologies. Also, it is necessary to broaden an engineer's education and to direct basic research so that the potentials and restrictions of product modeling approaches can be evaluated and realized by future engineers who will be responsible for developing new products in industry.

One of the most significant approaches towards the implementation of integrated product models is the development of the ISO Standard 10303 "Standard for the Exchange of Product Model Data" (STEP) which defines a neutral data format for the representation and exchange of product data Figure 11 [16]. The goal is the complete and system-independent representation of all product-related data during the product life cycle. The description of the data format provides a suitable basis for the definition of a neutral file format, the design of a product database and the conceptualization of a procedural interface. In regards to the information scope and possible forms of implementation, STEP goes far beyond the application potential of previously available interchange formats like IGES, SET or VDAFS which primarily facilitated the exchange of geometric information. STEP was the first approach which introduced the idea of complete product model exchange into standards and it is a methodology for the development of application oriented software. STEP development, even though it has already brought forth some standards, needs more elaboration. In the future, it could work as an information infrastructure for product model representation in order to integrate all product development processes.

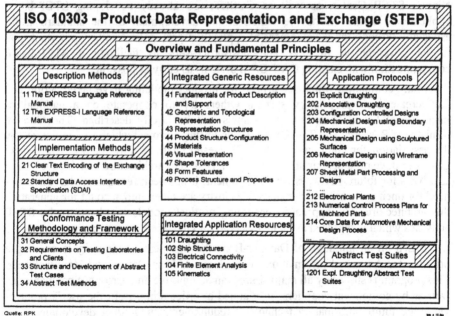

Figure 11. STEP - Standard

ISO 10303 can be broken down into a series of partial standards. At the core of the standard, information models for the representation of product data are defined. Among the various information models it is possible to distinguish between integrated resources and application protocols. Application protocols constitute implementable subsets of STEP. They define the specific segment of the entire information structure required for supporting a particular application. The integrated resources are further divided into generic resources and application resources. The generic resource models define application-independent basic elements which can be used in several application resources or directly in application protocols. Application-specific resources are tailored to a higher degree to the requirements of particular domains, but serve as a stock of elements for the definition of the application protocols.

To represent and develop the STEP information models a variety of tools for information modeling have been used. One of the tools, the data specification language EXPRESS, was developed to support the development of integrated information models. EXPRESS consists of language elements which allow an unambiguous object definition and specification of constraints on the objects to be defined. Implementation methods define the structure of physical files and the database access using the Standardized Data Access Interface (SDAI). The implementation of a database in the STEP format is based on data structures described in EXPRESS. In contrast to the physical file, the use of the SDAI allows data access independent from the physical representation of STEP data.

Although the standardization process of STEP is not yet finished, a growing number of companies and organizations, e.g. ProStep, PDES Inc., STEP Tools, Nippon STEP Center, as well as GOSET, are already developing STEP-based applications. These efforts are aimed at the development of the standard itself, especially through the development of application protocols. Another important goal is the development of pre- and post-processors to facilitate a STEP-based communication with existing CAD/CAM systems as early as possible.

Many companies which see a need for product modeling have taken action and they create their proprietary system. Others try to cooperate on a national or international basis. Three major examples will be briefly described. The German Ministry of Research and Technology is supporting a project which was initialized by the association for computer science (GI, Gesellschaft für Informatik). The project goal is to develop a description of a new CAD system architecture which could serve as a basis for future CAD system development. It has been recognized that there is a need for open systems. The demands of industry are taken into account for generating this architecture. The approach concerning product models is related to the STEP activities which is regarded as being essential for envisaged process chains.

Based on an initiative of Daimler-Benz the European Community is creating a program for Advanced Information Technology (AIT) in Design and Manufacturing. This program is shared by all major European automotive and aerospace companies as well as by component suppliers. It is aimed at articulating, from the user's point of view, the future information technology requirements in product development and manufacturing as well as in their integration. Process chains and product models are essential parts of these efforts. This will be the basis for the cooperative development of the needed software with IT industry. The consensus also should be an enabler for de facto standards.

Intelligent Manufacturing System (IMS) is a program which was initiated by MITI in Japan and is now a worldwide activity. The feasibility study which is performed has the goal to exercise this type of research and development cooperation on a global scale. Six subjects have been chosen for elaboration. At least two of the projects are affected by process chains and product models. These are GLOBEMAN 21 and GLOBAL CONCURRENT ENGINEERING. It can be expected that in the case of full IMS, the topics of product modeling will have an even stronger impact [17].

4 Enterprise Order Control

In a dynamic manufacturing environment the use of distributed, networked computers aiding to control orders is nowadys a standard. Adequate control software with powerful mechanisms for netwide interprocess communication as well as a dedicated manufacturing database are key elements.

The new factory has to be designed with an ability to follow rapid changes in the production system's environment. This involves flexible production equipment such as flexible manufacturing and assembly systems, a lean organization structure with a sufficiently agile management and highly skilled and motivated staff. Dependent on this, flexible and adaptive information technology capable of managing varying information flows is required, as the basis for what can be defined as a learning organization. Today's users of software systems therefore demand solutions which cannot only represent the organizational structure of their individual enterprise exactly and predictably but which can also easily adapt to changes the organization may undergo in the future. The factory's throughput is also changing. More and more information is processed in the workshop. It's main function, the processing of material, is increasingly enlarged and enriched by information processing tasks. Functionality and architecture of information systems for order control applications have to represent this.

Overdone division of labor and inflexible mass production concepts with layered hierarchies as practiced by F.W. Taylor and H. Ford describe yesterdays requirements. The appropriate enterprise model was supported by partially optimized application systems in functional organization units and by MRP systems, SFC functionality for factory coordination and the control tasks as well as production data acquisition (Figure 12).

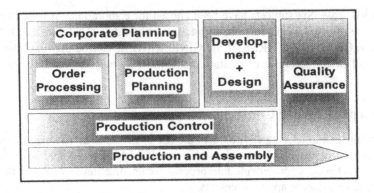

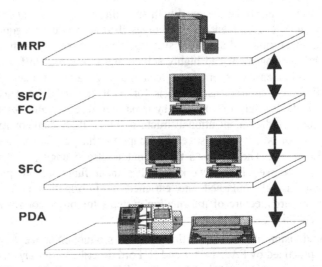

Figure 12. Conventional enterprise model and layered system support

Instead of conventional forms, self-regulating, decentral forms of work organization and production logistics better do represent the factory's new function determined by customer-orientation. Competent, responsive in-process decision making assigned to production units run by teams of motivated workers who are capable of entrepreneurial thinking has to be represented in adequate information systems for Order Control (OC).

Enterprise Order Control (EOC) systems have to manage complete process chains from the receipt of a customer order to the delivery of ordered products (Figure 13). The aims are to deliver quality products on-time, and to minimize disruptions of the production process.

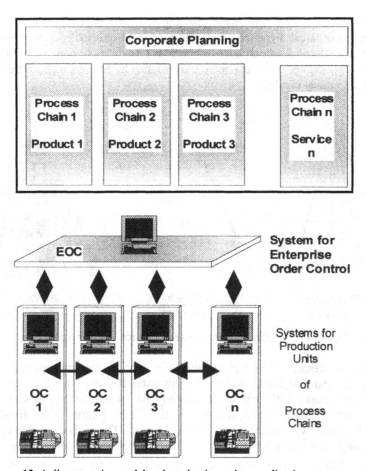

Figure 13. Agile enterprise model and production units coordinating system support

Interaction of different computer systems is required. It should only take a neglectable effort to define and implement control system features and the topology of the computer network. In communication with business administration information environments EOC has to support decision making and operation of more and more autonomous units of a modular production organization. Furthermore, it has to provide the link between and integration of all actions performed by such self-organizing and self-reflexive units.

Agile Organization and Information Technology

The remedy for an increasing amount of indirect work, augmenting lead times, rising inventories and increasing costs is to convert production units organized according to principles of Fordism - Taylorism into decentralized structures oriented at production process chains. In doing so, many items and structural design elements that seemed to be almost natural for industrial production will turn out as being inappropriate.. All attempts to achieve higher responsiveness of production systems by means of more technology for a better deterministic planning and control of big and intransparent production units failed. Pyramids of management levels erected to control workers in indirect functions and direct productive functions turn out to be contra-productive and inflexible. The attempt to achieve more transparency by forming organization islands ended in suboptimal results.

Figure 14. Complex business structures (Source: Küchler, syskoplan)

The following scenario of the typical situation of a producing company may serve as an example: Despite high investments in highly automated machines and facilities and a highly sophisticated MRP system the company did not succeed in achieving a higher degree of utilization of it's production capacities. In the contrary, the utilization rate decreased in some very capital-intensive departments. The analysis of these negative effects brought out two failure sources, the salary system and the work organization.

Wages had been differentiated into more than 50 different classes, allowing to identify and administer all kinds of unproductive work, however, the relative proportion of the only value-adding account indicating wages for productive work in manufacturing functions decreased.

Work organization is the second source of time spent unproductively. Problems of organization are the reason for an absence of production inputs like material, devices, for missing orders or unavailable personnel capacity, disproportional set-up times, breaks and pauses. These lay- and standstill-times which are not caused by technical requirements sum up to 90% of the total.

To support new system structures a new paradigm for information system solutions is required. Conventional factory models are based on complex mechanisms to control logistics. The pending Information Technology (IT) consists of partial systems that communicate almost exclusively in the vertical direction.

The application of already installed MRP systems and complementary systems, e.g. for shop floor control (SFC) and production data acquisition (PDA) has shown unsatisfactory results. This was caused by:

Deficiencies in timeliness and actuality of daily disposition plans,

- the development of inventories,
- the low degree of flexibility, especially in regards to decreasing lot sizes, and
- a steadily increasing expenditure on engineering and EDP capacities.

To cope with these problems one can follow two philosophies: The one is to apply more technology and to enhance efforts for supervision and monitoring. The idea would be to equip each machine with a PDA terminal in order to collect data on the when and why of any malfunctions and standstills, and to install an even more complex MRP system to fulfill requirements of Just-In-Time production and to cope with increasing inventories and work in progress. The second philosophy is quite different: Restructuring of the production process towards highly autonomous modular operational production units, and support for these units by IT solutions appropriate both in structure and functionality.

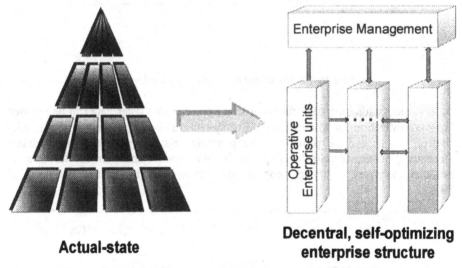

Actual-state

Decentral, self-optimizing enterprise structure

Figure 15. Optimized enterprise structures (Source: Küchler, syskoplan)

A suitable EOC system supports thinking in in-company customer/supplier relations. Centralist planning of time and quantities in this view has to be rejected. Each production unit has to be enabled to act as a business unit with proper responsibility for its business results, and with its own dedicated order control system. Roughly scheduled orders indicating the date of completion as the main control parameter can be held in an order pool and the workshops may pull appropriate production orders, feedbacking minimal information to the central coordination instance.

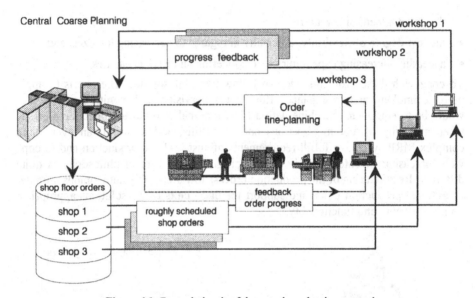

Figure 16. Control circuit of decentral production control

Basic business administration functions as well as certain master data can be retained there centralized in e.g. conventional MRP systems, and each production unit's order control system can be provided with planning and control data when needed. Depending on the organizational pattern or the product spectrum master data on products, machines, customers, suppliers and personnel can be retained in centralized form.

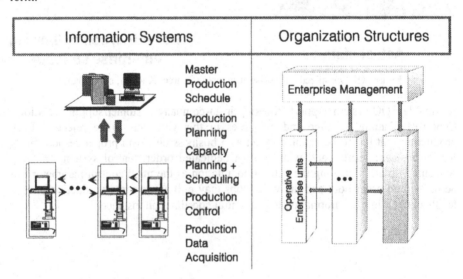

Figure 17. Local system support instead of dominating information systems

If for instance several production units can manufacture identical products or refer to the same supplier, the maintenance of the master data has to be performed by a centralized instance. But in principle order control systems of the production units should provide functions for the administration of such master data.

In case the order control system runs in an integrated mode together with a MRP system for business administration functions, provisions are required to deactivate the functionality for master data administration . EOC systems should consist of five functional blocks:

- Order take-over;
- Materials management;
- Order execution;
- Information system;
- Information system management.

Order take-over functionality has to process incoming customer orders and has to transform them into suitable orders for the production units. Net requirements, calculations, and pending reservations of materials are performed automatically. In the case of materials that have to be procured from out-company suppliers the system can initiate the procurement automatically.

Materials management functionality performs the acquisition of production inputs and outputs for the productions units, and the administration of all inventories; furthermore, it controls commission orders.

Order execution functionality performs order administration with functions for the display of orders, the tracking of orders, and the monitoring of order progress. For factory data and quality data acquisition it contains functions for the processing of such data, furthermore functions for transport and process control. For the process control a CAM interface is required, with provisions for monitoring, visualization and definition of process data.

Information system functionality provides functions for analysis and evaluation purposes concerning orders, inventories and quality. It contains functions for availability checks, monitoring all events of impact in regards to the availability and the status of resources.

System management functionality performs the administration of all master data required. Functions for a take-over and downloading of master data from business administration systems are also required.

The order control task requires a network of information exchange involving all functions of the factory. Chains of production coordinated by computer-aided decision making that involves engineering and production information plus business administration information need a high level of flexibility, which can be achieved by system support both for the mapping of organization units and the managing of events.

In the following some EDP and software architecture aspects of EOC systems will be discussed more deeply.

Manufacturing Database and Inter-Process Communication

For the given task - enterprise order control - the database appropriate to provide all the long-term data structures required for the control system and for the support of information flows between instances and functional blocks should be a dedicated manufacturing database (MFDB). It should contain all data for shop floor control tasks such as manufacturing order administration, scheduling, order arrangement, manufacturing resource management, DNC, and assembly.

As the development in the field of hardware makes it more and more likely that not one mainframe computer but also several smaller computers will perform order control tasks, there should be a possibility to distribute the EOC application network on a given hardware network. The way in which this distribution is done, and the question on which computer a part of a current application is running should be without visible consequences for the user. The functional blocks or application modules, i.e. executable programs or EDP processes forming an application system, have to communicate with each other in the frame of the enterprise order control network. Such programs may either run on the same or on different hardware systems concurrently, and they are not necessarily independent from each other. Information between programs is exchanged either directly - process-to-process - or programs take out information from the database which was written to that place by another process. Due to this fact of interdependency, certain sequences of processes are imperative, others are depending on events.

Event-Driven and Distributed Systems

From the requirement that an open order control system should not be restricted to a particular hardware class or enterprise organization pattern derives the need for a network distribution support. Such a service system is a main element of the order control software architecture, incorporating provisions for both time-driven and event-driven execution of functionalities in the system.

In event-driven systems the flow of time is described as a chain of discrete events. All events of importance for the software system and all necessary reactions on events should be described in the order control system. Obviously, the full complexity of a manufacturing system and every conceivable case or event can't be represented, however, many provisions can be implemented in an order control system. The possibility to specify important events and appropriate reactions together with some rules for the execution of a reaction in case of unpredictable events is already an interesting feature for a responding enterprise order control system.

Event-driven systems are more flexible than non-event-driven, subsequently, they are performing systems. The handling of events is performed by special program tasks getting information about reactions from configurable and therefore adaptable sources like files or database tables. Changes in the company's information flow can be realized by updating configuration files and reloading the new configuration.

The distribution of processes on different computers in a network follows the same logic. By this flexibility requirements on an EOC system can be met implementing company-specific "tailor-made" systems while keeping costs for the implementation and later adaptations of the system within budgetary and time constraints.

The modeling of a company's given and planned structure, its information flows, EDP processes, process chains and communication structures is an important step in implementing EOC systems. Data to be exchanged between EDP applications is modeled in this step. It is the input parameter for the system's configuration. Configuration data should be stored separately from the application to allow changes of parameters during the application's runtime without touching code. This is essential in regards to organizational aspects.

Representation of Organization

Projects introducing order control systems start-off with an analysis of internal and external information and materials flows of a choice of organizational units. Weak points in organization can be detected and eliminated. At the same time, the control system can be prepared and introduced. Usually however, analyses produce a static model of the organization without reference to moments when communication or data access occurs. Flexibility of an application system is given if it not only is easily adapted to existing organizations but if it also reflects the dynamic behavior well, and provides at low costs arrangements for changes the organization may undergo in future. To support system adaptation order control systems therefore need efficient methods and tools. Adaptation can be done on several places in a system:

- Variation of input and output masks as adding or deleting fields, changing titles or the language.

- Variation of the network topology, for example assignment of tasks to a new mode.

- Integration of new applications into the existing system.

Expensive adaptations within the application itself should be avoided. Therefore, the configuration data should be stored separately and apart from the application, so that it can be changed without greater efforts. User-friendly tools for the change of configuration data should support users to perform changes occurring during the software life cycle without involving software suppliers. Adaptability is important for:

1. Changes in organization, report writing, and communication.

2. User interfaces that must be adaptable to user- and task-specific requirements.

Ad 1. An in-plant communication structure is not fixed, and there are constellations for short periods of time. Connections between communication partners may be built-up depending on events:

- Work instructions may demand that a certain office has to be informed if a special type of malfunction occurs.

- A user decides to send a message (mail) to one or several other users within the system environment to inform them about new facts.

- The referential integrity of the database is endangered or has intentionally been violated by anyone. For instance, a rush order has to be scheduled and no NC-programs and tool definitions are available. Then, a request has to be sent to the appropriate instance to generate the missing data with highest priority.

- Sequence of time can also bring up events, for example if a requested worksheet did not arrive. After a defined waiting time the process planning office could be informed to generate an alternative process plan.

Ad 2. The second topic relates to the way a worker performs his work. This way is influenced to a high degree by personal preferences, varying from person to person. Often there is a difference between the actual way work is done and the way it was proposed to be done. If only the proposed way of performing a task is supported by the system, employees tend to apply additional informal means like personal card-indexes created to contain data obtained in informal ways. Merging computer work and manual work may end in inefficiency, in errors due to inappropriate data transfer and may therefore cause the discontent of employees (see also chapter 6).

The Way of Introducing Open Production Management Systems

As a summary it can be said that the implementation of EOC systems is a process induced by market requirements, benchmarking the enterprises' position as to cost structure and responsiveness of the operation.

As change is essential to recent and future production this has to be represented in open production management systems, involving, as a planning infrastructure, the dynamic representation of the company in appropriate models and, as an information basis, flexible IT.

5 CIM components

5.1 Data bases

With regard to the data processing components of a CIM architecture, a higher flexibility in manufacturing can be achieved by increasing the degree of data integration of the components. When connecting CIM components, it becomes necessary to install a transparent hardware- and software-independent access to all data previously held by the individual components.

Without data integration, the individual data objects have to be typed into computer systems several times manually, allowing for inconsistencies and faulty inputs. The flow of data between the CIM components is slowed down. Access control to data objects and backup of data cannot be guaranteed.

A first data integration step of CIM components has been realized exchanging data telegrams between the participating systems using definite protocols and storing data individually. Multiple data copies have led to different degrees of topicality. Access control and the surveillance of data distribution have not been taken into account. Evolutionary processes have not been represented.

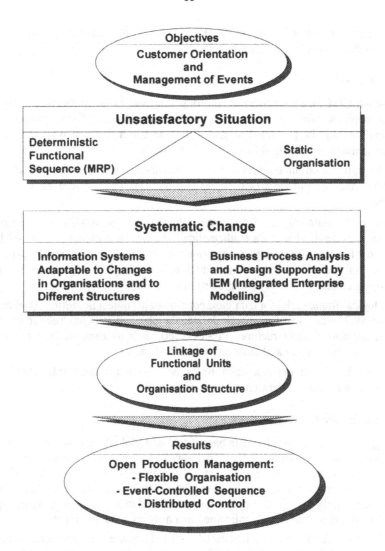

Figure 18. From corporate objectives to Open Production Management

Introducing a common data base system, either centralized or distributed, meets all requirements of consistency, accessibility and control of data. A transaction mechanism coordinates the concurrent access of several components to data objects. Previously linked applications become independent of each other, standardized access modules to data base objects offer a common information interface.

In an optimized production environment the customer order is typed in or transmitted from an order processing system and stored in the common CIM data base. This event triggers the processes of order scheduling and CAD. Afterwards CAP and NC program and bill of materials generation steps are initiated. Finally, the shop floor information and control systems make use of their individual view on the data objects. In each step all necessary input data are provided, and the results and the evolutionary progress of work on the data objects is registered. Consistency checks can easily be done before altering the object.

Integrated data model

A common CIM data base must store production data in an application independent form in order to satisfy all information demands of individual CIM components. The common data will be presented to each component in a different way. Views on the data are generated realizing different access methods.

The application independent data representation is based on models of the enterprise and the products (see chapters 2 and 3). They should be implemented directly.

The most commonly used data base data model is the relational model because it's easy to understand and it's mathematically based. Data objects are stored in two-dimensional tables describing the relations of the objects. Columns represent the attributes of the objects, rows individual objects. Logical views can be defined referencing columns or rows of single or joined tables.

A logical framework of a data base consists of all object definitions (fields, tables and views). The physical framework defines the physical storage structure of the data base. It contains access structures on all objects, their location on the storage media and the internal structure and location of data buffers.

Hierarchical and net-type data bases are nowadays used infrequently. They organize data objects in hierarchical trees or nets.

Centralized and distributed data

The logical view on a common data base is that of a globally accessible data pool with individual views. Physically the data pool may be located on one computer site in one data base, or it may be distributed on a local or wide area network of data base servers. In order to increase application performance, each server holds that part of data which mainly belongs to applications processed at the same site. Also, copies of data objects can be stored and maintained in a distributed environment.

The more distributed and replicated data objects are, the more complicated the management of the data base system is. Up to now there is no real distributed data base system available that meets all requirements of data partitioning. Research on the serviceability of distributed data bases in a production environment has revealed that the effort to maintain a permanent topicality of all distributed data objects is high and in most cases unnecessary.

Temporary and permanent data

Attributes of data objects often have multiple storage requirements in accordance with the durability of their values. Some attributes need to be stored permanently with their latest value, some with the complete history of all value changes. In contrast, temporary data are only of interest in their current state in the current context; they are not too numerous, change their values frequently and need not to be held permanently in a data base.

Loss of data is fatal for permanent data. They always have to be kept securely in all data processing steps throughout all systems. The loss of temporary data in case of system failures is not fatal but should be prevented because data have to be recollected resulting in a performance degradation of the involved facilities at that time.

Nowadays, most CIM system components store all their data in permanent data bases. The introduction of permanent and temporary storage methods, each with data protection mechanisms, leads to an increased availability of current production data for CIM systems with real-time components [24].

Temporary data platforms

In order to guarantee a high access speed for temporary data, it has to be stored in the main memory of CIM system components. To protect those data from loss in case of system failures, it has to be duplicated on a permanent storage media, such as magnetic disks.

Some operating systems support the permanent backup of main memory sections to a disk. After a system failure the memory contents can be reconstructed from the backup file.

Platforms for distributed temporary data make use of data interchange protocols over local or wide area networks to update replicated data objects. The protocols are quite similar to distributed data base protocols.

Data base mechanisms

The access coordination of multiple CIM components on a common data base has been realized using a transaction mechanism. Transactions consist of one or more single processing steps on common data. Either all modifications or none of them will persist, even after interrupts of the application or in case of system failures. During modifications the data objects are locked against access of other components. In distributed data bases the transaction mechanisms have been expanded to coordinate access to all copies of data objects.

In case of system breakdowns a rollback mechanism is used to undo all modifications of running transactions. Afterwards a recovery strategy is used to redo transactions as far as possible.

The transaction concept guarantees a consistent data base at every time. Consistency is defined as a state of the data base when no transactions are active and all data objects represent a meaningful model of reality.

Providing an application independent data definition and manipulation language, data objects can be accessed from different hardware and software platforms.

Object-oriented data base systems

Object-oriented data base systems have been developed improving the data storage capabilities in an object-oriented software design. The data base interfaces are included in the programming language as object class extensions.

One approach regards the data base system as a method server for permanent storage and retrieval methods which will take over the object for storage. Another approach expands the set of attributes of an object offering persistence. If persistence is chosen as a property of the object, a schematic definition in the data base system is automatically generated.

A view on data objects can be realized by using set-oriented retrieval methods or by regarding the inheritance property as a kind of view mechanism.

Integrity of the data base is defined and guaranteed by the access methods provided with the objects. Locking and recovery mechanisms work the same way as in conventional data base systems. A version control system is integrated in the majority of available object-oriented data base systems.

CIM components and their data requirements

Required data types and access methods vary very much regarding different CIM system components [25]. In a common data model all objects have to be represented adequately. *Table 1* lists a few important requirements of data base systems in relation to demanding CIM components.

CIM components Requirements	Production planning	Shop floor control (operative)	Shop floor control (dispositive)	CAD	CAP	CAQ
Independence of application views	●	●	●	●	●	●
Integrity	●	●	●	●	●	●
Security	●	●	●	●	●	●
Version control	●	●	○	●	●	◉
Complex structures	○	◉	○	●	◉	◉
Search for similar objects	◉	○	◉	●	●	◉
Evolutionary processes	○	◉	◉	●	◉	◉

● very important   ◉ important   ○ less important

Table 1: IM components and their requirements of data base systems [25].

5.2 Computer Communication

Over the last 10 years the central data processing was relieved by distributed systems. In the same period PC's tended to be linked together by networks and organized in company wide communication processes, instead of being used as stand alone desktop computers. Electronic interchange of information became a key position in the business structure. The public communication facilities belong to the most important factors when choosing a business location.

There are numerous concepts and technical possibilities that cover the area of computer communication. The fast growing processing capacity of modern computers demands more powerful networking facilities. The variety of nets and their more and more increasing capacities make the decision for a certain system difficult.

Computer Network Classification

Computer nets can be classified by their geographical extension. There are three major groups LAN, MAN and WAN.

The local area networks (LAN) is local to the company. And the company has the full responsibility for use and maintenance of the net.

The metropolitan area network (MAN) is used for fast communication between different systems in the same conurbanation.

The wide area network (WAN) is used to communicate between different systems over great geographical distances. The connections and services of the WAN are offered by public suppliers and communication societies.

To classify the capacity and the facilities of a data communication network it is necessary to introduce a systematic pattern. In 1977 the international standardization organization (ISO) developed the OSI (Open Systems Interconnection) reference model. It is the basis for standards that facilitate the communication between different teleconnections of various manufacturers. To realize the communication, it is necessary to synchronize the parallel processes on both sides of the connection and to guarantee a fault-free information interchange. The OSI reference model divides the different subtasks in seven layers. Each layer supports a certain class of services the higher layers use the services of the lower layers to realize their more complex services.

Physical Layer: Determines the electrical and mechanical requirements for the physical connection to a teleconnection and establishes the mode of transmission. This includes the specification of possible cabletypes and connection sockets for connecting computers with network cables.

Data Link Layer: Specifies the physical addressing of the communication partners and improves the data transfer by using a error recognizing and correcting encoding.

Network Layer: Creates the connection between two teleconnections and determines the optimal path between them (routing).

Transport Layer: Realizes the communication management for system independent data exchange between different processes and message addressing.

Session Layer: Creates, observes and terminates logical connections between different processes.

Presentation Layer: Transforms system dependent data into a standardized format.

Application Layer: Interprets data in the sense of the sender and supplies services for user processes in the network environment.

Local Networks

Commonly used LAN-networks are networks that follow the standard IEEE 802.3, also named CSMA/CD or colloquial ethernet, token bus networks or FDDI networks.

Ethernet was introduced in 1980 by DEC, Intel and Xerox. Since then it was further developed and became a standard. It uses the following technical transmission systems:

- transmission systems in bus topology that use a transmission rate of 1,5,10 and 20 Mbit/s on a co-axial cable also called 10 BASE 2.

- transmission systems in star topology that use a transmission rate of 1 and 10 Mbit/s on a twisted pair cable also called 10 BASE T.

- fiber optic transmission systems with a transmission rate of 10 Mbit/s on the basis of active or passive starcoupler also called 10 BASE F.

Each of these transmission systems uses certain sockets. Each System has its own restrictions regarding the length of the cable segments and the number of connected terminating equipment. The topology of the systems 10 BASE T and 10 BASE F supports error isolation much better then 10 BASE 2.

In winter 1985 the IBM-Token-Ring network was introduced by IBM. It was the basis of the standard IEEE 802.5 . Valid transmission cables are co-axial, twisted pair and fiber optic. It is possible to reach transmission rates up to 16 Mbit/s. The topology of the network is a set of ring-like connected stars.

The FDDI-protocol was designed for use on high frequency range fiber optic transmission systems. It permits a transmission rate of up to 100 Mbit/s with a maximum length of 100 km and up to 1000 connected stations. Originally designed for the use of two-way fiber optic cables, it was extended to support twisted pair cables too. This extension allows its use as a Backbone, which connects networks of different kinds.

The token bus network is described in the standard IEEE 802.4 that is used in production-oriented data processing. The token bus uses a wideband co-axial cable and can reach a transmission rate of up to 10 Mbit/s.

Layer 2 of the OSI reference model ends in a logical link control (LLC). The LLC supports the following services :

- non-confirmed connectionless service (LLC1): Exchange data units between network stations without a logical connection.

- connection oriented service (LLC2): Build up a logical connection, transmit the data units in a certain sequence, quit and close the connection.

- confirmed connectionless service (LLC3): As LLC1, except that delivery is acknowledged.

  Layer 1 and 2 describe the LAN transport service.

Furthermore, local networks include private branch exchanges and their further developments. Originally private branch exchanges were used for private telephone links. Today electronic digital technology allows the parallel use as private telephone links and the link of data connections. Their main advantage lies in the building of a uniform infrastructure for both, speech and data communication, and the optimal access of public communication facilities. Future developments in communication technology will further widen the possible use of the telephone network to allow video conferences - so called multi-media applications - in addition to the previously mentioned services. This service should be available not only for the internal company, but also for international communication. The CCITT issued a number of recommendations under the name of ISDN (Integrated Services Digital Network) to attain a standard that would help to make it international. The Wideband-ISDN (B-ISDN) specification of the CCITT includes the asynchronous transfer mode (ATM), a technology designed for long-distance traffic. This technology finds increasing use in local spheres. ATM uses a complex network of switches, that can move fast between different ports. With this method ATM reaches a transmission rate of 140 Mbit/s. An ATM network can connect any kind of system and network. ATM is used not only within the WAN area, were different ATM switches are connected by fiber optic cables, but also in the LAN area, were the ATM technology is used to connect different kind of networks.

The main parts that are used to connect different LAN networks are:

- Repeater regenerate the electrical capacities of the network and permit the extension of network components, which belong to the same kind of network. They work on layer 1 of the OSI reference model.

- Bridges have the same capacities as a repeater. Furthermore they can connect network components, which work with different transmission media, but use compatible network protocols. Bridges use address recognition for load balancing and to filter protocols. They work on layer 1 to 3 of the OSI reference model.

- Router are protocol dependent. They hold information about other router thus making it possible to locate the optimal path through a global network. They work on layer 3 of the OSI reference model.

- Gateways connect networks with different network protocols. They are mostly application-dependent and work on the upper levels of the OSI reference model.

Because of the development of independent quasi-standards, it is difficult to classify LAN-networks on the upper levels of the OSI reference model. The most important quasi-standards are the TCP/IP-family protocols, protocols that where developed especially for PC networks, the IBM system network architecture (SNA) and the Digital DECnet.

TCP/IP has components of the layers 3 to 7 of the OSI reference model but is not as hierarchic structured as the OSI reference model. TCP/IP includes a network and routing protocol, as well as the following applications telnet (remote terminal), SMTP (Simple Mail Protocol for message exchange), ftp (file transfer protocol) and SNMP (system network management protocol), a protocol that supports network management. Besides their own protocol stacks IBM and DEC developed support for TCP/IP.

PC networks were developed to facilitate the sharing of file and print services. Furthermore they support e-mail (electronic message interchange). PC networks allow two different methods of cooperation between the network resources peer to peer networking (equal cooperation) or client server computing (unequal cooperation). In the case of peer to peer networking each network resource puts other applications and services at the disposal of the other net participants. This is possible only in small groups of up to 10 work stations. In the second case a group of clients uses the facilities of one or more servers. A server may offer file and print services, backup service database services as well as the use of large applications or a WAN connection. Common network operating systems are NetWare from Novell, IBM LAN Server, Microsoft Windows NT Advanced Server and VINES from Banyan.

IBM SNA was developed as a hierarchical protocol that described the communication between a host and simple resources like terminals or printers. With the development of APPC (advanced program-to-program communication) a data transmission interface for the communication between transaction oriented programs, SNA opened to the needs of office communication, local networks and distributed systems.

DECnet was developed corresponding to the digital network architecture (DNA). Today it is developed up to phase V. DNA resembles the ISO model and all important protocols of the OSI reference model will be included in the realization of the higher levels.

In further developments the manufacturers strongly refer to layer 7 of the OSI reference model. Basically this layer includes the following protocols:

- FTAM (File Transfer, Access and Management) includes file transfer facilities and manages the access on contents and attributes of a file.
- VTP (Virtual Terminal Protocol) controls the online access to distant computer systems.
- MHS (Message Handling System) describes a standard for exchanging messages between persons. It follows the reference of the CCITT (X.400 - X.430).
- Directory Service is an information system, that saves the addresses, names, and other properties of people and objects that participated in a communication. It follows the reference of the CCITT (X.500).
- TP (Transaction Processing)
- JTM (Job Transfer and Management)
- RDA (Remote Database Access)

Most of the previous mentioned aspects stretch wide beyond the area of LAN networks.

MAN

The task of a MAN is to connect LANs or digital private branch exchanges over hundreds of miles with a minimum transmission rate of 100 Mbits/s. For that purpose the IEEE developed the standard 802.6, the DQDB (double queue double bus) in 1987. The transmission is usually made with fiber optic cables . In the USA this service is called SMDS (Switched Multimegabit Services and in Germany it is offered by Telekom under the name of Datex-M, with a protocol converter for the current LANs. It is expected, that these facilities will be included by ATM/B-ISDN.

WAN

Today the telephone system, the PSDN-network (CCITT X.25) and ISDN are at the participants disposal for connections that use public lines.

The telephone system constitutes a worldwide network. Teleconnections can be connected with the telephone system with the help of an acoustic coupler (transmission rates up to 1200 bit/s) or a modem (transmission rate up to 24000 bit/s depending on the type of modem). In this way it is possible to transfer data between hosts from nearly all over the world, if they are connected to the public network.

Another world-wide spread network is the PSDN-network (Packet Switched Digital Network). It works on the layers 1 to 3 of the OSI reference model and broadcasts data in form of packets. Because of the so-called PAD (Packet Assembler-Disassembler) from the Telekom, this network is within reach of the telephone system. After dialing a PAD the participant can be contacted by all users of the PSDN-network.

The slim band ISDN-network provides the user defined interfaces. The data transmission to a connected ISDN station uses two channels for user data (64 kbit/s) and one control channel (16 kbit/s). The data processing is handled by the connected terminating equipment. Higher requests demand a primary multiplex connection with the interface S2m, 30 channels for user data and one D-channel with 64 kbit/s as control channel. There are two different standards for the primary multiplex connection: the 1TR6, that is used by Telekom so far and the CCITT Q.931, also called Euro-ISDN.

A WAN-connection makes it possible to connect LANs of different permanent establishments, to transmit data to other enterprises and to use public or private information services, for example Datex-J as Telekom service for tele-banking or flight reservation and CompuServe for e-mail. Besides numerous databases offer various information. Internet is a world-wide network that covers more than two million computers. It provides services of e-mail, netnews, ftp (file transfer protocol) and supports the interactive, dialog-oriented use of distant computers. A new development is the WWW (world-wide web), a hypertext-oriented information system developed by the center of European nuclear research (CERN), that supports the search for information in the internet. Meanwhile big enterprises use WWW for representation and as information system for their clients.

Internet started as a project of the US Department of Defense 25 years ago and was extended by science and research all over the world. Commercial suppliers offer internet access for industrial users.

6 Human Factors

Computer integrated manufacturing aims at an increased effectiveness in customer service by production. To reach this aim, organization and technology have to be designed at the same time and tools for planning and factory management have to be provided.

Human factors in CIM design

CIM covers information processing in a manufacturing environment where people interact with computers. Considering human factors reaches far beyond the man machine interface design of information processing workplaces as for secretaries, scientists, or aircraft pilots.

Information systems have to follow organizational and task design and therefore shaping work contents can not be separated from CIM design in both directions. All aspects of how people manage their tasks with information processing tools individually and in groups and teams are subject to "human factors in CIM".

Scientific research contributing to Human Factors

The results of forty years of sociotechnical research and design have proved the effectiveness of a human oriented approach. Sociotechnical system design (STSD) denotes the simultaneous shaping of task, organization and technology in a way that a major impact of the user on the solution and a long term growth of personality is possible. Productivity and quality levels of the Japanese transplants in the United States of America and in the United Kingdom have demonstrated that human oriented management techniques are not limited to a certain culture but can be applied, with modifications, globally.

Basis for sociotechnical design is a knowledge about the specific human characteristics where the following are relevant for the design of information systems [26]:

- integrated body and mind, multisenorial,
- learning, creative, associative, individual,
- communicative and social,
- ready to take over responsibilities and
- goal oriented.

Each point has to be recognized and addressed to develop and use effectively human resources. Human factors as design principles are well known when it comes to ergonomic machine design and workplace design. This addresses mostly the human being as a system of bones and muscles in respect to chair or keyboard design or to a system of eyes, brain and muscles in respect to control panel design. The firstly named effort aims at preventing illness and longterm absenteeism, the second at preventing accidents in a power plant or in a plane cockpit.

The aim of any reengineering or CIM design task is a coherence of task design and the specific person, organization, technology and process of reengineering (Figure 19).

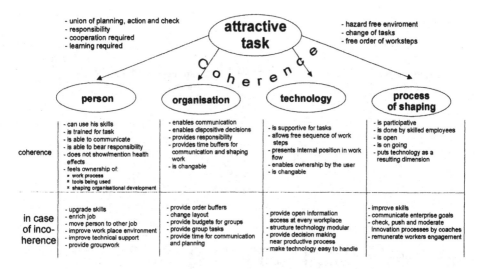

Figure 19. Coherence of task, organization and technology [27]

These interdependencies and coherences are the scientific basis for human CIM design and it makes no difference whether the task is a major organizational structure design or a minor user interface design.

The fundamental organizational design enables a higher level of effects because more human characteristics are addressed specially cognitive, communicative and goal and responsibility oriented needs of skilled staff.

Within a given turbulent market and with a skilled work force the following concept is most promising. Providing a maximum of *responsibility* and entrepreneurial challenge to small operational units results in organizational structures of profit centers (Figure 20) who operate independently on the market. This stimulates customer orientation, cost consciousness and creativity.

This approach goes beyond the already successfully implemented structures of segments or production islands. It is known that an increase of mental stress, group pressure, uncertainties and intensified work is connected to this approach. In spite of that the increased direct influence on the work process and the visibility of the agreed own business strategy and results in positive votes from the majority of the involved workforce. Although it seems to be a pure business approach on the first glance, it has considerable technical implications because it requires powerful decision support, open information access and distributed information handling on shop floor level. Increasingly any workplace uses networked application software so that company wide and focused views have to be applied at the same time.

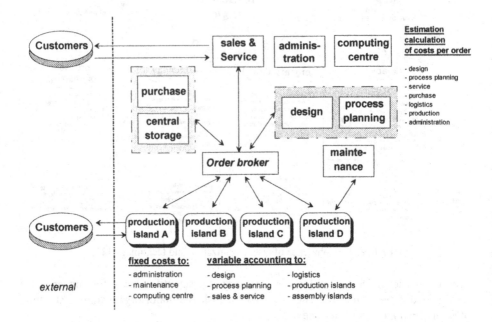

Figure 20. Reference structure of an extremely profit center oriented organization [29]

Workplace and software design

A traditional hierarchical menu forces the user to follow a certain sequence to reach the final application screen. This contradicts the human characteristic of creativeness and individuality. An associative hypertext-oriented navigation and a free selection of worksteps would serve to avoid this contradiction.

Modern NC-machine controlers and NC programming systems are devoted to support chains of action and provide the following features [28]:

- graphical process feedback,
- programming at machine or by the operator near to the machine ,
- interlinked software for order management, tool management, diagnosis etc.

This is a more expensive technology on the shop floor level instead of a DNC system with a CAD/CAM interface. Nevertheless, these systems can be more productive because they support workers to use body *and* mind, be creative by selecting tools, arranging cuts, use their tacit knowledge and select order sequences.

Handling of formal and implicit knowledge

Enterprises can be seen as sociotechnical organisms. It is often overlooked though, that the real workflow and decision flow as well as informal structures can not be completely described in words. Three broad projects in the framework of CeA (computer supported experience guided work) with on-site-studies in companies pronounced the importance of the influence of implicit knowledge to reach best performance. /see 29]

Decision making and empowerment

When technology is used for decision making, some companies aim to replace skilled workforce. Thus they face the danger to loose the above mentioned implicit knowledge.

On the other hand, knowledge bases and expert systems offer the chance to distribute the decision making process to a greater extend than before, if there is a decision support device. The configuration of a shop floor control system in Figure 21 offers feedback to the machine operators and allows them to simulate and test possible schedules. It also provides communication links between groups and uses CSCW features (computer supported cooperative work) to enable distributed and fast decision making.

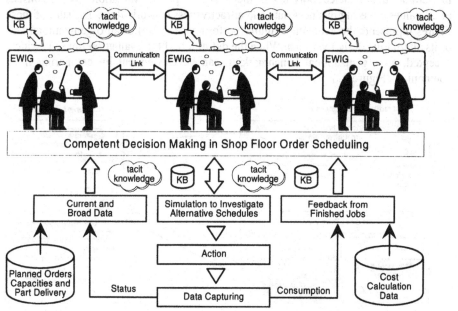

Figure 21. Skill enhancing decision support system [29]

The examples indicate that considering human factors in a profound way, this mostly means to shape tasks and user interfaces differently than if a pure functional approach is followed. Attractive worktasks and organization forms such as groupwork and/or simultaneous engineering and networked decision making requires an advanced, flexible, and fault-tolerant information technology that does not yet exist to its full extent.

Quantifying multidimensional improvements

Organizational Benchmarking

A profit or loss at the 'bottom line' still interests managers most, but this only represents the results of a period of operation. Engineers have to shape company structures during a reengineering process with CIM introduction that help to produce 'black figures' in the future.

It is well known that e.g. flexibility is one of the non-monetary characteristics that ease good business results in a turbulent business environment. Since 15 years, the method to determine the 'work-system-value', is one of the methods of the extended business evaluation. Here these aspects as flexibility including also Human Factors lead to the comparable work system value, a factor that combines 9 values, their weight and degree of fulfillment for alternative options according to cost-benefit analysis principles [30].

It is easier to assemble a countless list of comparable figures that are documented in literature than to select only few of them. The proposed evaluation scheme consists of 28 values covering business data and attractiveness of work alike. The sum of these figures represents the evaluation figure (performance value) of an organizational unit or its alternative design variants. When the definition of the evaluation figure is widely accepted, an organizational benchmarking can be carried out by comparing more companies. (Figure 22)

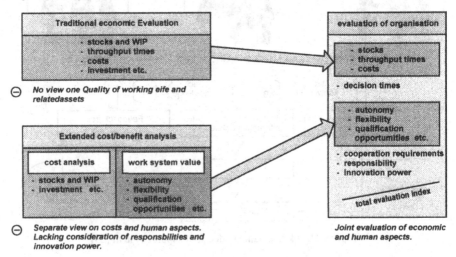

Figure 22. Organizational evaluation [31]

The usefulness of building human oriented work systems is widely accepted. There is also a growing number of design rules and embedded procedures being applied by engineers to shape technology and organizational structures according to human needs. Still managers are not sure how far they should go, what structure is appropriate, what technical specifications should be set up and what effort is reasonable. The above sketched methods could be used to guide a multidimensional reengineering process.

7 Conclusion

CIM is far more than just technical requirements of integrated data handling and data transfer standardization. CIM consists of an interrelated design of organization and technology with databases and networks being a crucial point for an unrestricted access to information.

Human characteristics are to be respected and are aimed at to be further developed instead of eliminating the human control of production and information processing. Evaluation methods that scan foreseeable business results as well as attractiveness of work, help to find long term solutions that ease the animation of the employees' hidden assets such as motivation and creativity.

The international use of the term 'agreeable factory', or 'agile', 'holonic' or 'organic factory' reveals that similar ideas about small business units and human orientation are pursued worldwide. Europe has a long tradition of developing human resources and thus should make better use of these assets by combining these with the "best practice" business approaches available.

A systematic planning and factory management means to use models for enterprises and products, to relate enterprise structures to control structures and to use appropriate software tools that support human activity. These systematic and interdisciplinary procedures have to be supported by integrated methods resulting from joint efforts of different disciplines, countries and schools.

References

1. Coad, P., E.Yourdon. 1990. Object-oriented analysis. Englewood Cliffs, NJ: Jourdon Press/Prentice Hall.
2. Flatau, U. 1986. Digital's CIM-architecture. Rev. 1.1. Marlboro, MA: Digital Equipment Corporation.
3. Harrington, J.R. 1984. Understanding the manufacturing process. Key to successful CAD/CAM implementation: Marcel Dekker.
4. ISO TC 184]SC5 DOC N 148, Technical Report. Reference Model for Shop Floor Production, Part 1.
5. Spur, G., K. Mertins, R. Jochem. 1993. Integrierte Unternehmensmodellierung. Berlin: Beuth Verlag.
6. Mertins, K., W.Süssenguth, R. Jochem. 1994. Modellierunsgmethoden für rechnerintegrierte Produktionsprozesse, Hrsg. G.Spur. München/Wien: Carl Hanser Verlag.
7. Süssenguth, W., R. Jochem, M. Rabe, B. Bals. 1989. An object-oriented analysis and design methodology for computer integrated manufacturing systems. Proceedings Tools '89. Paris: CNIT.
8. Mertins, K., R.Jochem. 1993. Integrierte Unternehmesnmodellierung - Basis für die Unternehmensplanung. DIN-Tagung.

9. Süssenguth, W. 1991. Methoden zur Planung rechnerintegrierter Produktionsprozesse. Dissertation. TU Berlin.

10. Mertins, K., R.Jochem. 1992. An object-oriented method for integrated enterprise modelling as a basis for enterprise coordination. International Conference on Enterprise Integration Modeling Technology (ICEIMT). US Air Force-Integration Technology Division. Hilton Head, SC.

11. Mertins, K.,W. Süssenguth, R. Jochem. 1991. Integration. Information. Modeling. Proceedings of Fourth IFIP Conference on Computer Applications in Production and Engineering (CAPE '91). Bordeaux, France: Elsevier Science Publisher B.V. (North Holland).

12. Mertins, K., R. Jochem. 1992. Planning of enterprise-related CIM-structures. Proceedings of 8th International Conference CARS and FOF. Metz, France.

13. Spur, G. 1986. CIM - Die informationstechnische Herausforderung an die Produktionstechnik. Produktionstechnisches Kolloquium. Berlin.

14. Kimura, F. 1993. Product and process modeling as a kernel for virtual manufacturing environment. Annals of the CIRP, 42[1. 147-151.

15. Krause, F.-L., B. Ochs. 1992. Potentials of advanced concurrent engineering methods. Manufacturing in the era of concurrent engineering. North Holland.

16. Grabowski, H., R. Andul, L.A. Polly. 1993. Integriertes Produktmodell. Berlin: Barth-Verlag.

17. Krause, F.-L., F. Kimura, T. Kjellberg, S.C.-Y.Lu. 1994. Product modeling. CIRP Annual Conference.

18. Mertins, K., R. Albrecht, H. Krause, W. Müller, V. Steinberger. 1992. A flexible architecture for distributed shop floor control systems. Third International Conference on Computer Integrated Manufacturing at Rensselaer Polytechnic Institute Troy, NY. Proceedings in: IEEE. Los Alamitos, CA: Computer Society Press.

19. Mertins. K., A. Salisch, F. Duttenhofer. 1991. Software architecture supporting tailor-made look and feel of customized applications. Proceedings of the international working conference on "New Approaches Towards One-Of-A-Kind Production". Bremen.

20. Reisch, S., F.-W. Lutze, K. Mertins, R. ALbrecht. 1991. Industrielle Softwareproduktion für die Fertigungsleittechnik. ZwF/CIM 86[2, 65 - 69.

21. Mertins, K., R. Albrecht, V. Steinberger. 1992. Werkstattsteuerung, Werkstattmanagement. Hrsg. G. Spur. München, Wien: C. Hanser Verlag.

22. Mertins, K., W. Süssenguth, R. Jochem. 1994. Modellierungsmethoden für rechnerintegrierte Produktionsprozesse. Hrsg. G. Spur. München, Wien: C. Hanser Verlag.

23. Spur, G., P. Liedtke, K. Mertins, R. Albrecht, N. Vieweg. 1993. Organisatorische Neustrukturierung in einer Manufaktur. Einführung der Gruppenarbeit mit Hilfe der Simulationstechnik. ZwF 88, 303 - 306.

24. Duttenhofer, F., et al. (to be published) 1994. Echtzeit-Produktionsmanagement. Entscheidungen treffen mit aktueller Fertigungsinformation. Tagungsband zu PEARL '94, GI-Workshop über Realzeitsysteme.

25. Duttenhofer, F. et al. 1992. Datenbanken für CIM. Hrsg. G. Spur.- Reihe CIM-Fachmann. Hrsg. I. Bey. Berlin: Springer-Verlag.

26. Mertins, K., B. Schallock. 1994. Menschliche Eigenschaften als Bestimmungsgrößen in der Fabrikwissenschaft. ZwF 89[12, 613-616.

27. Mertins, K., B. Schallock, P. Heisig 1994: Participative design of decentralized Structures , IPK Berlin.

28. Hohwieler, E., B. Schallock, V. Weiß u.a. 1994. Handlungsorientierte und gruppenfähige Maschinen- und Steuerungskonzepte. Hrsg. G. Spur. Stuttgart: IRB-Verlag.

29. Mertins, K., M. Carbon. 1994. Shop floor control systems supporting teams. Hrsg. P.T. Kidd, W. Karwowski. Advances in Agile Manufacturing. Amsterdam, Oxford, Washington, Tokyo: IOS Press.

30. Mertins, K., H. Edeler. 1994. Reengineering auf der Basis von Geschäftsprozessen. Hrsg. K. Mertins, S. Kempf, G. Siebert. Benchmarking '94. Tagungsband zur gleichnamigen Tagung. Berlin: IPK.

31. Schallock, B. 1993. Anforderungen an die technische Unterstützung von Gruppenarbeit. Hrsg. P. Binkelmann, H.J. Braczyk, R. Seltz. Entwicklung der Gruppenarbeit in Deutschland - Stand und Perspektiven. Frankfurt, New York: Campus Verlag.

Shortcomings and extensions of relational DBMS

Kroha, Petr

Technical University of Chemnitz,
Department of Computer Science,
P.O.Box 964, Chemnitz, Germany

Abstract. This contribution explains the reasons leading to the development of object-oriented databases. In the fist part, drawbacks of the relational model will be described including the lack of modelling power and the lack of performance. In the other part, the proposed solutions and extensions of the RDBMS will be introduced.

1 Introduction

Relational database systems have many advantages and they are very efficient for simply structured data. In the next sections, we will show disadvantages and shortcomings of relational database systems which led to the development of object-oriented systems.

1.1 Problem of splitting complex objects

Using the relational model requires that domains of attributes can contain only single values in the sense that these values may not have any internal structure (they must be atomic), i.e., we cannot work with a domain of lists, tuples, etc.

In practice, it means that if it is incidentially possible to represent each object in the database (i.e., each instance of each entity) by a single row in a table representing a relation whose attributes have only atomic values, then we have a very simple model and we can get a very efficient implementation. If a domain of at least one attribute is a set of objects which have some internal structure like lists, tuples, etc. then we must start a process called normalization of relations which causes splitting attributes of one object into more than one table.

Normalization and decomposition A relation whose attributes domains are sets of atomic values (type: integer, real, Boolean, character, date, string) is called a relation in the first normal form (1. NF). Now we will discuss the problem of what to do when the reality which is modelled contains domains with structured values.

To get relations in the 1. NF, we must split the given relation into more relations. This process is called decomposition. Each new relation constructed during the decomposition must contain the key of the original relation. We get one relation by projection of the former relation on the columns which have domains with atomic values, and then we get one relation for each domain with structured values. These relations are created from the keys and the values stored in the structures of their domains. Relations in the 1. normal form are called flat relations.

The problem of structured values in domains is not the only one. Other problems are caused by semantics of data, exactly by so called functional dependencies. For example, we desribe attributes of a product and know that PRICE depends on tuple $< SUPPLIER, PRODUCT >$, and we use it as a key in this table. On the other hand, we know that the customs-duty depends only on the product. It partially depends on only part of the complete key. These facts result from the semantics of the real world.

Another case of functional dependency is a transitive dependency. We know that the attribute BUILDING does not depend on the identification key of an Employee directly but transitively, i.e., $CODE- > DEPARTMENT- > BUILDING$.

- insertion anomaly,
- deletion anomaly,
- update anomaly.

Insertion anomaly means that we can insert only a complete row into the table. If we found, for example, PRODUCT and CUSTOMS- DUTY, we cannot insert this information into the table PRODUCTS, because we do not know the SUPPLIER, which is part of the key. We cannot insert a row into a table when we do not know its key, i.e., when we do not know its position in the table.

Deletion anomaly can be illustrated in a university database if the last teacher of a department were be fired, we would lose the information about the location of the department.

Update anomaly is caused by the redundancy. If a department has 100 teachers, then we repeat in each case the information about the location of the department. If the department changes its building, or if the building changes its name, we must update many items. The solution to these problems is again in decomposition.

We described reasons for splitting an entity description from one table into several tables, and we have to note that the idea that every instance of every entity (i.e., every object of the real world) should be represented by one row in one table - cannot be followed any more.

In fact, the reason of normal forms derived from the functional dependencies (2. NF and 3. NF) is that more than one entity was represented by the split relation. In other words, our estimation of what an entity is was not correct.

Using normalization we can transform our primary model to another model with another system of entities which can be represented as a system of two-dimensional tables.

In many cases, we can get many hundreds of 3. NF tables for one table representing a non-first normal form relation.

Example:

In CAD-databases, we can represent 3-D objects in the following way: Each object will be decomposed into a combination (supported by set operations) of transformed primitive basic objects.

#

It is not difficult to imagine a description of a relatively simple part which would be dispersed in some hundred tables. This phenomenon is very useful when we need to process one property of all instances of one entity in the normalized model but it is a burden when we need to process all properties of one whole object, i.e., one complete instance of an entity in the real world model.

Example:

Let us imagine a model of a relational database for a store of Christmas trees. Each Christmas tree will be decomposed into stem, branches, and needles. Each part of the tree will get the tree identifier as one of its properties (as a part of a key). Then, we put all stems in a STEM-box, all branches in a Branch- box, and all needles in a NEEDLE-box.

This organization is very suitable if we have to answer a query: "How many 5-inch branches do we have in our store alltogether?" but it is not suitable for selling whole Christmas trees. If somebody wants to buy a Christmas tree then we must go through all the boxes and look up parts with the same STEM-key as the stem selected by the customer has.

#

The problem is how to get a description of a whole complex object from the database when necessary. The structure of the whole object is not stored directly in the data dictionary of the database. In the data dictionary of a relational database, there can be found everything about all the stored tables (representing relations) and their columns (representing attributes), but there is nothing stored about how to collect all properties of some object of the real world. In the data dictionary, the primary real world model with its complex objects having structured attribute values is not stored, but what is stored is the secondary normalized model, which was derived using the normalization process. We can describe this problem as a problem of modelling power.

The information about reassembling objects from tables is stored indirectly in procedures which must be called up to get a complex object together from different tables. Running these procedures can be time-consuming and it often causes a problem with performance.

1.2 Problem of object synthesis

There are in principle two ways to collect the description of an object, i.e., all attributes of an object, from many tables:

- a synthesis using join operations,
- a synthesis using navigation through foreign keys.

In the first approach, we follow relational algebra, where the join operation is used for this purpose. However, the join operation is a subset of the Cartesian product and it is a very time-consuming operation.

In the second approach, we give the programmer the responsibility for writing the right logical operations for all tables, which will be accessed or connected by pointers.

The first approach is used in databases systems based on SQL, the second approach is used in database systems based on dBASE-like languages. We will describe both of these briefly in the next subsections.

An SQL-synthesis of an object When using SQL in its pure form we have no other tool for synthesis of an object description than the join operation.

```
SELECT      <list of all object properties>
FROM        <list of tables where the properties are
              stored>
WHERE       <conditions of the joins>
```

We must emphasize again that the list of tables where the properties of the object are stored is not stored in the data dictionary but must be given in the clause FROM in the SQL query statement.

Example:

To synthesize the object Nr. 177 in the Christmas tree database:

```
SELECT    STEM_BOX.*, BRANCH_BOX.*, NEEDLE_BOX.*
FROM      STEM_BOX, BRANCH_BOX, NEEDLE_BOX
WHERE     STEM_BOX.STEM_KEY   = 177
    AND   BRANCH_BOX.STEM_KEY = 177
    AND   NEEDLE_BOX.STEM_KEY = 177
```

#

Often it will be an equi-join operation, but not necessarily. There are many clever optimizing techniques but the complexity of the join operation is still quite considerable.

When using SQL in embedded mode we can simulate the navigation through foreign keys described in the next subsection.

A dBASE-synthesis of an object Database systems using a dBASE-language began with dBASE II on small computers under CP/M without a hard disk. In principle, it was not possible to use a join operation. In the set of operations of dBASE, you can find a JOIN statement, but its purpose is to create a new table (i.e., a new file) containing desired data. If you use it for big tables, it will probably run the whole night.

For composing an object with parts from many tables (better to say for accessing an object's parts stored in many tables), there is the SET RELATION TO statement. It should be noted that each table in dBASE is represented by a file, and that we can use altogether 10 working areas in the main memory for 10 opened data files. We will describe some properties of the statement SET RELATION TO for the purpose of illustrating navigation in the relational model. The main idea is that we have some tables stored in working areas, and for each such table there is a pointer at the current row. Using SET RELATION TO, we describe the logical connection under tables for the purpose of collecting properties of the given object. We get a chain of tables. This chain of tables can be considered here as a representation of the whole object structure. If we set the pointer of the first table in the chain on the key value identifying the given object, the statement SET RELATION TO provides the movement of all pointers in the chained tables to rows representing properties of the given object. For connected columns of tables, we must have index tables.
This method seems to be better in performance but:

- The whole responsibility is tranferred to the programmer who has to write all connections and logical operations.
- It was probably never tested for big tasks because of limitations concerning the number of opened files, used working areas, etc.

At this point, it should be noted that in the dBASE family we can include the database system Clipper which offers for this purpose 255 working areas. In each of them, there can be one data file and 15 index files opened. Each datafile in each working area can be connected by a SET RELATION TO statement to 8 other data files.

1.3 Problem of recursive querying

In the previous section, we described how to get a flat tuples description of complex objects and how to get complex object from flat tuples. Another problem is in the limitation of traditional query languages. It has been proved that the relational query languages do not allow expression of the computation of the transitive closure of a relation. The transitive closure of a binary relation R is the smallest relation S that includes R and is transitive, i.e., $S(X,Y)$ and $S(Y,Z)$ imply $S(X,Z)$.

In the following example we will show a typical application and explain the motivation of transitive closure of queries.

Example:

To investigate the hierarchical structure of an office trying to find if Mr. Y3 is a subordinate to Mr. X, we describe the hierarchical structure in the following table:

Name	List of direct subordinates
Z	{ X, V, W }
X	{ A, B, C, D }
...	
B	{ E, F, G}
...	
G	{ Y1, Y2, Y3, Y4 }

From the table, we see that Mr. Y3 is not directly subordinate to Mr.X but we must investigate if he is not directly subordinate to any of A,B,C,D, etc. This means that we must investigate if Mr. Y3 belongs to a subtree with a root built by Mr.X.

#

We see that the query has to be repeatedly applied and the number of steps is not known beforehand. In the relational model with underlying relational algebra, i.e., in SQL, tools are not available for writing a loop except for a quantifier. That is why we cannot express this type of query in the relational model.

The problem is we need to ask them. It was found that more than 15% of all queries to an airport database were of the following kind: " I am searching the cheapest connection from X to Y. It is not important how many times I will have to change."

1.4   Problem of inefficiency

As shown above, relational databases have two main shortcomings [Maier89]:

- lack of modelling power,
- lack of performance.

Both of these are created by decomposition. This is the reason why most CAD systems either perform their own data management on top of the operating system's file system or use only some specific services of a relational database system. However, the file processing system concept has disadvantages in not understanding the structure of the data items. This is important for integrity and for storage mapping driven by context. If a CAD system is built atop a relational database, then it uses its operations exclusively for selection, because of the efficient indexing routines, or for projection. Further, features are used which concern recovery, concurrency control, and buffer management.

The main reason that the commercial database systems are not fast enough to support the interactive design tools of today is that they are built for other tasks. Commercial database systems were developed for use in bank agendas where the actions mostly required are:

- to get a few tuples from a relation and update them,
- to select large groups of tuples from a relation and perform the same operation on them.

In the case of bank agendas, the structure of the real world entities fits very well with the logical (and so far with the physical) structure of the database.

When designing an integrated circuit, the situation is quite different. Rerouting a wire can induce a process taking up to half an hour because this one change must be propagated to many tables.

Summarising, we can stress the following main reasons for the unsuitability of commercial relational databases for manipulating structured complex objects:

- Normalization and its requirements for the splitting of complex objects increases the level of indirection between an object and its representation in the normalized relational model. Reassembling the pieces of the description of an object requires many join operations, and it is too time consuming.
- Commercial systems are optimized for high transaction throughput, i.e., for big amounts of simple transactions. Demands of transactions are collected together and the disk accesses, which are to be made for this collection of transactions, are optimized. For CAD systems, a small amount of complex transactions must be processed and a fast response to individual transactions is required. Design transactions can be long-lived, i.e., the designer can work on the update the whole day.

To overcome these shortcomings we have two possibilities:

- to extend the relational model,
- to substitute for the relational DBMS something which represents objects in their entirety within a single data structure because, as we have seen, the splitting (the necessary decomposition) of the representation of an object causes most of the problems. Such a representation is offered by an object-oriented DBMS.

Both possibilities will be discussed in this book.

2 Extensions of the relational model

As shown in the previous chapter, there are some shortcomings in the use of the relational model. Commercial and experimental database systems problems are answered by some extensions of the pure relational model. The answer to the problem of splitting a complex object is:

- nested relations,
- surrogates.

The answer to the problem of recursive querying is the embedded mode in SQL.

Further, we will discuss the concept of integrity constraints and the concept of triggers, because it should be stressed that the goal of the development of DBMSs is to store as much semantic information in descriptive form as possible in the data dictionary.

2.1 Nested relations

As shown in previous sections, properties of real-world objects differ from the idea of storing objects of one type simply into one table and the solution by decomposition has its shortcomings. It is quite natural to ask about possibilities of a relational algebra that would accept a relation as a value of an attribute.

The idea of the nested relational model is simply to allow relations at any place where attributes occur. Hence, attribute values may either be atomic or relations. This (hierarchical) nesting of relations may be repeated for an arbitrary (but fixed) number of levels.

When defining nested relations where attributes may be relations valued, obviously the "flat" definition becomes recursive. We can speak about a schema of a nested relation, which is a set of attribute descriptions each of which consists of an attribute name and an attribute schema. If an attribute schema is empty as a set, then this attribute is an atomic one, otherwise the schema describes the relations contained in the attribute values. Accordingly, the domain of an attribute is either a set of primitives (for an atomic attribute) or the powerset of the Cartesian product of the domains of the subattributes. Hence, every valid value of a non-atomic attribute is a set of tuples over the corresponding domains, which reflects the idea of "nested" relations. Schemata of nested relations can be represented by trees, and values by nested tables.

A suitable language is necessary for manipulation of nested relations. It should follow a simple idea: whenever a relation is encountered at some place in the language, we want to allow the application of queries expressed in that language. Such a language we will call a nested SQL.

Example:

We can describe a nested relation in an SQL-like notation:

```
CREATE TABLE Departments
    (DeptNumber    NUMBER,
    DeptName      CHAR(20),
    (TABLE Teachers
            (TeacherNumber      NUMBER,
            TeacherName         CHAR(25),
            (TABLE Teaches
```

```
        (SubjectNumber        NUMBER,
         SubjectName          CHAR(25),
         Day                  CHAR(2),
         Time                 CHAR(8),
         Room                 CHAR(8)
     )
   )
 )
)
```

How to query and manipulate data in nested relations, we will show in the next example.
We will define

```
CREATE TABLE Courses (    CourseNo  : NUMBER,
                          DeptName  : CHAR(25),
                          CourseName: CHAR(25),
                          Type      : CHAR(4),
     (TABLE Personal (    Person    : CHAR(15),
         (TABLE History(Time         : CHAR(10),
                          Students  : NUMBER
                       )
              )
                  )
     ),
     (TABLE RoomDemand(   Room      : CHAR(5),
                          Priority  : NUMBER
                       )
     )
                  )
```

Before querying, the nested relation must be filled with data:

```
INSERT INTO Courses(CourseNo, DeptName, CourseName, Type,
     (INSERT INTO Personal( Person,
          (INSERT INTO History( Time, Students)),
          (INSERT INTO RoomDemand( Room, Priority))
                       )
     )           )
     VALUES (113,"Math","Algebra","Lect",
              ("Brown",("1.Q.85",100),("A115",1))
          )
```

In the INSERT statement we name all columns, nested tables and their columns which have to be inserted - a list of structures - and then we write a list of values. The structures of the two lists must match. We need not express the structure of a nested relation, if not necessary:

```
SELECT      DeptName, History
FROM        Courses
```

We can use nested relations in the FROM-clause in the following way.

Example:

Query: For each course display teachers who were teaching
            it in the past at the same time, and display how many
            students preferred their lectures or exercises.


```
SELECT    CourseNo
      (SELECT    P1.Person
(SELECT   P2.Person, Time, P1.Students, P2.Students
 FROM     P1.History, Courses.Personal P2
 WHERE    P1.Person <> P2.Person
          AND
     P1.History.Time IN (SELECT    P2.History.Time
                         FROM      P2.History
                         )
)
      FROM      Courses.Personal P1)
FROM   Courses;
```


#

Many nested SQL-like languages have been proposed but there has been no standard until now. We have used only the main idea here, not specifying details.

By using nested relations with a nested SQL we solved only one of the shortcomings of the relational model - its lack of modelling power. The additional modelling power made available by nesting allows the representation of a higher-level user interface. At the high level, objects that have set-valued fields can be represented in a single tuple rather than several tuples. The correspondence between a tuple and an entity in the user view leads to a more natural interface.

Nesting is also useful for physical organisation of data. At the lower level, such implementation techniques as clustering or repeating fields can be represented using the formalism of the nested relational model. This added power is due to the fact that a nested relation expresses more from the semantic because it includes the structure which cannot be represented in the 1. NF relations.

A difficulty with the use of any relational language for that matter, within a programming language interface, is the fact that typical programming languages are record-oriented, while relation languages are set-oriented.

The extension of SQL/NF is a step towards the incorporation of the class-subclass semantics of object-oriented databases within a nested relational framework as we will explain later. Still lacking within the nested relational model is a means of representing executable code within the data model. Such a representation is used for methods in the object-oriented approach.

2.2 Object identity and the surrogate concept

In the relational model, every row of a table will be uniquely identified by a user-defined key. It can cause the following problems:

- The user-defined key is unique only within the table.
- In different tables, different types and different combinations of attributes are used as keys.
- During development, there may be a requirement that the identification system chosen by the user should be changed (e.g., two companies using a different identification system merge).
- In CAD applications, a mouse will be used for manipulation with an interface and it will not be convenient to force the user to identify each part of the constructed equipment.

As a solution a surrogate will be defined. This is a system-defined and system generated, globally unique identifier for each object in the database. It will not be changed during the life-time of the object. System date and time are often parts of this identifier. It is persistent and independent of object state or object address. The user has read-only access to it.

With use of the surrogate, the following extensions can be made:

- nested relations,
- hierarchical relations,
- more-dimensional relations.

Hierarchical relations are built from dependent relations which contain not only their own surrogate but the reference surrogate too. The reference surrogate often represents the IS_PART_OF relationship and will be used for constructing the hierarchical structure.

In ORACLE, there is a ROWID system attribute which has a unique value for each row. The first problem is that it has a constant value only within one query and not during the life-time of the row. The second problem is that one row does not represent the complete object very often.

In dBASE, the concept of indirect file name can be used. To build a hierarchical structure of relations (tables), we can store the names of child-tables into the fields of parent table. Then, we can use the mechanism USE(file_name) or USE &file_name. This concept can be used as surrogate if we would represent every object by a special file containing one record only. This approach would be not very efficient, of course.

Concepts of object identity will be deeply explained in chapter describing object-oriented paradigms.

2.3   Complex objects

The concept of complex objects has been introduced in attempts to release the
first normal form restriction of the relational model. A complex object can be
thought of as an abstraction of a structure or a record in a programming language. It allows repeating groups, record nesting, and using a tree structure for
storing its components. The simplest complex object can be an aggregate of components (also objects), each of which may have a different type. In later chapters
about object-oriented programming, we will introduce this as a collection.
The concept of complex object contains the following extensions:

- sets of atomic values,
- sets of tuples (nested relations),
- tuple-valued attributes,
- general sets and tuple constructors,
- object identity,
- bitmap-valued attributes for storing multimedia data (storing pictures, finger-
  prints, X-ray photographs, audio signals, etc.)

In practice, we also use complex objects in simple applications.

Example:

A foreigner is discussing his telephone number with the immigration officer
when filling in a form.

```
Question: "Do you have a telephone number?"
Answer:   "Well, I work for a building firm and every
          morning I am in my room on NumberA. I am at a building
          site on NumberB every afternoon. I have a telephone in
          my car - NumberC. On working days I am in this
          town, where I have no telephone, but my neighbour John
          has NumberD. At weekends, I live in the country
          where I can be reached on NumberE."
```

The officer supposed to get one number only but he has got a complex data
structure.

#

2.4   Iterative and recursive querying in nested relations

In the previous section, we described the research area concerned with removing
the limitation of flat tuples by supporting structured objects instead. Another
area of research deals with the limitation of traditional query languages. It has
been proved that the relational query languages do not allow expression of the
computation of the transitive closure of a relation. The reason is that the query
has to be repeatedly applied and the number of steps is not known beforehand.

In the relational model with underlying relational algebra, i.e., in SQL, tools are not available for writing a loop (except for a quantifier).

Consider about an airport information system containing data of direct connections between airports. Using an SQL-like language we can define these connections as follows:
We define the direct connections:

```
CREATE TABLE Connection ( Start          CHAR(40),
                    (TABLE Destination   CHAR(40) ) );
```

```
Connection

Start           Destination

London          Paris, Frankfurt
Paris           Frankfurt, London
Frankfurt       London, Paris, Vienna
Vienna          Frankfurt, Rome
Rome            Vienna
```

Query: Which cities can be reached directly from London?

```
      SELECT    Dest
      FROM      St   IN   Connection, Dest IN St.Destination
      WHERE     St.Start = 'London';
```

We used two pointers. The first pointer St points to an element of the column Start and the second pointer Dest points to an element of the list Destination.

If we ask an inverse query, i.e., we search all cities from which Frankfurt can be reached, we can write:

```
      SELECT    St
      FROM      St   IN   Start
      WHERE     EXISTS   'Frankfurt' IN St.Destination;
```

We can find all cities which are indirectly reachable with at most one change by using the following approach:

```
SELECT X.Start, (SELECT   YY.Dest
      FROM      Connection Y, Y.Destination YY
      WHERE     Y.Start IN (SELECT   Destination
                            FROM   X.Destination)
      FROM Connection X;
```

Obviously, the problem of computing all change connections cannot be solved by using the SQL because there may be arbitrarily long chains whose lengths have to be known in the compile-time.

Some theoretical approaches have been published, but the embedded mode of query languages will be used in applications.

2.5 Integrity constraints

One of the objectives of database technology is to provide means to maintain control over and preserve the integrity of the database. Database integrity expresses a true mapping between the data and the real world objects and their properties. The properties and relationships represented by data must be the same as in reality. In database programming, the term integrity has evolved to refer to the qualities of validity and consistency. Integrity is also concerned with illegal data accessing and modification.

The intent is to set up a number of integrity constraints on the data via special facilities available to them, and then for every update to make sure that the integrity constraints are satisfied.

It has to be mentioned that integrity is like security as far as the extent to which it is achieved and the cost of achieving it. The higher the degree of integrity, generally the higher the cost of achieving it in terms of the necessary software, access time, and storage cost. Thus, the intent is to achieve a satisfactory degree of integrity at a satisfactory cost. In fact, it may be practically impossible to avoid some invalid data values.

Example:

If a new employee is hired and the secretary makes a spelling error when writing his name then there is no reasonable method of checking it.

\#

There is no industry-wide standard for integrity, although a number of proposals have been suggested. In most applications, designers have implemented at least some degree of data validation at input time for some critical fields, e.g., checking if an entered value is in a given range of values (MONTH must be in the range 1..12). This level of protection is obviously important, and difficult to achieve reliably. It has been suggested that the specification of integrity rules could account for as much as 80 percent of a database description. In some cases the validation mechanisms that have had to be developed are exceedingly specialized and sophisticated. The intent of database technology is to provide a wide range of easy to use generalized integrity mechanisms and avoid much of the traditional effort of implementing the mechanisms themselves, using them and enforcing them.

Integrity mechanisms must be provided with a set of integrity rules that define:

- what errors to check for,
- what to do if an error is detected.

Integrity rules can be divided into two categories:

- data validity,
- consistency.

We will now outline some of the most important data validity and consistency constraints.

Data validity constraints are also called domain integrity rules. They concern acceptability of a value considered independent of its relationships to other values in the database.

- Data types and format
  A data field may be allowed to contain only values whose type and format must exactly match the characteristics declared for the data field in the schema.

- Value ranges
  The value of a particular field may be required to fall within a certain range. For example, the MONTH field must have values between 1 and 12. A field may be required to take only specified values. For example, DEPT_NO can take only one of the five given values. It is usually implemented as a main part of the input data validation step prior to processing.

- Uniqueness of a key field
  Practically all DBMSs provide means to ensure the uniqueness of a key of an inserted record.

Consistency constraints also concern the case in which two or more values in different tables in the database are required to be in agreement with each other in some way. The values of a data field in a given record may be required to appear also as values in some field of another record (or even perhaps in the same record). This type of constraint is also called referential constraint.

- The value of a data field in a record may be required to be the result of a procedure using actual values of other fields. Such a data field may be called a calculated field. A calculated field is materialized only at execution time and the value is not actually stored in the database.

Maintenance of data validity and consistency constraints entails much data checking. All update, insert and delete operations on the portion of the database with the constraints will have to be monitored. It may be very expensive for extensive constraints. Thus, the appropriate checking may be done more cost-effectively by means of a special utility program run periodically to perform the checking, rather than whenever the update operations occur. An important concern of the database administrator is setting up all these checking and monitoring methods.

2.6 Triggers

Occasionally, it is necessary to incorporate side-effect-type mechanisms into the system. For example, it may be required that access to one table leads to changes in another table; alternatively it may be that the addition of a row to a table set results in the production of a report. A DBMS can support those facilities which are associated with a table and defines that when a particular condition involving information from that table occurs, then an action is induced. When such an event is triggered then a procedure is called to carry out the necessary processing. Parameters passed to the procedure can include a copy of the information which caused the event to trigger. In order to reduce the complexity of the system, the DBMS allows any number of triggers to be associated with an entity set, but imposes a restriction that only one event can be associated with each trigger. The condition for checking is called a trigger-condition, and the procedure representing the expected action is called a trigger-procedure. Syntactically, we can write:

```
AFTER / BEFORE    <event>   CALL <procedure>
```

Events can be specified as STORE, DELETE, MODIFY, GET.

Triggered procedures have applications in the area of maintaining integrity and also in computing values of calculated fields, in enforcing authorization constraints, in selection of fields for insertion, in providing data encryption, in monitoring and tracing, in debugging, in compressing and decompressing data, in exception reporting. In dBASE-like systems, there are not really any triggers but we can use the mechanism of the statement ON.

2.7 Simple extensions in commercial DBMSs

As shown, the use of the pure relational model is associated with serious short-comings. The commercial DBMSs are not designed and implemented in the purely relational way. We will show the extensions on examples from the dBASE family (representing simple database systems) and from the SQL family (representing complex database systems). All these systems suppose data in the first normal form.

Extensions in dBASE The dBASE system was developed as a generalization of a file processing system. Some of its former features are still included in it. It is important to say that these features are often very useful. We will discuss:

- variables,
- heterogeneous arrays,
- user-defined functions.

Variables and arrays are used to store data found in files (in dBASE, a table is implemented as a file). Variables are destined for storing single values. One-dimensional heterogeneous arrays can have the same structure as a record of a data file. This implies that two-dimensional arrays can be used for storing sets of the records found.

User-defined functions (UDF) are very powerful instruments which, however can also be dangerous. A user-defined function can be called in an expression in a dBASE statement. This means that such a function call interrupts the execution of the dBASE statement and can cause some changes in the data structures used by the interrupted dBASE statement (e.g., the UDF can change the current record pointer). This can result in an inconsistency.

Embedded SQL and extensions in ORACLE When using SQL we must remember that the relational model is set-oriented whereas the reality is object oriented. When we describe each object as a record, we can say that the data processing is record oriented. All records in a file can be identical as far as their contents are concerned but they are still distinct objects having different physical addresses. Set orientation does not take into account the order of elements in a set. These concepts are important to remember when using the clauses DISTINCT and ORDER BY.

In data processing, the order of records in a file is very important from the efficiency aspect. The use of indexing and clustering techniques is purely record oriented and has nothing to do with relational algebra.

In those data processing features which are not supported in the relational model, we need to perform operations on sets prepared by SQL. The concept of embedded SQL opens up the operational capabilities of the DBMS, i.e., in a C program, we can only ask SQL to select specified rows from specified tables.

The fundamental idea underlying embedded SQL is that SQL statements (with various differences) can be used embedded in a program in some host language (C, Pascal, COBOL, FORTRAN, etc.). They can cooperate with host language variables (these variables have a preposition ":" in the SQL environment). The hybrid program consisting of host language statements and SQL statements will be translated by a preprocessor. SQL statements are usually introduced by EXEC SQL, i.e.:

```
EXEC SQL <SQL statement> ;
```

The preprocessor replaces SQL statements with specific procedure calls matching the host language syntax. After that, the generated program (completely in C, Pascal, etc.) will be compiled. In the following examples, we will use a Pascal-like language without EXEC SQL (there is no standard).

In the simplest case, the result of a query will be a single number. We can store the number found using the INTO clause in the SELECT statement.

```
Example:
```

```
Query: How many first names do the department teachers have?
```

```
program Names;
```

```
            var
                  Counter : integer;
      begin
            SELECT    COUNT( DISTINCT NAME )
            INTO      :Counter
            FROM      DEPARTMENT;
            if SQLCODE = 0 then { OK }
                  write('Different first names :', Counter)
            else
                  write('Error');
      end.
```

#

If we can be sure that only one row of the table (record) will be found, we can use the following construction.

Example:

Query: Who is heading the given department?

```
      program Head;
            var
                  Dept_number : integer;
                  TeacherName          : string;
      begin
            read(Dept_number);
        SELECT  NAME
        INTO    :TeacherName
        FROM    EMPLOYEE
        WHERE   TEACHER_CODE IN
                      (SELECT    TEACHER_CODE
                       FROM      DEPT
                       WHERE     DEPT_NR = :Dept_number);
      if SQLCODE = 0 then
                  write('The name is : ', TeacherName)
                        else
                  write('Error');
            end.
```

#

In the following example, we will show the case of a SELECT statement that selects a whole set of rows (records). These records can be made accessible successively by using a mechanism called cursor. The explanation of statements DECLARE, OPEN, FETCH, CLOSE is given in program comments.

Example:

Query: What are the names and the numbers of all departments?

```
    program List;
        var
                DeptNumber : integer;
                DeptName    : string;
    begin

    { declares a cursor DEPT_CURS for a given table and
projection }              DECLARE DEPT_CURS CURSOR FOR
            SELECT    DEPT_NO, DEPT_NAME
            FROM      DEPT;

    { prepares the cursor DEPT_CURS for using }

        OPEN DEPT_CURS;

    { makes accessible the first tuple produced
      by the previous SELECT }
        FETCH DEPT_CURS INTO :DeptNumber, :DeptName;

    { repeat fetching records until
      the last record not reached}
        while SQLCODE = 0 do
            begin
                    print(DeptNumber,'    ',DeptName);
                    FETCH DEPT_CURS INTO :DeptNumber, :DeptName;
            end;

    { cancel using the cursor DEPT_CURS }

        CLOSE DEPT_CURS;

    end.
```

#

The next SQL extension, which we explain here, is so called dynamic SQL. The principal idea is that we can construct the text of the SQL statement dynamically, i.e., at run time. This text has to be assigned to a string-variable defined in the host language environment which is often named SQLSOURCE. In the SQL environment a variable (often called SQLOBJ) will be declared into which the SQL statement will be translated. For this purpose the statement PREPARE exists in the SQL environment. Execution will be triggered by the

SQL statement EXECUTE.

Example:

The following program will ask for a statement and then perform it as if we input the statement from the keyboard in the interactive mode.

```
program Inter;
    var
        SqlSource : string;
begin
    DECLARE SQLOBJ STATEMENT;
    write('Query: ');
    read(SqlSource);
    PREPARE SQLOBJ FROM :SqlSource;
    EXECUTE SQLOBJ;
end.
```
#

References

[Abit86]   Abiteboul,S., Bidoit,N.: Non first normal form relations: An algebra allowing data restructuring. Journal of Computer and System Sciences, 33, pp.361-393, 1986.

[Banc90]   Bancilhon,F., Bunemann,P.(Eds.): Advances in Database Programming Languages. ACM Press Frontier Series. Addison-Wesley, Reading, MA,1990.

[Codd79]   Codd,E.F.: Extending the database relational model to capture more meaning. ACM Transactions on Database Systems, 4(4), pp. 397-434, 1979.

[Kroha93]   Kroha,P.: Objects and Databases. McGraw-Hill, 1993.

[Roth88]   Roth,M.A., Korth,H.F., Silberschatz,A.: Extended algebra and calculus for nested relational databases. ACM Transactions on Database Systems, 13(4), pp.389-417, 1988.

[Schek86]   Schek,H.-J., Scholl,M.H.: The relational model with relation-valued attributes. Information systems, 11(2), pp. 137-147, 1986.

[Stone87]   Stonebraker,M.,Rowe,L.A.: The POSTGRES Papers. Mem.No. UCB/ERL M86/85, University of California, Berkeley, 1987.

[Date86]   Date,C.J.: An Introduction to Database Systems. Fourth Edition, Addison-Wesley, Reading, 1986.

[Korth86]   Korth,H.F.,Silberschatz,A.: Database System Concepts. McGraw-Hill, 1986.

[Maier89]   Maier,D.: Making Database System Fast Enough for CAD Applications. In: Kim,W., Lochovsky,F.H.(Eds.): Object-Oriented Concepts, Databases, and Applications. Addison-Wesley, 1989.

[Ullman82] Ullman,J.D.: Principles of Database Systems. Second Edition,Computer Science Press, Rockville, Maryland, 1982.

Evaluation of Object-Oriented Database Systems

C. Huemer†, G. Kappel‡, S. Rausch-Schott‡, W. Retschitzegger‡, A Min Tjoa†,
S. Vieweg† , R. Wagner*

†Institute of Applied Computer Science and Information Systems. Department of Information
Engineering. University of Vienna, Austria

‡Institute of Computer Science. Department of Information Systems.
University of Linz, Austria

*FAW. Research Institute for Applied Knowledge Processing.
University of Linz, Austria

Abstract. Despite the fact that object-oriented database systems (OODBS) have
gained potential as promising database technology for non-standard applications
such as computer integrated manufacturing there does not yet exist broad experience with the use of OODBS in real-world applications. One reason is that the
features of OODBS, both functional and performance, haven't been exposed to a
broader audience. The goal of this chapter is to shed some light upon evaluating
the features of OODBS. In the first part of the chapter we discuss an extensive
evaluation catalogue for advanced database systems which has been developed
during the course of a real-world project. A discussion of the pros and cons of
using such evaluation catalogues summarizes our experience. The second part of
this chapter surveys existing performance benchmarks for OODBS. Special
emphasis is given to the requirements a benchmark has to fulfill, and to the practical applicability and usefulness of the proposed benchmarks.

Categories and Subject Descriptors: C.4 [Performance of Systems]; H.2.8
[Database Management] Database Applications; K.6.2 [Management of Computing and Information Systems] Benchmarks

General Terms: Database Systems, Performance Evaluation, Advanced Database Applications

Additional Key Words and Phrases: database requirements for CIM, functional evaluation of object-oriented database systems, application benchmarks

1 Introduction

Advanced database applications such as computer integrated manufacturing have
emerged over the past decade [Cattell91, Encarnação90, Gupta91]. Object-oriented
database systems are designed to meet their database requirements.

The first prototype implementations of OODBS were hardly usable in production
environments. Recent developments and improvements of existing systems increase
the usability of OODBS in practice. Consequently, the demand for means to evaluate
this new technology has emerged.

In this chapter we describe the functional and performance evaluation of OODBS. The functional evaluation of OODBS concerns the assessment of criteria that form the core of object-oriented database technology whereas performance evaluation provides means to measure database systems with respect to a given workload in a given application domain. In the following we will present approaches to both functional and performance evaluation of OODBS.

This chapter is organized as follows. In the next section we will describe a functional evaluation framework for object-oriented database systems. Section 3.3 gives an overview of performance benchmarks for object-oriented database systems. We conclude with a survey of the most popular approaches.

2 Evaluation Criteria for Object-Oriented Database Systems

Object-oriented database systems were designed to meet the database requirements of advanced database applications. The functionality of OODBS can be characterized by the features depicted in Table 1. These features are part of the evaluation framework as developed by the authors in the course of an evaluation project [Kappel92]. The framework serves as the basis for our further investigations of the functional evaluation of OODBS. The developed criteria catalogue was designed to allow the qualitative and quantitative assessment of object-oriented database systems. The presented criteria have been developed through an extensive study of the literature [Ahmed92, Atkinson89, Encarnaçao90, Kappel94, Stonebraker90] and from experience gathered in previous projects. Our analysis of object-oriented database systems is made up of a detailed list of questions. Note that our approach is not restricted to object-oriented technology, but that it provides a framework for evaluating advanced database technology in general. These questions are structured into sections representing the main features of database technology. Each section is further refined into two or more levels of subcriteria. The subcriteria allow a more detailed assessment of the evaluated systems. The evaluation catalogue comprises 20 sections each with about 25 subcriteria. The amount of information items (sections and subcriteria) totals to more than 500 questions.

OODBS Features	
• Data Model	• Concurrency Control
• Constraints & Triggers	• Recovery
• Persistence	• Authorization
• Data Dictionary	• Architecture
• Tools	• Storage Management
• Query Management	• Query Optimization
• Host Programming Languages	• Operational Conditions
• Schema Evolution	• Distribution
• Change Control	• Interfaces
• Version Management	• Business Criteria

Table 1. Features of OODBS

A functional evaluation of the OODBS represents the starting point of any database evaluation. We will therefore describe the evaluation features in more detail.

An object-oriented database system is a database system with an object-oriented **Data Model**. At present, there exist several different object-oriented data models. They are either based on existing object-oriented languages, like C++ and Smalltalk, or they have been newly developed. There exists no single object-oriented data model as it was the case for traditional (hierarchical, network, relational) database systems [Maier89]. Nevertheless, there is consensus that a data model to be called object-oriented has to exhibit the following core features [Atkinson89]: *complex object modeling, object identity, encapsulation, types, inheritance, overriding with late binding, extensible types,* and *computational completeness.* Complex object modeling deals with the specification of objects out of other objects which may also be complex by nature. Object identity provides a system-defined unique key and thus, allows the sharing of objects. The principle of encapsulation defines data (= structural information) and accessing operations (= behavioral information) as one unit called object type (or object class). It allows to distinguish the interface of the object type, i.e., the set of operation signatures, from its implementation, i.e., the implementation of the operations and the underlying data structure. An operation signature consists of the name of the operation, the types and names of its input parameters, and the type of its return value. An object type defines structural and behavioral information of a set of objects called *instances* of the object type. Inheritance supports incremental type specification by specializing one or more existing object types to define a new object type. Overriding means redefinition of the implementation of inherited operations, and late binding allows the binding of operation names to implementations at run time. The extensible type requirement permits new object types consisting of structural and behavioral information to be defined by the user and, furthermore, to be treated like any other predefined type in the system. Finally, computational completeness requires that any computable algorithm can be expressed in the data definition language (DDL) and/or the data manipulation language (DML) of the database system. The features described above have all been incorporated into the evaluation catalogue.

Constraints and Triggers represent an advanced technique to specify integrity constraints for a database. Constraint and trigger specification is important in the context of active database systems (see chapter „Active Object-Oriented Database Systems for CIM Applications). We therefore included it in our evaluation. The questions mainly focus on how constraints and triggers are incorporated into the database system. This includes the kinds of triggers and their enforcement.

Persistence is one of the main features of any database system. Object-oriented database systems follow different approaches to reach persistence. Persistence can be reached by declaration, by reachability from other persistent objects or by collection membership. The way how objects may become persistent is a crucial issue in the context of porting applications between OODBS and is therefore evaluated in detail. Furthermore we included the deletion semantics, the orthogonality to types, and the homogeneous handling of persistent and transient objects in our investigations.

Usually, advanced database applications operate in a complex environment. A **Data Dictionary** supports the management of such an environment by providing access to schema information. Furthermore, **Tools** for application development, user management, report generation, database archiving, database integrity checking, and for accessing the data dictionary were investigated.

The flexible access to the database and thus **Query Management** is one of the most important requirements for advanced database applications. With flexible database access we address the need for varying and demanding database manipulation requirements. The database access strategies differ from task to task. Some of the tasks, such as engineering design may require the access to objects via inter-object references (navigational access) while others such as the manipulation of incoming orders require associative access via the specification of predicates (associative access). In order to allow flexible access to the database, both access strategies must be supported within the same language. [Bancilhon89] discusses several requirements for an advanced (object-oriented) query language. These include ad-hoc facilities to state queries and updates in a high level language without using a programming language. In addition, multi-database queries and recursive query processing were considered important. Thus, the query language must have equal expressive power as conventional programming languages. Therefore, we also investigated the supported **Host Programming Languages**. Object-oriented programming languages such as C++, Smalltalk, Eiffel, and CLOS and non-object-oriented programming languages such as C, Modula-2, and Ada were included in this section.

Besides query facilities a flexible data management should also include the support for **Schema Evolution**. Schema evolution describes the mechanisms to cope with changes in the data definition of a given database. The database schema is the result of requirements specification and conceptual (and logical) database design. It thus reflects the 'real-world' entities elaborated in these design steps. These entities are subject to more or less frequent changes due to the evolving nature of 'real-world' situations. Schema evolution provides a framework to manage these changes in a controlled manner. Our analysis of schema evolution features included conceivable changes in the database schema and invariants in order to preserve the database schema's structural consistency. We assume that the database is populated whenever changes occur. With **Change Control** we denote the mechanisms that are used for the database conversion in order to conform to the new schema resulting from the schema evolution process. Thus, we investigate mechanisms for preserving the behavioral consistency in the course of schema changes. This includes strategies for the adaption of instances, methods, and queries.

The section **Version Management** investigates how the evolvement of database instances is supported. Our analysis of version models includes the specification of the structure and behavior of versioned objects. The structure describes the way in which versionable objects are composed and the basic granularity of the versioning mechanisms. The dynamic component of version models describes the intra- and inter-object relationships of versioned objects. Intra-object relationships define how object versions are created and how they relate to each other. Inter-object relationships characterize the versioned design object as referenced from other objects. The referencing objects may be other versioned objects or non-versioned objects.

Concurrency Control, **Recovery**, and **Authorization** provide means for the management of multiple users accessing the same database under restricted resources. The scheduling of these resources requires a measure of consistency and correctness. The concept of transactions provide a framework to ensure this. Transactions are a collection of actions that access the database. They are logical units that group together operations to form a complete task and they are atomic units preserving the consistency of the database. They also serve as unit for recovery to a consistent database state in case of failures. Our analysis in these sections therefore includes lock types, lock granularity, deadlock detection, the logging concept, access control, and authorization.

The achievements in workstation and network technology forced the shift from the host based computing paradigm to the client/server computing paradigm. The **Architecture** of the database system and the **Storage Management** are strongly connected to this computing paradigm. Our analysis is based on the supported client/server concept, disk management, and the memory architecture. We investigated which paradigm is followed in the client/server communication; whether pages, files, or objects are transferred from the client to the server. Furthermore, we investigated the main memory layout, and how disk replicas, indexing and clustering are supported. These issues mainly address performance considerations at the physical level.

Query Optimization addresses the efficient execution of queries. Our investigations in this section mainly focus on the use of the above mentioned features (indexes, clustering) as well as the management of the query optimizer (tuning and interrogating).

In the section **Operational Conditions** we analyzed the hardware and software requirements for each of the database systems. Object-oriented database systems mainly operate in a distributed hardware and software environment. The section **Distribution** addresses how data and control may be distributed. We investigated whether data distribution is transparent to the user. Furthermore, remote-database access, multidatabase queries, and the heterogeneous environment were investigated.

Advanced database applications do not operate in an isolated environment. They are integrated into various communication and information services. Therefore, we included the supported **Interfaces** in our analysis. This comprises the database interface to CASE tools, and standardized description languages such as STEP-Express.

Business Criteria such as reference installations, customer support, documentation, pricing, and the support of standards (ANSI C++, ODMG [Cattell94], etc.) are of great importance in a production environment. However, it is quite difficult to assess most of these features.

As mentioned above, all of these features are relevant to the evaluation of any OODBS. The presented evaluation schema is used as a starting point for the evaluation framework. The evaluation is carried out in three phases:

(a) Functional Evaluation of the OODBS

(b) Assessment of the Database Requirements

(c) Rating of the Database System

In the first step the evaluation catalogue is filled in by database experts in order to rate the evaluation features. The functional evaluation is carried out either by means of interviews with product experts or by an extensive study of the product literature. In addition to the 'on-paper' evaluation simple benchmark programs that emphasize a particular feature may be implemented to judge controversial evaluation features. In the second step the evaluation features are assigned weights corresponding to the importance of the evaluation features as rated by the application domain expert. This task is highly dependent on the intended use of the OODBS and eminently relies on the database requirements of the application domain. In the last step, the results of the functional evaluation and of the assessment of the database requirements are joined. This task may be accomplished either with qualitative or quantitative methods. In the former case the decision is taken on the basis of informally stated evaluation ratings such as 'suitable/non-suitable' and 'high/low'. In the latter case the overall rating of the evaluated products is computed by summing up the evaluation ratings weighted with the importance of the evaluation features. In both cases, the result represents a sound basis for the final decision process.

Besides functional requirements performance considerations may influence the decision process as well. In the following we will describe the issues that are relevant to performance evaluation in non-standard application domains. We will then discuss the approaches to benchmark advanced database applications; namely, the *Sun* benchmark, the *HyperModel* benchmark, the *OO1* benchmark, the *OO7* benchmark, and the *SEQUOIA 2000* benchmark.

3 How to Benchmark Object-Oriented Database Systems?

The need for new approaches to database management systems for advanced applications has been widely reported. Besides the advanced functionality, one of the central issues for the acceptance of advanced database systems is *performance*. Thus, measuring performance is an important topic. In this section we will present some of the main issues that effect on the evaluation of OODBS and thus, should be included in any OODBS benchmark.

Domain-specific benchmarks should follow some key criteria. In general, an application benchmark should be

- relevant,
- portable,
- scalable, and
- simple.

Furthermore it should provide a clear measure and both vendors and users should embrace it [Gray91, Gray93]. *Relevance* means that a benchmark should measure the performance of a system when performing typical operations within that problem domain. The benchmark should be *portable*, i.e. easy to implement in different systems and architectures. The benchmark should apply to computer systems in any size. *Scaling* the benchmark must be possible. The measurement of benchmark operations should result in a clear metric that allows insight into the performance of the system under test. Finally, the benchmark should be *simple* and *understandable* in order to be accepted by the audience. Within this framework, the main characteristics of an application domain must be mapped onto a benchmark specification.

When given the complexity of advanced database applications, it is obvious that traditional approaches to benchmarking DBS [Bitton83, Turbyfill91, TPCA92, TPCB92, TPCC92] lack relevance when measuring the performance of advanced database applications. Relevance can be expressed in terms of database requirements for a given application domain. These requirements include both functional requirements and performance requirements. The requirements form the basis of the database schema and the operations included in a benchmark. Consider CIM as an example for a complex application domain. Database systems represent a key technology for the realization of CIM. The database requirements for CIM applications can be grouped into three basic requirement clusters: *data modeling issues, querying and manipulation issues*, and *integration issues* (see chapter „Database Requirements of CIM Applications" in this book). Table 2 gives an overview of these requirements. The data modeling requirements comprise extended attribute domains, complex object support, relationships and dependencies between the modeled objects, and active consistency checking and knowledge-base support. The querying and manipulation requirements are widespread and include advanced transaction management issues, flexible database access structures, change management issues like versioning and schema evolution, and interfaces to various data formats. Due to the distributed nature of manufacturing systems the integration aspect of computer integrated manufacturing deserves particular emphasis. Distributed data management and multi data management are the key requirements for distributed computing in this field. Reverse engineering and integrated data and process modeling are rather concerned with the logical integration of manufacturing and business applications.

Data Modeling Requirements	• Extended Attribute Domains • Complex Object Support • Relationships and Dependencies • Active Consistency Checking and Knowledge-Base Support
Querying and Manipulation Requirements	• Advanced Transaction Management • Flexible Database Access • Change Management • Interfaces

Table 2. Database Requirements for CIM Applications

Integration *Requirements*	• Distributed Data Management • Multi Data Management • Reverse Engineering of Data • Integrated Data and Process Modeling

Table 2. Database Requirements for CIM Applications

A database requirements analysis as presented above forms the basis of any database application benchmark. The characteristic functional requirements, typical database operations, and the database size requirements are then merged into a benchmark suite. We will now investigate the approaches that have been chosen for advanced database application benchmarks. Therefore, we have identified the following issues as a framework for our analysis:

- Database Schema
- Database and its Generation Process
- Benchmark Operations
- Measurement Guidelines
- Reporting of Benchmark Results

The database schema describes the structure of the database and represents the complexity of the application domain under investigation. With generation process we denote the way how the benchmark database has to be constructed, and the restrictions to be met. Additionally, we describe the database size which is relevant for an application domain. The benchmark operations describe the operations that form the operational basis for the benchmark suite. They usually represent a mix of the most common and most characteristic queries and manipulations of an application domain. Moreover, we describe the measurement guidelines of the benchmark suite. They indicate how the results of the benchmark must be derived and which metric is used. As described above, the full mentioning of the system environment is very important in order to get comparable results from any benchmark run. Therefore, we also included the disclosure requirements for the reporting of the benchmarking results. In the following, each benchmark approach is investigated with respect to the above described characteristics. We will survey the Sun Benchmark [Rubenstein87], the HyperModel benchmark [Anderson90], the OO1 benchmark [Cattell92], the OO7 benchmark [Carey93a], and the SEQUOIA 2000 storage benchmark [Stonebraker93]. For all these approaches we will analyze the underlying database schema, the generation of the database, the operations, how the measurements must be performed and finally, how the benchmark results must be reported.

3.1 The Sun Benchmark

The Sun benchmark is one of the first approaches to measure database performance for OODBS. It thus represents a starting point for further approaches presented below.

Database Schema

The Sun benchmark database schema (Figure 1) describes a simple library database and consists of three basic concepts. The Person class describes persons in terms of `person_ID`, `name`, and `birthdate`. The Document class consists of `document_ID`, `title`, `page_count`, `document_type`, `publication_date`, `publisher`, and a document `description`. The Authorship relationship associates each person to 0 or more documents and each document to exactly 3 authors.

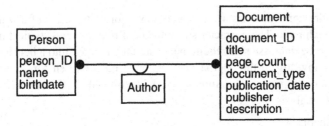

Figure 1. The Sun Benchmark Database Schema

Database Contents and Generation

The benchmark database is generated in a small and a large version. The small database consists of 20000 Persons, 15000 Authors, and 5000 Documents. The large database comprises 200000 Persons, 150000 Authors, and 50000 Documents.

Operations

The Sun benchmark consists of seven operations. These operations can be grouped into lookups (NameLookup, RangeLookup, GroupLookup), inserts, database scan, and database open. The NameLookup operation selects a person with a randomly selected person_ID. The RangeLookup operation selects all persons with birthdate within a randomly generated range of 10 days. The GroupLookup query finds the person which authored a given document. The Reference Lookup finds a person object that is referenced by a randomly selected author object. RecordInsert inserts a new person in the database. SequentialScan scans all document objects in the database. Furthermore, the time for the database initialization is measured.

Measurements

The measurement for the operations described above work as follows. The operations are repeated 50 times. This operation set is then repeated 10 times. This results in 500 queries and 50 open database operations. The average elapsed time for these operations is then reported.

Disclosure Report

No disclosure report for benchmark results is specified. The authors "hope" that anyone will report the used indexes, the access methods, and the total space required for the database.

OMT is an object-oriented modeling technique [Rumbaugh91]. For further details the interested reader is referred to the literature.

3.2 The HyperModel Benchmark

The HyperModel benchmark is based on the Sun benchmark described above. The extensions include a more complex database schema and additional benchmark operations. As pointed out in [Gray91], the benchmark provides more than a single performance measure. The operations cover a wide spectrum of database operations, each focusing on special queries.

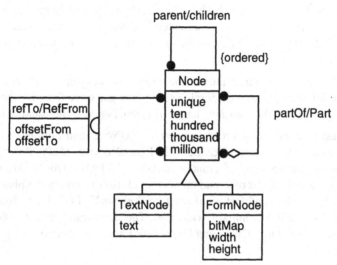

Figure 2. The HyperModel Database Schema

Database Schema

The conceptual schema of the HyperModel Benchmark (Figure 2) is based on an aggregation and generalization hierarchy of HyperModel documents. A HyperModel document consists of a number of sections each of which is represented by an object of type Node. Node has two sub-types: TextNode and FormNode. The instances of type Node have 5 attributes (uniqueID, ten, hundred, thousand, and million). In addition, a TextNode has a text attribute and a FormNode has a bitmap, width, and height attribute.

Nodes are interrelated by three relationships: the parent/children relationship (1:N), the partOf/parts relationship (M:N), and the refTo/refFrom relationship (M:N). The parent/children relationship is used to model the recursive aggregation structure of sections within a document. The children are ordered. The partOf/parts relationship is constrained to be hierarchical. Parts may share subparts but the relationship forms an acyclic graph. The refTo/refFrom relationship has been designed to model hypertext links, when attributes are attached to each link to describe the offset of each ending point within a node. The two attributes offsetFrom and offsetTo describe the starting/ending point of these links.

Database Contents and Generation

A HyperModel database consists of a single document composed of a network of nodes with the relationships described.

The parent/children relation forms a hierarchy that is made up of 7 levels (0-6). Each non-leaf node has 5 successors. The nodes are numbered starting with 1 (level 0) and ending with 10471 (last node at level 6). The total amount of nodes is 19531.

The partOf/parts relation is created by relating each node at level k to five randomly chosen (uniformly distributed) nodes of level k+1 in the tree structure built up in the parent/children relation.

The refTo/refFrom relation is built up by visiting each node in the parent/children hierarchy and connecting the node to a randomly chosen node of the entire hierarchy. The values of the attributes (offsetFrom, offsetTo) are initialized randomly (between 0 and 10).

The nodes in the parent/children hierarchy are numbered sequentially. The attributes are initialized with a randomly selected integer value from the interval 0 to the number specified by the "attribute name" (i.e. ten, hundred, thousand, million).

The lowest level in the hierarchy consists of TextNodes and FormNodes. There is one FormNode per 125 TextNodes (125 FormNodes, 15500 TextNodes). TextNodes contain a string of a random number (10-100) of words. Words are separated by a space and consist of a random number (1-10) of lowercase alphabetic characters. The first, middle and last word must be "version1". Each FormNode has a bitMap attribute which is initially white (0s). The size varies between 100x100 and 400x400 (always a square). Clustering (if possible) has to be done on the parent/children relationship.

Operations

The HyperModel benchmark consists of 7 groups of operations (20 operations). The operations can be grouped into Name Lookup Operations (nameLookup, nameOIDLookup), Range Lookup Operations (rangeLookupHundred, rangeLookupMillion), Group Lookup Operations (groupLookup1N, groupLookupMN, groupLookupMNAtt), Reference Lookup Operations (refLookup1N, refLookupMN, refLookupMNAtt), Sequential Scan (seqScan), Closure Traversal Operations (closure1N, closureMN, closureMNAtt, closure1NAttSum, closure1NAttSet, closure1NPred, closureMNAttLinkSum), and Editing Operations (textNodeEdit, formNodeEdit).

- *Name Lookup Operations*
  The nameLookup selects a node instance with a randomly chosen uniqueId value and returns the value of the hundred attribute. The nameOIDLookup operation finds the node instance given a random reference to a node.

- *Range Lookup Operations*
  The rangeLookupHundred query selects 10 % of the nodes with the hundred attribute in the range of 10. The rangeLookupMillion selects 1 % of the nodes with the million attribute in the range of 10000.

- *Group Lookup Operations*
  All group lookup operations follow the parent/children relationship, the partOf/parts

relationship and the refTo/refFrom relationship. The groupLookup1N selects the children of a randomly chosen node and returns a set of five objects. The groupLookupMN selects all part nodes of a random node. The groupLookupMNAtt selects the related node of the refTo/refFrom relationship.

• *Reference Lookup Operations*
The reference lookup operations (refLookup1N, refLookupMN, refLookupMNAtt) operate like the group lookup operations, except that they select the nodes in the reverse direction.

• *Sequential Scan*
The seqScan operation visits each node object in the database and accesses the `ten` attribute value of each node. The sum of all `ten` attribute values is returned.

• *Closure Traversal Operations*
The closure traversal operations retrieve a node from the database and transitively visit related nodes. All operations start off with a randomly chosen node on level 3 of the node hierarchy.
The closure1N operations selects a random node and follows the parent/children relationship in pre-order to the leaves of the tree. The closureMN selects a random node and traverses the partOf/part relationship recursively to the nodes of the tree. The closureMNAtt selects a random node and follows the refTo/refFrom relationship 25 times. The closure1NAttSum sums the `hundred` attribute values for all nodes reachable from a random node via the parent/children relationship. The closure1NAttSet visits all nodes reachable from the starting node via the parent/children relationship and updates the `hundred` attribute value to (99 - the actual value). The closure1NPred visits all nodes reachable by the parent/children relationship. Only those nodes (and their children) are retrieved that are out of the range 10000 of their `million` attribute values. The closureMNAttLinkSum performs the closureMNAtt operation and sums up the offsetTo attribute values of the nodes.

• *Editing Operations*
The editing operations test the interface to other programming languages and the updating of a node. The textNodeEdit operation selects a random TextNode and updates all three occurrences of the string "version1" to "version-2". The formNodeEdit operation inverts a 50 x 50 subrectangle (starting at position (25, 25)) in a randomly chosen FormNode. The operation is performed 10 times per node in order to emulate interactive editing.

Measurements

The time of each operation is measured in seconds. In order to achieve correct results in case of extensive cache use of the OODBS the operations are run with empty memory (cold run) and full cache memory (warm run). Basically, the benchmark suite consists of three phases, the setup, the cold run, and the warm run. The following sequence must be followed. First, perform the database setup. This incudes the preparation of the inputs to the benchmark operations. Then perform the "cold run". The operations must be run 50 times. If the operation is an update operation, the changes must be committed once for all 50 operations. The entire cold run is timed and divided by 50. This number is reported as the "cold run" result. The "warm run" repeats the operations 50 times with the same inputs. If the operation is an update operation, then the changes are committed once for all 50 operations. The total time is divided by 50 and reported as the "warm run" result.

Disclosure Report

The specification of the HyperModel benchmark does not require a disclosure report.

3.3 The OO1 Benchmark

As in the case of the HyperModel benchmark, the OO1 benchmark is based on the Sun benchmark. The improvements in the OO1 approach rather focus on simplifying the earlier approach. The benchmark is designed to measure the performance in the domain of engineering database applications.

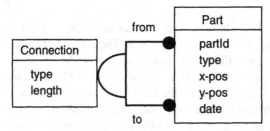

Figure 3. The OO1 Database Schema

Database Schema

The database schema of the OO1 benchmark (Figure 3) consists of two basic components: the part relation and the connection relationship. Parts are built up by `partID`, `type`, `x-pos`, `y-pos`, and `date`. The connection relationship itself contains information about the connected parts (`type` and `length`).

Database Contents and Generation

The database consists of N parts and 3*N connections. Parts have a unique identifier (`partID`) ranging from 1 to N. There are exactly three connections to other parts. The random connections between parts are selected to produce some locality of reference (90 % of the connections are randomly selected among the 1 % of the "closest" parts, the remaining connections are made to any randomly selected part. Closeness is defined by using the parts with the numerically closest `partIds`).

The algorithm for the database generation is provided by the authors. Three benchmark databases are provided: small, large, and huge (see Table 3). The unformatted small database comprises approximately 2 megabytes of data. The authors claim the database to be a good representative of engineering databases. The benchmark results are mostly reported for the small database only. It is assumed that the database resides on a (remote) server and that the application runs on a workstation. There are no restrictions on indexes, cache-sizes, and network architectures.

Database	Parts	Connections	Scale
small	20000	60000	1
large	200000	600000	10
huge	2000000	6000000	100

Table 3. The OO1 Database Sizes

Operations

The OO1 benchmark suite consists of three operations: LOOKUP, TRAVERSAL, and INSERT. The LOOKUP operation generates 1000 random Part-IDs and fetches the parts from the database. For each part a null procedure is called (in any host programming language) passing the x,y position and type of the part. The TRAVERSAL operation scheme finds all parts connected to a randomly selected part, or to a part connected to it (7-level closure = 3280 parts with possible duplicates). For each part a null procedure is called passing the x,y position and the type. The time for REVERSE TRAVERSAL (swapping the from and to directions) is also measured. The INSERT operation inserts 100 parts and three connections from each to other randomly selected parts into the database. Time must be included to update indices or other access structures used in LOOKUP or TRAVERSAL. A null procedure to obtain the x,y position for each insert must be called. The changes must be committed to the disk.

Measurements

The benchmark measures the response time of a single user from the instant when a program calls the database system with a particular query until the results of the query have been placed into the program's variables. Each measure is run 10 times, and the response time is measured for each run to check consistency and caching behavior.

General order of execution

The general execution order in the OO1 benchmark is similar to that of the Hyper-Model benchmark. The benchmark run also consists of a "cold run" and a "warm run". The cold run results report the time of the first execution of the benchmark operations starting off with an empty cache memory. The warm run results are derived from 9 subsequent executions of the operations without clearing the database cache memory. For the insert operation, the database has to be restored to its original state.

Disclosure Report

The work does not define specific requirements. The information which should be included in a benchmark report is listed in Table 4:

Area	Types of reportable information
Hardware	CPU type, amount of memory, controller, disk type/size
Software	O/S version, size of cache
DBMS	Transaction properties (whether atomic, level of supported concurrency, level of read consistency, lock modes used etc.); recovery and logging properties; size of database cache; process architecture (communication protocol, number of processes involved, etc.); security features (or lack thereof); network protocols; access methods used
Benchmark	any discrepancy to the implementation described here; real "wall-clock" run-time, CPU time and disk utilization time are also useful; size of the database

Table 4. The OO1 Disclosure Report

3.4 The OO7 Benchmark

The OO7 benchmark has been designed to overcome the limitations of the OO1 and the HyperModel approach for benchmarking OODBS. This includes complex object operations, associative object access and database reorganization.

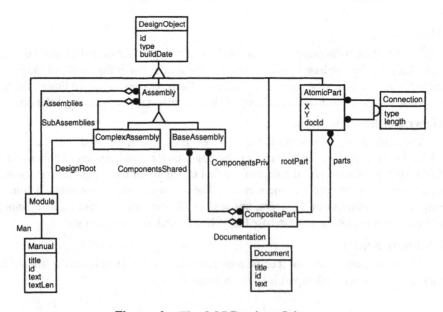

Figure 4. The OO7 Database Schema

Database Schema

The OO7 benchmark [Carey93a, Carey93b] database consists of a set of composite parts. The composite parts have a number of attributes (id, buildDate, type) and are connected to a set of atomic parts. The number of interconnected atomic parts depends on the database parameters. The degree of connectivity varies between 3, 6, and 9 connections to other atomic parts. Furthermore, a document object contains documentation information (title, string) for each composite part. Composite parts are referred to by base assemblies that form a 7-level complex assembly hierarchy. Complex assembly hierarchies are compiled to modules each of which has an associated manual object that contains additional information. Figure 4 shows the complete database schema of the OO7 benchmark specification.

Database Contents and Generation

The size of the OO7 benchmark database varies between small, medium, and large scale. The connectivity of the atomic parts, the assemblies, and the modules are shown in Table 5.

Parameter	Small	Medium	Large
NumAtomicPerComp	20	200	200
NumConnPerAtomic	3, 6, 9	3, 6, 9	3, 6, 9
DocumentSize (bytes)	2000	20000	20000
ManualSize (bytes)	100K	1M	1M
NumCompPerModule	500	500	500
NumAssmAssm	3	3	3
NumAssmLevels	7	7	7
NumCompPerAssm	3	3	3
NumModules	1	1	10

Table 5. OO7 benchmark database parameters

Operations

The OO7 benchmark includes three clusters of operations; traversals, queries, and structural modifications.

• *Traversals*

T1: Raw traversal: This traversal query scans each assembly and visits the associated composite parts. A depth-first search is performed on the atomic parts. The number of the visited atomic parts is returned.

T2: Traversal with updates: The traversal query T2 operates like traversal T1. In addition, the objects are updated during the traversal. Three types of updates are specified: (T2a) update one atomic part per composite part, (T2b) update every atomic part, (T2c) update each atomic part of a composite part four times. The traversal query returns the number of update operations that were performed.

T3: Traversal with indexed field updates: T3 performs the same steps as T2, except that the updates are performed on the indexed field buildDate (increment if odd, decrement if even).

T6: Sparse traversal speed: This traversal query scans each assembly and visits the associated composite parts. The root atomic parts are visited. The number of the visited atomic parts is returned.

T8: Operations on manual: T8 scans the manual object and counts the number of occurrences of the character "I".

T9: Operations on manuals: T9 checks the manual text if the first and the last characters are the same.

TCU: Cached update: This traversal operation first performes T1, then T2a. Both traversal operations are performed in a single transaction. The total time minus the T1 hot time minus the T1 cold time is reported.

• *Queries*

Q1: Exact match lookup: The exact match lookup query selects atomic parts by a lookup of their randomly generated id fields. An index can be used for the lookup. The number of atomic parts processed is returned.

Q2: Range query: Q2 performes a 1 % selection of the atomic parts via the buildDate field.

Q3: Range query: The range query Q3 performs a 10 % selection of the atomic parts via the buildDate field.

Q7: Range query: Query Q7 scans all atomic parts.

Q4: Path lookup: Query Q4 generates 100 randomly selected document titles and performs the following lookup query. For each title retrieve all base assemblies that belong to the composite part associated with the corresponding document object. The number of base assemblies is reported.

Q5: Single-level make: Q5 performs a selection of base assemblies that have a component part of a more recent buildDate than that of the base assembly. The number of qualifying base assemblies is reported.

Q8: Ad-hoc join: Q8 performs a join over document ids between documents and atomic parts.

• *Structural Modification Operations*

SM1: Insert: The insert operations create five new composite parts (with the correponding number of atomic parts) and inserts them into the database. References from base assemblies to these composite parts are randomly generated.

SM2: Delete: SM2 deletes the five previously created composite parts (and its associated atomic parts and document objects) from the database.

SM3: Database reorganization: All composite parts are scanned. For each composite part 50 % of its atomic parts are deleted and then newly inserted.

SM4: Database reorganization: This reorganization operation deletes and reinserts all composite parts and their associated atomic parts.

Measurements

The benchmark measures the elapsed time for each operation. Measurement include a "cold" run with empty cache memory and the average execution time of three further "warm" runs.

Disclosure Report

The authors do not give an explicit procedure for a disclosure report, but they describe the testbed configuration of the hardware and software that were being used in the benchmark runs.

3.5 The SEQUOIA 2000 Benchmark

The SEQUOIA 2000 Storage Benchmark addresses the application domain of engineering and scientific databases. It has been developed in the SEQUOIA 2000 research project that searches to investigate DBS support for Earth Scientists. Earth Scientists are mainly investigating issues that have effects on the condition of our environment. These investigations can be divided into three areas: field studies, remote sensing, and simulation. The SEQUOIA 2000 benchmark has evolved from those areas and addresses these application domains by specifying a set of databases and queries in order to measure the performance of databases for this application domain. Note that the benchmark data is not synthetic data as in the previous approaches but real data collected for scientific use. In the following we will describe the SEQUOIA 2000 benchmark according to the benchmark characteristics we identified above.

Database Schema

The kind of data that is mainly used by Earth Scientists can be divided into four categories: raster data, point data, polygon data, and directed graph data. The benchmark database is thus made up of these kinds of data sets. Figure 5 shows an OMT diagram of the SEQUOIA 2000 benchmark database schema; Figure 6 shows the corresponding Postgres schema.

The raster data represents data from the Advanced Very High Resolution Radiometer (AVHRR) sensor on NOAA satellites. The observed data is divided into so-called tiles, each of which is 1 square km. Two integers represent the relative position of a point within the corresponding tile. The satellite sensors observe 5 wavelength bands for each tile. The benchmark database consists of 26 observations per year.

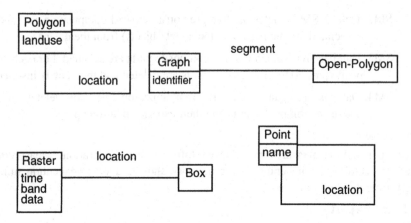

Figure 5. The SEQUOIA 2000 Benchmark Database Schema

The point data consists of names and locations taken from the United States Geological Survey (USGS) and from the Geographic Names Information System (GNIS). Each name in the database (with an average of 16 bytes) and its location (two 4-byte integers) are recorded.

The polygon data consists of homogeneous landuse/landcover data (available from USGS). Each item in the polygon database consists of a variable number of points (two 4-byte integers) and a landuse/landcover type (4-byte integer). An average polygon has 50 sides.

The graph data represents USGS information about drainage networks. Each river is represented as a collection of segments. Each segment is a non-closed polygon. For each segment the segment geometry and the segment identifier are recorded.

```
Raster(time=int4,       location=box,        band=int4,
data=int2[][])
Point(name=char[], location=point)
Polygon(landuse=int4, location=polygon)
Graph(identifier=int4, segment=open-polygon)
```

Figure 6. Postgres Schema of the SEQUOIA 2000 Benchmark

Database Contents and Generation

Usually, geological, geographical, and environmental database systems operate on a huge amount of data. This fact was taken into account in the SEQUOIA 2000 benchmark. The scales of the benchmark data cover the regional database, the national database, and the world database. Table 6 gives an overview of the benchmark sizes for each of the benchmark databases. The regional database consists of data of a 1280km x 800km rectangle. The national database comprises the benchmark data for the whole United States (5500km x 3000km). The world database covers all the world data for each of the types of data described above. The overall database size for the regional database is 1.1 GBytes, for the national database 18.4 GBytes, and more than 200

TBytes for the world database. The authors claim this approach to be durable with respect to technological progress in hardware. Up to now, SEQUOIA 2000 benchmark results are only available for the regional database [Stonebraker93].

Database Scale	Benchmark Data				
	raster data	point data	polygon data	directed graph data	
regional database	1 GB	1.83 MB	19.1 MB	47.8 MB	1.1 GB
national database	17 GB	27.5 MB	286 MB	1.1 GB	18.4 GB
world database	200 TB	numbers not available			

Table 6. The SEQUOIA 2000 Database Size

There are no restrictions on the layout of the database schema in the target DBS as long as AVHRR elements are 16 bit objects and point objects are pairs of 32 bit objects. In addition, the users are free to decompose the data for storage needs. Any necessary indexing or clustering technique may be used.

Operations

The SEQUOIA benchmark queries can be grouped into 5 collections of benchmark queries. Data loading, raster queries, polygon and point queries, spatial joins, and recursion. We will now describe each of these queries that form the basis of the benchmark suite.

• *Data Load*

Q1: Create and load the data base and build any necessary secondary indexes: This benchmark measure includes the loading of the data from the disk into the database system and the construction of the necessary indexes. It provides a measure for the efficient loading of bulk data into the DBS.

• *Raster Queries*

Q2: Select AVHRR data of a given wavelength band and rectangular region ordered by ascending time

Q3: Select AVHRR data of a given time and geographic rectangle and then calculate an arithmetic function of the five wavelength band values for each cell in the study rectangle.

Q4: Select AVHRR data of a given time, wavelength band, and geographic rectangle. Lower the resolution of the image by a factor of 64 to a cell size of 4 square km and store it persistently as a new object.

This type of queries is dedicated to the analysis of raster data. Q2 retrieves raster data of a specified region within a specified time. Q3 emphasizes the retrieval of raster data followed by an arithmetic analysis of the data. Q4 retrieves raster data, computes a lower resolution image of the data, and stores it in the database.

• *Polygon and Point Queries*

> Q5: Find the POINT record that has a specific name.

> Q6: Find all polygons that intersect a specific rectangle and store them in the DBS.

> Q7: Find all polygons of more than a specific size and within a specific circle.

The polygon and point queries represent queries about geographic points or polygons. Q5 represents a simple lookup operation given a point name. Q6 computes the intersection of polygons. Q7 combines spatial and non-spatial queries; by retrieving the polygons that satisfy spatial and non-spatial restrictions.

• *Spatial Joins*

> Q8: Show the landuse/landcover in a 50 km quadrangle surrounding a given point.

> Q9: Find the raster data for a given landuse type in a study rectangle for a given wavelength band and time.

> Q10: Find the names of all points within polygons of a specific vegetation type and store them as new DBS objects.

The collection of spatial joins emphasizes the database system's ability to perform joins to different spatial data types. Q8 joins polygon and point data. Q9 joins raster data and polygon data. Q10 combines point and polygon data.

• *Recursion*

> Q11: Find all segments of any waterway within 20 km downstream of a specific geographic point.

Q11 represents a restricted recursive query of the graph data. It computes all affected regions of a waterway that are 20 km downstream of a specific point.

Measurements

The metric of the benchmark results is the elapsed time in seconds for each of the queries described above. In addition, a performance/price ratio is reported. The overall performance of the database system is computed as the total elapsed time divided by the retail price of hardware. Software and maintenance costs etc. are neglected.

The elapsed time of the benchmark operations does not include the display of the retrieved data on the screen. The measurement includes the time from the start of the query until the placement of the results in the application memory. There are no restrictions as to the language in which the benchmark queries are coded. In case of using low-level DBS interface routines, the results must also be reported for the high-level interface. Furthermore, the DBS and the application must operate within different system domains in order to ensure the minimum security requirements.

Disclosure Report

The authors do not explicitly give a procedure for a disclosure report, but describe the testbed configuration of the hardware and software that were being used in the benchmark runs.

4 Conclusion

In the previous sections we described approaches to the evaluation of OODBS. We presented an approach to the *functional evaluation* based on the presented criteria catalogue. This catalogue comprises about 500 criteria that allow the functional assessment of OODBS. We also surveyed the most popular *object-oriented database benchmarks*, namely the Sun Benchmark, the HyperModel benchmark, the OO1 benchmark, the OO7 benchmark, and the SEQUOIA 2000 benchmark. We classified the approaches according to the database schema, the database generation, the benchmark operations, the measurement guidelines, and the reporting of the results. A summary of the approaches described above is presented in Table 10.

	Sun Benchmark	HyperModel Benchmark	OO1 Benchmark	OO7 Benchmark	Sequoia 2000 Benchmark
Benchmark Description	The Sun benchmark database describes a simple library database. It consists of a person class that describes authors and a document class that describes documents. The authorship concept describes the relationship between documents and persons. In the small version the database consists of 20000 persons, 15 000 authors, and 5000 documents. The large database scales these numbers by the factor ten.	The HyperModel database consists of an aggregation hierarchy and a generalization hierarchy of Nodes. Each Node has 5 successors (7 levels); thus forming a hierarchy of 19531 Nodes. The lowest level consists of 15500 TextNodes and 125 FormNodes (subtypes of Node). In addition to the parent/children relationship, a partOf/parts and a refTo/refFrom relation is generated.	N parts form the basis of the OO1 database. Each part is connected with exactly 3 other parts (N*3 connections). Small (20000 parts, 60000 connections), medium (200000 parts, 600000 connections), and large (2000000 parts, 6000000 connections) database size is specified.	The OO1 database consists of composite parts which are connected to a documentation object and a set of interconnected atomic parts. The composite parts form the leaves of a complex assembly hierarchy. Each of the connections is based on the benchmark database parameters. Small, medium, and large database size is specified.	The SEQUOIA 2000 database consists of four sets of data: Raster Data, Point Data, Polygon Data, and Graph Data. The data represents geological and geographical data. Three scales of the database sets are specified: regional database (1.1 GB), national database (18,4 GB), and world database (over 200 TB)
Database Schema					

Table 10. An Overview of Advanced DBS Application Benchmarks

	Sun Benchmark	HyperModel Benchmark	OO1 Benchmark	OO7 Benchmark	Sequoia 2000 Benchmark
Restrictions	There are no restrictions on the database schema, indexing, clustering for the benchmark measurements.	There are no restrictions on the location of the database (local or remote). Clustering has to be done on the parent/children relationship (if possible).	The database is assumed to reside on a remote server, the application runs on a workstation. There are no restrictions on indexing, clustering, cache-sizes etc.	There are no restrictions on the database schema, indexing, clustering for the benchmark measurements.	There are no restrictions on the database schema, indexing, clustering for the benchmark measurements.
Operations	7 operations: • Name Lookup • Range Lookup • Group Lookup • Reference Lookup • Record Insert • Sequential Scan • Database Open	7 groups of operations (20 operations): • NameLookup • RangeLookup • GroupLookup • ReferenceLookup • Sequential Scan • Closure Traversal • Editing Operations	3 operations: • Lookup • Traversals • Inserts	3 groups of operations: • Traversals • Queries • Structural Modification Operations	5 groups of operations (11 operations): • Database Loading • Raster Queries • Polygon and Point Queries • Spatial Joins • Recursion
Measurements	The time (response-time) is measured for each operation. Cold and warm times are reported.	The time (response-time) is measured for each operation. Cold and warm times are reported.	The time (response-time) is measured for each operation. Cold and warm times are reported.	The time (response-time) is measured for each operation. Cold and warm times are reported.	The elapsed time for each benchmark operation is reported. An overall performance number may be reported (elapsed time for all benchmark queries / retail cost of hardware).
Reporting	no disclosure report required	no disclosure report required	The authors specify types of reportable information.	no disclosure report required	no disclosure report required

Table 10. An Overview of Advanced DBS Application Benchmarks

References

[Ahmed92]   S. Ahmed, A. Wong, D. Sriram, R. Logcher; Object-oriented
                   database management systems for engineering: A Compari-
                   son; JOOP, June 1992

[Anderson90]   T. Anderson, A. Berre, M. Mallison, H. Porter III, B.
                   Schneider; The HyperModel Benchmark; Proc. of the EDBT
                   Conf., 1990

[Atkinson89]   M. Atkinson, F. Bancilhon, D. DeWitt, K. Dittrich, D. Maier,
                   S. Zdonik; The object-oriented database manifesto; Proc. of
                   the Conf. on Deductive and Object-Oriented Databases; 1989

[Banc89   A. F. Bancilhon; Query Languages for Object-Oriented Data-
                   base Systems: Analysis and a Proposal; Datenbanksysteme in
                   Büro, Technik und Wissenschaft, Springer, IFB 204, Zürich,
                   March 1989

[Bitton83]   D. Bitton, D. DeWitt, C. Turbyfill; Benchmarking Database
                   Systems, A Systematic Approach; Proc. of the VLDB Conf.;
                   1983

[Carey93a]   M. Carey, D. DeWitt, J. Naughton; The OO7 Benchmark;
                   Proc. of the ACM SIGMOD Conf. 1993

[Carey93b]   M. Carey, D. DeWitt, J. Naughton; The OO7 Benchmark;
                   Tech. Report, CS Dept., Univ. of Wisconsin-Madison; 1993

[Cattell91]   R. Cattell; Object data management: object-oriented and
                   extended database systems; Addison-Wesley; 1991

[Cattell92]   R. Cattell, J. Skeen; Object Operations Benchmark; ACM
                   TODS; Vol. 17, No. 1; 1992

[Cattell94]   R. Cattell, The Object Database Standard: ODMG-93, Morgan
                   Kaufmann Publishers, San Mateo, California, 1994

[Encarnação90]   J. Encarnação, P. Lockemann; Engineering Databases, Con-
                   necting Islands of Automation Through Databases; Springer
                   Verlag; 1990

[Gray91]   J. Gray; Standards are a Prerequisite for Interoperability and
                   Portability; Tutorial Notes; held at the EDBT Summer School,
                   Alghero/Italy; 1991

[Gray93]   J. Gray; A Tour of Popular DB and TP Benchmarks; Tutorial
                   Notes, held at the ACM SIGMETRICS Conf.; 1993

[Gupta91]   R. Gupta, E. Horowitz (eds.); Object-Oriented Databases with
                   Applications to CASE, Networks, and VLSI CAD; Prentice-
                   Hall; 1991

[Kappel92]   G. Kappel, S. Rausch-Schott, W. Retschitzegger, M. Schrefl,
                   U. Schreier, M. Stumptner, S. Vieweg; Object-Oriented Data-
                   base Management Systems - An Evaluation; Tech. Rep. ODB/
                   TR 92-21; Institute of Applied Computer Science and Infor-
                   mation Systems; Univ. of Vienna; 1992

[Kappel94]   G. Kappel, S. Vieweg; Database Requirements of CIM Appli-
                   cations; in this book; 1994

[Maier89]   D. Maier; Why isn't there an object-oriented data model?
                   Information Processing 89 - IFIP World Computer Congress;
                   G.X. Ritter; North-Holland; 1989

[Rubenstein87]   W. Rubenstein, M. Kubicar, R. Cattell; Benchmarking Simple
                   Database Operations; Proc. of the ACM SIGMOD Conf.,
                   1987

[Rumbaugh91]	J. Rumbaugh, M. Blaha, W. Premerlani, F. Eddy, W. Lorensen; Object-Oriented Modeling and Design; Prentice Hall; 1991
[Stonebraker90]	M. Stonebraker, L. Rowe, B. Lindsay, J. Gray, M. Carey, M. Brodie, P. Bernstein, D. Beech; Third-Generation Database System Manifesto; SIGMOD Record, Vol. 19, No. 3; 1990
[Stonebraker93]	M. Stonebraker, J. Frew, K. Gardels, J. Meredith; The Sequoia 2000 Storage Benchmark; Proc. fo the ACM SIGMOD Conf.; 1993
[TPCA92]	TPC Benchmark™ A, Standard Specification, Revision 1.1, Transaction Processing Performance Council (TPC); March 1, 1992
[TPCB92]	TPC Benchmark™ B, Standard Specification, Revision 1.1, Transaction Processing Performance Council (TPC); March 1, 1992
[TPCC92]	TPC Benchmark™ C, Standard Specification, Revision 1.1, Transaction Processing Performance Council (TPC); August 13, 1992
[Trapp93]	G. Trapp; The emerging Step standard for production-model data exchange; IEEE Computer; Vol. 26, No. 2; 1993
[Turbyfill91]	C. Turbyfill, C. Orji, D. Bitton; AS^3AP: An ANSI SQL Standard Scaleable and Portable Benchmark for Relational Database Systems

Active Object-Oriented Database Systems
For CIM Applications

G. Kappel[‡], S. Rausch-Schott[‡], W. Retschitzegger[‡], A Min Tjoa[†], S. Vieweg[†],
R. Wagner[*]

[†]Institute of Applied Computer Science and Information Systems. Department of Information
Engineering. University of Vienna, Austria

[‡]Institute of Computer Science. Department of Information Systems.
University of Linz, Austria

[*]Research Institute for Applied Knowledge Processing (FAW).
University of Linz, Austria

Abstract. Object-oriented database systems (OODBS) have gained wide attention as the most promising database technology for non-standard applications like computer aided design and computer integrated manufacturing. Among the most important requirements of these applications which are supported by OODBS are complex object modeling, advanced transaction management, and version management, to mention just a few. Active database systems (ADBS) have reached equal potential for non-standard applications as they offer event driven and constraint driven behavior of the database system necessary for implementing time critical reactions. For example, manufacturing plants need timely support of raw material and assemblies to keep going the manufacturing process. Active object-oriented database systems (AOODBS) provide the best of both worlds. They are based on one hand on an object-oriented data model for representing the objects of the problem domain, and on the other hand on a rule-based knowledge model for representing the event driven and constraint driven behavior. In the first part of this chapter we survey various approaches to AOODBS. In the second part we report on our own experience in extending the commercially available OODBS GemStone[TM] with active capabilities. TriGS (= Trigger system for GemStone) focuses on several concepts that originate both from shortcomings of existing active object-oriented database systems and from advanced requirements of Leitstand systems, for which TriGS has been developed in the first place. TriGS makes explicit use of objects, message passing, inheritance, and overriding to provide for a seamless integration between rules and an object-oriented data model. Rules consisting of events, conditions, and actions monitor the behavior of objects and can be attached to specific classes (local rules) or defined independently of any class hierarchy (global rules). Concerning the execution of rules, TriGS uses a flexible event based mechanism defining the points in time when conditions are evaluated and actions are executed. An evaluation of TriGS and a discussion of future research issues concludes this chapter.

[TM]GemStone is a registered trademark of Servio Corporation

Categories and Subject Descriptors: I.2.4 [Artificial Intelligence]: knowledge representation formalisms and methods - representation languages; representations; H.2.1 [Database Management] logical design - data models

General Terms: active database systems, object-oriented database systems, triggers, rules

Additional Key Words and Phrases: rule-based systems, overriding of rules, local message rules, global message rules, production planning and control system

1 Introduction

Non-standard application areas like computer integrated manufacturing (CIM) systems require timely responses to critical situations as well as sophisticated constraint management. A Leitstand, for example, constitutes the short-term production scheduling and control component of a CIM-system [Adel91]. For the purpose of monitoring messages concerning the actual production status, which are continuously received from the shop floor, a Leitstand has to be equipped with real-time capabilities [Sche91]. The real data has to be compared with the corresponding scheduled data and, depending on critical deviations, the production controller has to be informed and corrective actions have to be undertaken. Furthermore, whenever an operation is scheduled, a Leitstand has to generate high-quality schedules [Adel91] considering a large number of constraints, such as restricted availability of multiple resources, limited capacity buffer space, and time dependencies.

Traditional database systems which are used to store, retrieve, and manipulate non-standard application data do not provide satisfying solutions to these tasks. They are passive in their behavior, which means that they perform only operations when they are actually required to do so. Thus, in passive systems knowledge concerning the management of time-critical as well as constrained data is encoded in the application program. But this approach distributes and replicates knowledge into several application programs. As a consequence, it interferes with independent application development and impedes optimization. Another approach is to poll the database system periodically to detect whether critical data has arrived or whether constraints have been violated. This solution leads to wasted resources and problems concerning the frequency of polling [Daya88a, Chak90]. If polling is too frequent the database is overloaded with queries that most of the time will fail leading to decreased system performance. On the other hand, if the polling frequency is too low events will be missed [Bern92].

With respect to these shortcomings of passive systems, database systems are required to be "active" in the sense that they are able to react automatically to certain events. The database is no longer seen as a slave to the application but it cooperates with the application in a peer-to-peer communication based on knowledge stored inside the database system. This knowledge determines when and how to react [Diaz91]. In almost all active systems the representation of knowledge is based on the Event/Condition/Action paradigm (ECA paradigm) [Daya88]. Hence, the database system is able to monitor the situation represented by an event and by one or more conditions and to execute the corresponding actions when the event occurs and the conditions evaluate to true [Beer91]. Event/condition/action triplets are synonymous with triggers and rules. The terms will be used interchangeably in the following.

Object-oriented database systems are equally important for non-standard application areas, as they support the modeling of complex objects like production schedules and bills of materials, and they provide an open environment necessary to adapt to changing requirements of the application [View94]. Integrating active concepts with object-oriented concepts provides advantages from both approaches. Firstly, it allows full exploitation of object-oriented features in the modeling of rules. Secondly, it allows extension of the object-oriented paradigm with (declarative) mechanisms to specify event-driven behavior and constraints, both, local to one object and global to a system of objects.

The remaining sections are organized as follows. In Section 2, we will give a short overview of several approaches to active object-oriented database systems. Section 3 surveys our approach to an AOODBS called TriGS. The discussion of TriGS comprises its knowledge model (section 4), its execution model (section 5), and architectural aspects of the prototype implementation of TriGS (section 6). Finally, in Section 7 an example illustrates the applicability of TriGS.

2 Approaches to Active OODBS

There are a number of approaches to active OODBS differing according to various criteria. Concerning *expressiveness of the knowledge model* of an active system which defines how rules and its execution environment can be specified, various kinds of events, conditions and actions are supported by different systems. For example, some models support only primitive events whereas others provide a powerful event specification language allowing to specify composite events. Another issue is the binding between event, condition, and action, i.e., which parameters, if any, are passed from event to condition to action. The second criterion is the *execution semantics* of active systems, i.e., how rule execution relates to database operations including those that mark transaction boundaries. Active systems differ in whether condition evaluation and/or action execution is done in the application's transaction or in different transactions, how multiple concurrently triggered rules are processed, and whether rule cascading is supported. Finally, active systems differ in their *architecture*. In this respect, two alternatives for the integration of active concepts with OODBS are proposed. On one hand, the underlying OODBS can be augmented with a layer responsible for providing active capability (layered approach). All rule specifications and - in order to detect events that are able to trigger rules - all transactions are routed through this layer. Additionally, triggered rules are also processed within this layer. One advantage of the layered approach is that the additional layer may perform some optimizations. For example, it can decide whether to rewrite a transaction to include condition monitoring code or to use polling or aperiodic checking depending on the meta-data used by the system. Another example is caching some data for condition monitoring in this layer. On the other hand, the kernel functionality of a passive DBS can be extended to support event detection, condition monitoring and extended transaction management for concurrent rule processing (integrated approach).

In the following, we will survey an illustrative sampling of approaches to active object-oriented database systems with respect to the above mentioned criteria as far as they are known from literature.

2.1 HiPAC

The pioneering work in the area of AOODBS has been done within the development of the HiPAC system (High Performance ACtive Database System) [Daya88a, McCa89], which integrates active concepts into the OODBS PROBE [Daya87]. Central to the HiPAC knowledge model is the concept of Event-Condition-Action (ECA) rules. Rules are first-class database objects, i.e., subject to the same operations as user-defined objects. They can be created, modified and deleted like any other database object. The attributes of a rule specify its event, condition, action, E-C coupling, and C-A coupling.

The *event* specifies when the rule is triggered. HiPAC supports both primitive and composite events. Primitive events include database operations (data definition, data manipulation), transaction events, temporal events, and external (application defined) events. Primitive events can be combined using disjunction and sequence operators to specify composite events. The *condition* is a collection of queries expressed in an object-oriented data manipulation language. A query may refer to actual arguments bound at event detection. The condition is satisfied if all of these queries produce non-empty results.The results are passed on to the action together with the argument bindings obtained from the event signal. The *action* is a sequence of database operations and calls to application programs. A rule can be enabled or disabled at runtime. If a rule is disabled, it is not automatically triggered if its event occurs. Rules can be triggered manually, too, irrespective of whether they are enabled or disabled.

E-C coupling describes the relationship between triggering event and condition evaluation, relative to transaction boundaries. Three different coupling modes are supported:

- **immediate:** immediately after an event triggers a rule the condition of that rule is evaluated in the same transaction. The transaction that signalled the event (triggering transaction) is preempted during processing of the rule.
- **separate:** when the triggering event occurs the condition is evaluated in a separate transaction without interrupting the triggering transaction.
- **deferred:** the condition is evaluated at the end of the triggering transaction

If a rule's E-C coupling mode is immediate or deferred, condition evaluation is done in a subtransaction of the triggering transaction. The parent transaction is suspended while this subtransaction executes. If the rule's E-C coupling mode is separate, then condition evaluation takes place in a top level transaction that executes concurrently with the triggering transaction. The same modes are available for *C-A coupling*, which specifies the relationship between condition evaluation and action execution.

2.2 SAMOS

The overall goal of SAMOS (Swiss Active Mechanism-Based Object-Oriented Database System) [Gatz91, Gatz92, Gatz94] is the combination of active and object-oriented characteristics within one coherent system by means of a layered approach. Since current OODBS differ in their data models and functionalities only characteristic properties provided by almost all OODBS, like inheritance, user-definable types and operations, and encapsulation, are assumed in SAMOS. Like in HiPAC, the knowledge model of SAMOS is based on ECA-rules, which are represented as objects themselves. Rules may be permitted or prevented to operate on or to access object values by supporting two kinds of rules; class-internal rules and class-external rules. *Class-internal rules* are part of the class definition encapsulated within objects and, thus, visible to the class implementor only. Events, conditions and actions of class internal rules may operate directly on the object's instance variables. This leads to a high level of object autonomy since specific tasks relevant to a certain object may be kept completely local to that object. *Class-external rules* are defined independently of a class definition, preventing the direct use of instance variables in events, conditions and actions. Since a class-internal rule is attached to a specific class it is subject to inheritance and propagated to subclasses like methods.

A specific contribution of SAMOS to rule specification is its extensive collection of event specific features. Primitive and composite events are specified using an event language. A primitive event in SAMOS is always regarded as a specific point in time. The way how this point in time is specified (explicit time definition, beginning or end of a database operation) leads to various event classes:

- **Time event:** specified as an explicitly defined point in time, which can be absolute, periodical, or relative to occurring events.

- **Method event:** Each message gives rise for two events, the point in time when the message is arriving at the object, and the point in time when the object has finished executing the corresponding method. A method event can be attached to one or more classes, which means that the event is signalled before or after the execution of the appropriate method on any object of these classes. In addition, a method can be attached to a particular object only.

- **Value event:** This event class is related to the modification of (parts of) the values of instance variables of an object, which can take place in various methods. Due to encapsulation, appropriate rules have to be regarded as part of the class definition and are definable/visible by the class implementor only. Thus, value events can be used in class internal rules only.

- **Transaction event:** The start and end of arbitrary transactions can raise an event. If transactions are named, a transaction event can be restricted to transactions with a specific name.

- **Abstract event:** Users can define and name their own events, which may be used in rules like any other event. Unlike other event classes, abstract events are not detected by SAMOS but explicitly raised by the user or the application, respectively.

Composite events can be defined by combining primitive events by means of six event constructors: A *disjunction* of events (E1|E2) is signalled when either E1 or E2 is signalled. *Conjunction* of events (E1,E2) is signalled when E1 and E2 occur, regardless of order. A *sequence* of events (E1;E2) is signalled when first E1 and then E2 is signalled. With the *-constructor* an event is composed which is signalled after the first occrrence of E only once, even if E occurs several times within a specified time interval. A *history* event (TIMES(n,E)) is signalled when the event has occurred with the specified frequency n during a specified time interval and finally, *negation* (NOT E) is signalled when the event did not occur within a specified time interval. For the last three event constructors the default time interval is the time between event definition and infinitive. Event classes can be parameterized by environment parameters (e.g., point in time of the event occurrence and identifier of the triggering transaction) and parameters depending on the event class. The constructor *same(parameter_kind)* can be used to denote parameter binding between different event parts of composite events. For example, (E1;E2):same(tid) is signalled when E1 occurs before E2 in the same transaction.

Like HiPAC, SAMOS uses three different coupling modes (immediate, deferred, decoupled) to specify the points in time when condition evaluation and action execution have to take place. The execution of triggered operations is based on multi-level transactions and semantic concurrency control at method level. Since class-internal rules behave comparable to methods, conflict relations comparing condition transaction and action transaction have to be provided. Class-external rules call methods which are synchronized with other methods and rules by themselves. The execution of multiple rules triggered by the same event is handled by means of priorities.

2.3 Ode

Ode [Geha92a, Geha92b] is a database system based on the database programming language O++, which is an extension to the object-oriented programming language C++. O++ extends C++ by providing database facilities such as creating and manipulating persistent and versioned objects, and associating constraints and triggers with them. Constraints are used for maintaining database integrity whereas triggers are used for automatically performing actions depending on the database state. The key differences between constraints and triggers are the following:

- Constraints ensure consistency of the database state. If this consistency cannot be maintained the surrounding transaction is aborted. Triggers are not concerned with object consistency but are fired whenever an appropriate event occurs and the specified conditions are satisfied.

- Constraints are active from the moment the appropriate object is created to the moment it is deleted. Triggers have to be activated explicitly.

- All objects of a given type obey the same constraints. However, different triggers may be activated for different objects of the same type.

Since in most other systems constraints are defined by using triggers without introducing an additional concept, in the following both Ode's trigger and constraint mechanisms are surveyed.

Constraints

Constraints in Ode are associated with class definitions. Thus, they are inherited together with all other class properties. All objects of a class must satisfy all constraints associated with that class. Constraints exist of a predicate, which is a boolean expression, and an action (handler), which is executed when the predicate is *not* satisfied. Constraint checking can be performed immediately after accessing the object (*hard constraints*) or deferred to the end of the transaction in which the object is accessed (*soft constraints*). Thus, the granularity of checking hard constraints is at member function level. The constraints of a class are checked at the end of execution of every constructor and every member function of the class. They are not checked by the destructor or if (public) attributes of an object are directly accessed. The reason is that each public member function must leave the object in a consistent state. The granularity of checking soft constraints is at transaction level. Soft constraints may be used when multiple objects are involved in a constraint allowing for a temporary violation of constraints.

Ode distinguishes between *intra-object constraints* and *inter-object constraints*. A constraint is said to be intra-object if it is associated with a (single) specific object, and the condition associated with it is evaluated only when this object is updated. Otherwise, a constraint is said to be inter-object. Most inter-object constraints can be converted into one or more intra-object constraints. There are two particularly important cases of inter-object constraints, namely reference validation and relationship integrity. Suppose, that an object to be deleted still has a reference to it. In order to ensure reference validation, Ode supports three standard maintenance options. The reference can be set to NULL as part of the transaction deleting the object (nullify), the referencing object can be deleted (ripple), or the deletion of the object can be disallowed (abort). Relationship integrity is the proper maintenance of inverse attributes constituting a binary relationship between two classes. If one of the inverse attributes is updated, either the value of the other inverse attribute is fixed (ripple), or the transaction is aborted (abort).

Triggers

A trigger, like a constraint, is specified at the level of class definition and consists of two parts: an event predicate and an action. It applies only to specific objects for which it is activated. Activation and deactivation for a certain object has to be done explicitly by sending an appropriate message to that object. However, trigger activation can be automated by putting the activation code in constructors. In this way, triggers are activated for a specific object as soon as it is created. Triggers are parameterized and can be activated multiple times with different parameter values.

A very interesting feature of Ode is its powerful event specification language, called COMPOSE [Geha93]. In an object-oriented system most of the events are local to a particular object. In addition, Ode supports events within other scopes, such as the database (e.g. the creation of an object type, schema modification, etc.). Events are classified into basic events, logical events, and composite events. *Basic events* are *object state events* (creation, deletion, modification of an object), *method execution events* (immediately before or after the specified member function is applied to an object), *time events*, and *transaction events* (immediately before/after transaction commit/abort). Time events and transaction events are global events, but are considered to be local to all objects. A *logical event* is a basic event that optionally can be qualified with a *mask*, i.e. a predicate used to restrict the set of appropriate events. Logical events can be combined to create *composite events* using logical operators (conjunction, disjunction, negation) and a variety of special event specification operators. These include partial and total ordering (prior, relative, sequence), limited and unlimited repetition of (composite) events (relative n, relative+), and periodically occurring events (every). For a complete description the reader is referred to [Geha93].

The action of an active trigger is scheduled as soon as the corresponding event predicate is satisfied. By default, an action is executed as a separate commit dependent transaction. Other coupling modes supported by Ode include execution in a separate independent transaction and execution in the same transaction, either immediately or at the end of the transaction. Most of these coupling modes have been introduced for ease of expression, since the only coupling mode required is immediate. All other coupling modes can be obtained by appropriate event composition.

Concerning activation, Ode supports two kinds of triggers: *once-only* (default) and *perpetual* triggers. A once-only trigger is automatically deactivated after the trigger has fired, whereas a perpetual trigger remains activated after each firing.

2.4 O_2

This section surveys a production rule management system which has been integrated into the OODBS O_2 [Mede91]. O_2 rules also build on the ECA-paradigm. They are conceptually considered to be properties of an O_2 database schema and can be attached to classes, methods or named objects. Rules attached to a given class are inherited to all its subclasses. Rules can be added, deleted or modified and, unlike other schema components, enabled or disabled dynamically. They are implemented as objects with priorities and access rights. To define rules the O_2 system-defined class hierarchy rooted at class RULE is employed. Two levels of rule support are distinguished, namely *External_rule* and *Internal_rule*. External rules correspond to rules seen and manipulated by users other than system programmers. They are transformed by the system into lower level internal rules.

An external rule object is specified as seven-tuple <rule name, event, query, action, type, priority, status>. *Event* is an expression describing the set of events to which the rule applies. O_2 supports message events and time events. Message events can be signalled before or after sending any message to any object of a given set of classes or by sending it to a specified named object. Time events are signalled periodically according to a given time interval. *Query* is the name of an O_2 query representing the condition. *Action* refers to a sequence of operations that can be coded as a CO_2 method (O_2C is a C-like language enhanced with an object-oriented layer and database functionality for O_2). An operation may be an external user intervention or some database action, such as undoing the triggering operation. If multiple rules are triggered by one event they are ordered according to their *priorities* which can be changed dynamically. The *type* attribute indicates whether the rule is message-related or time-related. The *status* indicates whether the rule is enabled or not.

Each external rule is transformed into a set of hidden internal rules stored in priority queues. This is done by decomposition, i.e. each atomic event of an event expression is mapped to an internal rule. Every rule attached to a class is propagated to all subclasses of that class, i.e., it is inherited to the class's subclasses by a *flattening mechanism*.

2.5 GOODSTEP

Another attempt for integrating rules into the O_2 system is done in the course of the GOODSTEP project [Coll94] aiming at the development of a software engineering environment by enhancing and improving the O_2 system. Rules have been introduced mainly for (i) notifying users, i.e., programmers of Software Development Environments or end-users, (ii) application access logging, (iii) organizing related application programs, (iv) tools communication, (v) change propagation, and (vi) maintaining data consistency.

GOODSTEP supports only primitive events which are signalled on manipulation of entities, on execution of programs, or on transaction begin and end, respectively. These events are detected by the O_2 engine. Rules are defined as schema elements not violating encapsulation, i.e. only authorized operations can produce valid events. Rules associated with only one class are subject to *inheritance*. Conditions (predicates stating O_2SQL queries on objects and values) and actions (executable O_2C code) of a rule are also inherited but cannot be redefined.

Concerning coupling modes, an *immediate E-C / immediate C-A* combination is used for rules responding to operations on a single entity, whereas a *deferred E-C / immediate C-A* combination is provided for set-oriented rules. Concurrently triggered rules are ordered by specifying priorities between rules using a *precedence relationship*. Cascading rule execution is based on the notion of *execution cycles*. An execution cycle describes the execution of a sequence of operations belonging to a program, a user-defined transaction or a rule. Immediate triggered rules are executed in a depth first order, deferred rules are executed in a breadth first order. Every execution cycle is associated with an event history used to build the execution environment of each rule considered in the cycle. Such an environment is represented by a *delta structure* containing data related to the triggering operation and the result of condition evaluation.

2.6 ACOOD

ACOOD (ACtive Object-Oriented Database System) [Bern92] is a prototype reactive object-oriented system built on top of ONTOS^{TM}, a commercial OODBS based on C++. ACOOD is an example for a layered approach. The data model of ONTOS provides ACOOD with persistent objects and nested transactions. Following HiPAC, ACOOD supports ECA rules. Primitive events can be signalled immediately before or immediately after any method of user-defined classes. Methods of system-defined classes cannot serve as primitive events. If multiple rules are triggered by the same event they are not evaluated on the same database state. Actions of already processed rules change the database state seen by subsequent rules, i.e., they are able to affect the evaluation of conditions of other rules. The processing order of concurrently triggered rules is determined by priorities. Rules with equal priority are processed in non-deterministic order.

2.7 ADAM

ADAM [Diaz91] is a prototype object-oriented database implemented in PRO-LOG. It provides support for metaclasses, multiple inheritance, and ECA-rules. Rules and events are first-class objects. Events reflect a variety of situations that can arise during message passing. Besides the usual start and end of method invocation situations like "method not found" and backtracking into a method - due to the underlying PROLOG execution mechanism - may be signalled by events. Being objects, events can be related to other objects, e.g., to rules triggered by that event or arranged in event hierarchies. Events can be generated by the database system itself or by any other external system such as a clock or an application program. Thus, an event object is instantiated from one of the predefined classes *DB-EVENT*, *CLOCK-EVENT*, and *APPLICATION-EVENT*, which are all subclasses of the more abstract class *EVENT-CLASS*. The condition is a set of queries to check that the database state is appropriate for action execution. The action is a set of operations that may have different aims, such as enforcement of integrity constraints, user intervention, or propagation of methods.

Condition and action definitions may refer to the current instance to which the rule is applied, and to the current arguments of the method firing the rule by two predicates *current-instance* and *current-arguments*. In addition, the action can make use of any values retrieved during condition evaluation using the condition-result predicate.

Rules are attached to classes by means of a binary relationship between rules and classes. Thus, rules are indexed by class reducing considerably the search for applicable rules. Rules are inherited without introducing any additional mechanism. Rules may be enabled and disabled at the level of the whole class (attribute *is-it-enabled*) as well as at the level of specific instances (attribute *disabled-for*).

---

^{TM}ONTOS is a registered trademark of ONTOS, Inc.

2.8 Sentinel

The Sentinel system [Chak93] is being developed on the basis of C++ using the OODBS Zeitgeist [Blak90].

Sentinel supports *message events* signaled before or after method execution and the construction of *composite events* by using a hierarchy of event operators (disjunction, conjunction and sequence) which are applicable to message events. A more expressive event specification language for Sentinel is currently under development within the Snoop project [Chak91].

Conventional C++-classes can be made *reactive* by deriving them from a pre-defined "Reactive Class" and specifying the methods which should generate a message event using the event interface in the public, private and protected sections of the C++ class definition. Event Detection is done by creating a wrapper member function for each event specification. Therefore, the user must explicitly generate events by invoking the newly defined method.

Instances of a reactive class are producers of events and know the consumers (*notifiable objects*) of these events. Rules and event operators are examples of notifiable objects, i.e. they are capable of receiving and recording the events propagated by reactive objects. In order to establish an association between an event and a notifiable object, a *subscription mechanism* is used allowing notifiable objects to dynamically subscribe to reactive objects. After subscription, a notifiable object will be notified of the events generated by the corresponding reactive objects and will react to those events. The distinction between reactive objects and notifiable objects clearly separates event detection from rules and leads to an *external monitoring viewpoint*. This viewpoint permits rule definition to be independent of the objects which they monitor.

The subscription mechanism supports, on one hand, *class level rules* which are subject to inheritance and applicable to all instances of a class, and on the other hand, *instance level rules*, applicable to specific instances, possibly from different classes and not subject to inheritance. Rules (regardless of their classification) and events are treated as *first-class objects*. Conditions and actions are methods of the rule class. Concerning rule execution, immediate and deferred coupling modes are supported.

2.9 Comparison of AOODBS

Summing up, a concise comparison of the systems described in this section is given in Table 1 [Röck93, Chak93]. The table entries have been discussed with each system. Concerning the table entries for our system TriGS, a detailled discussion is given in the forthcoming sections.

| Features / Systems | Expressiveness | | | | | | | | Execution Semantics | | | Archi-tecture |
	DB Events	External Events	Temporal Events	Composite Events	Rule Inheritance	Rule Overriding	Rules as first class objects	Events as first class objects	Coupling Modes	Rule Cascading	Multiple Rules	
HiPAC	✓	✓	✓	✓	*	*	✓	✓	i,d,s	✓	system-defined	integrated
SAMOS	✓	✓	✓	✓	✓	✗	✓	✓	i,d,s	*	user-defined	layered
Ode	✓	✓	✓	✓	✓	✗	✗	✗	i,d	✓	system-defined	integrated
O₂	✓	✗	✓	✗	✓	*	✓	✗	i,d	*	user-defined	integrated
GOOD-STEP	✓	✓	✗	✗	✓	✗	✓	✗	i,d	✓	user-defined	integrated
ACOOD	✓	✗	✗	✗	*	*	*	*	*	*	user-defined	layered
ADAM	✓	✓	✓	✗	✓	✗	✓	✓	*	*	*	integrated
Sentinel	✓	✗	✗	✓	✓	✗	✓	✓	i,d	*	*	layered
TriGS	✓	✗	✗	✗	✓	✓	✓	✓	i,d,s	✓	user-defined	layered

Legend: ✓ .... supported, ✗ .... not supported, * .... not explicitly stated in the literature
i .... immediate, d .... deferred, s .... separate

Table 1. Comparison of AOODBS

3 TriGS - Implementing Active Concepts on Top of OODBS

As surveyed in Section 2, several attempts exist to integrate active concepts into object-oriented databases. Current approaches to active object-oriented database systems, however, do not consider the various active concepts together with some of the basic object-oriented concepts. The system TriGS (=Trigger system for GemStone) [Kapp93, Kapp94] focuses on several concepts that originate both from shortcomings of existing active object-oriented systems and from advanced requirements of CIM applications. TriGS makes explicit use of objects, message passing, inheritance, and overriding to provide a seamless integration between rules and an object-oriented data objects. TriGS provides an extension of the object-oriented paradigm with (declarative) mechanisms to specify event-driven behavior and constraints, both local to one object and global to a system of objects. Since rules are activated at different levels of granularity ranging from the object instance level to the object class level, a powerful exception handling mechanism is also supported. Moreover, rules can be dynamically defined, modified, and extended independently of any application.

Concerning the execution model, TriGS, unlike most other systems, does not exploit coupling modes to specify the temporal relationships between rule triggering and condition evaluation, and between condition evaluation and action execution, respectively. This is because the expressive power of coupling modes is insufficient for specifying certain execution semantics required by non-standard applications. Instead, TriGS uses an event specification mechanism not only for defining the points in time for rule triggering but also for defining the points in time for condition evaluation and action execution. This idea has been described the first time by Beeri and Milo in [Beer91]. TriGS extends their approach in several directions, most importantly, conditions are considered in addition to events and actions, and a flexible event specification mechanism including both class level specification and object level specification is provided. The advantages of the approach compared to coupling modes are twofold. Firstly, fewer concepts are employed, thus higher uniformity and (hopefully) improved understandability is reached. Secondly, the event specification mechanism for describing the execution behavior of the rule scheduler is more general than using coupling modes, thus higher expressive power is reached.

TriGS is implemented on top of the object-oriented database system GemStone as part of a larger EC ESPRIT project aiming at the development of next-generation production scheduling and control systems (see also chapter "Database Requirements of CIM Applications" in this book). In the following sections TriGS is discussed in more detail.

4 Knowledge Model of TriGS

4.1 Specification of Rules

TriGS is based on the ECA paradigm [Daya88]. An event-condition-action triplet is denoted as rule. The event part of a rule is represented by a rule event selector determining the event(s) which is (are) able to trigger the rule. Events represent real-world situations that form the basis for subsequent rule execution. In our application environment, a typical event is the notification about a finished production process or a machine breakdown. In object-oriented databases, this is reflected by sending a message to some object. In TriGS, any message sent to an object may be associated with a *message event*. The condition part of a rule is specified by a condition event selector, which defines when to evaluate the condition (cf. Section 5.1), and a Boolean expression (e.g., is the machine available?), possibly based on the result of a database query (e.g., select all orders that are scheduled on the damaged machine). The action part (e.g., schedule the next operation) is specified by an action event selector, which (e.g., schedule the next operation) is specified by an action event selector, which defines when to execute the action (cf. Section 5.1), and a sequence of arbitrary messages. Furthermore, a rule is characterized by some properties, which define additional semantics of a rule:

- a *name,* which does not have to be unique
- an *invariance* flag for the purpose of restricted overriding
- a *priority,* defining the execution order in case of multiple rules
- two sets *activatedFor* and *deactivatedFor,* holding objects for which the rule is activated or deactivated, respectively, in order to handle exceptions
- two *transaction modes*, specifying transaction semantics for condition evaluation and action execution

Figure 1 defines the overall syntax of a rule specification (note that only the major parts are specified) using the Backus-Naur Form (BNF). The symbols ::= | [ ] { } are meta-symbols belonging to the BNF formalism.

```
<rule_definition>   ::=    DEFINE [VARIANT | INVARIANT] RULE <rule_name>
                           ON <rule_event_selector> DO /*event part*/
                               /*cond. part*/
                           ON <cond_event_selector> IF <boolean_expr> THEN
                               /*action part*/
                           ON <act_event_selector> EXECUTE <action>
                           [WITH PRIORITY <number>]
                           [ACTIVATED [FOR <set_of_instances>] |
                            DEACTIVATED [FOR <set_of_instances>] ]
                           [TRANSACTION (<transaction_mode>,
                                             <transaction_mode>)].
<rule_event_selector>::= {PRE|POST} ([<class_name>],
                           [CLASSMETHOD|INSTANCEMETHOD]
                           <method_signature>)
<cond_event_selector> ::=<rule_event_selector> |
                           {PRE|POST} (self[.<path_expr>],
                           <method_signature>)
<act_event_selector> ::= <cond_event_selector> |
                           {PRE|POST} (cond[.<path_expr>],
                           <method_signature>)
<path_expr> ::=            <instance_var> {.<instance_var>}
<set_of_instances> ::=     (<instance_ref> {,<instance_ref>})
<transaction_mode> ::=     SERIAL | PARALLEL
```
Figure 1. BNF of rule specification

In the following, examples of rule specifications are based on the syntax defined above. Almost all examples employ immediate event-condition coupling, i.e., rule triggering and condition evaluation take place at the same time. Thus the specifications of rule event selector and condition event selector are defined for the same triggering object. The same holds true for condition-action coupling. For a detailed explanation of the execution model we refer to section 5.

❏   The specification of a rule responsible for fetching the next operation from a machine's buffer and scheduling it each time this machine becomes available is shown in the following example (the whole example database schema is shown in section 7):

```
DEFINE RULE MachineRule_1
ON POST (Machine, changeState: newState) DO
ON POST (self, changeState: newState)
    IF newState = 'available' THEN
ON POST (self, changeState: newState)
    EXECUTE self schedule: (self getBuffer nextOperation).
```

To obtain tight integration with the underlying data model as well as for the purpose of dynamic rule definition and manipulation, the three components of the rule as well as the rule itself are first class objects. That is, new rules are defined by creating new instances of predefined object classes representing the rule and its components and equipping them with appropriate values.

4.2 Integration of Rules into the Class Hierarchy

Considering rules in the context of a class hierarchy, they can be attached to a specific class or defined apart from any class. According to this distinction, TriGS supports two categories of rules, called *local rules* and *global rules*, each of them having a specific scope determined by the rule event selector.

Local Rules

A *local rule* allows to monitor specific behavior of certain classes. Thus the event selector of a local rule consists of a triggering class in addition to a triggering method and to one of the keywords PRE or POST, respectively. The *triggering class* denotes the class which the rule is attached to. The *triggering method* denotes the method which raises the event when the corresponding message is received. The optional keywords CLASSMETHOD and INSTANCEMETHOD depict whether the triggering method - in terms of GemStone - should be a class method or an instance method (which is default). Finally, PRE or POST denote whether the event should be raised before or after execution of the triggering method.

Inheritance

Since a local rule is attached to a specific class it is subject to inheritance, consequently being propagated to the subclasses of the triggering class together with other structural and behavioral components by using the inheritance mechanism of the underlying object model. This implies that a local rule is triggered not only by invoking the triggering method on any object of the triggering class but also by invoking this method on any object of its subclasses. Therefore, the scope of a local rule is defined as the part of the class hierarchy rooted at the triggering class. To determine whether an event is able to trigger a local rule, both the triggering method and the triggering class of the selector have to be compared with the signalled event. Additional rules can be defined in subclasses, but are required to be assigned unique names. Regardless whether a triggering method is overridden in a subclass, the rules containing that method in their event selector are still inherited.

Let us assume that A, B and C denote object classes standing in a class-subclass relationship, and R_j denotes the name of a rule. For representation convenience we denote a local rule R_j defined on triggering class A with triggering method m_j as R_j (A, m_j). In Figure 2, on the left side this class hierarchy is shown. The corresponding rules are shown in the middle part of the figure and the inheritance of rules is depicted on the right side. Three rules R1, R2, and R4 are defined on the triggering method m1 and one rule R6 is defined on the triggering method m3. R1 and R4 are attached to the triggering class A, R2 is attached to B, and R6 is attached to subclass C of class B. As soon as an object of class B receives the message m1, rule R2 and - due to inheritance - rules R1 and R4, respectively, are triggered. Considering class C, rule R6 is triggered if m3 is sent to an object of class C.

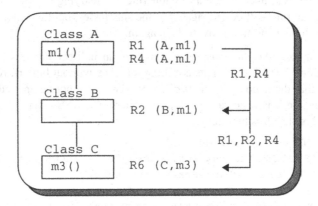

Figure 2.   Inheritance of local rules

❑   Considering the example, given in Section 4.1, `MachineRule_1` is inherited to all subclasses of `Machine`, namely `MillingMachine`, `Drilling-Machine`, and `MMWithoutBuffer`.

Overriding

Let us assume that local rule R1 (A, m1) is to be overridden in a subclass B by defining another local rule R1' (B, m1). For this purpose, firstly, the rule event selector has to be specified by using the same triggering method m1 but the triggering class B. Secondly, the new rule R1' is required to have the same name as the overridden rule R1. Rule R1' can be modified by changing the values of its properties (activation status, etc.) as well as by embedding another condition and/or action. Once R1 is overridden by R1', the latter is propagated to the subclasses of B until R1' itself is overridden (see Figure 3).

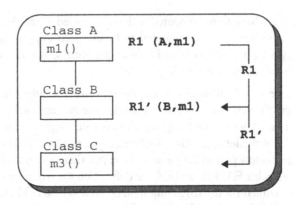

Figure 3. Overriding of local rules

Occasionally (e.g. if a triggering method is overridden) an inherited rule may be unsuitable for a subclass. For this purpose, we provide a selective inheritance mechanism. The trick is to override the unsuitable rule by falsifying the condition of the rule. The overridden rule will be propagated to the subclasses, but the action will never be executed until the condition is overridden again.

❑   In our example, the class Machine has an indirect subclass MMWithout-Buffer (which represents a milling machine without buffer), resulting in the fact that the action part of MachineRule_1 cannot be applied to that class. Therefore it is necessary to stop the inheritance of MachineRule_1 at this level of the class hierarchy:

```
DEFINE RULE MachineRule_1
ON PRE (MMWithoutBuffer, changeState: newState) DO
ON PRE (self, changeState: newState)
    IF FALSE THEN
ON PRE (self, changeState: newState)
    EXECUTE /* nothing */.
```

Moreover, TriGS provides the possibility to make a rule invariant by setting the invariance flag. An invariant rule cannot be overridden in any of the subclasses of the triggering class. This option represents a convenient way to enforce absolute constraints. By default, a rule is defined to be variant.

Activation and Deactivation of Rules

The activation status of a rule determines whether it can be triggered by a corresponding event. This provides a convenient way for experimenting with different rules, i.e., analyzing the impact of different rules on the system behavior by temporarily activating or deactivating them.

In TriGS, a local rule can be activated and deactivated at different levels of granularity. It is possible to (de)activate a rule at *class level* (cf. keywords ACTIVATED and

DEACTIVATED in Figure 1). By default, a rule is activated at class level. (De)activation at class level means that, if a message corresponding to the triggering method is sent to any instance of the triggering class or one of its subclasses within the rule's scope, the corresponding rule will (not) be triggered depending on the activation status of the rule. Figure 4a shows rule R1, which is attached to class A and overridden by rule R1' in subclass B. Both rules are activated by default. Therefore the scope of R1 covers only instances of class A (a1...a4 in Figure 4a), whereas the scope of R1' covers all instances of class B (b1...b3 in Figure 4a) as well as all instances of its subclass C (c1, c2 in Figure 4a).

As soon as rule R1' is deactivated at class level (see Figure 4b) the overridden rule R1 is propagated to all subclasses instead of the deactivated rule R1'. Note that the deactivation of rule R1' does not have the same semantics as falsifying the condition of a rule (see Section Overriding) since in the latter case R1' would be propagated to subclasses instead of R1 and thus no action would be executed.

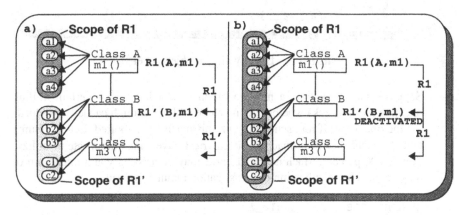

Figure 4. Activation and deactivation at class level

Besides (de)activation at class level, local rules can be (de)activated at *object level*. Thus exceptions can be specified for certain instances of the classes within the rule's scope. This is done by (de)activating the rule for the desired instances only by collecting them into the appropriate container (de)activatedFor (cf. keywords (DE)ACTIVATED FOR in Figure 1).

Figure 5 shows rule R1 attached to class A, which is activated at class level and explicitly deactivated for instance a3 of class A and instance b2 of A's subclass B. If message m1 is sent to object a3 or b2, rule R1 is not triggered, yet activated at class level for class A and class B.

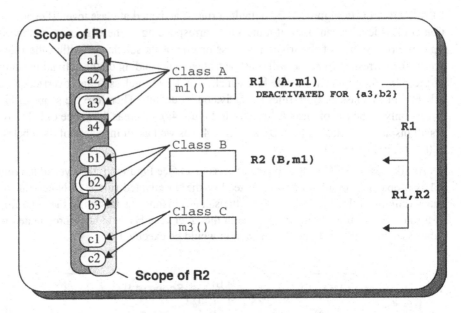

Figure 5. Deactivation at object level

❑ Suppose that every time a new operation is scheduled on a certain milling machine a rule checks whether the machine still has enough capacity to schedule the operation. If not, scheduling on this machine is aborted. Let us assume that all machines except machine `milling3` have a maximum rate of utilization of 100 percent. Machine `milling3` may be used only at a maximum of 80 percent. The specification of this situation requires two rules:

```
DEFINE RULE MillingRule_1
ON PRE (MillingMachine, schedule: operation) DO
ON PRE (self, schedule: operation)
   IF (self computeUtilization: operation) > 100 THEN
ON PRE (self, schedule: operation)
   EXECUTE self changeState: 'overloaded'. BREAK.
DEACTIVATED FOR {milling3}.
```

```
DEFINE RULE MillingRule_2
ON PRE (MillingMachine, schedule: operation) DO
ON PRE (self, schedule: operation)
   IF (self computeUtilization: operation) > 80 THEN
ON PRE (self, schedule: operation)
   EXECUTE self changeState: 'overloaded'. BREAK.
ACTIVATED FOR {milling3}.
```

MillingRule_1 is automatically activated for all instances of class Mill-ingMachine and explicitly deactivated for the instance milling3. In addition, MillingRule_2 is defined and activated only for that instance.

Global Rules

Local rules allow the definition of rules according to the behavior of certain classes only. But it would be also desireable to attach a rule to a certain triggering method without considering the class to which this method belongs. This is especially useful in an object-oriented system like GemStone which does not support multiple inheritance. Simulating multiple inheritance through single inheritance often results in the redundant specification of attributes and methods of classes. As local rules also belong to specific classes they would have to be defined redundantly, too. Global rules are a way to avoid this drawback. In contrast to a local rule, the event selector of a global rule consists of the triggering method together with one of the keywords PRE or POST, only.

As the triggering class is not specified, a global rule is not associated with a specific class and thus not subject to inheritance. To determine whether an event is able to trigger a global rule, the rule event selector has to be compared with the message name that signalled the event. The scope of a global rule covers all classes of the database schema that know the triggering method defined in the rule event selector.

Figure 6 illustrates the different scopes of global and local rules (denoted by different fill patterns). All rules in the example are defined on the same triggering method m1. Although method m1 has been defined twice due to the lack of multiple inheritance, only one global rule R5 has to be defined on m1 since the scope of R5 covers both the class hierarchy rooted at class A and the class hierarchy rooted at class Y. In contrast, the scope of local rule R1 attached to class A is defined only by the class hierarchy rooted at class A. Similarly, the scope of local rule R2 attached to class Y is defined by class Y and its subclass Z.

Assume that a message m1 is sent to an instance of class A. Since m1 matches the specified event selectors of local rule R1 and global rule R5, both rules are triggered. In case a global rule and a local rule with the same name are defined on the same triggering method, only the global rule is taken into consideration when an appropriate event occurs. This mechanism provides the possibility to override - in one blow with one global rule - a number of local rules defined on the same triggering method and located in different class hierarchies of the database schema. In Figure 6 the locally defined rules R5 on class A and R5 on class Y are overridden by a global rule with the same name.

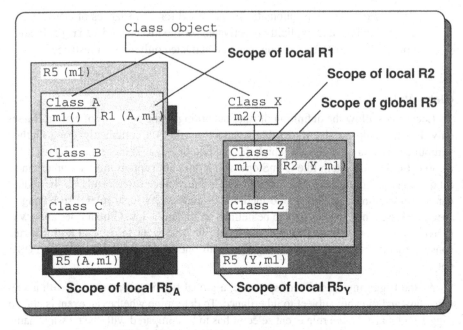

Figure 6. Scopes of local and global rules

❏ In our example, one could assume, that for the purpose of statistical analysis some change of the attribute `state` caused by method `changeState` is written to a log file. The attribute `state` is defined for classes `Resource` and `Schedule`. In every subclass of class `Resource` the attribute `state` can be set to different values. For example, the states of a worker are "ill/onHoliday/working/notWorking", whereas a machine may be "running/broken/available/overloaded". A schedule may have the states "inTime/delayed/notReleased". For each of these classes a local rule named `LoggingRule` is defined, which has its own log file. Now we want to unify all these local rules by writing to only one log file:

```
DEFINE RULE LoggingRule
ON POST (changeState: newState) DO
ON POST (changeState: newState)
    IF NOT ({'available','running','working','inTime'}
                includes: newState) THEN
ON POST (changeState: newState)
    EXECUTE globalLogFile put: newState.
```

This `LoggingRule` is globally defined for the triggering method `changeState` and overrides all locally defined logging rules, as they have the same name.

The (de)activation of global rules may be changed at class level covering all instances of all classes in the scope of the rule. Since a global rule is not subject to inheritance, no overridden version of the rule may be inherited instead of the deactivated one. (De)activation of global rules at object level has the same semantics as described for local rules.

5 Execution Model of TriGS

In this section we focus on relevant aspects concerning the execution of rules. These comprise the semantics of firing a single rule including the timing when its condition/action should be evaluated/executed, and how rule execution is embedded into our transaction model. In addition, we specify the semantics of multiple rule firing by analyzing their execution order.

5.1 Firing a Single Rule

TriGS uses an event-based approach to specify the behavior of rule firing similar to the concept proposed by [Beer91]. This approach allows to reuse the event selector mechanism originally introduced for rule triggering. *Condition event selectors* and *action event selectors* (cf. Figure 1) define the points in time when condition evaluation (*CEP*, i.e., Condition Evaluation Point) and action execution (*AEP*, i.e., Action Execution Point) have to take place. Condition event selector and action event selector are specified on the basis of the corresponding rule event selector.

In this context, we may distinguish three cases. Firstly, the objects which are involved in triggering a rule, in the corresponding condition evaluation, and in the corresponding action execution are independent from each other, being of the same or of different classes. In this case and if a local rule is defined the triggering class is specified for all three event selectors. Secondly, the keyword self is used for specifying that the object that is involved in triggering condition evaluation and/or action execution has to be the same (or reachable from the same) object which is involved in triggering the corresponding rule. Thirdly, the keyword cond provides a binding between the object involved in triggering condition evaluation and the one involved in triggering action execution.

❑ Suppose, for example, that a machine wants to fetch the next operation from its buffer which is already empty. In this case, a rule should be triggered which specifies that as soon as a new operation is inserted into the machine's buffer, this operation is scheduled on the machine. The insertion into the buffer is used as the event triggering the action. To ensure that the action is triggered only by an insertion into the same buffer which proved to be empty, the object triggering action execution has to be bound to the object triggering the rule. It is assumed that all machines are referenced by the persistent Collection Machines.

```
DEFINE RULE BufferRule_1
ON PRE (Buffer, nextOperation) DO
ON PRE (self, nextOperation)
    IF self isEmpty THEN
ON POST (self, insertOperation: anOp)
    EXECUTE (Machines detect:[:m|self=m getBuffer])
                schedule:self nextOperation.
```

Transaction Semantics

So far we have discussed when conditions may be evaluated and actions may be executed. We will now turn to the transaction semantics of rule processing. This effects the triggering transaction, the condition transaction, and the action transaction. The *triggering transaction* is the transaction in which the event is raised which triggers rule processing in turn. The *condition transaction* embodies condition evaluation, and the *action transaction* includes action execution. TriGS supports both serial and parallel execution of condition transactions and action transactions. In case of serial transaction execution (which is default) a parent transaction invokes a child transaction as nested transaction and blocks until the child transaction has finished. In case of parallel transaction execution a parent transaction invokes a child transaction as new top level transaction and continues its own execution immediately.

❑   Consider our example of a LoggingRule. It would be desirable to specify the action transaction to be parallel in order to avoid a delay of the application caused by writing to the log file. For this purpose, the statement TRANSAC-TION (SERIAL, PARALLEL) has to be appended to the definition of the LoggingRule.

Serial as well as parallel child transactions can be rolled back independently of the parent transaction. As GemStone provides neither nested nor parallel transactions these features have to be simulated. Nested transactions are realized by means of safepoints and parallel transactions are realized by starting a concurrent GemStone session [Butt91].

Granularity of CEP and AEP Definition

We will now turn to the question of how the event-based approach for specifying rule processing compares to the coupling mechanism originally introduced in [Daya88]. We will show that the event-based mechanism in conjunction with the specification of the transaction semantics provides a fine granularity for specifying event-condition coupling and condition-action coupling. These include immediate, deferred, and separate coupling of [Daya88]. In the following we will demonstrate these coupling modes for event-condition coupling. The same approach applies to condition-action coupling.

To simulate the semantics of *immediate* event-condition coupling the rule event selector and the condition event selector have to be defined by the same triggering method, and, in case of a local rule, by the same triggering class. In addition, the condition transaction has to be defined as serial. Note, that all our previous examples simulate immediate coupling. The *deferred* event-condition coupling mode may be simulated by using the commit or abort method of GemStone as triggering method of the condition event selector and defining the condition transaction to be serial. The *separate* event-condition coupling mode is different to the immediate coupling mode in the sense that the condition transaction has to be defined as parallel. It can be seen that the granularity of coupling modes of [Daya88] is restricted to an immediate or deferred CEP/AEP, defining the corresponding transaction to be serial and to a separate CEP/AEP defining the corresponding transaction to be parallel. That is, coupling modes determine not only CEP/AEP but implicitly also the transaction semantics for condition evaluation and action execution.

In TriGS, however, transaction semantics is defined orthogonal to CEP/AEP, and the granularity of CEP/AEP definition is not restricted to immediate, deferred, and separate coupling modes. Figure 7 illustrates the possibilities for defining CEP and AEP assuming that the methods m1, m2, and EOT (End Of Transaction, i.e. commit or abort method) are sequentially executed in the same transaction and that there exists a rule R1 with the triggering event being $m1_{pre}$. It is shown, that the start and end of any method can be used for specifying CEP and AEP (denoted by ✔). Of course, the event triggering the rule has to be signalled before or at the same time as the event triggering the rule's condition, and the latter one has to be signalled before or at the same time as the event triggering the rule's action. For this reason, some CEP/AEP combinations are denoted as "not applicable".

CEP \ AEP		m1 pre	m1 post	m2 pre	m2 post	EOT pre	EOT post
m1	pre	ii,is,si, ss,sd	*not applicable*				
m1	post	✔	✔				
m2	pre	✔	✔	✔			
m2	post	✔	✔	✔	✔		
EOT	pre	id	✔	✔	✔	di, dd, ds	
EOT	post	✔	✔	✔	✔	✔	✔

i ... immediate   d ... deferred   s ... separate

Figure 7. Granularity of CEP and AEP

Furthermore, it is shown which CEP/AEP combinations simulate the nine possible combinations of coupling modes between event-condition coupling and condition-action coupling. The appropriate transaction semantics necessary for the simulation is not illustrated in the figure.

❑   The following example illustrates how our event-based approach can be used to realize a deferred update of the class variable `totalNumOfOps` of class `Buffer`. Whenever the value of this variable is needed and the number of operations within any buffer had been changed by `nextOperation` the value of this variable has to be recomputed. The following rule checks if the contents of a buffer has changed and computes the actual `totalNumOfOps` in case of retrieval by use of the method `getTotalNumOfOps`. Note that `getTotal-NumOfOps` and `computeTotalNumOfOps` are class methods.

```
DEFINE RULE NumOfOpsRule
ON PRE (Buffer, nextOperation) DO
ON PRE (self, nextOperation)
    IF self notEmpty THEN
ON POST (Buffer, CLASSMETHOD getTotalNumOfOps)
    EXECUTE Buffer computeTotalNumOfOps
```

5.2 Firing Multiple Rules

Due to the fact that TriGS allows simultaneous firing of multiple rules, i.e. the evaluation/execution of two or more conditions/actions at the same CEP/AEP, it is necessary to analyze their evaluation/execution order. Two rules R1 and R2 have the same CEP/AEP if their conditions/actions are triggered by the same event. In Figure 8 CEP_{R1} is defined by the event selector (A, m1) whereas CEP_{R2} is defined by m1. Thus, CEP_{R1} is equal to CEP_{R2} only if m1 is sent to an object of class A, class B or class C.

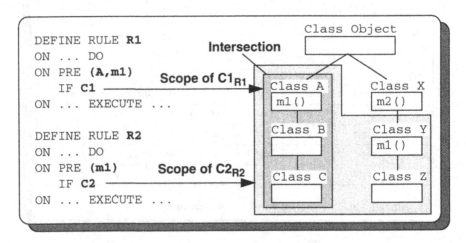

Figure 8.   Simultaneous firing of multiple rules

In case that two conditions have the same CEP, and the condition transactions of both are defined to be serial, their evaluation may be done in arbitrary order, since they access the database in read mode and therefore no conflicts may arise. Since actions are able to change the database state, they can influence each other if they have the same AEP and the action transactions of both are defined to be serial. In this case their execution order has to be specified, which is done by means of priorities similar to other systems [Kotz93, Mede91]. By default all rules have the same priority, which may be increased or decreased for any rule at any time.

6 Prototype Implementation of TriGS

6.1 Implementation Architecture of TriGS

The architecture of TriGS shown in Figure 9 consists of four main components, namely *Event Detector, Rule Scheduler, Condition Evaluator*, and *Action Executor*. The Event Detector is responsible for signaling detected events and passing them to the Rule Scheduler. The tasks of the Rule Scheduler are to collect rules (denoted by ①), conditions (denoted by ②), and actions (denoted by ③), all triggered by the signalled event, and to pass collected conditions and actions to the Condition Evaluator and Action Executor, respectively. Rules are transferred between components by means of persistent dictionaries.

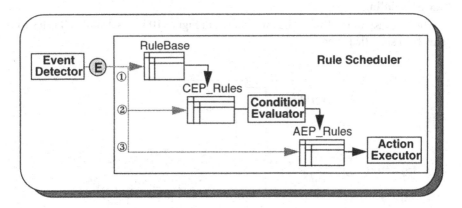

Figure 9. Overall architecture of TriGS

In the remainder of this section, a more detailed description of the tasks executed by the different components is given.

Event Detection

In TriGS, event detection is realized by means of method wrappers. At the first time a method m1 is defined to be part of an event selector, m1 is transformed into a triggering method by, firstly, renaming m1 to m1_original and recompiling it within the class where it has been defined in the first place and, secondly, adding a wrapper, i.e. a new method named m1, to that class. The body of the wrapper consists of a call of m1_original and a call of the scheduler responsible for scheduling all triggered rules, conditions, and actions. In case of pre method detection the Rule Scheduler is called before the original method is executed. In case of post method detection it's the other way round.

Rule Storage

Rules are stored in a persistent dictionary representing the *rulebase* of TriGS. Every entry in this dictionary consists of a key represented by the appropriate rule event selector and a value consisting of references to all rules containing that rule event selector. Thus rules can be indexed by their event selectors, which allows speeding up the process of scheduling all rules triggered by a specific event. Besides the dictionary **Rulebase**, two additional persistent dictionaries having the same structure as the rulebase are used to store rules waiting for condition evaluation (**CEP_Rules**) and action execution (**AEP_Rules**). The keys of these dictionaries are the event selectors of the conditions and actions, respectively. Finally, a transient dictionary (**triggered_Rules**) is used to implement overriding of rules by temporarily storing rule references indexed by their names.

Rule Scheduling

The first task of the Rule Scheduler shown in Figure 10 is to schedule all rules triggered by a signalled event.

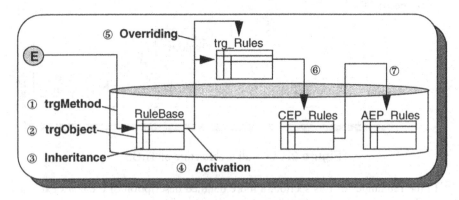

Figure 10. Rule scheduling

In a first step (denoted by ① in Figure 10) the Rule Scheduler looks for an entry in the rulebase corresponding to the triggering method to find global rules. This can be done by only one access since the dictionary is indexed by the triggering method. Next, the Rule Scheduler looks for local rules within the rulebase (denoted by ② in Figure 10). Since local rules are subject to inheritance, a lookup mechanism (denoted by ③ in Figure 10) based on metaclass information is used starting at the class of the object where the event was signalled, and ending at the class where the triggering method is defined (in the worst case at the root of the class hierarchy).

Figure 11 illustrates the lookup mechanism for collecting rules in response to the event m1 raised in object b1. After retrieval of the triggering method, the class of the triggering object is determined. The lookup process starts at this class, in our example class B, collecting all defined rules {R2B, R1', R4'}. Then, the superclass of B, class A, is determined and again all defined rules of class A, {R1, R4, R5A}, are collected.

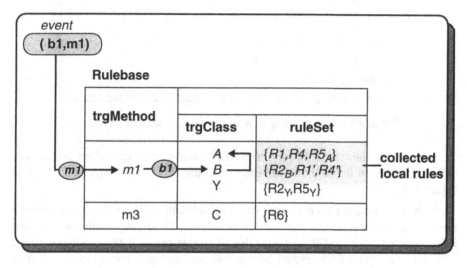

Figure 11. Collecting local rules

For each collected rule the Rule Scheduler has to check its activation status (denoted by ④ in Figure 10). If the rule is activated it is stored in the dictionary CEP_Rules if not overridden by another rule. To recognize overriding of rules (denoted by ⑤ in Figure 10) a mechanism based on the equality of rule names is used. Since overriding has to be checked for each specific event and since the dictionary CEP_Rules may still hold rules triggered by former events, the transient dictionary triggered_Rules having the rule name as key is used to filter overridden rules. Before making a triggered rule persistent within the dictionary CEP_Rules, its name is compared to the keys of rules already inserted into the dictionary triggered_Rules. Equality means that the rule has been overridden, and the more specialized rule has already been inserted into the transient dictionary triggered_Rules. Consequently the overridden rule is rejected, otherwise it is inserted.

Figure 12 represents the process of filtering overridden local rules collected in Figure 11 with regard to the global rule R5 which also has been collected due to event m1 (cf. Figure 6). The numbers denote the sequence of the filtering process. After global rule R5 is transferred into the dictionary triggered_Rules, all local rules are transferred in the same order as they were collected. Since R1' and R4' are collected before R1 and R4, the latter ones are rejected due to overriding.

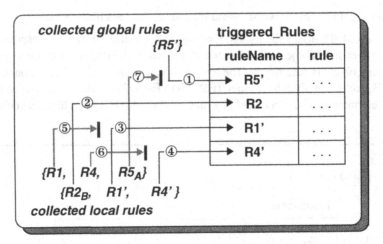

Figure 12. Overriding local rules using a transient dictionary

After having transferred all rules triggered by one specific event into the transient dictionary they are stored in the persistent dictionary CEP_Rules (denoted by ⑥ in Figure 10).

Right after a certain event has been used for scheduling all corresponding rules the same event is used to schedule conditions and actions. Firstly, and similar to collecting rules (denoted by ① and ② in Figure 10), it is determined if the conditions stored within the dictionary CEP_Rules are ready for evaluation or not, no matter whether their corresponding rules have been scheduled by the same or by a former event. Secondly, the Condition Evaluator is started to evaluate the conditions of the collected rules. The condition evaluator treats all conditions whose condition transaction is defined to be serial within the triggering transaction. All conditions whose condition transaction is defined to be parallel are evaluated in independent top-level transactions. If condition evaluation returns true, the rule is transferred into the persistent dictionary AEP_Rules (denoted by ⑦ in Figure 10). Finally, the Action Executor has to apply almost the same steps as the Condition Evaluator to all actions waiting for execution within the dictionary AEP_Rules. The only difference is that the Action Executor does not execute all actions defined to be serial in arbitrary order, as is the case with conditions, but according to the specified priorities of the corresponding rules.

6.2. Status of Implementation

Event Detector, Rule Scheduler and Rule Designer have been implemented in GemStone Vs. 3.2.5 [Butt91] and are already operational. GemStone provides a computationally complete data definition language and data manipulation language, called OPAL, which is a Smalltalk derivate. Event Detector and Rule Scheduler were imple-

---

[TM]OPAL is a registered trademark of Servio Corporation

mented directly in OPAL, i.e., they run in the database's address space. By extending a system-defined method responsible for compiling OPAL-methods they are automatically adapted for event detection at compile time. Since OPAL lacks direct terminal and disk I/O facilities the Rule Designer had to be implemented in one of the host languages Smalltalk or C++. We decided to use Smalltalk, which is more homogeneously integrated into the GemStone environment than C++. The GemStone Smalltalk Interface (GSI) is a set of classes installed in a Smalltalk image that permits a user to access and modify objects in the GemStone database. Rules are built by means of the Rule Designer within Smalltalk and transmitted to GemStone via the GSI. Within the database a rule is made persistent by referencing it from a globally known persistent dictionary.

7 TriGS by Example

Our universe of discourse is taken from the Leitstand environment. A schedule reserves a machine for the purpose of executing a certain operation within a specific start time and end time. There are three different types of machines, namely drilling and milling machines which have buffers with different capacities holding the operations to be executed, and some milling machines without a buffer. Machines and buffers are two kinds of resources that each have a certain state. The database schema representing this universe of discourse is depicted in Figure 13 and Figure 14.

Figure 13 illustrates the example database schema including its class hierarchy represented by plain lines, its class composition hierarchy represented by arrows, and its rules denoted by abbreviations of their names. The class composition hierarchy models the partOf relationship between composite objects and their component objects.

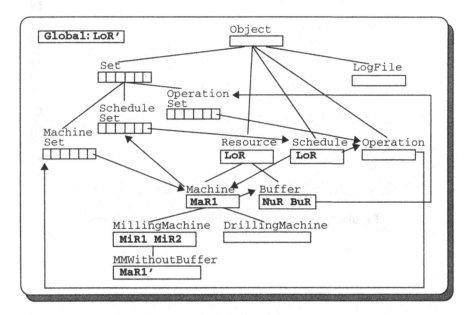

Figure 13. Example database schema

Figure 14 shows each class with its associated attributes (enclosed in square brackets) and methods (located below each class).

```
Object
 Set
  OperationSet
  ScheduleSet
  MachineSet

 Resource  [state]
  getState
  changeState: newState

  Buffer  [capacity, waitingOperations]
   nextOperation
   insertOperation: operation
   isEmpty
   notEmpty
   --- classmethods ---
   getTotalNumOfOps
   computeTotalNumOfOps

  Machine  [utilization, buffer, schedules]
   schedule: operation
   getBuffer
   computeUtilization: operation

   Drilling  [holeType]
   Milling   [profileType]
    MMWithoutBuffer  []

 Operation  [name, priority, applicableMachines]

 Schedule  [startTime, endTime, state, selectedMachine,
            scheduledOperation
  changeState: newState

 LogFile
  put: newState
```

Figure 14. Object classes with attributes and methods

In the following all rules used throughout the report are summarized.

```
DEFINE RULE MachineRule_1
ON POST (Machine, changeState: newState) DO
ON POST (self, changeState: newState)
   IF newState = 'available' THEN
ON POST (self, changeState: newState)
   EXECUTE self schedule: (self getBuffer nextOperation).


DEFINE RULE MachineRule_1
ON PRE (MMWithoutBuffer, changeState: newState) DO
ON PRE (self, changeState: newState)
   IF FALSE THEN
ON PRE (self, changeState: newState)
   EXECUTE /* nothing */.


DEFINE RULE MillingRule_1
ON PRE (MillingMachine, schedule: operation) DO
ON PRE (self, schedule: operation)
   IF (self computeUtilization: operation) > 100 THEN
ON PRE (self, schedule: operation)
   EXECUTE self changeState: 'overloaded'. BREAK.
DEACTIVATED FOR {milling3}.


DEFINE RULE MillingRule_2
ON PRE (MillingMachine, schedule: operation) DO
ON PRE (self, schedule: operation)
   IF (self computeUtilization: operation) > 80 THEN
ON PRE (self, schedule: operation)
   EXECUTE self changeState: 'overloaded'. BREAK.
ACTIVATED FOR {milling3}.


DEFINE RULE LoggingRule
ON POST (changeState: newState) DO
ON POST (changeState: newState)
   IF NOT ({'available','running','working','inTime'}
           includes: newState) THEN
ON POST (changeState: newState)
   EXECUTE globalLogFile put: newState
TRANSACTION (SERIAL, PARALLEL).


DEFINE RULE BufferRule_1
ON PRE (Buffer, nextOperation) DO
ON PRE (self, nextOperation)
```

```
    IF self isEmpty THEN
ON POST (self, insertOperation: anOp)
    EXECUTE (Machines detect:[:m|self=m getBuffer])
            schedule:self nextOperation.


DEFINE RULE NumOfOpsRule
ON PRE (Buffer, nextOperation) DO
ON PRE (self, nextOperation)
    IF self notEmpty THEN
ON POST (Buffer, CLASSMETHOD getTotalNumOfOps)
    EXECUTE Buffer computeTotalNumOfOps.
```

8 Conclusion and Future Work

After surveying existing approaches to active object-oriented database systems, we presented our own approach TriGS, an extension to the commercially available OODBS GemStone. TriGS focuses on a seamless integration of rules by making explicit use of object-oriented mechanisms like objects, message passing, inheritance, and overriding. As rules and their components are first-class objects, they can be specified dynamically and independent of any application. To support the specification of rules valid for all classes of an application's class hierarchy as well as the specification of rules attached to distinct classes and therefore subject to inheritance and overriding, global rules and local rules are distinguished. In order to enforce constraints on some subclass hierarchy overriding may be prohibited for specific rules. The possibility to activate and deactivate rules on different levels of granularity allows to experiment with different rules in a convenient way and, in addition, it provides a flexible exception handling mechanism. The most salient concepts of TriGS were illustrated by means of some examples out of the area of production scheduling and control. The first prototype of TriGS on the basis of the object-oriented database system GemStone is operational.

Further research is required for supporting other kinds of events like time events, and external events, for the purpose of communication between different application modules. In addition, further possibilities to integrate the trigger model with an advanced transaction model, besides that of GemStone, have to be explored. Last but not least, rules for designing applications based on active object-oriented database systems have to be investigated. Questions like "what is the right place to put rules?" and "what goes into rules and what into object behavior?" seek further research. One of these design guidelines should cover the "right" inheritance of rules. Since GemStone adheres to the principle of implementation inheritance, few restrictions have been placed on overriding rules so far. However, the inheritance and overriding of rules based on subtyping is a challenging problem to provide for smooth integration of the most important features of TriGS with (strongly) typed data models.

References

[Adel91]   H. Adelsberger, J. Kanet, *The Leitstand - A New Tool for Computer Integrated Manufacturing,* Production and Inventory Management Journal, First Quarter, pp. 43-48, 1991

[Beer91]   C. Beeri, T. Milo, *A Model for Active Object Oriented Database,* Proc. of the 17th Int. Conference on VLDB, Barcelona, pp. 337-349, 1991

[Bern92]   M. Berndtsson, B. Lings, *On Developing Reactive Object-Oriented Databases,* Bulletin of the IEEE Technical Commitee on Data Enginieering, Vol. 15, No. 1-4, December 1992

[Blak90]   J.A. Blakelely, C.W. Thompson, and A.M. Alashqur. *Zeitgeist query language (zql).* Technical Report TR-90-03-01, Texas Instruments, March 1990

[Butt91]   P. Butterworth, A. Otis, J. Stein, *The GemStone Object Database Management System,* Communications of the ACM, Vol. 34, No. 10, pp. 64-77, October 1991

[Chak90]   S. Chakravarthy, S. Nesson, *Making an Object-Oriented DBMS Active: Design, Implementation, and Evaluation of a Prototype,* Proc. of the Int. Conference on Extending Database Technology, Venice, pp. 393-406, March 1990

[Chak91]   S. Chakravarthy and D. Mishra. *An event specification language (snoop) for active databases and its detection.* Technical Report UF-CIS TR-91-23, Database Systems R&D Center, CIS Department, University of Florida, E470-CSE, Gainesville, FL 32611, September 1991

[Chak93]   S. Chakravarthy, E. Anwar, L. Maugis, *Design and Implementation of Active Capability for an Object-Oriented Database,* Database Systems R&D Center, CIS Department, University of Florida, E470-CSE, Gainesville, FL 32611, January 1993

[Coll94]   C. Collet, P. Habraken, T. Coupaye and M. Adiba, *Active Rules for the Software Entineering platform GOODSTEP,* 2nd Int. Workshop on Database and Software Engineering - 16th Int. Conf. on Software Engineering, Sorrento, Italy, May 1994

[Daya87]   U. Dayal, F. Manola, A. Buchmann, U. Chakravarthy, D. Goldhirsch, S. Heiler, J. Orenstein, A. Rosenthal, *Simplifying Complex Objects: The PROBE Approach to Modelling and Querying Them,* Proc. German Database Conference (BTW), Springer-Verlag, 1987

[Daya88]   U. Dayal, A.P. Buchmann, D.R. McCarthy, *Rules Are Objects Too: A Knowledge Model For An Active, Object-Oriented Database System,* Advances in Object-Oriented Database Systems, Proc. of the 2nd. Int. Wordshop on OODBS, K.R. Dittrich (ed.), Springer-Verlag, pp. 129-143, 1988

[Diaz91]   O. Diaz, N. Paton, P. Gray, *Rule Management in Object Oriented Databases: A Uniform Approach,* Proc. of the 17th Int. Conference on VLDB, Barcelona, pp. 317-326, 1991

[Gatz91]   S. Gatziu, A. Geppert, K.R. Dittrich, *Integrating Active Concepts into an Object-Oriented Database System,* Proc. of the 3rd Int. Workshop on Database Programming Languages, Nafplion, August 1991

[Gatz92]   S. Gatziu, K.R. Dittrich, *SAMOS: an Active Object-Oriented Database System,* Bulletin of the IEEE Technical Commitee on Data Enginieering, Vol. 15, No. 1-4, December 1992

[Gatz94]   S. Gatziu, K.R. Dittrich, *Detecting Composite Events in Active Database Systems Using Petri Nets,* Fourth International Workshop on Research Issues in Data Engineering, Houston, Texas, February 1994

[Geha92a]   N. H. Gehani, H. V. Jagadish, O.Shmuheli, *Composite Event Specification in Active Database Systems,* Proc. of the 18th Int. Conference on VLDB, August 1992

[Geha92b]   N. H. Gehani, H. V. Jagadish, O.Shmuheli, *Event Specification in an Active Object-Oriented Database,* Proc. of ACM-SIGMOD Int.'l Conf. on Managment of Data, 1992

[Geha93]   N. H. Gehani, H. V. Jagadish, O.Shmuheli, COMPOSE - A System for *Event Specification and Detection,* in Advanced Database Concepts and Research Issues, N.R. Adam, B. Bhargava, LNCS, Springer Verlag 1993

[Kapp92]   G. Kappel, M. Schrefl, *Local Referential Integrity,* Proc of the 11th International Conference on Entity-Relationship Approach, G. Pernul and A M. Tjoa (eds), Springer, pp. 41-61, 1992

[Kapp93]   G. Kappel, S. Rausch-Schott, W. Retschitzegger, S. Vieweg, *TriGS - Making a Passive Object-Oriented Database System Active,* to be published in: Journal of Object-Oriented Programming (JOOP), 1993

[Kapp94]   G. Kappel, S. Rausch-Schott, W. Retschitzegger, *Beyond Coupling Modes - Implementing Active Concepts on Top of a Commercial OODBMS,* submitted for publication, 1994

[Kotz93]   A. Kotz-Dittrich, *Adding Active Functionality to an Object-Oriented Database System - a Layered Approach,* Proc. of GI-Conference on Database Systems for Office, Technology and Science, Braunschweig (BRD), W. Stucky, A. Oberweis (eds.), Springer, pp.54-73, March 1993

[McCa89]   D. R. McCarthy, U. Dayal, *The Architecture Of An Active Data Base Management System,* Proc. of the ACM-SIGMOD Int.'l Conference on Managment of Data, Portland, Oregon, 1989

[Mede91]   C. B. Medeiros, P. Pfeffer, *Object Integrity Using Rules,* European Conference on Object-Oriented Programming (ECOOP), P. America (ed.), Springer LNCS 512, pp. 219-230, 1991

[Röck93]   B. Röck, M. Schrefl, Active Object-Oriented Database Systems, Technical Report, University of Linz, 1993

[Sche91]   A.-W. Scheer and A. Hars, *The Leitstand - A New Tool for Decentral Production Control,* in: Fandel, G. and Zäpfel, G. (eds.), Modern Production Concepts, Springer-Verlag, Berlin, pp. 370-385, 1991

[View94]   S. Vieweg, G. Kappel, A M. Tjoa, *Change Management in Object-Oriented Database Systems,* Institute of Applied Computer Science and Information Systems, Univerity of Vienna, 1994 (submitted for publication)

Databridge between RDBMS and OODBMS

Kroha,P.

Technical University of Chemnitz,
Department of Informatics,
P.O.Box 964, 09009 Chemnitz, Germany

Abstract. For applications, where some of used data is complex structured and some of data is simply structured, we can use an objct- oriented DBMS but the processing simply structured data also will also use the overhead of OODBMS necessary for processing complex structured data. This decreases the efficiency of the processing. In this contribution, we describe the hybrid approach where a relational DBMS will be called from within an object-oriented DBMS application to speed up the processing of simply structured data. We show the main idea illustrated on example of a cooperation between OODBMS GemStone and RDBMS Sybase.

1 Introduction

Currently used DBMSs based on the relational model have many advantages. They are very efficient for simply structured data. However, the data elements used in many applications have become too complex and diverse within a relatively short period of time to be represented with relational database technology. Unfortunately, it is not possible to use these systems in many engineering applications because of their inefficiency when applied on complex structured data.

Relational DBMS are appropriate for applications where data elements and the relationships among data elements are relatively simple. The primary strengths of the relational model are its generality in storing data in elementary tables and its flexibility in providing ad hoc views of the data by joining tables together. As relationships increase among data in a relational database, then the number of join operations begins to rise.

As a rule of thumb, when the number of tables that must be joined together to produce the derived information exceeds five, and when the data types become more complex it is time to consider choosing an object-oriented database instead of a relational database.

Object-oriented DBMSs operate on complex information efficiently because elements of information are "pre-joined" together and stored as a whole in the database. This is in sharp contrast to relational DBMS where elements of data are stored in their simplest, or normal forms in tables. To encover relationships among data in a relational database, these tables must be joined together. To uncover relationships in an object- oriented database, objects forming the relationships are simply fetched. In addition to the relational DBMSs, object-oriented

DBMSs must allow for navigation among objects and tracing of their interrelationships. An accessible repository of classes and objects is crucial to those who intend to cultivate a corporate framework which applications will be generated more efficiently.

Applications enabled by object-oriented DBMSs have been in use in the scientific and engineering communities for some time. More recently, the business community is seeing the benefits of object-oriented databases for certain types of aplications as well.

2 Databridge between RDBMS and OODBMS

Object-oriented DBMSs are necessary for processing complex structured data. In the contrary, OODBMS are not efficient for processing simply structured data. Important criterion is how many tables are necessary in relational model to store all attributes of an object of the real world, more precisely, how often these tables need to be joined in queries and how much time it takes. Often, applications use both complex structured data, and simply structured data. In this case, it is advantageous to work inside an object-oriented DBMS with complex structured objects and to "hire" services of a relational DBMS for efficient processing simply structured objects.

The main idea is to map RDBMS types to equivalent OODBMS types (classes) and to use a special library associated with the OODBMS offering the functionality of the databridge. We will show this process on example using RDBMS Sybase from OODBMS GemStone.

After starting the session in GemStone we create a persistent object of a predefined class sybaseSession. Functionality of this object enables to log in to the Sybase server. After log in to the Sybase, we can use some predefined classes and their methods supporting the databridge, e.g. by sending SQL commands from GemStone to the Sybase server. The results are then returned to GemStone.

This mechanism makes possible to process tables stored in the relational DBMS Sybase using all advantages of their implementation, i.e. without the overhead coming from the generality and the high modelling power of the object-oriented DBMS.

3 The databridge classes

The GemStone's query language OPAL which can be held for a dialect of Smalltalk contains classes (RdbmsSession, Tuple, RdbmsRelation) for accessing an external relational DBMS Sybase from GemStone. These classes have the following meaning:

– RdbmsRelation is a subclass of Set.

Instances of concrete subclasses are intended to hold the esults of SQL queries. For query results returned from the method dbexecute:, the system creates a subclass of RdbmsRelation and contains it to hold a system-created subclass of Tuple.

– Tuple is a subclass of Object.

Each instance variable in concrete subclasses of Tuple is intended to correspond to a row in a result relation.

– RdbmsSession is a subclass of Object.

Its methods represent the functionality of the databridge and will be described in the next section. SybaseSession used in the following examples is a concrete (already implemented) subclass of RdbmsSession.

4 The functionality of the databridge

The class RdbmsSession offers the following methods supporting the functionality of the databridge:

– dbLogin ... Logs in to Sybase.

– dbUse ..... Specifies a particular database.

– dbExecute: $< SQL - commandString >$

Executes the SQL-commandString on the database server, creates a new relation class as a subclass of sybaseRelation and a new tuple class as a subclass of sybaseTuple and uses instances of these classes to store the results of the query. Using options we explicitly can specify the tuple class and relation class in instances of which results of the SQL query will be stored and the global dictionary for storing these results as persistent objects.

– dbDo: $< SQLcommandString >$

Executes the SQL-commandString, but the result is pointed by a cursor and can be examined row-by-row using dbGetRow.

– dbGetRow ... Gets the next row from the query result pointed by the cursor.

– dbGetRowWithTupleName: $< aTupleName > inDictionary :< dict >$

Gets the next row from the query result, stores its value in an object of type aTupleName and stores this object into the given dictionary.

Example:

```
new dbLogin: <sybaseUserName>
        password: <password>
        server: <sybaseServerName>
new dbUse: 'uni'
new dbExecute: 'select * from teachers'
new dbExecute: 'select name from teachers where salary < 3000'
        relationClassName: 'oo-teachers'
        tupleClassName: 'oo-teacher'
new dbDo: 'select name from teachers where salary > 5000'
  resultClass: goodPaidTeachers
new dbGetRow
    .
    .
    .
new dblogout

#
```

References

1. Gemstone Databridge. Servio Logic Corporation, Beaverton, 1992.
2. Kroha,P.: Objects and Databases. McGraw-Hill,1993.

Database Requirements of CIM Applications

G. Kappel[‡], S. Vieweg[†]

[†]Institute of Applied Computer Science and Information Systems. Department of
Information Engineering. University of Vienna, Austria

[‡]Institute of Computer Science. Department of Information Systems.
University of Linz, Austria

Abstract. Changes in market and production profiles require a more flexible
concept of manufacturing. Computer Integrated Manufacturing (CIM) describes
an integrative concept for joining business and manufacturing islands. In this
context database technology is the key technology for implementing the CIM
philosophy. However, CIM applications are more complex and thus, more
demanding than traditional database applications like business and administrative applications. In this chapter we systematically analyze the database requirements for CIM applications including business and manufacturing tasks. Special
emphasis is given on integration requirements due to the distributed, partly isolated nature of CIM applications developed over the years. An illustrative sampling of current efforts in the database community to meet the challenge of non-
standard applications like CIM concludes this chapter.

Categories and Subject Descriptors: H.2.8 [Database Management] Database
Applications; H.2.m [Database Management] Miscellaneous; J.1 [Computer
Applications] Administrative Data Processing - business, manufacturing, mar-
keting; J.6 [Computer Applications] Computer-Aided Engineering - CAD, CAM

General Terms: Database Systems, Computer Integrated Manufacturing

Additional Key Words and Phrases: extended relational database systems,
object-oriented database systems, database requirements of CIM

1 Introduction

Computers have been widely used in business and manufacturing, but their use has
been limited almost exclusively to purely algorithmic solutions within the subtasks
design, engineering, production planning, and manufacturing. The data was organized
in file systems, application specific databases, and local databases. This lead to production, design and business islands. The same data was stored in different representations
and locations. The consequences of this approach are obvious. Database changes have
only local effects and the independent organization of various databases implies a high
and non-controllable degree of replication. Thus, the global consistency of the data can
hardly be managed and the interdependent support of the enterprise tasks is complicated.

Recent efforts are driven to integrate all of the technologies described above to
form an integrated and integrative model of business and manufacturing concepts. The

integration refers to the technical and administrative processes and the required data. This concept is known as Computer Integrated Manufacturing (CIM) [Harrington73]. The major components of a CIM system span from sales and marketing over production to storage and distribution. Computerization in sales, marketing, accounting, stores, and distribution mainly focuses on administrative tasks. This includes the analysis and presentation of sales data and the administration of customers and marketing data. In our analysis we will not cover these traditional issues explicitly. However, they are of great concern when discussing integration aspects (see Section '3.3 Integration Requirements'). Our investigation mainly focuses on engineering design (computer-aided design), manufacturing engineering, production planning and control, manufacturing, and quality assurance.

The realization of CIM requires very complex algorithms, suitable organizational structures, and integrated, distributed database systems. The decision-making processes cover the spectrum from high-level decisions at the strategic planning level down to the specific technical decisions at the machine scheduling level. Consequently, the database system's services and performance must be very compound in order to satisfy the information requirements of the several tasks.

Database systems were introduced in the sixties and seventies to support the concurrent and reliable use of shared data, mainly in the area of business applications. With the advance of so-called non-standard applications, like office information systems, multi-media systems, and computer integrated manufacturing, requirements for various database features have changed; for example, from simple-structured data to complexly structured data, from short single-user transactions to long multi-user transactions, and from stable long-lived data descriptions to evolving schema management, to mention just a few. This shift in requirements has also been reported in literature [Ahmed92, Encarnaçao90, Kemper93]. The various articles focus on engineering applications and their database demands. A comprehensive discussion of database requirements, mainly integration requirements, of all aspects of a CIM system, however, is still missing. The objective of this chapter is to fill this gap.

Based on a brief introduction to database systems we will discuss CIM's database requirements. Thereby, we distinguish between requirements concerning data modeling issues, querying and manipulation issues, and integration issues. Based on the discussed requirements we focus on the state-of-the-art of database technology and review to which extent CIM's database requirements may be already fulfilled nowadays and in the near future, respectively. The chapter concludes with a prospective outlook on database technology.

The work reported in this chapter is part of the ESPRIT project KBL (ESPRIT No. 5161), whose goal is the design and development of a Knowledge-Based Leitstand[1]. The authors are responsible for incorporating object-oriented database technology as an underlying information store, and as an integrative component between the Leitstand and various other CIM components.

---

1   The support by FFF (Austrian Foundation for Research Applied to Industry) under grant No. 2/279 is gratefully acknowledged.

2 Basic Features of Database Systems

In this section we will give a brief overview of the basic functionality of database systems. We will start with the discussion of the functions of a database system, give a definition of database systems, and review some basic data models. Due to the distributed nature of manufacturing we will also address issues of distributed data management.

A database management system (DBMS) is a collection of mechanisms and tools for the definition, manipulation, and control of databases (DB). A database system (DBS) consists of a DBMS and one or more databases. In the following, we will use the terms DBMS, DBS, and DB interchangeably if the meaning is unambiguous from the context. In general, a DBMS has to provide the following functional capabilities [Cattell91, Elmasri89, Ullman88]:

- management of persistent data (persistence)
- ability to manipulate the database, either with an interactive query language or from applications (querying and manipulation)
- independence between application programs and databases (data independence)
- consistent and concurrent access of multiple users to the database (transaction management)
- recovery from system failures and restoring of the database to a consistent state (recovery management)
- authorization of user groups and their view of the database, archiving and tuning of the database (database administration)

In order to support an independent view of the services provided by a DBMS, a three level architecture (ANSI/SPARC architecture [Tsichritzis78]) was introduced and is widely accepted. It consists of the *physical (internal) database level,* the *conceptual database level*, and the *view (external database) level.* The physical database (level) consists of a collection of files and access structures (the so-called internal schema). The conceptual database (level) consists of the conceptual schema of the universe of discourse, i.e., of the problem domain which has to be stored and manipulated in the database. The concepts and mechanisms used to specify the conceptual schema are part of the data model which comprises the logical foundation of each database system (see also below). The data definition language (DDL) is used to describe the conceptual schema. The view level (= external database) consists of subsets of the conceptual database. As the users' work with the database is mostly restricted to predefined functions and responsibilities, they do not need to know all the information stored at the conceptual level of the database. Views provide a mechanism to supply each user or each user group with his or her own database. The operations performed within the database are expressed by means of a data manipulation language (DML). The DML allows to query and manipulate the data and the data definition stored in the database.

DBMSs may be distinguished in terms of the data model they support. The first DBMSs followed the *hierarchical* and *network* model [Elmasri89]. Out of the deficiencies of those early data models which were mainly due to the interdependence between the logical data structure and the physical record and file structure the *rela-*

tional model [Codd70, Ullman88] emerged. The relational data model is based on the mathematical notion of a relation. The querying and manipulation of the data is set-oriented and follows the relational algebra. The de facto standard data definition and manipulation language for relational databases is SQL [Melton93]. SQL is based on the relational algebra. It fully supports the three level database architecture.

Relational databases have gained wide acceptance especially in the area of business and administration applications since they fulfil the demand of managing large amounts of simply-structured and equally-structured data and providing some intuitive querying and manipulation language. However, it was soon be realized that the modeling power of the relational model is strong and weak at the same time. It is strong in the sense that most problems may be expressed by a set of relations and by applying the relational operators. However, it is weak in so far as it lacks expressive power and extensibility to attribute domains. To cope with these insufficiencies various *extended relational data models* and systems have been proposed such as Postgres [Stonebraker91] and Starburst [Lohman91] (see also Section '4 Extended Relational and Object-Oriented Database Systems').

In spite of the higher expressiveness of extended relational data models some fundamental problems in working with (extended) relational database systems remain which are known as *impedance mismatch* [Cattell91, Shasha92]. The term stems from electrical engineering. When two attached wires have different impedances a signal traveling from one to the other will be reflected at the interface, which may lead to problems. This metaphor describes the problems caused by two different paradigms, the programming language paradigm and the database paradigm. Whereas traditional programming languages may be characterized as being procedural, supporting record-at-a-time processing, and complexly structured data types, traditional (relational) database systems can be characterized as being declarative, supporting set-at-a-time processing, and simple data types. To overcome the disadvantages of working with two different worlds object-oriented database systems were introduced.

An object-oriented database system follows an *object-oriented data model* [Dittrich90]. At present, there exist several different object-oriented data models and thus, different DDLs and DMLs. They are either based on existing object-oriented languages, like C++ and Smalltalk or they have been newly developed. Although there exists no single object-oriented data model there is consensus that a data model to be called object-oriented has to exhibit the following core features [Atkinson89]: complex object modeling, object identity, encapsulation, types, inheritance, overriding with late binding, extensible types, and computational completeness. *Complex object modeling* deals with the specification of objects out of other objects which may also be complex by nature. *Object identity* provides a system-defined unique key and thus, allows the sharing of objects. The principle of *encapsulation* defines data (= structural information) and accessing operations (= behavioral information) as one unit called *object type* (or object class). It allows to distinguish the interface of the object type, the set of operation signatures, from its implementation, the implementation of the operations and the underlying data structure. An operation signature consists of the name of the operation, the types and names of its input parameters, and the type of its return value. An object type defines structural and behavioral information of a set of objects called instances of the object type. *Inheritance* supports incremental type specification by specializing one or more existing object types to define a new object type. *Overriding* means the redefinition of the implementation of inherited operations, and *late binding* allows the binding of operation names to implementations at run time. The *extensible type requirement* permits new object types consisting of structural and behavioral information to be defined by the user and, furthermore to be treated like any other predefined type in the system. Finally, *computational completeness* requires that any computable algorithm can be expressed in the DDL and/or DML of the database system.

So far, we have not mentioned the architecture of a DBS. A database system architecture reflects the functionality described above. The core of a database system consists of the data itself (physical database). The physical database is managed by the file manager which handles requests from the database manager. The database manager operates at the conceptual level and coordinates concurrent access of multiple users to the database and it also manages authorization. The access to the database from the user's point of view is managed by the query processor. The query processor handles access to the database, either via application programs or interactively stated user queries. The resources involved (data and control) may be organized centrally or distributed. This leads us to both an important requirement of CIM (see Section '3.3 Integration Requirements') and a vital area of research in database systems, namely distribution.

Distributed database systems (DDBS) extend the notion of distribution of resources to the concept of *transparency* [Traiger82]. In a networking system the user has to explicitly access different nodes (data or services) in a distributed environment. In contrast to that, in a distributed database system the access control to subunits located at different nodes is kept completely invisible to the users of such a system. The user is not aware of the distribution of the resources such as data, machines, and services *(location transparency)*. The same is true in case of replicated data and services *(replication transparency)*. Database access and query submission to a distributed database can be performed from any node in the network *(performance transparency)*. Last but not least, data and schema changes should not affect distribution *(copy and schema change transparency)*.

Distributed database systems have recently been extended to so-called multidatabase systems (= MDBS = federated database systems) which incorporate issues of autonomy of local systems, their distribution, and their heterogeneity [Öszu91, Sheth90]. The *autonomy* of the database systems refers to the distribution of control over the manipulated data. This corresponds to independently operating database systems which cooperate within a multidatabase system. A *distributed* database system consists of a single distributed DBMS managing multiple databases. Distribution specifies the degree of distribution of data. The data either is physically distributed and occasionally replicated within the network, or it is located at a single site. *Heterogeneity* ranges from differences in the DBMS to differences in the operating system and hardware environment. The heterogeneity in the DBMS is due to distinct data models (structures, constraints, DDL, and DML), distinct management features (concurrency control, commit strategies, recovery) and semantic heterogeneity (= distinct schemas). The latter is probably the most difficult to cope with. Semantic heterogeneity occurs whenever various meanings or interpretations of the data have to be managed. Consider a local production database where the productivity of a worker is stored in terms of processed units per day. The management department may also store the worker's productivity, but computed as some ratio of working hour costs and cost of a bought-in part. Attempts to compare or analyze data from the two databases is misleading because they are semantically heterogeneous. To sum up, a multidatabase system supports access to multiple DBS, each with its own DBMS. Questions of implementing distributed database systems and multidatabase systems in terms of communication networks are beyond the scope of this chapter. The interested reader is referred to [Tanenbaum89].

3 Database Requirements of CIM

We will cluster the main database requirements for CIM into three areas which are *data modeling requirements, querying and manipulation* requirements, and *integration requirements,* respectively. There are also other requirements, mainly database administration requirements like archiving the data, tuning the database for better transaction throughput, and insuring data security. However, these requirements are neither restricted to CIM applications nor are there any special features CIM applications might need compared to traditional applications. These requirements are not further treated in this chapter. The interested reader is referred to the literature [Gray93, Shasha92].

We will now turn to a deeper analysis of the three requirement clusters mentioned above.

3.1 Data Modeling Requirements

In this section we will analyze data modeling requirements for CIM applications and introduce appropriate data modeling features. We recall that a data model comprises the logical foundation of a database system in terms of the concepts and mechanisms applicable to specify the conceptual schema and to query and manipulate the stored data. A feasible approach to select the "best" data model for an application is to analyze the data that has to be stored in the database. A data model is appropriate for modeling a given task if the information of the application environment can be easily mapped to the data model. Thus, we will first analyze the involved data and then, we will derive the requirements for a data model appropriate for CIM applications.

The application environments in CIM vary from business applications to machine control. The data encountered can be classified into the following kinds [Bedworth91]:

- Product data
- Production data
- Operational data

- Resource data
- Financial data
- Marketing data

Product data contain the description of the products involved in the manufacturing process. The data may consist of graphic, textual, and numeric information. The *production data* describe how the individual parts are to be manufactured. An example of production data are process plans with detailed instructions of how to make a part based on its design. This includes operations, machines, tools, feeds, stock removal, and inspection procedures. *Operational data* consist of all information about the production process itself. This includes production schedules, production planning data, and machine/part related data. *Resource data* describe the resources involved in production; the kinds and number of machines, the tools, and personnel. Beside these production related data *financial* and *marketing data* have to be managed as well.

Requirements for CIM applications differ a lot from those in pure business applications. Table 1 gives an overview of the differences between business and CIM databases with respect to data modeling issues. Data types describe the conceptual entities stored in a database. Business applications usually operate on basic data types such as integer, floating point numbers, and character strings. CIM applications require more complex, predefined as well as user-defined data types because of the complexity of the objects that have to be designed and manufactured. These objects are far more interrelated than data in conventional applications. The relationships between objects of business applications are simple compared to those in CIM applications where they are manifold and may imply dependencies between objects. There are lots of objects involved in business applications as well as in CIM applications; most probably there are more objects involved in the latter.

	Business Databases	CIM Databases
Data Types	few, simple	many, complexly structured
Relationships	simple	complex, highly dependent
Amount of data	high	high

Table 1. Data modeling issues in Business Databases and CIM Databases

To accomplish a unified view of data in a heterogeneous environment such as CIM a data model has to deal with a variety of aspects: data must be kept in different representations, data domains are heterogeneous, relationships between data are very complex, data changes may require multiple changes to other data, and consistency of data must be kept in the whole environment. In data modeling these requirements refer to the following features of data models [Encarnaçao90, Gupta91]:

- Extended Attribute Domains
- Complex Object Support
- Relationships and Dependencies
- Active Consistency Checking and Knowledge-Base Support

In the following, we will discuss these features in turn.

Extended Attribute Domains

Extended attribute domains address the data types of the basic units that can be stored in the database. In a data model the attributes mirror the properties of an object in the real world. The attribute domains reflect the potential set of states an object can be in. In business and administration applications the *attribute domains* are mostly restricted to numerical and character data types. In CIM applications the set of attribute domains must be *extended*. Extensions include application-specific data types such as dimensions, part descriptions, and various measures which are all built out of basic data types such as numerical data and character strings. Additionally, there may be geometric data types which define the structure and the dimensions of products and assemblies. Matrices and lists will be used for production specific information. Unstructured data such as long character string attributes may serve as references to user manuals and maintenance manuals. In addition to these extended attribute domains user-defined attribute domains must also be supported. Out of existing attribute domains and type constructors, such as listOf, arrayOf, and recordOf, new attribute domains may be constructed and used like predefined ones further on. Multi-media data types like voice, image, and video, complement the demands on extended data types. The use of extended data types is not only restricted to the description of product related information. Office information systems and workflow models also require extended attribute domains.

Complex Object Support

Complex object support allows the composition of complexly structured objects by using existing types and thus, the creation of user-defined object types. An object type extends a data type in so far as it has an implicitly defined attribute modeling the object identifier of the instances of the object type. We use both terms interchangeably if the meaning is clear from the context. In a manufacturing enterprise the relevant objects are very complex and interdependent. Consider for example the product related data. Parts may consist of subparts; the subparts might themselves consist of subparts and so on. Complex object support is necessary to express the relationship between dependent data. Flat data models such as the relational model do not satisfy these requirements.

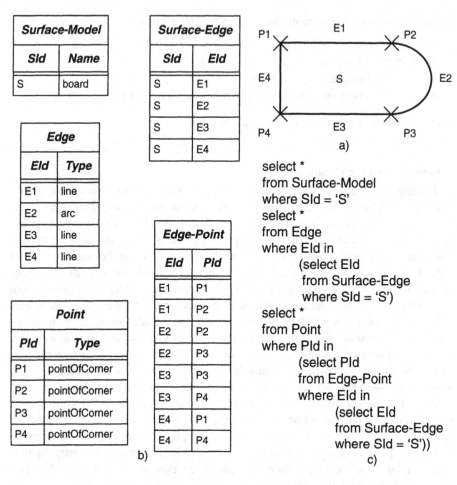

Surface-Model	
Sid	**Name**
S	board

Edge	
Eid	**Type**
E1	line
E2	arc
E3	line
E4	line

Point	
Pid	**Type**
P1	pointOfCorner
P2	pointOfCorner
P3	pointOfCorner
P4	pointOfCorner

Surface-Edge	
Sid	**Eid**
S	E1
S	E2
S	E3
S	E4

Edge-Point	
Eid	**Pid**
E1	P1
E1	P2
E2	P2
E2	P3
E3	P3
E3	P4
E4	P1
E4	P4

a)

b)

```
select *
from Surface-Model
where Sid = 'S'
select *
from Edge
where Eid in
        (select Eid
         from Surface-Edge
         where Sid = 'S')
select *
from Point
where Pid in
        (select Pid
         from Edge-Point
         where Eid in
               (select Eid
                from Surface-Edge
                where Sid = 'S'))
```

c)

Figure 1. 2D model a) with data definition based on the relational model b) and retrieve statement in SQL c), a) and b) are taken from [Scheer90, p. 269])

Example: Consider a 2D model as it is shown in Figure 1.a) and its specification using the relational data model shown in Figure 1.b). The information concerning just one graphical object is scattered over five relations. The situation gets worse if one wants to retrieve the complex object, i.e., the surface object with all its direct and indirect component objects. The appropriate SQL statements are given in Figure 1.c). We will not go into details of SQL. However, the example already demonstrates the user's responsibility for consistent querying and manipulation. In our example, the user has to write three select-from-where statements to retrieve all information concerning just one complex object.

Object-oriented models capture these requirements and allow the modeling and processing of *complexly structured design objects* as logical units [Kim89]. This issue is also related to the extension of attribute domains since the definition of complexly structured objects equals the introduction of new object types.

Relationships and Dependencies

As these complex objects are highly interrelated to other objects, relationships and dependencies between these objects must be considered. There are two main kinds of relationships distinguished, generalization and aggregation. Generalization models the *isA*-relationship between object types. An object type *is a* subtype of another object type called supertype if it exhibits at least the same properties (= attributes and operations) as its supertype but possibly even more. *Generalization* may be used to classify a set of objects according to some classification attribute. The objects in the subset (= subtype) inherit all the properties from the superset (= supertype), thus, common properties are defined at a single place. Object types connected via an isA-relationship build up a generalization hierarchy or a generalization graph; the latter occurs in case of inheritance from more than one supertype.

Example: Parts may be produced inhouse or are delivered by external suppliers. Thus, the generalization hierarchy consists of the object types PART, INHOUSE-PART, and BOUGHT-IN-PART (see Figure 2). INHOUSE-PART and BOUGHT-IN-PART inherit from PART. (Note: the notation is taken from the object-oriented modeling technique OMT [Rumbaugh91].) In addition, parts may be classified according to their production status as RAW-MATERIALs, INTERMEDIATE-WORKPIECEs, and PRODUCTs. The hollow triangle represents the isA-relationship. As already noticed in the example CIM applications comprise various generalization hierarchies which should be expressible in the data model.

Aggregation models the *partOf*-relationship between objects. An object called component object is a *part of* another object called composite (= complex) object if the latter is made out of the former. Component objects may again consist of (other) component objects. The resulting hierarchy is called aggregation hierarchy.

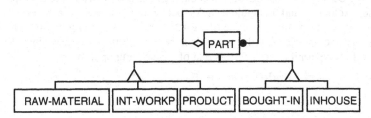

Figure 2. isA- and partOf-hierarchies involving part objects

Example: Figure 2 models a recursive partOf-relationship between part objects. It is represented by an edge outgoing from and ingoing into the object type PART with a hollow rhomboid at the complex object side and a circle with a black fill at the component object side. The circle models the fact that a part may consist of zero or more component parts.

The concepts of generalization and aggregation have to be extended by cardinality constraints, i.e., the amount of relationships a single object may participate in, and by constraints to the object's existence. Objects may be shared by several objects or they may belong exclusively to one object. They may be independent from or dependent on the existence of other objects. To support the consistent querying and manipulation of complex objects and their relationships cascading operations are necessary. Cascading operations insure, for example, the retrieval of dependent component objects if the corresponding complex object is accessed.

Example: Consider the 2D model of Figure 1. Its specification using the ORION object-oriented data model is given in the following. (ORION is an object-oriented database system prototype [Kim89, Kim90]. It was developed at MCC in the late eighties and captures the main advanced features of object-oriented database technology. If not stated otherwise ORION is used as an example system throughout the chapter.) The object class SURFACE-MODEL comprises the attributes SId, Name, and Edges. The attribute Edges models a set of component objects of class EDGE (:composite true) where each edge belongs exclusively to one surface-model (:exclusive true) on which it is also dependent (:dependent true). The object classes EDGE and POINT are specified accordingly.

```
make-class SURFACE-MODEL
  :superclasses GEOMETRIC-MODEL
  :attributes (SId :domain String)
            (Name :domain String)
            (Edges :domain (set-of EDGE)
                   :composite true
                   :exclusive true
                   :dependent true)

make-class EDGE ...

make-class POINT ...
```

The specification of surface models consists of three object classes. The attributes Edges and Points (= attribute of object class EDGE not shown in the example) model the exclusive and dependent links between these object classes. For retrieving one complex object of type SURFACE-MODEL including its direct and indirect component objects only one select statement is necessary since the semantics of the operation select is that of a cascading operation.

```
select Surface-Model where SId = "S"
```

Active Consistency Checking and Knowledge-Base Support

Due to the constraint-based behavior of a manufacturing enterprise and its complex decision processes active consistency checking mechanisms and knowledge-base support should be provided by the data model. In a manufacturing as well as a business application the need for consistent data is obvious. *Consistency* may refer to data types, valid ranges and access conditions. In real-time systems such as manufacturing control consistency mainly refers to timing constraints, and triggering and automatic invocation of actions under specific conditions. Manufacturing plants need timely supplies of raw material and assemblies to keep the manufacturing process going. Some manufacturing process inherently relies on breaks between subprocesses, for example for cooling the material. Current approaches which transfer the problem to the application programs distribute consistency checking and constraint maintenance over several places in a complex application and thus are error-prone and not efficient. The definition of consistency constraints and active monitoring of these constraints must be part of the data model and of the DBS considered in a CIM environment. Active database management systems (ADBMS) are designed to support the definition of consistency constraints in terms of production rules tightly coupled to the entities in the database. A production rule may be specified as event/condition/action rule [Dayal88, Kappel93] where the event specifies an operation performed within the database, the condition part describes some additional constraint, and the action part specifies operations on the database. ADBMSs also provide active monitoring of the production rules. If an event takes place it triggers the evaluation of the corresponding condition and if the condition is satisfied the corresponding action is executed.

Example: In an automated factory there may occur the task of monitoring the quantity on hand for each item in the production cycle. If the quantity on hand is below a certain threshold then a reorder procedure has to be initiated automatically. A production rule for this problem may be stated as follows [Dayal88]:

```
Event:      update quantity_on_hand(item)
Condition:  quantity_on_hand(item) +
            quantity_on_order(item) <
            threshold(item)
Action:     reorder(item)
```

For an in-depth discussion of active database management system features in the realm of CIM applications the interested reader is referred to chapter „Active Object-Oriented Database Systems For CIM Applications" in this book. Production rules provide declarative mechanisms for specifying event-driven behavior. In addition, mechanisms for specifying incomplete information, heuristic knowledge and derived knowledge are necessary. Many tasks in a CIM application cannot be described in a deterministic way or their exact solution is not achieved with respect to time and computing resources. Consider the assignment of work orders to machines or manufacturing cells. Varying arrival rates of work orders, missing materials, and machine breakdowns complicate this task. Traditional optimization methods are mostly inapplicable. The automation of such tasks requires a method to handle uncertain data and some inference mechanism which still works under uncertain situations. Expert system and rule management system techniques are vital to accomplish this task.

3.2 Querying and Manipulation Requirements

In this section we will address the requirements for querying and manipulation of data. We will start with an overview of the tasks that have to be performed in the CIM environment. The tasks are analyzed according to their need for database services in the course of the life-cycle of a product. In the second part we will investigate the functional units of database systems which should handle these requirements.

Task Level	Data	Response Time
Engineering and Planning	Mbytes	minutes
Manufacturing Control	Kbytes	seconds
Cell Management	bytes	0.1 second
Machine Control	bits	ms

Table 2. Size and time requirements of control tasks in CIM databases [Hammer91]

The management and control tasks in a manufacturing enterprise are built up hierarchically. The requirements with respect to the amount of data, and response time differ from level to level. Table 2 gives an overview of this control hierarchy including a sketch on the amount of data involved and the required response time.

At the engineering and planning level the acceptable response time is within the range of minutes and the amount of processed data is relatively high. At the machine control level the response time has to be very short; small amounts of data gathered with sensors have to be transformed immediately. In our analysis we focus on the engineering and manufacturing control levels.

	Business Databases	CIM Databases
Queries	associative	associative and navigational
Updates	short	long, highly interactive

Table 3. Querying and manipulation issues in business databases and CIM databases

In addition to size and time requirements, accessing a CIM database differs a lot from accessing a business database. Table 3 provides an overview. Access to the highly interrelated objects in the database is more complex. The underlying query language must support complex querying of data both in terms of query predicates supporting associative access and in terms of path expressions supporting navigational access. Usually the modification of data takes longer than in traditional transaction processing applications. Consider the process of designing a new product as opposed to the transaction processing system of an automatic teller machine. The design of the geometric object is a long and highly interactive process whereas bank transactions are mostly predetermined in the amount and the way they access the database.

The processing of engineering problems covers qualitative and quantitative analysis of design data. If applied separately none of the techniques can address all problems in manufacturing successfully, thus, both qualitative and quantitative analysis must be applied [Rao91]. Qualitative analysis is based on symbolic and graphical information while quantitative analysis mostly relies on the processing of numerical data. Database systems in the CIM environment must be capable of supporting the processing and retrieval of both types of data, numerical and symbolic.

The functions of a manufacturing enterprise that have to be supported during the life cycle of a product are manifold. The assemblies have to be designed and a tolerance analysis has to be performed for those assemblies. The engineering drawings of assemblies, individual parts, fixtures, and manufacturing facilities have to be prepared. Analytical models of parts have to be carried out with respect to structural and thermal analysis. The weights, volumes, and costs of manufacturing have to be calculated. The parts have to be classified in order to be reused or to find similar parts already existing in stock. Before the manufacturing may start the bill of materials has to be computed, usually a very resource intensive task. Production plans for the parts and assemblies have to be designed and NC programs must be implemented and assigned to the individual production of the parts. In case of robot use in work cells the movements and tools must be programmed. Furthermore, changes in the design of assemblies and their effects on manufacturing have to be controlled.

All phases in manufacturing require administrative support. Thus, business computing is not only important per se but also as a supporting task of the manufacturing process. As an example consider cost accounting. Product costs have to be computed in any step of the production process. Cost planning, resource management, and marketing strategies heavily rely on this data. Although the requirements for these applications are covered by traditional transaction processing they are noteworthy to mention just to stress the integrative character of CIM applications once more. In our requirements analysis we put emphasis on advanced transaction processing issues.

CIM requires all the tasks during the product life-cycle to be supported by database services. The major features for data querying and manipulation by multiple users are presented in the following. These include:

- Advanced Transaction Management
- Flexible Database Access
- Change Management
- Interfaces

In the following we will discuss these features in turn.

Advanced Transaction Management

Advanced transaction management is concerned with database support for applications that demand consistent database access beyond the traditional scope. Traditional data processing in business applications is known as on-line transaction processing (OLTP). Relatively simple operations such as querying and updating of huge amounts of small records are performed. CIM applications such as engineering design and production control have different access patterns and thus need advanced transaction management concepts, such as *semantics based concurrency control, cooperative and long transactions, real-time transactions,* and *hypothetical transactions.*

In an engineering environment the work profiles differ a lot. Firstly, the types of data are different, as pointed out in the previous section. Furthermore, the processing of designs or manufacturing schedules in production planning is a highly interactive task. The processing times are much longer than in traditional data processing. A single engineer may check out a design object from a public database into his private database and may work on it for days without any requests for concurrent updates from other engineers. Thus, the level of concurrency may be different. Additionally, collaborative engineering requires the same data being available for different users interacting during the design phase. This implies that the results within a transaction have to be promoted to the database before the end of the transaction. This may violate traditional transaction properties - atomicity, consistency, isolation, and durability, known as ACID properties [Härder83]. *Atomicity* requires that a transaction either commits entirely or not at all. *Consistency* requires that a database be transformed from one consistent state into another. The *isolation* property forces the operations of concurrent transactions to yield the same results as if they were executed serially in some order. *Durability* guarantees the preserving of the effects of committed transactions after recovery from system failure or loss of memory.

CIM applications demand a redefinition of the ACID properties. Design transactions may last for weeks, thus, a complete rollback in case of memory failure is inadequate. Consistency checking using the no read-write conflict paradigm has to be extended to *semantics based concurrency control* [Garcia83]. The conventional approach to the problem of concurrency control is based on the synchronization of database reads and writes. The concurrent execution of transactions is allowed (consistency and isolation) if they have the same result as if executed serially (serializability). Semantics based concurrency control concepts relax this requirement by using knowledge about the application domain, the application process, and the access patterns of the users that concurrently use the database. Non-serializable transactions thus allow the interleaving of transactions in the course of their execution.

Cooperative and long transactions play an important role in computer-aided engineering. Consider a collaborative design environment where designer groups must coordinate their designs before the actual design decision takes effect within the database and serves as the basis for further decisions. Non-serializable transactions are one way to accomplish this task, versioning is another (see subsection on change management).

Manufacturing control and factory cell management are closely related to the actual manufacturing process, not only logically but also with respect to time. This requires immediate reaction to changes reported from the machine control sensors. Machine breakdowns or problems with the part supply system may require the rescheduling of work orders. The response time requirements in cell management and machine control reflect real-time conditions. *Real-time* databases guarantee that a transaction commits until a certain deadline [Stankovic88]. Hard real-time databases secure this requirement for each transaction, soft real-time databases guarantee it for a definable percentage of the database load. In any of these two cases the scheduling of the transactions has to cope with resource constraints of any kind (availability of material, machines or workers) and an efficient system-wide management of time.

Up to now we have assumed that transactions are designed to change the state of a database, thus, they always take effect in the database. This is not true for *hypothetical transactions* [Garza88]. Hypothetical transactions provide a framework to perform "what-if" analyses. What-if analyses (and thus hypothetical transactions) can be used in any situation when effects on the database state must be checked experimentally without the transaction management overhead of normal transactions. As an example consider a what-if analysis for the scheduling of additional work orders in an interactive production planning and control system.

For further information on advanced transaction management the interested reader is referred to [Barghouti91].

Flexible Database Access

With *flexible database access* we address the need for varying and demanding database manipulation requirements. In a CIM organization the database access strategies differ from task to task. A design engineer may require the parts that are connected to the part currently under manipulation. The clerk in the production planning center may demand the incoming orders of the last few weeks. In the first case the user wants to access structured data through inter-object references. This is also known as *navigational database access*. The latter requires *associative access* to the database, i.e., the selection of those entities in the database whose field values satisfy a certain condition such as looking up of order objects by their arrival date. A further issue in query languages is *recursive query* processing. A typical application of recursive query processing is the computation of bills of materials.

Example: Consider the part hierarchy modeled in Figure 2. We would like to retrieve all (direct and indirect) component parts of a specific part named "bicycle". The specification of the part hierarchy as well as the recursive query statement is given in the following.

```
make-class PART
:superclassesNil
:attributes (PartNr :domain Int)
            (PartName :domain String)
            (HasParts :domain (set-of PART)
                :composite true
                :exclusive true
                :dependent Nil)

select PART (recurse HasParts) where PartName =
"bicycle"
```

Note that the semantics of (recurse HasParts) is such that if a part object named "bicycle" is found, all values of the attribute HasParts are retrieved recursively.

Standardized high level query languages like SQL are set-oriented and only support an associative access. The navigational part of the database access must be performed within some (host) programming language. In order to allow flexible access to the database, however, both access strategies must be supported within the same language. [Bancilhon89] discusses several requirements for an advanced (object-oriented) query language. These include ad-hoc facilities to state queries and updates in a high level language without using a programming language. Thus, the query language must have equal expressive power than conventional programming languages. Recent efforts in this direction either address the extension of high level query languages (various dialects of ObjectSQL are under development; see also example in ORION above) or the extension of programming languages with database functionality.

Change Management

Let us further elaborate on the example of the production plan. It is not only desirable to compute the production plan but also to handle several versions for comparison and optimization. For example, the production planning and control system optimizes the timely production of products. The optimization process heavily relies on constraints based on the availability of resources and based on administrative and marketing needs. Different versions of the production plan may exist due to partly conflicting goals which have to be satisfied. In addition to version management evolving applications such as CIM applications must cope with changing requirements which imply changes in the description of the data in the database, known as schema evolution and schema versioning. Both version management and schema evolution/versioning have to cope with changes either of data values or of data descriptions. These tasks are commonly referred to as change management [Joseph91]. In the following, the requirements of both aspects of change management will be discussed in turn.

Version management allows to keep track of evolving data values, i.e., the values of the attributes, in a controlled way. *Versioning* of data is very important in the course of the design process of product configurations. The design process is highly interactive and consists of several abstraction levels. Each level introduces new constraints to be satisfied, thus causing different versions of the design object. A version represents a meaningful snapshot of a design object at a point in time [Katz90]. Versions of the same design object either may exist in sequence and in revisions, or in alternatives. These are also referred to as linear and branching versions. Revisions represent improvements of a specified version. Alternatives describe different ways of manipulating objects effecting the same result. Revisions and alternatives should be combined in any order. Together the representations form a *version tree* where a node is linked to its parent node via a *derivedFrom*-relationship (= *versionOf*-relationship). The manipulation operations of the version tree include the merge of branching versions and the way the version tree is referenced by other objects. Version merging is imperative during the design phase when the versions of subunits have reached a stable state and have to be put together to form a consistent design object. Thus, a version tree becomes a directed acyclic version graph. Referencing versioned objects addresses the problem of version states. Let us again consider the design process. It can be divided into several phases. During the initial phase the designers tend to experiment with design options; the versions are in a *transient* state where they may be updated and deleted. Later, the versions of some design object may turn into a more stable or *working* state where they can only be deleted but not updated any longer. Finally, when the design is finished, a stable or *released* version of the design object which can neither be deleted nor updated is available for further processing.

Consider the development of an engine. Different designers may develop distinct approaches for the transmission of the engine. Several designer groups develop their own solutions to that problem. Each of the designers start off with a requirement specification of the transmission element, and an initial design is *checked out* from the group database. During the design process the versions evolve from a transient state to a more stable state. Versions reaching a state that could be advantageous for other design groups are promoted to working versions and are added to the version graph as alternative versions. This is performed with a *check-in* of the design object from the local database to the group database. Finally, a design alternative from all alternative versions is selected as released version. The need for version management is not only limited to engineering design tasks. Manufacturing tasks and group work in office information systems may also require version management. An example of the former application are versions of production plans, an example of the latter is cooperative editing of documents.

Let us now turn to changes in the description of the data in the database, to *schema evolution* and *schema versioning*. The database schema describes the structure and behavior of the entities in a database, also referred to as the description of a set of object types. Changes to the schema description may occur at any time. Consider an engineering environment; usually, the schema descriptions are likely to be modified as soon as designers arrive at a better understanding of their problem. Attribute names and attribute domains may change, the structure of composite objects may change, new attributes may be added, and existing attributes may become obsolete. An important issue is that the DBS must be able to check and resolve inconsistencies due to schema changes dynamically, i. e., without database shutdown. The database services must be available during the reorganization of the database [Banerjee87, Nguyen89, Zicari91]. The reorganization of the database, i.e., of the stored data, to conform to the new schema can be achieved in three ways: (a) conversion, i.e., the objects are immediately changed to follow the new schema, (b) screening, i.e., the changes are deferred to the time when the data is accessed, and (c) schema versioning. In the first approach the contents of the database is immediately converted after changes of the schema have occurred. This implies that whenever the conversion takes place the database is not available. This might not be desirable. The second approach starts the conversion only when the data is accessed and thus, does not require a database shutdown. Schema versioning [Skarra90] is the most flexible way to cope with schema changes. In addition to the changed schema the old schema is available. Existing entities which were created within the old schema need not be converted but may be processed as usual. Schema versioning might also build up a version tree or a version graph.

Dynamic schema evolution and schema versioning must be supported in CIM applications. Consider the example of the Leitstand toolkit with an underlying database system as it is developed within KBL [Adelsberger92]. The Leitstand toolkit consists of a set of generic classes implementing an electronic planning board. It is operational but at the same time not customized to specific user needs. When working with the Leitstand toolkit at the customer's site it might become necessary to change and extend the existing schema to make it more suitable. It should be possible to carry out changes not only at the client's site but even during normal work. Alternative approaches are too time-consuming and might be unacceptable for the production process.

Interfaces

So far, we have discussed querying and manipulation of data which has been defined using the DDL of the underlying database system. However, it is of equal importance for CIM applications that information be automatically transferred to/from the particular format of the underlying database system from/to other data formats. On the one hand, this requirement is due to various software products used within the very same manufacturing enterprise with syntactically and semantically different data formats for the same information. On the other hand, this is due to information coming from outside the system, for example from suppliers of raw material and intermediate workpieces. Up to now, the information has to be processed manually and stored in the database system.

Considering the above mentioned problems of translation from and to various data formats it would be highly desirable to work with one standard neutral format used by every software to communicate its data. One of the mainstream ISO standardization efforts in this area is STEP, STandard for the Exchange of Product model data [Trapp93]. Within STEP the specification language Express was proposed for formal specification of product data [Schenk90]. Express already provides features of an object-oriented specification language in terms of user-defined types, single and multiple inheritance, and local and global rules. Local rules are defined for the objects of a single object type (called entity type in Express), whereas global rules may incorporate objects of several types. Extensions to Express, for example to even incorporate operations as part of the object type definition, are currently under investigation and have also been proposed by the recently finished ESPRIT project IMPPACT (Integrated Modeling of Products and Processes using Advanced Computer Technology [Bjørke92]).

Database systems for CIM applications should support at least the import and export of STEP/Express data descriptions to justify their role as common and integrating information store. The question of interfaces is highly related to that of integration which will be discussed in the next section.

3.3 Integration Requirements

The "I" of the acronym CIM already points to the importance of integration in CIM applications. Database systems seem to be the natural solution for providing an "integrated" data storage and manipulation facility since they support a concurrent and consistent use of the data. However, there are specific requirements concerning integration which stem from the distributed, partly isolated nature of CIM applications developed over the years. Thus, database solutions have to cope with two facts. Firstly, there exist several isolated databases, for example for CAD, PPC, CAM, and business tasks, but no integrated one. However, the implementation of CIM concepts requires, in addition to the local databases, a unified consistent view of the data in a manufacturing enterprise. Secondly, due to the geographically distributed nature of various CIM tasks, a minimum requirement for "integrated" solutions is a client/server architecture, which is realized via some communication network.

Based on the current situation in manufacturing enterprises, four main features of database technology may be distinguished to overcome the integration problem:

- Distributed Data Management
- Multi Data Management
- Reverse Engineering of Data
- Integrated Data and Process Modeling

In the following, we will discuss these features in turn.

Distributed Data Management

We already mentioned the need for distributed data processing in the previous sections. The distribution in a manufacturing enterprise results from the fact that in general the tasks in such an enterprise are geographically distributed. Even if we assume that there are no local databases (this assumption will be relaxed when discussing multi data management issues), a central database solution must be based on a client/server architecture, where several client applications are distributed over different nodes of a communication network and access data from a single central database via a database server process. This *m clients / 1 server* architecture bears several disadvantages. Firstly, the server might soon become the performance bottleneck since all clients have to access the data via this process. Secondly, the availability of the database services depends on a single process. Thirdly, the system architecture does not reflect reality in a manufacturing enterprise, where certain information is only used in a single place, i.e., by a single client application, and thus should be stored together with the application to avoid communication overhead. These problems may be overcome by distributed database systems which are based on a *m clients / n servers* architecture. Although there exists a single global database schema and all users (client processes) may query and update the data as if there were one central database the data is physically distributed into several databases over a network. For each database a single server process exists. The different kinds of transparency, mostly data transparency which hides the physical distribution of data, were already discussed in Section '2 Basic Features of Database Systems'. A distributed database management system should provide these kinds of transparency.

Multi Data Management

The assumption that there are no local (isolated) databases is not very realistic in today's manufacturing enterprises. In general, some parts of the enterprise may have been computerized before others, or not at all. Business databases most probably were among the first which to be installed in an enterprise. Special CAD databases have been in use since several years [Ketabchi87]. Consequently, database systems, data models, and database technology may vary a lot within an enterprise. The integration of local and independent databases to factory-wide information systems requires tremendous efforts but is a prerequisite for the success of CIM. Data migration, schema integration, and data model integration may have to be performed. This requires open system architectures and data modeling flexibility. In the following we will discuss these issues in more detail.

The integration of multiple databases into a single enterprise-wide information system, also called multidatabase system, requires multiple approaches. From the user's perspective, a single view of the multidatabase must be provided. The user may access multiple databases without knowing the local structure of these databases. This requires two services of the MDBS, firstly, to provide a unified view of the local databases and secondly, to allow consistent access to these databases. Different schemas originating from local databases (islands of automation) must be integrated. Schema integration is one of the most challenging problems in database theory [Sheth92]. In general, two approaches are suitable: a single global schema or multiple federated schemas.

In the first case all of the local schemas are integrated in order to represent all the data in the distributed environment. The local user's views are then derived from the global schema. Developing a global schema of an enterprise is a difficult, yet commonly used approach in schema integration. With the advance of reference models for enterprise-wide information systems [Hars92, Scheer92] this task might go through certain standardizations and thus, become easier. Such a reference model aims at being generic enough to capture the information of various enterprises at least at some abstract level. The advantages of using a reference model or of starting with a reference model which is customized in further steps are evident. One does not have to start from scratch but may rely on existing expertise. For example, depending on the level of abstractions, sales and marketing information, product and production information, and manufacturing control data may have equal structures in different enterprises. Thus, it is advantageous to reuse existing models (schemas) and customize them for the specific requirements at hand.

In the second approach the local schemas are split into multiple federated schemas, and the user views are derived from these federated schemas [Thomas90].

Things get more complicated when multiple data models are involved. In this case, the DBS must translate between different local data models. This might occur in two ways. One approach is to find a single common data model and to match all local schemas onto that global model. Obviously, this requires a global model being at least as expressive as the locally used models. The second approach is to allow different data models and to provide translation mechanisms for accessing the local data models. It is the database management system's task to invoke these translations when different databases are accessed simultaneously. In any of these cases the access to multiple databases has to be managed in a consistent way. Distributed concurrency control mechanisms and distributed commit protocols take care of this requirement. Distributed concurrency control allows to coordinate the access to multiple distributed databases. Distributed commit protocols provide mechanisms to achieve atomicity in the distributed case of database access. A detailed analysis of concurrency control in distributed databases and multidatabases can be found in [Bernstein87].

Reverse Engineering of Data

Existing data but not existing systems are the driving forces behind integrating local databases. No enterprise can afford to throw away its data and build everything from scratch. Since schema integration is still a very demanding task an alternative solution might be the reverse engineering of existing data. Reverse engineering attempts to extract abstract, semantically meaningful information from the concrete database contents to gain a conceptual understanding of the stored information [Batini92]. The target model of the abstraction process is in general a semantic data model (like the object-oriented data model introduced in Section '2 Basic Features of Database Systems') which has more expressive power than the implementation model on the basis of which the data has been stored.

Reverse engineering of data might quickly become a very important task in CIM applications since integration is a real issue. Consider, for example, early database use in manufacturing enterprises. If it has been decided to change to new database technology without using a multidatabase approach, it will be crucial to automatically incorporate the existing data of the old databases into the new system.

Integrated Data and Process Modeling

So far, we have mainly concerned ourselves with the integration of existing data and systems, leaving aside the potential of integrating passive data information with active process information. The importance of process modeling together with data modeling has been recognized also for CIM applications. Firstly, it is highly advantageous that the whole manufacturing process including the business activities is scheduled and monitored by some integrated database system. If there occur any deviations from the schedule, correcting actions can be taken immediately (see also Section '3.1 Data Modeling Requirements' on active consistency checking and knowledge-base support and chapter „Active Object-Oriented Database Systems For CIM Applications" in this book). Secondly, modeling and storing operations together with the data which they access reduces repeated development of the same operations and supports the integrity-preserving use of the data. For example, several activities during the manufacturing process such as "computing the price" and "shipping" are concerned with the total weight of a product. The total weight has to be computed since the weight of a product is the sum of the weights of its components. If the operation "computeWeight" is stored together with the products in the database, it does not have to be implemented by each activity requesting this information. In addition, since the operation is implemented at a single place and all activities asking for the total weight of a product have to use this operation it is less error-prone.

Storing data and operations together obeys the principle of encapsulation. Allowing access to the data exclusively via its operations obeys the principle of information hiding. Both principles have emerged in the software engineering area to aid in writing more maintainable, more extensible, and less error-prone codes. With the advance of object-oriented modeling and object-oriented databases these principles are also of great concern in the database area. An object-oriented database supports encapsulation and information hiding through the specification of object types integrating data (= structural information) and operations on the data (= behavioral information).

The discussion of integrated data and process modeling issues concludes the analyses of database requirements. The next section will deal with current efforts in the database area to meet the requirements discussed in this section.

4 Extended Relational and Object-Oriented Database Systems Revisited

The previous section included several arguments why database technology is crucial for the success of CIM applications. In addition, we already argued why traditional (mostly relational) database systems fall short of meeting the database requirements of CIM. The goal of this section is to provide an illustrative sampling of current database systems to meet the database challenge of non-standard applications like CIM. We will shortly mention existing prototypes and products in this area but will deliberately omit thorough descriptions thereof. Instead, we will refer to the existing literature where necessary.

The last decade has witnessed several efforts in extending traditional database technology along the lines discussed in the previous section. These efforts may be classified according to the underlying technology into extensions to relational database systems, extensions to object-oriented programming languages, and new developments.

Extensions to relational database systems extend the relational model with object-oriented data modeling features like object identity, user-defined attribute domains, stored procedures, and inheritance. The advantage of this approach is an evolutionary transition from traditional to extended relational database systems including (easier) portability of existing data, non-procedural interfaces like SQL and extensions thereof, and data independence. The last issue is concerned on the one hand with the strict distinction between conceptual data definition and manipulation and physical storage, and on the other hand with the distinction between user views, i.e. derived data, and "real" data stored in the database. The disadvantages of this approach are mainly due to the still existing impedance mismatch between the application programming language and the database definition and manipulation language. Representatives of this approach are the prototypes Postgres [Stonebraker91], and Starburst [Lohman91]. In addition to the features mentioned above both systems provide active database capabilities. Currently, these prototypes fall short of meeting the requirements of advanced transaction management, change management, interfaces to various data formats, and distributed and multi data management.

Extensions to object-oriented programming languages incorporate database functionality like persistence, concurrency control, and recovery into existing programming languages like C++ and Smalltalk. The advantages of this approach are due to an (almost) seamless transition from an existing programming language to one with database functionality. These include the (easier) portability of existing applications. The disadvantage of this approach as may be observed in existing systems is a lack of end-user database functionality like some high-level, non-procedural query and manipulation language (such as SQL). Nevertheless, representatives of this approach are sufficiently known as object-oriented database systems (OODB). They may be distinguished according to the underlying programming language into C++-based systems like ObjectStore [Lamb91], Ontos [Andrews91], Objectivity/DB [Objectivity91], and Versant [Versant92], Smalltalk-based systems like GemStone [Butterworth91], and extensions to LISP like OpenODB (commercial counterpart of the Iris prototype [Fishman89]) [Ahad92], and Itasca (commercial pendant to the ORION prototype [Kim90]) [Itasca92]. These systems, each of them a commercial product, were developed with the requirements of advanced applications like CAD, CIM, and multi-media applications in mind. They are not used primarily by an end-user but by a (mostly interactive) application program. Thus, the stated disadvantage might be less important in this respect. The systems mentioned support (or have announced to support) advanced transaction management, change management, interfaces to and from the Express data format, distributed data management, and integrated data and process modeling. They fall short of active database capabilities not least because of the required changes to the syntax and semantics of the underlying programming language.

New developments also qualify as OODB and are characterized by the development of a new object-oriented data model together with a database management system. We are aware of only one commercial product qualifying as new development, O_2 [Deux91]. In addition to the basic object-oriented features O_2 provides an object-oriented extension to SQL, advanced schema evolution mechanisms, and a graphical interface both for the specification of database schemas and for querying and manipulation.

Overviews of the above mentioned systems may be found in [Ahmed92, Cattell91, Soloviev92]. Extended relational and object-oriented database systems are still in a premature phase compared to traditional (relational) database systems. New releases include major changes compared to their predecessors, and it is not yet predictable which systems will survive on the market. Nevertheless, these systems have gained in significance as underlying object store of advanced applications like CIM. This observation is also confirmed by a market survey of a major European consulting institute [Ovum91].

5 Conclusion

In this chapter we focused on the requirements of a key technology on which CIM systems have to be built to realize their potential, i.e. database technology. We identified three main clusters of database requirements for CIM applications, which comprise data modeling issues, querying and manipulation issues, and integration issues. Within data modeling we discussed extended attribute domains, complex object support, relationships and dependencies between the modeled objects, and active consistency checking and knowledge-base support. Within the realm of querying and manipulation we elaborated an advanced transaction management issues, flexible database access structures, change management issues like versioning and schema evolution, and interfaces to various data formats, mostly STEP/Express. The integration aspect deserved and still deserves special treatment due to the distributed, partly isolated nature of CIM applications developed over the years. We identified three main issues to overcome the seeming principle of isolation, which are distributed data management, multi data management, and reverse engineering of existing data. Last but not least, in addition to integrating existing data, we also emphasized the importance of integrating passive data modeling with active process modeling to reach more maintainable, more extensible, and less error-prone applications. We deliberately did not discuss requirements which are not restricted to CIM applications, like performance requirements. Concerning performance issues the interested reader is referred to [Cattell92, Winslett92] and to chapter „Evaluation of Object-Oriented Database Systems" in this book. We briefly surveyed current database system developments to conclude that most systems available today are still premature but gaining in significance as underlying object stores of CIM applications.

Acknowledgment

The authors are grateful to Michael Schrefl for developing the example depicted in Figure 1.

References

[Adelsberger92]   H. Adelsberger et al.; The Concept of Knowledge-Based Leitstand - Summary of First Results; Proc. of the 8th CIM-Europe Annual Conference; C. O'Brien, P. MacConaill, and W. Van Puymbroeck (eds.); Springer Verlag; Birmingham, UK; 1992

[Ahad92]   R. Ahad, D. Dedo; OpenODB from Hewlett-Packard: a Commercial Object-Oriented Database Management System; Journal of Object-Oriented Programming (JOOP); Feb. 1992

[Ahmed92]   S. Ahmed, A. Wong, D. Sriram, R. Logcher; Object-oriented database management systems for engineering: A Comparison; Journal of Object-Oriented Programming (JOOP); Jun. 1992

[Andrews91]   T. Andrews, C. Harris, K. Sinkel; Ontos: A Persistent Data-base for C++; Object-Oriented Databases with Applications to CASE, Networks, and VLSI CAD; Prentice-Hall; in: [Gupta91]; 1991

[Atkinson89]   M. Atkinson, F. Bancilhon, D. DeWitt, K. Dittrich, D. Maier, S. Zdonik; The Object-Oriented Database System Manifesto; Proc. First Int. Conf. on Deductive and Object-Oriented Data-bases; Dec. 1989

[Bancilhon89]   F. Bancilhon; Query Languages for Object-Oriented Database Systems: Analysis and a Proposal; Proc. of the GI Conf. on Database Systems for Office, Engineering, and Scientific Applications; Springer IFB 204; Mar. 1989

[Banerjee87]   J. Banerjee, W. Kim, H. Kim, H. Korth; Semantics and imple-mentation of schema evolution in object-oriented databases; SIGMOD Record, Vol. 16, No. 3; Dec. 1987

[Barghouti91]   N. Barghouti, G. Kaiser; Concurrency Control in Advanced Database Applications; ACM Computing Surveys, Vol. 23, No. 3; Sept. 1991

[Batini92]   C. Batini, C. Ceri, S. Navathe; Conceptual Database Design: An Entity-Relationship Approach, Benjamin/Cummings Publ.; 1992

[Bedworth91]   D. Bedworth, M. Henderson, P. Wolfe; Computer-Integrated Design and Manufacturing; Mechanical Engineering Series, McGraw-Hill NY; 1991

[Bernstein87]   P. Bernstein, V. Hadzilacos, N. Goodman; Concurrency Con-trol and Recovery in Database Systems; Addison-Wesley, Reading MA; 1987

[Bjørke92]   Ø. Bjørke, O. Myklebust (eds.); IMPPACT - Integrated Mod-elling of Products and Processes using Advanced Computer Technologies; TAPIR Publishers; 1992

[Butterworth91] P. Butterworth, A. Otis, J. Stein; The GemStone Object Data-base Management System; Communications of the ACM; Vol. 34, No. 10; Oct. 1991

[Cattell91]   R. Cattell; Object data management: object-oriented and extended database systems; Addison-Wesley, Reading MA; 1991

[Cattell92]   R. Cattell, J. Skeen; Object Operations Benchmark; ACM Transactions on Database Systems (TODS); Vol. 17, No. 1; 1992

[Codd70]   E. Codd; A relational model for large shared data banks; Com-munications of the ACM; Vol. 16; No. 6; 1970

[Dayal88]   U. Dayal et. al.; The HiPAC Project: Combining Active Data-bases and Timing Constraints; SIGMOD Record Vol. 17, No. 1; 1988

[Deux91]   O. Deux; The O_2 System; Communications of the ACM; Vol. 34, No. 10; Oct. 1991

[Dittrich90]   K. Dittrich; Object-Oriented Database Systems: The Next Miles of the Marathon; Information Systems; Vol. 15, No. 1; 1990

[Elmasri89]   R. Elmasri, S. Navathe; Fundamentals of Database Systems; Benjamin/Cummings; Redwood City CA; 1989

[Encarnaçao90]   J. Encarnaçao, P. Lockemann; Engineering Databases, Connecting Islands of Automation Through Databases; Springer Verlag; 1990

[Fishman89]   D. Fishman et. al; Overview of the Iris DBMS; in: W. Kim, F. Lochovsky (eds.); Object-Oriented Concepts, Databases, and Applications; ACM Press; 1989

[Garcia83]   H. Garcia-Molina; Using Semantic Knowledge for Transaction Processing in a Distributed Database; ACM Transactions on Database Systems (TODS), Vol. 8, No. 2; Jun. 1983

[Garza88]   J. Garza, W. Kim; Transaction Management in an Object-Oriented Database System; Proc. of the ACM SIGMOD Conf.; 1988

[Gupta91]   R. Gupta, E. Horowitz (eds.); Object-Oriented Databases with Applications to CASE, Networks, and VLSI CAD; Prentice-Hall; 1991

[Gray93]   J. Gray, A. Reuter; Transaction Processing: Concepts and Techniques; Morgan Kaufmann; 1993

[Hammer91]   H. Hammer; CAM-Konzepte - am Beispiel flexibler Fertigungssysteme; in: CIM Handbuch; U. Geitner (ed.); Vieweg Verlag; 1991

[Harrington73]   J. Harrington; Computer Integrated Manufacturing; Krieger Publishing, Malabar Fla., 1973

[Hars92]   A. Hars et al.; Reference Models for Data Engineering in CIM; Proc. of the 8th CIM-Europe Annual Conference; C. O'Brien, P. MacConaill, and W. Van Puymbroeck (eds.); Springer Verlag; Birmingham, UK; 1992

[Härder83]   T. Härder, A. Reuter; Principles of Transaction-Oriented Database Recovery; ACM Computing Surveys, Vol. 14, No. 4; 1983

[Itasca92]   ITASCA Systems, Inc.; Technical Summary for Release 2.1; 1992

[Joseph91]   J. Joseph, S. Thatte, C. Thompson, D. Wells; Object-Oriented Databases: Design and Implementation; Proc. of the IEEE, Vol. 79, No. 1; Jan. 1991

[Kappel93]   G. Kappel, S. Rausch-Schott, W. Retschitzegger, S. Vieweg; TriGS - Making a Passive Object-Oriented Database System Active; accepted for publication in: Journal of Object-Oriented Programming (JOOP); 1993

[Katz90]   R. Katz; Toward a Unified Framework for Version Modeling in Engineering Databases; ACM Computing Surveys, Vol. 22, No. 4; Dec. 1990

[Kemper93]   A. Kemper, G. Moerkotte; Object-Oriented Information Management in Engineering Applications; Prentice-Hall; 1993

[Ketabchi87]   M. Ketabchi, V. Berzins; Modeling and Managing CAD Databases; IEEE Computer, Vol. 20, No. 2; Feb. 1987

[Kim89]   W. Kim, E. Bertino, J.F. Garza; Composite Objects Revisited; Proc. of the ACM SIGMOD Conf.; 1989

[Kim90]   W. Kim, N. Ballou, H. Chou, J. Garza, D. Woelk; Architecture of the ORION Next-Generation Database System; IEEE Transactions on Knowledge and Data Engineering; Vol. 2, No. 1; Mar. 1990

[Lamb91]	C. Lamb, G. Landis, J. Orenstein, D. Weinreb; The Object-Store Database System; Communications of the ACM; Vol. 34, No. 1; Oct. 1991
[Lohman91]	G. Lohman et al; Extension to Starburst: Objects, Types, Functions and Rules; Communications of the ACM, Vol. 34, No. 10; Oct. 1991
[Melton93]	J. Melton, A.R. Simon; Understanding the new SQL, Morgan Kaufmann; 1993
[Nguyen89]	G. Nguyen, D. Rieu; Schema evolution in object-oriented database systems; Data & Knowledge Engineering, Vol. 4, No. 1; Jul. 1989
[Objectivity91]	Objectivity, Inc.; Objectivity/DB System Overview; Version 1.2; Jan. 1991
[Ovum91]	J. Jeffcoate, C. Guilfoyle; Databases for Objects: The Market Opportunity; Ovum Ltd.; 1991
[Özsu91]	T. Özsu, P. Valduriez; Principles of Distributed Database Systems; Prentice-Hall International, Inc.; 1991
[Rao91]	M. Rao, J. Luxhoj; Integration framework for intelligent manufacturing processes; Journal of Intelligent Manufacturing; No. 3; 1991
[Rumbaugh91]	J. Rumbaugh, M. Blaha, W. Premerlani, F. Eddy, W. Lorensen; Object-Oriented Modelling and Design; Prentice-Hall; 1991
[Scheer90]	A. Scheer; Wirtschaftsinformatik; Springer Verlag; 1990
[Scheer92]	A. Scheer, A. Hars, Extending Data Modeling to Cover the Whole Enterprise; Communications of the ACM, Vol. 35, No. 9; 1992
[Schenk90]	D. Schenk: EXPRESS Language Reference Manual, ISO TC184/SC4/WG1 N466; 1990
[Shasha92]	D. Shasha; Database Tuning, A principled approach; Prentice-Hall; 1992
[Sheth90]	A. Sheth, J. Larson; Federated Database Systems for Managing Distributed, Heterogeneous, and Autonomous Databases; ACM Computing Surveys; Vol. 22, No. 3; Sept. 1990
[Sheth92]	A. Sheth, H. Marcus; Schema Analysis and Integration: Methodology, Techniques, and Prototype Toolkit; Technical Memorandum TM-STS-019981/1, Bellcore; Mar. 1992
[Skarra90]	A. Skarra, S. Zdonik; Type Evolution in an Object-Oriented Database; in: A. Cardenas, D. McLeod (eds.); Research Foundations in object-oriented and semantic database systems; Data and Knowledge Base Systems, Prentice Hall; 1990
[Soloviev92]	V. Soloview; An Overview of Three Commercial Object-Oriented Database Management Systems: ONTOS, ObjectStore, and O₂ SIGMOD Record, Vol. 21, No. 1; March 1992
[Stankovic88]	J. Stankovic; Real-Time Computing Systems: The Next Generation; in: J. Stankovic, K. Ramamritham (eds.); Hard Real-Time Systems; IEEE Computer Society Press; 1988
[Stonebraker91]	M. Stonebraker, G. Kemnitz; The Postgres Next-Generation Database Management System; Communications of the ACM, Vol. 34, No. 10; Oct. 1991
[Tanenbaum89]	A. Tanenbaum; Computer Networks; 2nd edition; Prentice-Hall; 1989

164

[Thomas90]   G. Thomas, G. Thompson, C. Chung, F. Carter, M. Templeton, S. Fox, B. Hartman; Heterogeneous Distributed Database Systems for Production Use; ACM Computing Surveys, Vol. 22, No. 3; Sept. 1990

[Traiger82]   I. Traiger, J. Gray, C. Galtieri, B. Lindsay; Transactions and Consistency in Distributed Database Systems; ACM Transactions on Database Systems (TODS), Vol. 7, No. 3; Sept. 1982

[Trapp93]   G. Trapp; The emerging STEP standard for production-model data exchange; IEEE Computer; Vol. 26, No. 2; Feb. 1993

[Tsichritzis78]   D. Tsichritzis, A. Klug; The ANSI/X3/SPARC DBMS Framework Report of the Study Group on Database Management Systems; Information Systems, Vol. 1; 1978

[Ullman88]   J. Ullman; Principles of Database and Knowledge-Base Systems; Vol. I; Computer Society Press, Inc.; 1988

[Versant92]   Versant Object Technology; System Reference Manual; Versant Version 1.7.3; Jan. 1992

[Winslett92]   M. Winslett, I. Chu; Database Management Systems for ECAD Applications: Architecture and Performance; NSF Design and Manufacturing Conf., Atlanta; 1992

[Zicari91]   R. Zicari; A Framework for Schema Updates in An Object-Oriented Database System; Proc. of the IEEE Data Engineering Conf.; 1991

Planning and Scheduling

Gerd Finke

University Joseph Fourier
Grenoble, France

Abstract. We present briefly the theoretical framework of planning and scheduling in automated manufacturing systems. Emphasis is on the algorithmic approach to sequencing and scheduling of products through a machining center. Recent developments of the classical theory are described. These include issues of resource constrained scheduling, dealing for instance with tool management aspects. The objective is to provide a "tool-box" of scheduling algorithms and models which would form an important part of a production scheduling module in computer-integrated manufacturing systems.

1 Introduction

A successful implementation of automated production systems requires the solution of many interconnected problems. In general, these problems address issues of organizational, economical and technical nature. In its final and ideal form, as flexible manufacturing systems (FMS), these automated systems offer many benefits: high quality products, manufactured in small and medium size volumes, with quick transition times in order to respond rapidly to customer demands. But these systems are extremely complex (and also expensive), both in hardware and software, and careful management and control strategies greatly influence their performance.

In this section, we are interested in production planning and scheduling. There exists a vast literature on general scheduling theory, developed since the early 1950. Our intention is to outline in a very selective way the basic concepts. Then we want to describe in particular the applicability of this theory to manufacturing systems, point out its limits and indicate some extensions that are necessary so that the theory provides a useful algorithmic base for computer-integrated manufacturing.

2 Production Planning

Although an integrated solution approach to all levels of the production planning is required, one should not attempt to develop a single model to deal with all problems simultaneously. One has to decompose in some way the problem into smaller manageable modules. Usually, some kind of hierarchical framework is used. Decisions at the higher level define the constraints for the next lower level. Following [8, 14,

17], one may distinguish four main levels in such a hierarchy: design or strategic planning, tactical planning, production scheduling, and control and monitoring.

Strategic planning is concerned with long term planning. Management has to select the parts to be produced. One has to decide on the type and number of NC-machines, their layout in the factory, the transportation system. Tools, pallets and fixtures have to be chosen. The system performance is to be analyzed carefully, for instance by simulation and queueing network techniques [13].

At the tactical level, issues of the medium term production planning are considered. Depending on the actual orders, parts are split into batches, depending on due dates and resource availability. In this way, a master production plan is obtained. At this level, one is still dealing with aggregate measures. One selects the part types and their quantities, the production ratio, and allocates the resources for a given production horizon. This is above all a static problem of selection and allocation. Mathematical programming tools are very useful in this phase. Usually, linear programming models are formulated, solved and rounded to integers in order to give an indication of the cumulated time parts and resources are to be allocated to the various machines. With these informations, the short term production may be scheduled. It is precisely at this level where the scheduling theory is applicable. Finally, there remains the control and monitoring of the real-time production process. Here one is faced with the day by day perturbations, tool failure and nonavailability, machine breakdown etc.

3 Scheduling

3.1 Some Concepts of Scheduling Theory

Recent references to the classical scheduling theory are [1, 3, 9]. The problems are described in terms of two basic sets: the set of m machines $M=\{M_1, M_2, \ldots, M_m\}$ and the set of n jobs $J=\{J_1, J_2, \ldots, J_n\}$. Although applicable to many situations (e.g. computer systems), in production scheduling the machines M_i refer in most cases to the NC-machines and jobs J_j are the parts to be produced. A schedule is the specification of the production time intervals on the various machines that are required by the parts. There are two fundamental underlying assumptions: at any instant, a machine cannot process more than one job (part), and at any instant a part cannot be located and processed on more than one machine. Many other technological constraints have to be satisfied for a feasible (realistic) production schedule. The most useful tool to represent the schedules are the Gantt charts (Fig. 1). An incredible number of very different scenarios may occur on the factory floor. Nevertheless, the classical scheduling models catch many useful practical situations.

Job processing and machine environment

For each job J_j , one has to specify the processing requirement. We are assuming

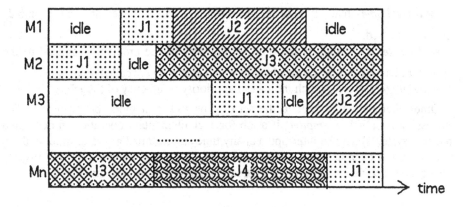

Fig. 1 Gantt Chart: Allocation of time periods on machines M1, M2, ...
for multi-operation jobs J1, J2, ...

deterministic processing times and consider first a set of jobs that are produced by a single machine operation. If all machines can carry out the same function, we obtain a system of *parallel machines*. In general, job J_j requires p_{ij} time units of production on machine M_i. The machines are *identical* if $p_{ij} = p_j$ for all M_i. The system is *uniform* if the parts are produced with different speeds on the machines, i.e. $p_{ij} = p_j/s_i$ where s_i is the *speed factor* of M_i. Finally, we have *unrelated* parallel machines if the speeds also depend on the particular job.

The second class consists of dedicated machines that are specialized on certain production operations. In this environment, a job is a part to be produced as a succession of operations to be carried out on different machines. One distinguishes three classical multi-operation models: The so-called *open shop*, where each job J_j consists of elementary operations $J_{1j}, J_{2j}, \dots , J_{mj}$. The operation J_{ij} is to be done on machine M_i during p_{ij} time units, but the order of execution is arbitrary. In a *flow shop*, the job J_j consists again of operations $J_{1j}, J_{2j}, \dots , J_{mj}$. However, their order is given. Operation J_{ij} is the i-th operation to be processed on machine M_i during p_{ij} time units. The third case is a *job shop*, in which each job J_j is a given sequence of operations to be processed on a job dependent sequence of machines. Here also the length of the sequence, i.e. the number of machine operations, may vary with the jobs.

Job characteristics

In addition to the processing requirement, one has further parameters associated with each job J_j:

- An *arrival time* (ready time or release time) r_j , which is the earliest time job J_j can be processed.
- A *due-date* d_j , which is the time for delivery of the product promised to the customer . Due-dates may be absolute, in which case they are called *deadlines*.
- A *weight factor* w_j , which indicates the priority or urgency of the job.

Other characteristics refer to the set of jobs and their mode of execution. The processing is called *preemptive*, if a job (or each elementary operation in dedicated machine systems) may be interrupted at any time and restarted at a later time, without extra cost and eventually on another machine. Allowing preemptions will usually reduce the schedule length when compared to nonpreemptive systems (Fig. 2).

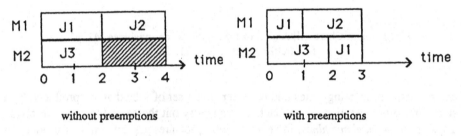

without preemptions   with preemptions

Fig. 2 Three-job processing $p_1=2$; $p_2=2$; $p_3=2$

Very often there exist *precedences* among the jobs. For a pair of jobs J_i and J_j, the precedence $J_i < J_j$ means that the processing of J_j can only be started after the completion of job J_i . A set of jobs is called *dependent* if certain precedences exist, otherwise the jobs are *independent*.

Performance criteria

For a given feasible schedule, visualized by its Gantt chart, one can calculate the following quantities for each job J_j :
- the *completion time* c_j ;
- the *flow time* $f_j = c_j$-r_j, which gives the time the part is in the system, being processed and waiting;
- the *lateness* $l_j = c_j$-d_j ;
- the *tardiness* $t_j = \max\{0, l_j\}$, where the earliness of the job, the completion before its due-date, is ignored.

The schedules are most commonly evaluated by one of the following criteria:
- the *schedule length* or makespan $C_{max} = \max\{c_j\}$;
- *mean flow time* $F = \Sigma f_j$;
- *maximum lateness* $L_{max} = \max\{l_j\}$;
- *maximum tardiness* $T_{max} = \max\{t_j\}$.

The optimal schedule refers to the one that minimizes the chosen performance criterion. These are the main criteria. In some applications, one also considers the *mean tardiness* $T = \Sigma\ t_j$ or the number of late jobs. One also may add the weights w_j to obtain for instance the *mean weighted flow time* $\Sigma w_j f_j$.

The problems are specified by the so-called three-field classification $\alpha\ /\ \beta\ /\ \gamma$, where the first field indicates the machine environment, the second describes the job parameters, and γ stands for the performance criterion. Without providing all the details (which can be found in [1, 9]), we only give some examples.

1) *Problem* $Pm\ /\ /\ C_{max}$

We have n jobs with processing times $p_1, p_2, \dots , p_n$ to be processed on m identical machines. The jobs cannot be interrupted once they are started (nonpreemptive system) and there are no precedences among the jobs (independent jobs). Execution for all jobs may start at time $r_j = 0$ and the objective is the minimization of the schedule length.

2) *Problem* $Q2\ /\ pmtn,\ tree,\ r_j\ /\ T_{max}$

Here we have two uniform parallel machines. On one of the machines the processing times are $p_1, p_2, \dots , p_n$ and these times are changed by the other machine by a certain speed factor. The jobs may be interrupted at any time (preemptive system) and there are precedences among the jobs which have the form of a tree. The jobs arrive at different times r_j . The performance criterion is the maximam tardiness which automatically requires as input a list of due-dates d_j .

3.2   Complexity of Scheduling Problems

Scheduling problems, as a special class of combinatorial optimization problems, are classified according to their computational complexity. The precise theoretical framework of complexity can be found in [6]. For our purpose, we can rely on the extensive literature of analyzed scheduling models, select the appropriate one and find out from the existing catalogue whether or not this particular problem is 'easy' or 'difficult'.

For easy problems, an algorithm with polynomial complexity exists. Avoiding any details on the exact mathematical model (the Turing machine), one may imagine that the complexity function is in fact an upper bound on the number of elementary computational steps in an actual implementation by a computer program. The complexity usually is of the form $O(f(n))$ or $O(f(n,m))$, depending only on the number of jobs and eventually on the number m of machines. The functions f are in most cases low order polynomials like n, nlogn, n^2 and n^3 .

For the difficult (NP-complete , NP-hard) scheduling problems, no polynomial time algorithms are known till now. Although it is still an undecided question, there is a very strong indication that no polynomial algorithm can exist for such a problem. But one has to keep in mind that the complexity refers to the worst case and only gives the asymptotic behavior for large numbers of jobs. In a practical context, one may have an automated production system with few machines and not too many parts to be produced in a given production horizon. In such a case, an NP-completeness result is of less importance and enumeration techniques can solve the problem exactly.

Within the class of NP-complete problems, certain problems possess so-called pseudo-polynomial algorithms. Their running time is polynomial, but depends in addition on the size of certain job characteristics, for instance the processing times. In a real production environment, the parts may require nonexessive execution times so that a pseudo-polynomial algorithm may well be used. In general , however, one has to rely on approximation schemes and heuristic methods and be content with suboptimal solutions.

3.3   Analysis and Results in Scheduling Theory

The parts that have to be processed in a given production horizon are allocated to a particular machining centre. In the simplest case, the production unit consists of one NC-machine. Single machine scheduling problems have been studied very extensively (see for instance [3]). This is also due to the possibility of decomposing more complex centres into a set of one-machine components on which groups of parts are sequenced separately. According to the previous section, the problems are classified as easy and an efficient (polynomial-time) algorithm is known or they are difficult and a heuristic has to be used.

Problems $1 / ... / C_{max}$

Here we consider the scheduling of jobs on a single machine in order to minimize the schedule length. If all jobs are independent and ready at time t=0 , one may sequence in any order. No machine idle time is inserted and $C_{max} = \Sigma p_j$. This is also true for arbitrary precedences in which case one has to use any admissable (topological) ordering. The problem with different arrival times, $1 / r_j / C_{max}$, is solved by ordering the jobs with increasing r_j.

In more realistic production systems, even these problems may become very complicated. If jobs arrive at different times and have to be delivered at different (absolute) due-dates d_j , the problem $1 / r_j, d_j / C_{max}$ is difficult (NP-complete) and is usually solved by a branch and bound method (compare [3]). In some processes, the change-over from a job to the next may be very time consuming and may last much longer than the actual production phase. Machines have to be set up for the new product and long regulations and adjustments are required. For such problems we want to incorporate transition times $c_{ij} = c(J_i, J_j)$ from job J_i to J_j. These times really depend on the two jobs involved since certain jobs may be quite similar so that only little retuning is necessary, but other pairs may be less compatible. Let (1), (2), ... , (n) be any ordering or permutation of the n jobs. Since the pure machining time Σp_j is constant, the best schedule is the one that minimizes the total transition times: $c_{(1)(2)} + c_{(2)(3)} + \cdots + c_{(n-1)(n)}$. However, this is equivalent to the traveling salesman problem, which is a notoriously hard problem.

Another interesting industrial application is given by the following process. The jobs arrive at different times and have to be processed p_j time units on the given machine. The product is terminated with a finishing operation (coating, painting or chemical bath) of length q_j. This operation can be done simultaneously on several jobs. The objective is to find the best sequence on the machine so that the total schedule length, including the finishing process, is minimized. Also this problem is difficult [6].

Other performance criteria on single machines

The second important measure is the mean flow time $F=\Sigma f_j$. Since f_j indicates the time the part j stays in the system, from arrival to completion, this criteria minimizes the mean response time to the customer's orders. The simplest problem, 1 / / F, is solved by the *SPT* or *Shortest-Processing-Time* principle: sequence the jobs in increasing order of the durations p_j. Adding different arrival times or precedences makes the problem difficult.

The third class concerns the lateness and tardiness measure. In the absence of any other constraints, one has a simple optimal ordering principle, namely Jackson's *EDD* or *Earliest-Due-Date* rule: sequence the jobs in increasing order of their due-dates. Again, further restrictions tend to make the problems difficult.

We have introduced previously the concept of preemption. With preemptions, certain (one-machine) scheduling problems become easy. This model, applied to manufacturing systems, means that the production process of a part is interrupted, the semi-finished part is taken off the machine, stored and mounted again onto the machine at a later time. This is a rather unlikely scenario, except in cases where a very urgent job has to be done immediately and no other machine is available. The concept of preemptions is quite natural for instance in computer systems where the execution of computer programs is interrupted and restarted repeatedly.

Parallel machine systems

We consider machining centers that consist of identical or similar machines. All machines can carry out the same operations. Either one has copies of the same machine or the machines can process the jobs with varying processing times (uniform and unrelated cases), due for instance to a more advanced technology of recently purchased machines.

The complexity results of most of these scheduling problems are discouraging. Already the simplest problem to sequence n jobs on m≥2 machines with no further constraints, i.e. problem $Pm \, / \, / \, C_{max}$, is difficult (NP-complete). Preemptions would trivialize the problem (applying McNaughton's rule). However, we had to discard this possibility as nonrealistic in manufacturing, except if one wants to generate lower bounds through a problem relaxation. In fact, preemptions in parallel machine systems are even more complicated since one might be forced to restart a semi-finished product on another machine, thus requiring the transfer of the part between the machines.

Nevertheless, problem $Pm \, / \, / \, C_{max}$ possesses a pseudo-polynomial solution approach with $O(nC^m)$; $C = \Sigma p_j$; elementary calculations. The method is a dynamic programming algorithm and may be used for a small number m of machines. The most popular heuristic methods, applicable also in the presence of general precedence constraints, are priority list procedures: One imposes a priority list of the jobs. Whenever a machine becomes available, the most urgent job from the list that is ready (i.e. all predecessors have been processed) is executed next without delay. It is known that the resulting schedule may be very poor, twice as long as an optimal schedule, and certain anomalies may occur (Graham's anomalies, 1966). However, list schedules belong to the most investigated procedures and good ordering strategies for the priority lists yield acceptable results.

Only very special scheduling problems on parallel machines with the C_{max} criterion are solvable exactly in polynomial time, for instance if all processing times are equal or the precedences have the form of a tree (Hu's level algorithm, 1966, and the label algorithm of Coffman and Graham, 1972, for two machines).

Scheduling problems for the other performance criteria F, L_{max} and T_{max} are in general even more difficult. Some special cases are solved by extensions of the SPT and the EDD rules, known fom the one-machine case. Considering in addition uniform or unrelated processing times further complicates the problems, and only few exact methods are known for nonpreemptive scheduling.

Dedicated machine systems

We turn now to jobs that require a multi-operation production sequence to be executed on different machines. We have previously distinguished three forms: open-shops, flow-shops and job-shops. Only the C_{max} criterion is considered. Polynomial algorithms exist for the two-machine shops. The most famous is Johnson's algorithm for the flow-shop problem $F2 \, / \, / \, C_{max}$ which is extremely simple: first arrange the jobs with $p_{1j} \leq p_{2j}$ in order of increasing p_{1j}, and then arrange the remaining jobs in order of decreasing p_{2j}. The result is a *permutation schedule* : the processing order is the same on the two machines. The three-machine flow-shop problem is difficult, but there still exists a permutation schedule that is optimal. This is no longer true in general for four and more machines. Approximation methods usually construct permutation schedules and are related to Johnson's algorithm [12].

The three types of shops do not seem to be of equal importance in actual manufacturing systems. Open-shops are not very realistic. One might imagine a sequence of minor operations , to be added to a product, that can be executed in any order. Job-shops are more suitable. However, one seems to have very often some kind of flow pattern in the factory so that the flow-shop is perhaps the most important form. This also includes all production lines. But a number of extensions are necessary. One may have a *no-wait* situation, as for instance in some chemical processes: once a job is started, the passage to all machines is to be done without any delay. The most important modification are so-called *flexible flow-lines* : at each stage of the flow-shop one may have the choice of several parallel machines. The classical flow-shop model requires unlimited storage space for the parts between the machines. One may instead impose capacity limits.

A large number of such models have been described in the literature and many heuristics have been proposed. We have only given a brief description. The suggestion is to set up a tool box of scheduling models and algorithms, exact and approximate, one of which should be activated in a given production environment.

4 Scheduling under Resource Constraints

4.1 Manufacturing Resources

In the classical scheduling theory, one solves the problem of assigning and sequencing the jobs on the available machines in the factory. Machines are, therefore, the main resource. But in automated manufacturing systems, a number of important additional resources are involved in the production process; like tools, fixtures, pallets and material handling devices.

There exists, in fact, an extension of the theory - the so-called *resource constrained scheduling* theory - that incorporates such supplementary resources [2]. Resources may be of quite general nature. They may be *discrete* (like tools) or *continuous* (like energy); *renewable* or *nonrenewable*.The raw material required by a part is a nonrenewable resource. Fixtures, pallets, tools and *automated guided vehicles* (AGV) are renewable. But the equipement may wear out or break after a certain (unpredictable) period of usage. One is therefore faced with problems of maintainance, monitoring and replenishment of such resources.

In resource constrained scheduling, the job-machine system is augmented by s types of supplementary resources $R_1, R_2, \ldots , R_s$ available in $m_1, m_2, \ldots , m_s$ units. Each job J_j requires for its processing a machine M_i and certain amounts of resources. These are usually specified by a resource requirement vector $R(J_j) = [r_1(J_j), r_2(J_j), \ldots , r_s(J_j)]$, where $r_k(J_j)$ denotes the number of units of resource type R_k that is needed for the processing of J_j . In case of flow-shop and job-shop scheduling, each operation is characterized by its resource requirement vector.

Resources are scarce and have to be shared by several jobs. The main problem is, of course, to get the required resource at the right time to the right place. Tools and fixtures that are necessary to hold the part are blocked throughout the processing of the job. Other resources, like AGVs, have only to be available at the beginning (to deliver the raw material) and after termination (in order to pick up the part). If $J(t)$ is the set of all jobs allocated to the machines at time t, then the resource constrained schedule is only admissable if

$$\sum_{T_j \, \varepsilon \, J(t)} r_k(T_j) \leq m_k$$

for all times t of the production horizon and for all resources R_k.

Scheduling theory under resource constraints has been originally initiated to incorporate a few additional resources, for instance a supplementary operator or a limited storage space. However, in manufacturing systems, one is faced with a huge variety of resource types. Already every different tool is counting as one new resource category. It is not surprising that most of these scheduling problems are difficult so that one has to rely on heuristics. Let us just mention one such approximation algorithm - the *first-fit-decreasing* or *FFD* method. The ratios of the resource requirement to available amounts, i.e. $r_{max}(J_j) = \max \{ r_k(J_j) / m_k ; 1 \leq k \leq s \}$, is determined. The jobs are arranged in decreasing order of $r_{max}(J_j)$ and a list scheduling algorithm is used. Obviously the idea behind this approach is to schedule early the potential resource bottlenecks.

A sample FMS

In order to get an idea of the number of machines and tool resources, we consider a recent FMS of Pratt &Whitney for the production of helicopter engine parts (Fig. 3).

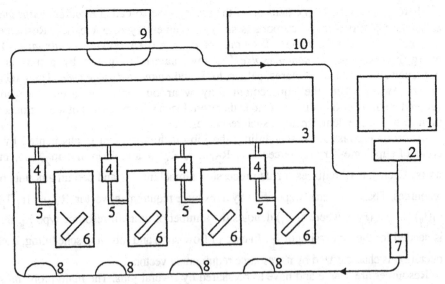

Fig. 3. Schematic view of an existing FMS

The raw material is taken from the automated storage area (1), loaded onto a pallet (2), and transported by one of the AGVs (7) to the allocated NC-machine (6) where it is unloaded automatically at (8). Each of the four machines is capable of processing the same parts. This flexibility is achieved by a large number of different tools that can be mounted onto the machines. In fact, the tool magazines (4) possess a capacity of up to 130 tools. Tools are arranged in two layers so that longer tools can occupy two vertical slots. Almost 100 quite different parts can be produced by this system. Small parts require about 30 different tools and operations, the complicated parts need up to 80 tools. The tools are loaded from the automated central tool reservoir (3) which contains nearly 2000 tools. Finished parts are transported to the inspection section (9) and then stored at (10).

As pointed out by this example, there is a tendency in modern FMS technology to design more and more versatile NC-machines. In the metal cutting industry, the parts are often produced on a single machine (as in our example). In such a case, the system is described by the scheduling theory of parallel identical machines. In addition, the number m of machines is often small (only 4 in Fig. 3). However, this simplification of the job-machine scheduling goes along with a significant increase of additional resources (categories of tools, fixtures, pallets, etc).

In more general processes (job-shops or flow-shops), the production of a part is carried out on several machines. Again, the tendency is to use few machines in order to reduce the calls for material handling, transfers and machine setups. It was found in the survey [7] that modern systems hardly consist of more than five stages (machines), but there may exist multiple choices at each stage (*flexible flow-lines*).

4.2 Scheduling with Resource Management

It appears that resource constrained scheduling theory provides the appropriate framework . But because of the large numbers and varieties, these supplementary resources are of dominating importance. There are problems of resource allocation, competition and sequencing. A careful management greatly influences the system performance and scheduling strategies. There are many interesting resulting algorithmic problems [4]. This is in fact currently a very active area of research.

We shall present one such strategy in the tool management. Many technical details can for instance be found in [16]. Let us consider the very restricted case of a single machine. Suppose the family of parts to be produced on this machine is given and in addition their sequence is known.We would like to determine the optimal tool flow between the crib and the magazine. Since the tools, required by a part, fill only partially the tool magazine, one has to decide what to keep in the unused slots (assuming that the magazine remains completely filled at all times). Consider the example in Fig. 4. Six jobs are to be processed in the order 1, 2, 3, 4, 5, 6. There are altogether eight tools and the magazine has the capacity of four tools. The optimal tool management strategy is the one that minimizes the total number of tool switches (tool exchanges between crib and magazine). The solution is given by the *KTNS*-strategy [15]: *Keep the tools in the magazine that are needed soonest.*

Tool-Part Incidence Matrix
Part 1 2 3 4 5 6

```
          0  1  1  1  1  0
          1  1  0  0  0  0
          0  0  0  0  1  1
Tool  0   1  1  0  0  0
          0  0  0  1  1  1
          1  0  0  0  0  0
          1  0  0  1  0  0
          1  0  1  0  0  1
```

Optimal Loading of the Magazine (5 tool switches)
Part 1 2 3 4 5 6

```
            2    2   2---5  5  5
tool        6---1   1   1   1
magazine 7---4   4---7---3  3
            8    8   8   8   8  8
```

Fig. 4. Optimal Tool Management

Finding also the best production sequence makes the problem even more complicated than the traveling salesman problem [5, 10, 11] . In this example, the number of tool exchanges has been the performance criterion. One may also consider the number of times the production has to stop since tools (in any number) have to be inserted. The appropriate model depends on the available tool exchange technology and the ratio of tool changing time to production time.

5   Decentralized Intelligent Scheduling Modules

We have outlined a collection of theoretical methods and models of scheduling in manufacturing systems. One has to keep in mind, that these can only provide a toolbox in an actual production environment. The pratical situations in the factories, ranging from the metal-cutting industry to assembly lines and also including complicated chemical processes, are far too manyfold to be modeled by a single universal system. Careful analysis and dialogues at the production sites are essential to design an adapted model.

We have presented the scheduling procedures as they apply to automated manufacturing systems. In particular, the extensions to the resource management are vital. For instance as described in [16], the initial investment in cutting-tools and fixtures may reach up to 25% of the total FMS investment. Cases are mentioned where about 16% of the production time may be missed because tools are not available; half of the tool inventory contains obsolete tooling and up to 30% of the tool inventory is just "lost".

The scheduling module is part of a hierarchical production system. The aim is to reduce the number of levels in this hierarchy and increase their autonomy as much as possible. In such a partially decentralized system, the modules have to be more "intelligent". We think that our tool-box can provide the basis for an intelligent scheduling module. The models are simple , flexible, easy to implement and, above all, quite sufficient to describe correctly the process (if properly chosen). The algorithms (exact or heuristic whenever necessary) have low-order polynomial running-times. They can be executed in real-time and can be rerun very quickly in the presence of new circumstances (disturbances, new urgent orders, etc).

Scheduling with resource management is a basic ingredient for frequent production changes. The possibilities of the system have to be explored in order to change rapidly and intelligently the tools (this is known under the term *SMED*). This flexibility of the production is imposed by the frequently changing demands of the market. The quick reaction and effective handling, and at the same time avoiding all wasteful operations and unnecessary stocks in the factory, is one of the centerpieces of the so-called *"lean production"*.

6 Final Remark

The present chapter has been based on a series of lectures given to students at the Czech Technical University in Prague in June 1993.

References

1. J. Blazewicz: Selected Topics in Scheduling Theory. Annals of Discrete Mathematics 31, 1-60 (1987)

2. J. Blazewicz, W. Cellary, R. Slowinski, J. Weglarz: Scheduling under Resource Constraints. Basel: Baltzer 1987

3. J. Blazewicz, K. Ecker, G. Schmidt, J. Weglarz: Scheduling in Computer and Manufacturing Systems. Berlin-Heidelberg: Springer 1993

4. J. Blazewicz, G. Finke: Scheduling with Resource Management in Manufacturing Systems. To appear in EJOR

5. Y. Crama, A.W.J. Kolen, A.G. Oerlemans, F.C.R. Spieksma: Minimizing the number of tool switches on a flexible machine. To appear in Int. J. of FMS 6 (1994)

6. M.R. Garey, D.S. Johnson: Computers and Intractability: A Guide to the Theory of NP-Completeness. San Francisco: Freeman 1979

7. R. Jaikumar: Postindustrial manufacturing. Harvard Bus. Rev. 1986

8. A. Kusiak: Intelligent Manufacturing Systems. Englewood Cliffs, New Jersey: Prentice-Hall 1990

9. E.L. Lawler, J.K. Lenstra, A.H.G. Rinnooy Kan, D.B. Shmoys: Sequencing and scheduling: Algorithms and complexity. CWI Amsterdam; Report BS-R8909

10. C. Privault, G. Finke: Tool management on NC-machines. Proc. Int. Conf. on Ind. Eng. and Production Management, Vol. II, Mons 1993, pp. 667-676

11. C. Privault, G. Finke: Gestion on-line des changements d'outils dans un système flexible. Proc. Congrès biennal de l'AFCET, Vol. I, Aide à la Décision et Recherche Opérationnelle. Paris 1993, pp. 175-184

12. C. Proust: De l'influence des idées de S.M. Johnson dans la résolution des problèmes d'ordonnancement de type Flowshop. Report, Laboratoire d'Informatique, Ecole d'Ingénieurs en Informatique pour l'Industrie. Tours 1992

13. P. Solot, H. Van Vliet: Analytical Models for FMS Design Optimization: A Survey. Technical Report, Lausanne 1993

14. K.E. Stecke: Formulation and solution of nonlinear integer production planning problems for flexible manufacturing systems. Management Science 29, 273-288 (1983)

15. C.S. Tang, E.V. Denardo: Models arising from a flexible manufacturing machine. Part I: Minimization of the number of tool switches. Part II: Minimization of the number of switching instants. Operations Research 36, 767-784 (1988)

16. D. Vaaramiani, M. Upton, M.M. Barash: Cutting tool management in computer-integrated manufacturing. Int. J. of FMS 3/4, 237-265 (1992)

17. A.J. Van Looveren, L.F. Gelders, L.N. Van Wassenhove: A review of FMS planning models. In: A. Kusiak (ed.): Modelling and Design of Flexible Manufacturing Systems. Amsterdam: Elsevier 1986, pp. 3-31

Operational Research Models and Methods in CIM

Marino Widmer

Université de Fribourg, IIUF - Regina Mundi, Faucigny 2
CH - 1700 Fribourg

Abstract : Many models and methods of Operational Research can be adapted for industrial applications. In this chapter, we show on the one hand the main problems of a manufacturing system and, on the other hand, how they can be ranged in an hierarchical order, derived from a CIM architecture (from the strategic decisions to the production constraints). Then, we present an Operational Research tool for solving each of these problems.

1 Introduction

Flexible Manufacturing Systems (FMS) are nowadays installed in the mechanical industry, especially in the factories in which cars are produced. However, the market constraints impose to improve the production system and the whole production organization.

The concepts developed by Taylor and applied at the beginning by Ford are progressively abandoned and replaced by the Just-In-Time concept and the Computer Integrated Manufacturing philosophy (CIM). One of the aims of the CIM philosophy is to provide an integrated information system which avoids the rigid separations between the different functionalities of a complete production system. With such integrated information systems, the loss of time on the one hand between the customer order and the part delivery, on the other hand between the product design and its manufacture will be drastically reduced.

To understand the complete production system, it is relatively easy to find, in the scientific literature, excellent general books explaining the different aspects of the Production and Operations Management (POM) ([1], [2], [3], [4], [5]). It is more difficult to discover a writing dedicated to use of Operational Research (OR) models and methods in the industrial context [6]. And it is quite impossible to find a book which offers a good balance between POM and OR ...

In this chapter, we will show how a CIM architecture can be partially decomposed along two main axes : the production management aspect (from the customer order to the bill) and the logistics aspect (from the supply to the distribution). Then, we will show how to integrate Operations Research models and methods in this CIM architecture.

A basic CIM architecture defined by Scheer [7] is presented in Section 2 and its limits are highlighted.

In Section 3, a production planning decision hierarchy is proposed, containing five main phases. The two last phases are then developed in the different components, describing a two-axes structure, the vertical axe representing the production control and the horizontal one dealing with the logistics.

Section 4 is dedicated to the application of different OR models and methods in the CIM context. These one are mainly useful for solving the following problems : plant sizing and location, equipment type and amount, production allocation among plants, order management, product design, planning, scheduling, shop floor control, supplying logistics and inventory management, and distribution logistics.

Finally, Section 5 concludes this chapter suggesting new research topics for Operational Research in CIM context.

2 A CIM architecture

Chase and Aquilano give in their book an interesting description of different aspects of the Production and Operations Management, based on a life cycle approach [1]. To define a good CIM-architecture, Scheer takes a step which is quite similar [7]. He considers on the one hand the organizational planning functions and on the other hand the technical functions (fig. 1). Of course, these two sets of functions are tightly linked in the implementation phase, which corresponds to the real time control of the manufacturing system.

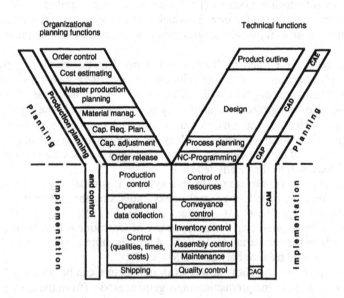

Fig. 1. Information systems in production : the "Y" of Scheer [7]

This architecture described by Scheer is relatively complete for an existing manufacturing system, where the production resources are globally known and where the production strategies have been defined. But if the manufacturing system has to be created, some important decision steps do not appear in the "Y". We propose in next Section a multiple phase approach which takes also into account the strategic decisions.

3 The Integrated Production Management

3.1 The production planning decision hierarchy

When an industrial company decides to built news factories, it has to solve a set of problems, which are ordered according to a production planning decision hierarchy (fig. 2). It is important to solve each problem, one after the other, to avoid an economical disaster !

The first step consists in selecting the optimal **locations** for the new plants, taking into account constraints like economical factors (taxes, wages) and management facilities (proximity of the suppliers and the customers).

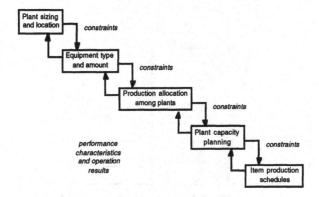

Fig. 2. The production planning decision hierarchy [1]

The **equipment type** and their **amounts** are defined for each plant in the second phase. When the shop floor descriptions are complete for the different resources (human and material), the next step consists in defining the **production allocation** among the plants.

At this point, all the conditions to use the CIM architecture of Scheer are satisfied. However, we prefer to describe a concept of integrated production management, where it is possible to highlight easily some basic problems which can be solved thanks to Operational Research tools.

3.2 The integrated production management

To define the integrated production management, we will describe the different modular notions which appear in every production system, independently of the size or of the functionality of the factory (fig. 3).

To depict the *production control*, the process between the **customer order** and its delivery is first studied.

When a customer gives an order to the factory, this order belongs to one of the two following categories :

- the ordered product is described in a catalogue (like a car with its several options). That means that the manufacturing process is known. In such a case, the customer order is recorded in a module of **order management** ;
- the ordered product is new (the manufacturing process is unknown). Before being able to insert the customer order in the module of order management, it is necessary to design the product : this work is done in a module called **product design**.

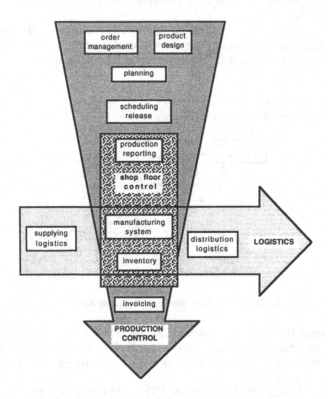

Fig. 3. The Integrated Production Management [8]

The order management module, eventually associated with a **forecasting** module (depending of the chosen production type), is in constant dialogue with the module of **inventory management** to define the quantities to produce and the delivery delay for the customer.

Naturally, to determine the delays, a **planning** module has to be used, taking into account the work in process or the foregone manufacture. This module answers to the question : when should we produce ? These delays are also important to determine the moments when orders have to be given to the suppliers : that means that it exists a dialogue between the planning module and a **supplying management** module.

When the quantities of the different products to manufacture during the next planning period are known (we dispose of the list of parts to be processed during a production period), we have to determine the **scheduling** of the products which should be manufactured respecting some criteria (to avoid too many set ups of the machines, to minimize the transfers, to minimize the waiting times,...), to avoid a non-optimal utilization of the resources. So, this module answers to the question : how should we produce ?

Then, when the production is **released**, we have to respect as faithfully as possible the sequence established by the scheduling. This is the task of the *shop floor control* module : it takes care of the production reporting and of the production resource management in real time (machines, conveyors,...). A module of **quality control** could be associated to it, as well as a module of **maintenance management**. If the firm does not work in *just-in-time*, the shop floor control module has to dialogue with the forecasting module.

After this phase of real time control, the parts are finished products (or considered as finished products by the factory which manufacture them). The last operations consist in sending the goods to the customer and to submit the bill (**distribution logistics** module and **invoicing** module).

This concludes the main description of the production control.

Let us now explain the components of the *logistics aspect*, which define the product in term of material, from the draw material to the finished product.

Before the arrival in the shop floor, the draw material has to be ordered : this task is devolved upon a module of **supplying logistics**. To apply successfully the *just-in-time* concept, the suppliers as well as the firm have to work in *just-in-time*.

Then, when we decide to manufacture the products, we have to manage the **inventory**. This could be an inventory of draw material, in-process parts, semi-finished or finished products. Even if the *just-in-time* concept recommends the notion of zero-inventory, a minimum inventory is always necessary (like a stabilization stock).

Finally, when the products are finished, they have to be delivered to the customers. This last action is done by the **distribution logistics**. Thus, the ordered product arrives to the customer and this complete the **logistics** aspect.

To conclude with this two-axes structure, we highlight the fact that some modules have an evolutive aspect. For example, the planning module loses the main part of its importance when the firm works in *just-in-time*. If we add to this aspect the fact that the life cycle of the products is more and more shorter, it is easy to understand the non negligible consequences on the inventory management module.

Notice that we have deliberately omit to talk about a **marketing** module, which is marginal in the production process. However, this marketing aspect has not to be neglect, seeing that it has an influence on the forecasted order management and on the new products, which are susceptible to interest some potential customers.

3.3 An hierarchico-cyclical structure

In the integrated production management, the aspects of planning, scheduling and shop floor control could be joined in a global decision approach : the hierarchico-cyclical structure ([9], [10]). To describe this last one, it is useful to consider the life cycle of a product (fig. 4).

To produce parts, it is necessary to know the kind of available resources. The **shop floor description** defines the physical structure of the shop floor, this means the number of components of each type (machines, conveyors), the material handling system, the pallet quantity and eventually the fixture quantity. The resources described here are *permanent* elements of the system.

The **part description** gives informations on the process plan of the different parts (sequence of operations, processing time, set of machines on which each operation can be executed).

The **planning** takes into account the list of the customer orders, the inventory level (if the *just-in-time* concept is not applied) and the work in process. This production planning helps the seller when he negotiates the delivery lead times.

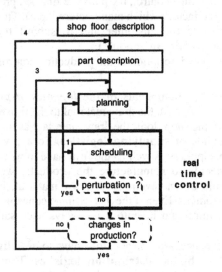

Fig. 4. The hierarchico-cyclical structure [10]

When the quantities of parts (to produce during the planning period) are known, the part *scheduling* has to be determined. It must respect some criteria to avoid an under-optimal utilization of the resources (minimize the transfer durations, minimize the set ups on the machines, ...). Then, when the manufacturing is released, it is necessary to respect as strictly as possible the defined scheduling. This task is dedicated to the *system control* and to realize that, one has to take into account the machine occupation, the pallet and fixture availability and the tools loaded on each machine. These two levels (scheduling and control) make up the **real time control** of the shop floor.

When the production system is working, breakdowns and maintenance operations disturb the smooth working of the factory. If the maintenance operations can be kept down (they can be considered as fictitious parts), the breakdowns can generate complex problems. According to the importance of the perturbation, the reactions are fundamentally different. In the case of a momentary interruption, a new allocation of the workload among the remaining resources can be done thanks to a rescheduling (1). This means that the perturbation is *absorbed* by the real time control system. When the perturbations are more serious (2), it is necessary to establish a new planning.

If there are no changes in the set of the products, the planning process restarts at the beginning of the next production period (3). Else, when new parts have to be produced, for which the process plans are not defined, it is necessary to introduce these informations in the part description (4).

Let us now explain how the **technical evolution** can be taken into account : if a product is modified in its process plan, one has to record it as a new part, whose production increases progressively in comparison with the old version of the product. When the old process plan is no more used, the associate product disappears from the *part description.*

This hierarchical and cyclical structure has huge advantages. It allows above all a global view on the whole control of the shop floor. It offers the advantage to be enough general to suit for different types of shop floor, classical or flexible. Naturally, it must be adapted to be efficient for a determined factory : for example, it would be useful to couple an inventory management module to the planning module in some cases.

In the conditions of the real time control, when some events occur, it is easy to see at which level one has to intervene. Thus, this structure helps the managers to take optimal decisions.

Finally, notice that Solot has proposed a quite similar basic hierarchico-cyclical structure, including some elements of Artificial Intelligence like expert systems [11] (fig. 5).

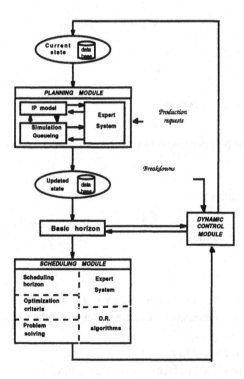

Fig. 5. The model proposed by Solot [11]

4 OR models and methods in CIM context

In the last Section, we have decomposed the CIM architecture into a set of problems. Each one should find a solution thanks to Operational Research tools. We present in this Section a few models and methods, mainly based on linear programming. Of course, they are basic answers for general problems : for solving real-life problems, the models have to take into account more accurate informations to be suitable to the reality.

4.1 Plant sizing and location

The selection of the optimal locations for the new plants has to take into account constraints like the establishment costs, the proximity of the suppliers and the customers, the production costs, etc. We present here a simplified formulation of this problem.

$$min \ \sum_i f_i . y_i + \sum_i \sum_j c_{ij} . x_{ij}$$

subject to :
- demand constraints :

$$\sum_i x_{ij} \leq b_j \qquad \forall \ j$$

- production constraints :

$$\sum_j x_{ij} \leq a_i \qquad \forall \ i$$

- location and capacity constraints :

$$x_{ij} \leq a_i . y_i \qquad \forall \ i, \forall \ j$$

- the number of new plants is fixed to N :

$$\sum_i y_i = N$$

- y_i are Boolean variables :

$$y_i \in \{0,1\} \qquad \forall \ i$$

- x_{ij} are non negative variables :

$$x_{ij} \geq 0$$

where :
- f_i is the establishment cost of a new factory in i
- c_{ij} is the unit transportation cost from the factory i to the region j
- b_j is the demand in region j
- a_i is the capacity of factory in i
- x_{ij} is the volume produced in factory i for the region j
- y_i is the decision variable if a factory is located in i or not

4.2 Equipment type and amount

Simulation is here an efficient tool. If we consider the discrete event simulation, some performant softwares could help the decision maker to select the suitable quantity of each resource type. But he should waste plenty of time before he arrives to the adequate solution !

Hopefully the analytical simulation avoids this tedious research. The queueing network theory gives all the elements to model the resources of the manufacturing system as well as the behaviour of the parts (fig. 6).

Dallery and Frein have proposed an efficient method to determine the optimal configuration of a FMS [12]. Knowing the costs of the machines and the pallets, their algorithm finds the optimal quantity of resources allowing to reach a given production rate.

Solot has developed some interesting models also based on queueing network theory dealing with pallet changes and resource reliability ([13], [14], [15],[16]).

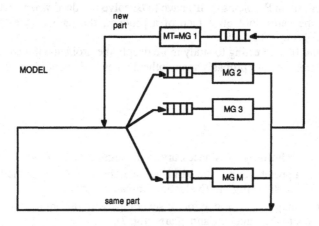

Fig. 6. Queueing networks for shop floor description

4.3 Production allocation among plants

This problem is quite similar to the plant location. As the two previous steps have defined the geographical location of the plants and their resources, we know :
• the ability of each factory to manufacture the parts ;
• the production capacity of each factory.

In these conditions, it is easy to adapt the formulation presented in 4.1 to take into account the production costs and the transportation ones.

4.4 Order management

For a long time, when the demand was greater than the offer, it was primordial to forecast the demand, in order to release the production of the parts before receiving the customer order ! Such an anticipation was possible thanks to forecasting methods, as those presented by Chatfield ([17], [18]) : exponential smoothing, regression analysis, Box-Jenkins technique, ...

However, in the today's situation, the forecast is not so important, carrying away the **order management** with it : indeed, if it is necessary to know the quantity of basic parts which will be useful in the next production period, the specific components of the product are done only when the customer order is real.

4.5 Product design

The **product design** is one of the topics neglected by the OR tools. Some Traveling Salesman Problems (TSP) have to be solved. Indeed when several holes are made with the same tool on a face of a product, the path of the tool could be minimized solving a TSP [19].

But it would be interesting to study more deeply the problems linked to the product design, because they are probably many others cases which could be solved with OR methods.

4.6 Planning

Concerning the **planning**, it is necessary to distinguish two cases :
• if the part is a prototype or if the part is unique, the most efficient method is to deal with a PERT model (Program Evaluation Research Task) ;
• else, the planning of the production of several parts in different quantities could be solved thanks to the famous Gantt charts (fig. 7).

As explained before, some new concepts have appeared in the last twenty year, like Just-in-time (JIT) or Optimized Production Technology (OPT), which give less importance to the planning module [20].

Notice that it exists some interesting models combining planning and scheduling problems, which could be applied in mechanical industry [11] or in chemical industry [21].

Let us present an integer programming model to plan the production of different parts over several time periods, in a manufacturing system where it exists an unique critical resource.

$$min \ \sum_i \sum_t (c_{it} \cdot x_{it} + f_{it} \cdot y_{it}) \ + \ \sum_t (g_t \cdot u_t + h_t \cdot v_t)$$

subject to :
- the satisfaction of the demands:

$$x_{it} + y_{i,t-1} - y_{it} = d_{it} \qquad \forall \ i, \forall \ t$$

- capacity constraints :

$$\sum_j k_i \cdot x_{it} = u_t + v_t \qquad \forall\, t$$

- time constraints :

$$0 \le u_t \le \bar{u}_t \qquad \forall\, t$$

$$0 \le v_t \le \bar{v}_t \qquad \forall\, t$$

- y_i are Boolean variables :

$$y_i \in \{0,1\} \qquad \forall\, i$$

- x_{it} and y_{it} are non negative variables :

$$x_{it}, y_{it} \ge 0 \qquad \forall\, i, \forall\, t$$

where :
- x_{it} is the quantity of products i manufactured during period t
- y_{it} is the quantity of products i in inventory at the end of period t
- u_t is the quantity of normal working hours during period t
- v_t is the quantity of extra working hours during period t
- c_{it} is the production cost of one product i during period t
- f_{it} is the inventory cost of one product i during period t
- g_t is the cost of one normal working hour during period t
- h_t is the cost of one extra working hour during period t
- d_{it} is the demand of product i in period t
- k_i is the processing time to manufacture product i
- $\bar{u}_t$ is the quantity of normal working hours dedicated to period t
- $\bar{v}_t$ is the maximal quantity of extra working hours allowed during period t

We do not give more details on this topic, as this book contains a complete chapter dedicated to the planning and the scheduling.

4.7 Scheduling

The **scheduling** problems have held the attention of an huge amount of searchers during the last forty years, since Johnson has proposed optimal two and three-stage production schedules. Many heuristics have been developed to solve the problems of flow shop, job shop and open shop. Unfortunately, none of these heuristics are general models : they deal with some special cases, each one with specific constraints. A good review of some of these models is described in the books of Blazewicz ([22], [23]).

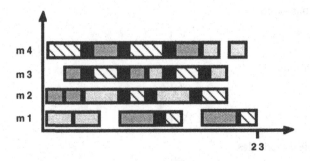

Fig. 7. A scheduling solution taking into account tool switches
(the black boxes represent the tool switches)

New developments are given in the topic of the scheduling considering compact cylindrical scheduling (repetitive production) [24] or dealing with the tool switches on one machine [25] or on several machines [26] (fig. 7).

Let us present an integer programming model to minimize the number of tool changes in a flexible manufacturing system containing m machines and n mono-operation products [27].

$$min \sum_{k=0}^{p} \sum_{j=1}^{m} \sum_{z} u_{zjk}$$

subject to :

- the number of tools allocated to machine j is less than or equal to the capacity C_j of the tool magazine of machine j :

$$\sum_{z} y_{zjk} \leq C_j \qquad \forall j, \forall k$$

- the number of tools of type z allocated to all the machines is less than or equal to the limited number nb_z of tools of type z :

$$\sum_{j=1}^{m} y_{zjk} \leq nb_z \qquad \forall z, \forall k$$

- a_i parts of type i have to be produced :

$$\sum_{k=1}^{p} \sum_{j=1}^{m} x_{ijk} = a_i \qquad \forall i$$

- the right tools have to be in the right place at the right moment :

$$x_{ijk} \leq a_i \cdot y_{zjk} \qquad \forall z \in Z_i, \forall i, \forall j, \forall k$$

- s_k is positive if at least one type of part is produced in time-shift k :

$$\sum_{i=1}^{n} \sum_{j=1}^{m} x_{ijk} \leq s_k \cdot \sum_{i=1}^{n} a_i \qquad \forall k$$

- s_k is a non increasing variable :

$$s_k \geq s_{k+1} \qquad \forall\, k \in \{1,..,p-1\}$$

- y_{zjk} and s_k are Boolean variables :

$$y_{zjk} \in \{0,1\} \qquad \forall\, z, \forall\, j, \forall\, k$$

$$s_k \in \{0,1\} \qquad \forall\, k$$

- x_{ijk} is a non negative integer variable :

$$x_{ijk} \in \mathbf{N} = \{0, 1, 2, ... \}$$

- no tools are loaded if no parts are processed in time-shift k :

$$y_{zjk} \leq s_k \text{ and } y_{zj0} = 0 \qquad \forall\, z, \forall\, j, \forall\, k \in \{1,..,p\}$$

- and finally :

$$-u_{zjk} \leq y_{zjk} - y_{zj,k+1} \leq u_{zjk} \qquad \forall\, z, \forall\, j, \forall\, k \in \{1,..,p\}$$

where :
- x_{ijk} is the number of parts of type i produced on machine j in time-shift k
- y_{zjk} is 1 if a tool of type z is allocated to machine j in time-shift k , 0 otherwise
- s_k is 1 if at least one type of part is produced in time-shift k , 0 otherwise
- u_{zjk} is 1 if a tool of type z is added to or removed from machine j at the end of time-shift k, 0 otherwise
- a_i is the quantity of parts i which have to be manufactured
- C_j is the capacity of the tool magazine of machine j
- nb_z is the limited number of tools of type z

As the reader can realize, even "simple" scheduling problems have a complex formulation !

4.8 Shop floor control

This is probably the topic where Operational Research models and methods have to be progressively developed. The **shop floor control** is a really complex problem : this fact explains partially the lack of efficient OR tools. However, it seems to be possible to introduce some algorithms or some heuristics methods in this context [28], but the search is at the beginning in this topic.

4.9 Supplying logistics and inventory management

These logistics aspects are two of the basic elements in a firm and they have focused the attention of several searchers. A good synthesis dealing with **supplying logistics** models and **inventory management** methods is presented by Peterson and Silver [29] and by Giard [30]. We do not give here more detailed explanations, seeing that this topic is one of the most prevalent in the scientific literature.

But as we have mentioned it previously, the *just-in-time* concept is now implemented in many factories and it recommends the notion of zero-inventory. That means this topic is less important nowadays, even if a minimum inventory is always necessary.

4.10 Distribution logistics

If we refer to Semet [31], the **distribution logistics** becomes a new topic of research. Indeed due to the reduction of production costs observed during the recent years, distribution costs have increased significantly as a percentage of the cost price of a product. Therefore the minimization of these costs is a new challenge for many companies which can addressed through solving numerous combinatorial problems, e.g., location problems or vehicle routing problems.

Semet considers problems dealing with the vehicle routing problem when different means of transport are used (trains and trucks) and defines the accessibility constraints. He proposes a new feature of traveling salesman problem when these constraints have to be taken into account and can define a model using an integer programming formulation. Then he tackles the vehicle routing problem under accessibility constraints, modelled also thanks to an integer programming formulation.

Notice that this type of problems is not easy to solve, because of the variety of real-life situations (time-windows for the delivery, accessibility constraints, rest time, capacity of the truck (weight and volume), ...) [32].

5 Conclusion

We have shown how the complete production process can be defined as a production planning decision hierarchy, including an integrated production management, which can be decomposed in several modules. Then we have shown that it exists some OR models and methods which can be applied to each module.

A good survey of the applications of Operational Research models and techniques in flexible manufacturing systems is presented by Kusiak [33], in a special issue of the European Journal of Operational Research dedicated to the FMSs.

To conclude this chapter, we think that two topics could provide some interesting research fields :
- the product design
- the shop floor control.

For this last point, it will be important to develop several dynamic algorithms, to react quickly to the behaviour of the manufacturing system.

6 Bibliography

[1]   R. Chase, N. Aquilano : Production and Operations Management : A Life Cycle Approach. Irwin 1989

[2]   J. Dilworth : Operations Management : Design, Planning, and Control for Manufacturing and Services. McGraw-Hill 1992

[3]   N. Gaither : Production and Operation Management. The Dryden Press 1992

[4]   J .Heizer, B. Render : Production and Operation Management. Allyn and Bacon 1993.

[5]   W. Stevenson : Production / Operations Management. Irwin 1993

[6]   E. Turban, J. Meredith : Fundamentals of Management Science. Irwin 1991

[7]   A.-W. Scheer : CIM Computer Integrated Manufacturing : Towards the Factory of the Future. Springer Verlag 1988.

[8]   J.P. Wermeille, M. Widmer : Les composantes de l'intégration. Marché Suisse de Machines 21, 46-49 (1991)

[9]   K.E. Stecke : Design, planning, scheduling, and control problems of flexible manufacturing systems. Annals of Operations Research 3, 3-12 (1985)

[10] M. Widmer : Modèles mathématiques pour une gestion efficace des ateliers flexibles. Presses Polytechniques et Universitaires Romandes 1991

[11] P. Solot : A Concept for Planning and Scheduling in an FMS. European Journal of Operational Research 45, 85-95 (1990)

[12] Y. Dallery, Y. Frein : An Efficient method to Determine the Optimal Configuration of a Flexible Manufacturing System. Annals of Operations Research 15, 207-225 (1988)

[13] P. Solot, J. Bastos : MULTIQ : A Queueing Model for FMSs with Several Pallet Types. JORS 39, 811-821 (1988)

[14] P. Solot : A Heuristic Method to Determine the Number of Pallets in a Flexible Manufacturing System with Several Pallet Types. IJPR 2, 191-216 (1990)

[15] P. Solot : Nouvelles approches mathématiques des problèmes de conception et de pilotage des ateliers flexibles. Thèse de l'Ecole Polytechnique Fédérale de Lausanne no 975, 1991

[16] M. Widmer, P. Solot : Do not forget the breakdowns and the maintenance operations in FMS design problems ! IJPR 28, 421-430 (1990)

[17] C. Chatfield : The analysis of times series : theory and practice. Chapman and Hall 1975

[18] C. Chatfield : What is 'Best' Method of Forecasting ? Journal of Applied Statistics 15 (1988)

[19] V. Schaller : Ordonnancement d'opérations sur une machine à commande numérique (industrie mécanique). Projet de semestre EPFL-DMA, Lausanne 1986

[20] J. Browne, J. Harhen, J. Shivnan : Production Management Systems : a CIM perspective. Addison-Wesley 1988

[21] H. Groeflin, H. Schiltknecht : PEPI : Ein innovatives Produktionsplanungs instrument in der chemischen Industrie. Output 12, 59-64 (1989)

[22] J. Blazewicz, W. Cellary, R. Slowinski, J. Weglarz : Scheduling under resource constraints : deterministic models. Annals of Operations Research 7 (1986)

[23] J. Blazewicz, K. Ecker, G. Schmidt, J. Weglarz : Scheduling in Computer and Manufacturing Systems. Springer-Verlag 1993

[24] D. de Werra, P. Solot : Compact Cylindrical Chromatic Scheduling. SIAM Journal on Discrete Mathematics 4 (1991)

[25] Y. Crama, A.W.J. Kolen, A.G. Oerlemans, F.C.R. Spiekma : Minimizing the number of tool switches on a flexible machine. IJFMS 6 (1994)

[26] A. Hertz, M. Widmer : A new approach for solving the job shop scheduling problem with tooling constraints. Submitted for publication

[27] D. de Werra, M. Widmer : Loading problems with tool management in flexible manufacturing systems : a few integer programming models. IJPR 3, 71-82 (1990)

[28] Y. Mottet, M. Widmer : Dynamic Scheduling and Tool Loading. In : V.C. Venkatesh and J.A. McGeough (Eds.) : Proceedings of the 7th International Conference on Computer-Aided Production Engineering. Cookeville USA : Elsevier (1991), pp. 325-332

[29] R. Peterson, E.A. Silver : Decision Systems for Inventory Management and Production Planning. John Wiley and Sons 1985

[30] V. Giard : La gestion de la production. Economica 1988

[31] F. Semet : Elaboration de tournées de véhicules sous contraintes d'accessibilité. Thèse de l'Ecole Polytechnique Fédérale de Lausanne no 1163, 1993.

[32] D. de Werra, F. Semet, P. Solot : La distributique. Output 9, 10, 11, 1990.

[33] A. Kusiak : Application of operational research models and techniques in flexible manufacturing systems. EJOR 24, 336-345 (1986)

Production Planning and Control Systems
- State of the Art and New Directions

V Sridharan and John J. Kanet

Department of Management, Clemson University
Clemson, SC 29634-1305, USA

Abstract: This chapter begins with a description of the role of production planning and control (PPC) within the manufacturing function. After discussing the impact of the operating environment on the choice a system for PPC, we describe some recent empirical evidence regarding the use and performance results of various PPC systems. This is followed by a brief overview of the two most widely used systems for production planning and control. We then describe a recent development in the area of short-term detailed scheduling exploiting the latest developments in computing technology. The chapter concludes with a discussion of an emerging paradigm for exploiting the advancements in computing technology for developing sophisticated state-of-the-art PPC systems capable of satisfying the needs of tomorrow's market place.

1. Introduction

The Production Planning and Control system is a major component of the infrastructure which supports the manufacturing process selected for the specific environment faced by the firm [27]. Production planning and control (PPC) is a major function essential for the successful operation of every manufacturing company. It is a complex task because of the large data volume involved [18]. Villa [26] characterizes it as a control optimization problem for a large-scale dynamic system. He argues that the evolution of a PPC system is influenced largely by two types of events: (1) events to be planned and (2) events to be controlled.

A PPC system is mainly concerned with ensuring efficient and effective use of capacity for satisfying anticipated demand. Four major tasks comprise production planning and control: Aggregate Capacity Planning, Material Planning, Production Activity Planning, and Production Activity Control. Aggregate Capacity Planning focuses on matching demand and supply (i.e., capacity) over the short or medium horizon. Material Planning deals with making sure that the needed material is available at the right time. Production Activity Planning generates detailed plans ensuring that the available capacity is consumed in an effective and efficient fashion. Production activity control is concerned with execution of the plans.

Successful companies plan and control their operations using a formal system. Systems designed to support companies in planning and controlling their operations are called Production Planning and Control Systems. Many PPC systems are being used by companies around the world. Earlier this century firms often used what has come to be known as Reorder Point Systems (ROP). With the advent of computers, during the early sixties, the use of Material Requirements Planning Systems became widespread. Beginning in the 1960's, many traditional reorder point systems were replaced by what are known as Manufacturing Resource Planning (MRP) systems. According to some estimates, more than 20,000 companies in various parts of the world deploy MRP systems to plan and control their operations. In the early 1980's the "MRP Crusade" ran up against the "JIT Crusade" built around the Kanban system. In the last decade, largely due to some Japanese firms operating under the Just-In-Time philosophy, Kanban-based systems became popular. At about the same time, another new system called Optimized Production Technology based upon the "drum-buffer-rope" approach [5] was introduced and is used by some companies. Thus, the four most common PPC systems used in practice and discussed in the literature are the following: MRP-based push systems (MRP), Kanban-based pull systems (Kanban), constraint theory based systems that identify and schedule according to bottleneck resources (OPT), and traditional reorder point based systems (ROP).

Computer integrated manufacturing (CIM) links all vital functions of a manufacturing organization through an integrated computer system thus providing the capability to consistently produce the desired products at a low cost and high quality, and in a timely fashion. The CIM concept spans the entire firm. For the purposes of this chapter we limit attention to only the planning and control function. Within CIM, the PPC system is computerized. CIM requires that the PPC system be tightly integrated and be capable of speed and sophistication. This is so because of the increased product variety typically associated with a CIM environment. Furthermore, the enhanced capability to supply products at short notice (reduced customer leadtimes), the need for accurate planning of capacity and materials, and tighter control of the shop floor activities in greatly increased. In short, under CIM the PPC system plays a more central and important role in ensuring effectiveness and efficiency.

2. Manufacturing Environment and PPC Systems

Recent market-trends indicate that manufacturing firms are being required to excel in a variety of dimensions [6]. Low cost manufacturing, quicker product development, faster delivery, wider variety of products, wider range of efficient production volumes, and steadily increasing quality standards have all become important. Demand for capabilities that would have been impossible to meet under the more dichotomous strategies of the not too distant past have become the norm for competition in today's manufacturing environment [2, 25, 28].

The environmental conditions faced by the manufacturing function can be characterized by (1) product volume and variety, (2) competitive priorities, and (3) process technologies and infrastructure available within the firm. The volatility of demand, the level of product design changes, and the rate of new product introduction define the product volume and variety mix. In terms of competitive priorities, firms are faced with the need for holding the line on costs while meeting demand for more frequent and smaller lot deliveries of an increasing variety of products. The process technology available within the firm (e.g., Numerically Controlled Machines, Flexible

Manufacturing Systems, etc.) determines its flexibility and ability to support the competitive priorities.

Given the multidimensional strategic objectives of manufacturing firms and the increasingly complex environments within which they operate, the selection and implementation of a suitable PPC system is an important concern for manufacturing firms worldwide. Consequently, many managers are concerned with the fit between their infrastructure support system and manufacturing environment [22]. However, historically, the selection of PPC systems has been influenced more by the latest system developments, internal knowledge, and information processing constraints of the firm than by environmental factors faced by the firm. Some might argue that a PPC system does not in itself add value. A good fit between the manufacturing environment and PPC system, however, can facilitate better execution of activities that add value. Others might argue that the choice of PPC system is simple since the systems are mutually exclusive in terms of the environments they fit. However, manufacturing environmental factors such as differences in complexity, information processing burden, implementation discipline requirements, perspective (local Vs global) or focus of operation (top down Vs bottom up) make one system more attractive than another. Clearly, when demand is both stable and predictable there is very little need for a sophisticated system for production planning and control. It is when faced with a highly unstable or a highly unpredictable demand that choosing the right kind of PPC system becomes crucial for achieving both effectiveness and efficiency.

For example, one could probably make a strong case for selecting a system such as MRP to support a functionally laid out process that produces (in batches) a large variety of products, that are in different stages of their product life cycle. Given the associated volume fluctuations and the necessity for frequent design changes, MRP provides sufficient planning and replanning ability to accommodate the dynamic nature of such an environment. In general, MRP based push systems are likely to perform well in a complex environment and in the presence of high demand variability as long as the predictability of demand is high. Its performance may deteriorate rapidly as demand uncertainty is increased (i.e., predictability decreases).

On the other hand, products further along their product life cycle may have consistently higher volume with minimal design changes. For example, consider a firm producing a small variety of products using a product layout. Under these more stable conditions, a simple information system such as Kanban may well serve the process more effectively. Overall, pull systems, such as Kanban, may work well when the demand variability is low. The level of uncertainty in demand (i.e., not knowing when the next Kanban card will arrive) may not be as critical to system performance. Thus, pull systems may not be as sensitive to predictability of demand as, for example, push systems are likely to be. When the demand is highly unpredictable and demand variability is also high, systems such as OPT are likely to produce superior results.

Based on such considerations, for each process type we superimpose a PPC system that might represent the "best" for the conditions that define the firm's environment. Table 1, taken from Newman and Sridharan [17], presents a mapping of the four types of PPC systems commonly found in use in terms of the environments in which they are likely to perform well. However, the empirical validation of such a taxonomy may be cloudy. While a good fit between the PPC system and the firm's environment may be necessary, it alone may not be sufficient to assure success. For example, consider a firm producing a wide range of custom products in small batches using a make-to-order policy. While MRP may seem like a logical choice, performance may be clouded by the use of dedicated process technologies or overly centralized decision making. Alternatively, a similar firm producing the same product mix with general purpose technologies may still find performance below expectations if competitive priorities are based upon price and delivery speed.

Demand Predictability	Demand Variability Low High	
Low	Kanban	OPT
High	All	MRP

Table 1. PPC Systems Suitable for a Given Environment

Clearly, in a stable and predictable environment there is very little need for a complex and sophisticated planning and control system. In such cases a simple PPC system such as ROP may provide acceptable performance. However, when the environment is not stable the planning and control system should be able to cope with the volatility and thus become very important. Under such conditions, a simple system like ROP is often inadequate. When the production process is streamlined and simple, it is often flexible and, thus, is able to handle much of the uncertainty in demand. In such cases Kanban systems are the most effective. A complex production system facing an unstable environment needs a sophisticated planning and control system such as MRP or OPT. When demand is <u>not</u> stable, i.e., it is highly fluctuating, but largely predictable, MRP systems may be the appropriate choice. MRP systems are particularly well suited for firms facing an environment where the process is complex (i.e., product variety is large, production process is batch oriented, and demand is dynamic). When demand is <u>unpredictable</u> but is steady, and the production process is simple, Kanban systems appear to be the best choice. Finally, OPT based systems appear to be best equipped to handle the case of highly variable and unpredictable demand. Granted that external predictability of demand is, at least in some sense, a function of forecasting, yet a great deal of internal demand predictability is a function of internal stability. However, variability of demand is external to the firm and is often not fully controllable.

3. Performance of Existing PPC Systems

A recent survey of companies covering a wide spectrum of manufacturing industries ranging from machine tools, automobile components, furniture, plastics, and medical equipment to computers and defense electronics provides some indication regarding the effectiveness of the four traditional PPC systems [16, 17]. The results show that, of the four PPC systems described above, MRP system was found to be the most widely used system (56%). This is followed by ROP based systems (22%), Kanban (8%), and OPT based systems (5%). Roughly 9% of the firms reported using some "in-house" system (categorized as "other") created to meet their unique needs.

The results indicate that MRP based systems appear to be used in firms belonging to each of the three different process types (job shops, repetitive, and process) and across all size firms. ROP based systems were implemented mainly in smaller firms and were evenly divided between process and job shop situations. Most Kanban implementations were also found in smaller shops. OPT users, on the other hand, appear to be concentrated in larger process industries. Table 2 provides a summary of the survey results in terms of the distribution of firm performance within each PPC system user group.

	Percentage of respondents using			
	MRP	**Kanban**	**OPT**	**ROP**
Inventory Turns				
> 100	5.45	0.00	16.67	0.00
10 - 100	20.00	9.09	16.67	30.77
5 - 10	56.36	54.55	16.67	53.85
< 5	18.18	36.36	50.00	15.38
On-Time Delivery (%)				
> 95	27.27	27.27	50.00	50.00
90 - 95	16.36	18.18	0.00	19.23
80 - 90	30.91	54.55	33.33	15.38
< 80	25.45	0.00	16.67	15.38
Lead Time (Weeks)				
< 2	14.55	18.18	16.67	46.15
2 - 5	27.27	18.18	16.67	15.38
5 - 10	30.91	36.36	16.67	23.08
10 - 15	14.55	9.09	16.67	7.69
> 15	12.73	18.18	33.33	7.69
Lateness (Weeks)				
< 1	41.82	36.36	33.33	65.38
1 - 2	38.18	36.36	50.00	23.08
2 - 5	16.36	9.09	0.00	7.69
> 5	3.64	18.18	16.67	3.85
Utilization (%)				
> 95	3.64	18.18	33.33	19.23
90 - 95	16.36	18.18	16.67	15.38
80 - 90	25.45	27.27	33.33	23.08
< 80	54.55	36.36	16.67	42.31

Table 2. Performance Distributions

These results provide some valuable insights into the relative performance of alternative PPC systems. Based on the results, it appears that MRP systems are most versatile and are able to cope with increased complexity. The increased information processing requirements entailed by MRP systems, however, may prove to be a hindrance for firms that do not need a complex system (i.e. where product variety is low and demand is steady). Under such cases the use of ROP systems may be appropriate. ROP systems are cheaper to implement and are simpler compared to MRP systems. They, however, are inferior in handling situations where demand is unsteady and product variety is high. Kanban and OPT systems are used by so few firms and under restrictive conditions that it is difficult to interpret the results. The most significant finding, however, is that none of the existing system appears to be capable of providing consistently superior performance across all criteria important to a firm from a strategic/competitive point of view.

4. Traditional PPC Systems Described

Given the above results and the predominance of ROP and MRP -based systems in use, we restrict our discussion to these two systems only. Readers interested in learning more about Kanban systems may refer to any of a number of textbooks such as Schonberger [19]. See also Karmarkar [13] for an exposition concerning situations where Kanban-based systems are not likely to be effective. Likewise, for learning more about OPT-based systems please refer to Goldratt and Cox [5]. It is important to recognize that almost all of the existing systems deploy a hierarchical decision architecture. The differences between the systems stem mostly from implementation details rather than conceptual underpinnings. They essentially adapt a 'divide and conquer' approach to decompose the problem based on either the product structure or the process structure. See Villa [26] for an exposition and analysis of alternative design approaches, via a unifying mathematical formulation of the production plan optimization problem, to recognize the main features of the existing PPC systems. The formulation also serves to compare the usefulness of alternative PPC systems in different manufacturing processes.

4.1. Reorder Point Systems

A reorder point system (ROP) is also known as a replenishment system. They are mainly concerned with determining answers to two basic questions: when to order (order timing) and how much to order (order quantity). Typically, such systems answer these two questions by first determining an 'economic' quantity for ordering and then determining a reorder point based on (1) estimates of lead time required to replenish the item and the variability in demand during the lead time and (2) a target customer service level specified by the management. Two basic types of re-order point systems are commonly found in practice: Fixed Quantity System or Fixed Interval System. In a fixed quantity system the order quantity is predetermined and order timing is allowed to vary depending on actual demand. In a fixed interval system, orders are placed once every P units of time, where P is predetermined, and order quantity is allowed to vary based on some predetermined target inventory level. It is important to recognize that several hybrid systems based on these two basic systems have been proposed in the literature.

It is clear that re-order point systems are simple in nature and are most suited for situations experiencing independent demand. One major flaw of such systems concerns their lack of forward visibility. This severely curtails their effectiveness in situations where a part of the demand is "dependent" as in most manufacturing operations. Second, they assume a uniform and continuous demand. This is unrealistic in most manufacturing operations. Demand for components is often lumpy, time-phased, and not amenable to the traditional forecasting techniques. Third, placing replenishment orders solely based on inventory levels tend to generate work load that is highly erratic. Thus, often the factory is unable to handle them efficiently.

4.2. MRP-based Systems

Figure 1, taken from Kanet [8] illustrates what is meant here by the term MRP-based PPC systems. Central to the MRP-based approach is an MRP component inventory planning system, surrounded by other logistics modules such as master production scheduling, capacity planning, shop scheduling and control, and the like. Typically, the approach takes a set of forecasted customer orders and develops a master schedule of production. The master scheduling task is often aided by a "rough-cut" capacity planning module. The MRP "explosion logic" then orchestrates the release of production orders based on planned lead time and predetermined lot sizes. Planned order releases from the MRP inventory system are used to conduct "machine load" analysis for capacity planning. As orders are released to the production system, the factory scheduling module uses the MRP due date as a means for providing priority to orders as sequenced through the factory in competition for limited resources.

There are a number of fundamental flaws in the MRP-based approach to production planning and control. A central weakness is MRP's modus operandi of sequential, independent processing of information. The approach attempts to "divide and conquer" by first planning material at one level and then utilization of manpower and machines at another level. The result is production plans which are often found to be infeasible at a point too late in the process to afford the system the opportunity to recover. A second flaw concerns the use of planned lead times. Planned lead times are management parameters which are provided prior to the planning process and represent the amount of time budgeted for orders to flow through the factory. This can result in a tremendous amount of waste in terms of work-in-process inventory. Thirdly, MRP-based systems do not provide a well-designed formal feedback procedure instead depend on ad hoc, off-line, and manual procedures. When a problem occurs on the shop floor, or raw material is delayed, there is no well-defined methodology for the system to recover. Thus, the firm depends on and actively promotes safety buffers, leading to increased chances for missing strategic marketing opportunities. Fourth, MRP systems often produce schedules that are extremely nervous which, in turn, leads to reduced productivity and increased costs [23]. Firms resort to either freezing a portion of their master schedule for combating schedule nervousness or keeping safety stock of end products and, thus, incur the penalty of reduced customer service or increased costs [20, 21].

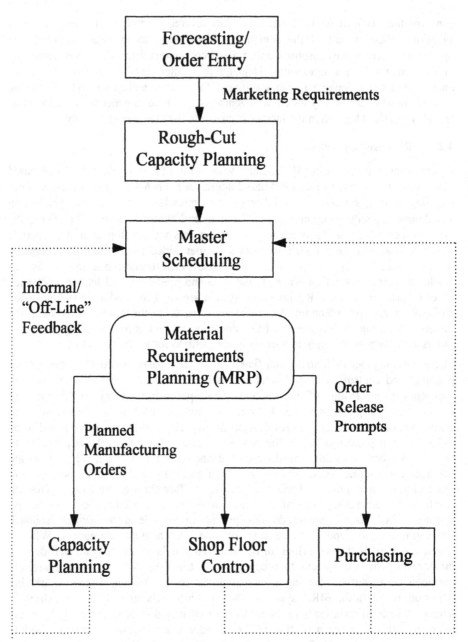

Fig. 1. MRP-Based Logistics System Architecture

Several empirical studies dealing with the practical issues surrounding efficient and effective implementation of PPC systems, in particular MRP-based systems, have appeared in the literature. See, for example, Monniot, J.P., D.J. Rhodes, D.R. Towhill, J.G. Waterlow, D.H.R. Price, and A.K. Kochhar [15]; Duchessi, Schaninger and Hobbs [4], and Kochhar and McGarrie [14]. Kochhar and McGarrie [14] report seven case studies and face-to-face meetings with senior managers and identify key characteristics for the selection and implementation of PPC systems. They conclude that (1) the operating environment significantly impacts the choice of the system and (2) the existing framework for an objective assessment of the need for individual control system functions is largely inadequate in serving the needs of managers. This result demonstrates the need for a modular design and a decentralized architecture for PPC systems, thus providing individual companies the maximum flexibility in tailoring the system to meet their needs within a common framework. Such an architecture and design, in our view, should automatically preserve the best features of in all variants of the system and, thus, be able to guarantee efficiency and effectiveness.

5. Recent Advances

In spite of 20 years of advancement in computer capabilities, there has been little change in the basic design of production and inventory planning systems. Since 1970, the de facto standard for such systems has been material requirements planning (MRP). As discussed earlier a major problem with MRP is its use of fixed "planned lead times". Given a customer's date of need, planned lead times are used for planning the arrival of materials and the release of orders to a production facility. In MRP systems design, planned lead times are fixed parameters of a given inventory item and use no information regarding the sequence in which orders for items are eventually processed through the facility. Because sequencing knowledge is not used, planned lead times must be made large enough to accommodate the worst case situation. The end result is unnecessarily large inventories at every stage of the production process. In effect, under the current MRP regimen, the ultimate timing of how orders pass through the shop is treated as a random variable and an order's lead time is a conservative point estimate of its eventual flow time through the shop. This has led to a concerted effort to develop an alternative to MRP and it is heartening to note that some significant progress has been made in this arena.

In particular, a number of advances in applying computer technology have been made in the area of detailed scheduling and short term control of open manufacturing orders. Very notable is the rapid development of "leitstands" (computer graphics-based scheduling support systems) in Germany. Adelsberger and Kanet [1] characterize the emergence of electronic leitstand as one of the most exciting new developments in computer-integrated manufacturing. Leitstands make extensive use of recent advances in computer graphics to support human decision makers in manufacturing settings and fit perfectly into the CIM concept by connecting the planning module with the shop floor.

A leitstand is a computer-aided decision support system for interactive production planning and control [1]. The word "Leitstand" is German for command center or directing stand, and it is in Europe, and Germany in particular, that the leitstand technology is most highly developed. [9]. As Figure 2 shows a leitstand is comprised of the following five components:

1. a *graphics component* for providing visual representation of schedules (i.e., Gantt charts) for production resources;

2. a *schedule editor* for assisting a human scheduler to manipulate and edit a schedule;

3. an *evaluation component* for helping to measure and analyze the performance characteristics of a given schedule;

4. an *automation component* for automatically generating schedules (or partial schedules) and for performing complex schedule edits;

5. a *data base manager* (a computer program) for managing the data base within a leitstand and for communication and interaction with other computer-based systems and data bases.

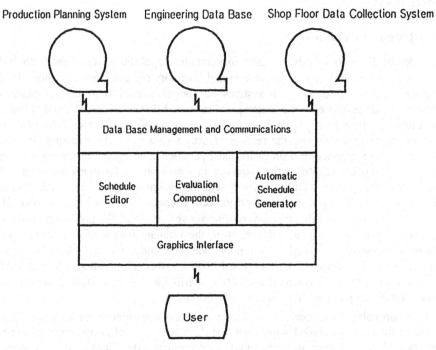

Fig. 2. Conceptual View of an Electronic Leitstand

It is clear from Figure 2 that a leitstand uses information from at least three external sources:

(a) a *production planning system*, where it receives information concerning customer demand (such as due dates, order quantities, order priorities, etc.);

(b) an *engineering data base*, where it receives information concerning process technology (such as process times, setup times, routing alternatives, etc.);

(c) a *shop floor data collection system* which provides information regarding the status of resources and orders within a production system (such as available machine capacity, current queue sizes, tooling status, etc.).

A leitstand helps users in getting to the details quickly by providing the capability to focus on a specific job and its schedule throughout the shop or an individual operation. It also provides the ability to quickly obtain information on machine utilization, machine loading pattern (i.e., input-output analysis), work run-out times, and job lead times. While making available the detailed information, it also provides the option to selectively access, view, present, or skip information concerning a portion of or the entire schedule, thus avoiding information overload. In a leitstand, the schedule editor is useful for manually constructing a production schedule or modifying an existing schedule. It is important to note that the schedule editor does not automatically alter the entire production schedule every time a schedule is edited. Factory scheduling typically involves many jobs with each job having a set of requirements. In performing the schedule editing functions, one need not be concerned with violating any hard technical constraints (e.g., machine-operation match, operation precedence relationship, etc.) because leitstand systems typically have a built-in constraint checking capability. Partial schedule generation and automatic constraint checking are features of a leitstand that make the editing function quick and accurate.

Choosing a good production schedule requires one to choose a performance measure. The leitstand concept provides the flexible capability of using any standard or custom-built criterion for evaluation. Summary statistics using each of the pre-defined criteria are computed and presented as demanded by the user. The evaluator may offer the capability of a hierarchical evaluation: operation, job, machine or resource, and shop level evaluation. This feature is important because there may be a variety of management objectives in production scheduling. A leitstand can also furnish a variety of information such as machine loading pattern for a given schedule, work-in-progress inventory diagrams, machine run-out times, and job and operation leased times.

Most importantly a leitstand can be intelligent. We visualize intelligence in production scheduling occurring in at least three ways. One is the capability of performing complex editing functions. A second way is through detailed constraint checking as described earlier. A third way that leitstands can provide intelligence is by their ability to conduct clever searches for scheduling alternatives. For example consider the JOBPLAN system developed at Siemens AG, Munich, Germany [9]. As shown in Figure 3(a), it first provides pop-up menus enabling the user to choose from among eight candidate schedule generation strategies. Given a schedule, the optimization routine is imbedded in JOBPLAN then helps the user search for better schedules. In doing this, it first leads the user into making a choice of the scheduling objective [see Figure 3(b)]. Such objectives may be to minimize total tardiness or flow time of the production schedule, or to maximize resource utilization for the given set of jobs. Other criteria can be built into the system: e.g., maximizing net present value of the production schedule. The system then conducts a search via a pair-wise exchange of neighbor operations strategy. Before the search begins, the user is prompted to specify the parameters of the search, such as total number of iterations [See Figure 3(c)]. Once specified, the underlying search algorithm quickly scans the solution space for an improved production schedule and presents the result for the user to examine (see Figure 4).

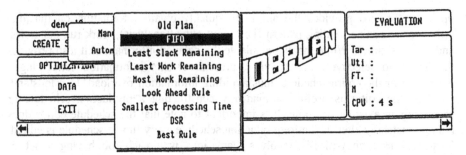

Fig. 3.a. Schedule Generation Strategies

Another attractive feature of leitstand systems is the ability to deploy advanced AI-based search methods. The types of approaches we refer to here include methods such as heuristic branch and bound, simulated annealing, beam search, tabu search, etc. Similar approaches that appear promising include application of genetic algorithms [10], application of basic decision theory [3, 12] and application of neural networks [7]. All these methods distinguish themselves from simple linear forward simulations in that they may include a (limited) capability for backtracking and/or the feature of dynamically changing the search path. Some of these approaches have the property that the more time they are given, the better the solution they produce. It is important to note here that such search methods can be made to run in the background so that the system is not tied up. This way, while the system is searching for a better solution the user can engage in other urgent/important work.

Current wisdom views the leitstand as a bridge between a firm's production and inventory planning system and its shop floor control system. A leitstand receives production orders from the PPC system (typically a manufacturing resource planning (MRP) system) and develops a feasible short term schedule for completing the orders as required. The leitstand uses information from the shop floor concerning the status of machines and open orders to determine the starting conditions for schedule generation. In attempting to develop a feasible schedule, a leitstand can uncover capacity problems and pinpoint exactly when and where future production bottlenecks will occur. In a limited sense, one can think of a leitstand as a tool for performing finite capacity planning, order scheduling, and production control -- all in one. We say these functions are limited because current thought on the role of leitstands limits the universe of orders which it must manage to those already released by, say, an MRP system. Leitstands can augment current MRP-based technology by providing a tightly-integrated method for short term capacity planning, production scheduling, and production control.

Perhaps the greatest virtue of the leitstand approach is that it facilitates rescheduling. In a typical manufacturing environment, any schedule, once determined, is almost immediately subject to new conditions, demands, and constraints. A leitstand starts with the schedule that was once valid (perhaps days, perhaps months ago) and helps a human planner process transactions (e.g., changes in machine availability, delivery of materials, changes in demand, etc.) to arrive at a new schedule.

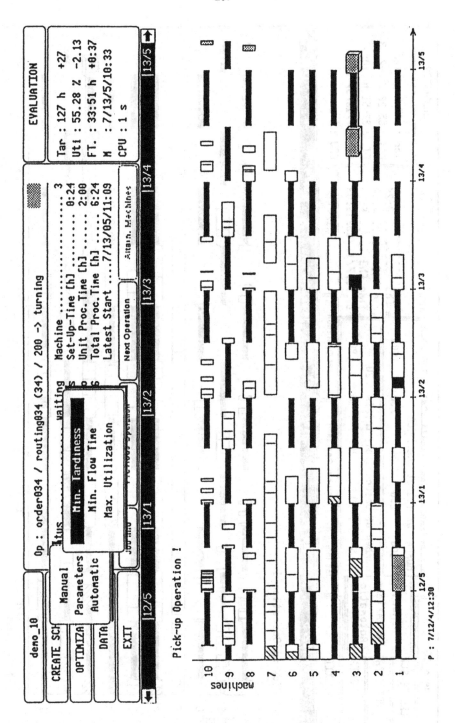

Fig. 3.b. Scheduling Objective

208

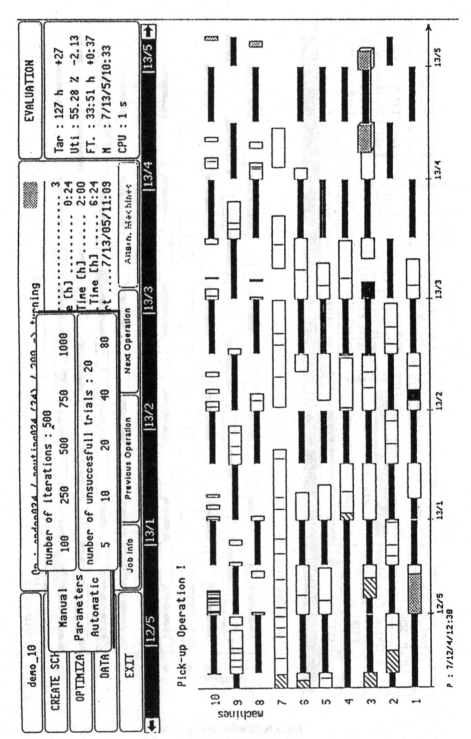

Fig. 3.c. Search Patters

209

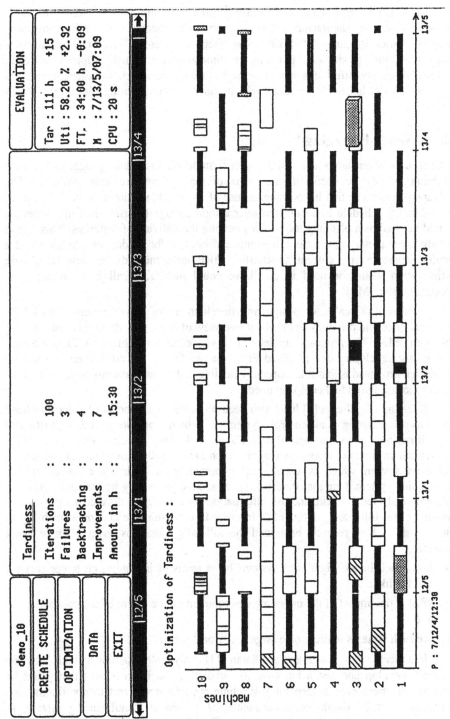

Fig. 4. Best Schedule Found

There is no doubt about the increasing demand for production planning methods which enhance a manufacturing unit's ability to reschedule. The entire movement toward "Quick Response," "Flexible Manufacturing Systems," etc., clearly indicates the need for production planning methods/systems which support efficient rescheduling. By offering a simple straight-forward approach to rescheduling, the leitstand model helps fill a critical void in current manufacturing planning methods [9].

6. A Newly Emerging Paradigm

The success of leitstands as a tool for short-term detailed scheduling suggests the same technology may be useful in the arena of material requirements planning. The advantage would be that in so doing the use of "planned lead times" would disappear. If a detailed schedule is maintained over a long enough horizon, then that schedule could be used as a planning tool -- for planning the delivery of materials. This would imply a completely new type of systems architecture for production planning and a major departure from current "stochastic" MRP-based methods. Because scheduling information would be used in detail we could justifiably call such an approach "deterministic" MRP.

Rolstads [18] calls for design and development of new computer based PPC systems since and provides an excellent overview of basic data structures and building blocks needed. The proposed model can be viewed as an architecture for a modular design for developing decentralized PPC systems. One of the advantages of the kind of design proposed is that it is extremely flexible and allows easy customization of the system to suit individual company needs.

Kanet and Sridharan [9] have also been promoting the concept of a decentralized production planning and control system in which computers and sophisticated algorithms play a key and central role. For example, they visualize a design in which individual workcenter managers (agents) each have a fully functional and integrated leitstand system which is capable of communicating with other agents in the production system. One major advantage of such systems may be that managers can electronically seek, obtain, and use information concerning the detailed shop plans for ensuring feasibility and 'optimality'. In order to determine the economic viability of such systems they recently examined the following questions using a single agent model:

a) how much better might such systems be in terms of inventory, customer service, and the like;

b) what is the impact of the operating environment on the extent of the improvement? and

c) are such systems feasible on a large real-world scale?

In a recent study, Kanet and Sridharan [11] analytically examine a single machine system; develop and test a heuristic for predicting the improvement that could be realized in more complicated multi-machine systems under certain conditions; and conduct a set of simulation experiments, of a general manufacturing system, to compare its operation under two fundamentally different policies:

Policy 1. Material deliveries planned using planned lead times as in MRP;

Policy 2. Material deliveries planned from a detailed schedule of operation.

For a single machine system operating under a policy of "forbidden early shipment," (where orders completed early are shipped only when they are due) and deploying the first-come-first-served priority rule and where every job has the same constant allowance A, the expected benefit ($\Delta\mathfrak{I}*$), in terms of flow time reduction, of using is

shown to be equal to $\Delta\mathfrak{I}* = \dfrac{\rho}{-\log f_T + f_T}$ where ρ is the expected utilization and f_T

is the fraction of orders expected to be tardy. From Little's Law it is clear that this reduction in flow time leads to a proportional savings in inventory related costs. Thus, the expected savings in raw material inventory, for various values of ρ and f_T, is as shown in Table 3.

Utilization	Fraction Tardy f_T		
ρ	.05	.10	.15
.75	25	31	37
.80	26	33	39
.85	28	35	42
.90	30	37	44

(The numbers shown in the table are in percentage)

Table 3. Expected Benefit of Using Scheduling Information: Single Machine Case

Extending their analysis to a multi-machine system where individual jobs are independent (i.e., no dependency between items as in an assembly) and when each job has a known probability p_i that it will require i operations they derive the heuristic expression to estimate the approximate expected benefit, in terms of reduced flow time:

$$Expected\ Benefit = \sum_i \frac{p_i}{i}\Delta\mathfrak{I}*,$$ where $\Delta\mathfrak{I}*$ is the expected benefit for the

corresponding single machine problem. The above expression assumes that (1) the jobs are independent (no assembly is involved) and (2) the order allowance is constant. Its validity was tested via a set of simulation experiments using an eight machine job shop. In the simulation experiments the average number of operations per job (n) was varied at four levels and the target shop utilization was varied at three levels. The four levels considered for the number of operations were: (1) every job has exactly one operation (n=1), (2) a job has one to five operations with equal probability (n=3), (3) a job has one to nine operations with equal probability (n=5), and (4) every job has exactly eight operations (n=8). The three levels of target shop utilization examined were 75%, 85%, and 95%.

The simulation model was designed to mimic a general "job shop." Jobs arrived according to a Poisson process with a specified mean interarrival time based on the target shop utilization level. Operation processing times were generated from a negative exponential distribution with a mean of 100. Job allowances (planned lead times) were identical for each job and set equal to a multiple (k) of the jobs total expected processing time.

For the system operating under policy 1, job system flow time was computed as max(d,C) - r and for the system operating under policy 2, job system flow time was computed as max(d,C) - s, where d=job due date, C=job completion date, r=job arrival date, and s=scheduled start time of the job's gateway operation. For each run the average flow times under both policies were computed based on the 5000 jobs beginning at the 301th job and ending at the 5300th job to avoid transient effects. The results of the simulation experiments are presented in Figure 5 in terms of the reduction in job flow time when policy 2 is used compared to policy 1.

Reduction in Flowtime

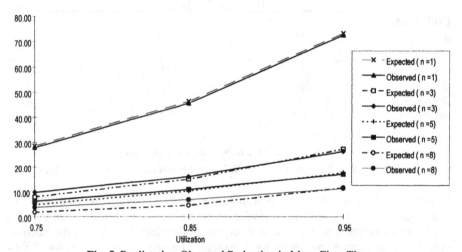

Fig. 5. Predicted vs Observed Reduction in Mean Flow Time

It is clear from these results that the actual reduction in flow time (improvement) is virtually identical to the value predicted by the heuristic for all cases considered. In addition to serving to validate the heuristic, these results demonstrate that it is possible to develop reasonable estimates of the value of using scheduling information for material planning.

Note that Policy 2 mentioned above only reduces the raw material inventory and has no effect on the work-in-process or finished goods inventory. Once we have the knowledge of a complete schedule, additional steps for reducing flow time further are possible. Such steps, however, are possible only under planning systems that implement Policy 2. Kanet and Sridharan [11] go on to illustrate one possible procedure, called 'SHIFTER', that uses scheduling information for reducing total system flow time and inventory. The 'SHIFTER' procedure is described below.

Given any general job shop and a complete schedule S0,

Step 1:   Let INTVL be the latest unmarked interval of machine idle time. Let I designate the operation just prior to the start of INTVL.

Step 2:   If INTVL = NULL then stop.

Step 3:   If operation I is early, shift as much of INTVL as possible in front of I preserving schedule feasibility so as to create schedule S1.

Step 4:   Mark INTVL and go to Step 1.

Note that in Step 3 shifting any idle time after the completion of an operation to before its start will only be possible if the operation is early. An operation i is considered early if it is completed before the start of its successor(s), if any, or in the event operation i is the last operation of a job, then it is early if it is completed before the due date.

The result of applying SHIFTER to schedule S0 is a schedule S1 with the same job permutation on each machine as in S0, but with a different allocation of idle time. The benefit of applying SHIFTER is that some of the idle time can be shifted into the position before the start of the first operation of jobs allowing further delay in raw material delivery dates.

SHIFTER considers each INTVL only once and moves it as long as it is beneficial to move. Suppose there are N total operations in S0. Then assuming preemption is not allowed, in the worst case there would be N INTVLs of idle time in the schedule. Therefore the algorithm is O(N). SHIFTER is a simple greedy algorithm, i.e., it is minimally intelligent. But it is only possible (and sensible) to apply SHIFTER after a detailed schedule is generated. So it represents one of the benefits of systems that are made possible by Policy 2. Note that application of SHIFTER does not change the total percent tardy and average tardiness values. The reason is that SHIFTER only reallocates the machine idle time and alters operation completion times only if the operation is early.

A simulation study of a general job shop was conducted to evaluate three systems: those operating under policies 1 and 2, and one operating under Policy 2 with SHIFTER (Policy 3). The simulation study examined the three policies under a wide range of operating conditions characterized by (1) three levels of utilization(75%, 85%, and 95%); (2) two operation sequencing rules (FCFS and OPNDD), (3) two order allowance methods (TWK and NOPS), and (4) four average number of operations per job (1, 3, 5, and 8).

Under each combination of sequencing rule, job allowance method, shop utilization level, and average number of operations per job the average flow time was computed based on the 5000 job sample for each of the ten replications. Thus, defining average flow time values for each combination as follows:

AFT_1: The average flow time under Policy 1;

AFT_2: The average flow time under Policy 2;

AFT_3: The average flow time under Policy 3;

The improvement in flow time (IMPRVMNT) when Policy 3 is used compared to Policy 1 was computed as $(AFT_1 - AFT_3)/AFT_1$.

Detailed statistical analysis of the results indicated that shop utilization has a dominant effect on IMPRVMNT (the reduction in flow time) for all values of average number of operations per job. Both order allowance method and sequencing rule have a significant impact on IMPRVMNT. However, their influence decreased as the average number of operations per job is increased. The results also showed that under all conditions examined the policy of using scheduling information for coordinating material delivery dates produces consistently lower average flow time values. For each combination of factors tested the actual benefit was significantly greater than zero. For the case where the number of operations per job ranged from 1 to 9 (average number of operations per job = 5), utilization ratio was 0.85, the job sequencing was done using OPNDD, and job allowances were set using the NOPS method, the 95% confidence interval for the expected benefit was found to be in the range 7.89% to 9.08%, indicating that systems which use the scheduling information for planning material requirements are likely to produce a significant amount of savings in inventory related costs.

These results are important not only because they show to what extent MRP-based systems can be improved, but also because they demonstrate how such improvement could be readily accomplished. What this preliminary study has in fact shown here is a simple "patch" to the inherent problems in MRP systems that arise from the use of fixed planned lead times. To expand on this idea, consider altering MRP systems in the following way: Change all planned lead times to some number greater than the length of the planning horizon. This will ensure that no shop order is ever un-released to the shop scheduling system. Let the shop scheduling system then establish a shop schedule and use this schedule to derive a schedule of planned order releases for all raw materials. The difference in this approach and the MRP-based approach is that MRP uses planned lead times to estimate the offset needed for manufacturing operations instead of deriving a schedule to see what offset is really needed for each manufacturing order.

Not only is the above patch straight forward, but the computational experience leads to the conclusion that such a procedure would be quite possible in even large scale practical settings. For example, Kanet and Sridharan [11] were able to completely rebuild entire schedules for their model factory in less than 5 minutes using relatively modest computational facilities (an 80386 PC running under DOS 6.0 and Windows 3.1 operating systems.) For example, at the extreme, this required the rescheduling of 5000 manufacturing orders with eight operations per order so that approximately 40,000 operations were scheduled. Moreover, the procedure we describe is polynomial in the number of operations to be scheduled.

7. Conclusion

Current thought on the role of so-called "finite" scheduling systems limits the universe of orders which it must manage to those already released by an MRP system. The next step would be to expand the scope of these systems so that they can supplant current MRP-based technology by providing a tightly integrated method combining order raw material planning with production scheduling and control. The above results represent a first verification that such integrated planning systems are beneficial and feasible on a practical scale. From here a number of questions remain open. First, in many real life factories the jobs are dependent (e.g., components and assemblies). So it would be important to study the behavior of the above approach in more complicated environments such as when jobs are dependent. A second question is what would be the impact of a rolling horizon with changing demands and new shop conditions. The concern here is whether or not such systems would introduce excessive "nervousness" in the raw material plans and thus prove to be counterproductive. Finally, testing the above approach in other types of shops (perhaps with different routing patterns and processing time distributions) are the next important questions to be addressed in this line of research.

References

1. Adelsberger, H. H. and J. J. Kanet, "The Leitstand -- A New Tool for Computer-integrated Manufacturing," *Production and Inventory Management*, Vol. 32, No. 1, 43-48, (1991).

2. Bower, J.L., T.M. Hout, "Fast Cycle Capability for Competitive Power", *Harvard Business Review*, Nov.-Dec., 1988.

3. Chryssolouris, G., Wright, K., Pierce, J., & Cobb, W., "Manufacturing systems operations: Dispatch rules versus intelligent control." *Robotics and Computer-Integrated Manufacturing*, Vol 4, 531-544, (1988).

4. Duchessi, P., C. M. Schaninger, and D.R. Hobbs, "Implementing a Manufacturing Planning and Control Information System," *California Management Review*, Vol. 31, No. 3, (1988).

5. Goldratt, E. and J. Cox, *The Goal*, North River Press, New York, NY, 1986.

6. Hill, T., *Manufacturing Strategy*, Irwin, Homewood, IL, 1988.

7. Johnson, M.D. and H. -M. Adorf, "Scheduling with Neural Networks -- The Case of the Hubble Telescope," Computers and Operations Research, Vol. 3, No. 4, 209-240, (1992).

8. Kanet, J.J., "MRP-96: Time to rethink Manufacturing Logistics", *Production and Inventory Management*, Vol. 29, No. 2, (1988).

9. Kanet, J.J. and V. Sridharan, "The Electronic Leitstand: A New Tool for Shop Scheduling," *Manufacturing Review*, Vol. 3, No. 3, 161- 170, (1990).

10. Kanet, J.J. and V. Sridharan, "PROGENITOR: A Genetic Algorithm for Production Scheduling," *Wirtschaftsinformatik*, Vol. 33, No. 4, 332-336, (1991).

11. Kanet, J.J. and V. Sridharan, "The Value of Using Scheduling Information in Planning Material Requirements," Working Paper, Clemson University, (1994).

12. Kanet, J.J. and Z. Zhou, "A Decision Theory Approach to Priority Dispatching for Job Shop Scheduling," *Production and Operations Management*, Vol. 2, No. 1, (1993).

13. Karmarkar, U., "Getting Control of Just In Time", *Harvard Business Review*, Sept.-Oct, 1989.

14. Kochhar, A. and McGarrie, B., "Identification of the Requirements of Manufacturing Control Systems: A Key Characteristics Approach," *Integrated Manufacturing Systems*, Vol. 3, No. 4, (1992).

15. Monniot, J.P., D.J. Rhodes, D.R. Towhill, J.G. Waterlow, D.H.R. Price, and A.K. Kochhar, "A Study of Computer Aided Production Management in UK batch Manufacturing," *International Journal of Operations & Production Management*, Vol. 7, No. 2, (1987).

16. Newman, W., Sridharan, V., "Manufacturing Planning and Control: Is There One Definitive Answer?", *Production and Inventory Management*, Vol. 33, No. 1, (1992).

17. Newman, W.E. and V. Sridharan, "Linking Manufacturing Planning and Control to the Manufacturing Environment," *Integrated Manufacturing Systems*, Vol. 6, No. 4, (1995). Forthcoming.

18. Rolstads, A., "Structuring Production Planning Systems for Computer Applications," *Annals of Operations Research*, Vol. 17, 33-50, (1989).

19. Schonberger, R.C., *World Class Manufacturing*, Free Press, New York, 1986.

20. Sridharan, V. and R.L. LaForge, "The Impact of Safety Stock on Schedule Instability, Cost, and Service," *Journal of Operations Management*, Vol. 8, No. 4, (1990).

21. Sridharan, V. and R.L. LaForge, "An Analysis of Alternative Policies to Achieve Schedule Stability," *Journal of Manufacturing and Operations Management*, Vol. 3, No. 1, (1990).

22. Stamm, C.L., Golhar, D.Y., and Smith, W.P. "Inventory Control Practices in Manufacturing Firms", *Mid-American Journal of Business*, Vol.4, No. 1, (1988).

23. Steele, D.C., "The Nervous MRP System: How to do Battle?" *Production and Inventory Management*, Vol. 16, No. 4, (1976).

24. Swamidass, P.M., *Manufacturing Flexibility*, OMA Monograph, #2, 1988.

25. Swamidass, P. M. and W. T. Newell, "Manufacturing Strategy, Uncertainty and Performance", *Management Science*, Vol. 33, No. 4, (1987)

26. Villa, A., "Decision Architectures for Production Planning in Multi-Stage Multi-Product Manufacturing Systems," *Annals of Operations Research*, Vol. 17, 51-68, (1989).

27. Vollmann, T.E., Berry, W.L., and Whybark, D.C., *Manufacturing Planning and Control*, Irwin, Homewood, IL, 1988.

28. Voss, C.A., *Managing Advanced Manufacturing Technology*, Springer Verlag, New York, 1986.

Simulation — A Tool for Developing Advanced Production Strategies

Heimo H. Adelsberger and Christian Hohendorf

University Essen, D-45117 Essen, Germany

Abstract. This paper is an introduction to simulation, especially covering the application of this methodology to production. At first, general aspects of simulation are presented, giving a survey of basic concepts and terminology. This is followed by a description of the main steps performed in a simulation study. Thereafter, a survey of currently available software is given. Finally, the application of simulation in different areas of production and manufacturing logistics is presented.

1 Introduction to Simulation

Today, simulation is considered to be one of the most important tools for the analysis of systems and processes, especially in the context of decision support. Simulation as a problem solving tool is used in any situation where analytical methods cannot be applied. Reasons for this are:

- The problem is too complex, or
- experiments cannot be performed, since using the real system is impossible or impractical, or
- these types of experiments are too expensive.

Whenever it is not necessary or feasible to study the system itself, the issue is to construct a model representing the relevant aspects of the system. By restricting the problem to the relevant aspects of the original system, a reduction of complexity is achieved. In many cases — without the complexity being reduced — it would not be possible at all to study the system. Simulation allows to study the model instead of the real or proposed system. Providing that the relevant aspects of the system are mirrored in the model, results gained from the study of the model can be applied to the system. It is always reasonable to use a model if the model is easier to manipulate (compared to the real system), or if the model is less complex and therefore the structure and the functions can be perceived more easily, or if the original system cannot be accessed without difficulty.

Simulation has its roots in statistics (sampling theory) and in system theory. Complex systems where many events are influenced by random processes are difficult to understand and to analyze.

1.1 Definitions

In the following we will give several definitions of simulation. The simplest definition of simulation which also covers the broadest meaning, is:

Simulation means making experiments on a model.

Shannon [19, p. 2] gives a more exact definition: *Simulation is the process of designing a model of a real system and conducting experiments with this model for the purpose either of understanding the behavior of the system or of evaluating various strategies (within the limits imposed by a criterion or set of criteria) for the operation of the system.*

Naylor [12, p. 2] defines simulation in a more mathematical or technical orientated way: *... simulation as a numerical technique for conducting experiments with certain types of mathematical models which describe the behavior of a complex system on a digital computer over extended periods of time.*

Two relevant aspects are to be considered: In the first place, the experimental use of a model, i.e., the simulation run, is understood by simulation, secondly the process of defining a model.

2 Basic Concepts

2.1 The Real System and the Simulation Model

Several scientific disciplines have dealt with definitions of systems and models, like system theory, mathematics, philosophy, psychology, and even linguistics, thereby focusing on different aspects of modeling. Bulgren [5] defines a system *... as an aggregation or assemblage of objects joined in some regular interaction or interdependence.*

Kreuzer [11] goes further, stressing the fact that modeling is a purpose-driven activity: *A system is defined as a collection of objects, their relationships and a behavior relevant to a set of purposes, characterizing some relevant part of reality.*

Zeigler's Approach. The basic elements and relations of the modeling and simulation enterprise are[20]:

Real System: Behavior. The Real System is a part of the real world which is of interest. The system may be natural or artificial, presently in existence or planned for the future. The real system refers to nothing but a source of observable data.

Model. The model is a set of instructions for generating behavioral data, expressed, e.g., by differential equations or some other formalism.

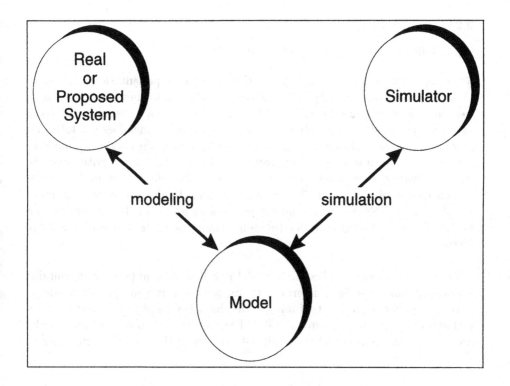

Fig. 1. Modeling and simulation

The Experimental Frame: Validity. The experimental frame characterizes a limited set of circumstances under which the real system is to be observed or experimented. The validity of a model is relative to an experimental frame and the criterion by which agreement of input-output pairs is gauged.

The Base Model: Hypothetical Complete Explanation. The base model is a model capable of accounting for all the input-output behavior of the real system. In any realistic situation, the description of the base model can never be fully known. It may be expected to comprise a huge number of components and interactions. The complexity is likely to be extremely great.

The Lumped Model: Simplification. Having specified the experimental frame of current interest, a modeler is likely to construct a relative simple model that will be valid within that frame. Such a model is called lumped model, because it is often constructed by lumping together components and simplifying interactions.

The Computer: Complexity. The computer is the computational device helping to generate the input-output pairs of the lumped model. Sometimes it is possible to work it out analytically using brain, paper, and pencil. Usually it must be done in a step by step fashion, from one simulated time instant to the next.

2.2 Terminology

In the following section, major terms are explained.

Static Versus Dynamic. A static simulation model is a representation of a system at a particular time or independent of time. Monte Carlo simulation models as well as scale models are typical for this type; static models can only show the values of system attributes in balance systems. A static model does not take into account the passing of time; e.g., a model, in which a certain weight is applied to a bridge at a given instant to determine whether the bridge can withstand the stress. Dynamic models consider the changes in the course of time that result from activities of the system. Therefore, a dynamic model is a representation of a system as it evolves; e.g., a model in which a variety of loads are applied to a bridge in a period of time to determine how long the bridge can withstand the stress.

Deterministic versus Stochastic. A model is called deterministic, if it contains no random variables. In a deterministic model, there is a unique set of model output data for a given set of inputs. On the other hand, a stochastic model includes one or more random variables. The output data of a stochastic model are random themselves and thus only estimations of the true characteristics in the real system.

Endogenous versus Exogenous. Activities occurring within the system are called endogenous; activities in the environment (outside the system) that affect the system are called exogenous.

Discrete-event versus Continuous. Continuous simulation describes systems by sets of equations to be solved numerically (e.g., systems of high order, non linear, non homogeneous differential equations). State variables may change continuously in the course of time. Typical application areas are systems which are based on state variables like electricity, liquid, or gas. Another example are traffic systems not based on single traffic units but on traffic flow.

Discrete-event simulation describes systems in terms of logical relations that cause changes of state at discrete points in time rather than continuously. Examples of problems in this area are most queuing situations: objects (customers in a bank, orders in a job shop, etc.) arrive and change the state of the system instantaneously. State variables can be the number of customers waiting to be processed, the number of machines busy, idle, or broken down, etc. State variables are changing only at a countable number of points in time.

Discrete Event Simulation World Views. Traditionally, discrete event simulation models are further distinguished by different „world views[1]":

– event approach

---

[1] The following summary follows closely the survey found in [10]

– activity scanning

– process interaction

Event Orientation. A system is modeled by defining the changes that occur at given points in time. The task of the modeler is to determine the events that can change the state of the system and then to develop the logic associated with each type of event. A simulation of the system is produced by executing the logic associated with each event in a time-ordered sequence.

Event routines (subprograms) encapsulate transformations to be applied at a specified instance in time, triggered under the supervision of some scheduler. For a long time, this framework has been the most widely used approach in discrete-event simulation. It is very easy to implement.

Activity Scanning Approach. Activities encapsulate transformations to be applied to model states at the start or end of some time consuming, state changing activity. This framework tries to capture the notion of causal connection of events. A model executor scans all activities and checks the conditions for starting or ending an activity.

The modeler describes the activities in which the entities in the system engage and prescribes the conditions which cause an activity to start or end. The events which start or end the activity are not scheduled by the modeler but are initiated by conditions specified for the activity. As simulated time progresses, the conditions for either starting or ending an activity are scanned. If the prescribed conditions are satisfied, then the appropriate action for the activity is taken. It is necessary to scan the entire set of activities at each advance of time.

Process Interaction Approach. Process-orientated schemes encapsulate sequences of events under a common name into textually closed descriptions. The principal of preservation of causal connections is better supported than in event-orientated models. Patterns of change, which in the mind of the analyst belong together, can also be textually grouped into one module. This permits the analyst to constrain complexity. The logic of the model becomes much easier to comprehend.

This approach is especially suited for modeling complex, highly interactive systems.

Transaction Flow or Network Approach. This is a specific form of the process interaction approach. Models of this type are driven by transient workload items, often referred to as transactions. Transactions are generated stochastically by some generator procedure and drive the model by flowing through their life cycle scenarios, waiting for resources and encountering service delays. Each flow of an transaction is an instance of a process, drawing one sample from the state space of the model. The static structure of the model is described by passive objects which include facilities and storage units.

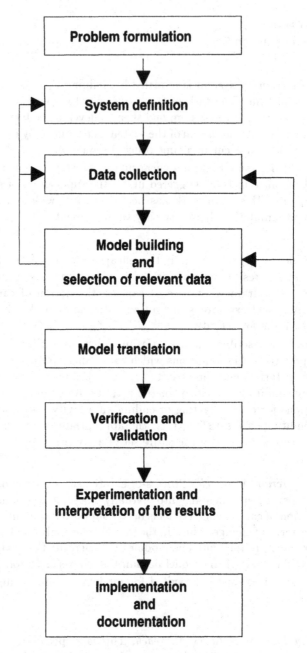

Fig. 2. Main steps of a simulation study

3   Steps Performed in a Simulation Study

The main steps to be performed in a simulation study are given in Fig. 2.

In the following, the single steps are explained in detail (after Pegden [14]):

Problem Formulation. The first task to be performed is an exact definition of the problem; this means that the goals and the reasons for performing the simulation study have to be formulated. The better this goal is understood and the more exact it is defined, the better the results will be.

System Definition. According to system theory, a very important and crucial task is to find the boundaries of the system to be studied, that means to stipulate which aspects of the system are relevant and which are not. This is most important for the definition of the state variables and all restrictions concerning these variables.

Data Collection. To perform a simulation study, the data for the system have to be collected. Depending on the system, these data can either be taken from an already existing or from a similar system. In case the real system is not yet in existence (e.g., simulation is used for designing a new factory), estimated data have to be used. The more exact the data are, the more reliable the simulation results will be.

Model Building and Selection of the Relevant Data. In accordance with the goal of the simulation study, the simulation model is formulated. The relevant objects are defined and relationships and behavior patterns are established to form the basic structure. This can be done in a rather simple representational form, e.g., by using flow charts or pseudo code. The relevant data will be selected and provided in a form needed for the model.

Model Translation. Depending on the selected simulation tool (simulation language or a simulator), the model will be formulated in form of a computer program or built interactively by using a simulator.

Verification and Validation. During the verification process, the simulation program is checked for correctness. This means, errors in the simulation code (or in the simulator) will be detected and corrected. After that, the validity of the model will be tested. This means that the correspondence between the model and the real system is checked.

Experimentation and Interpretation of the Results. Input data are collected and experiments are defined. The simulation runs are performed and the results are analyzed.

Implementation and Documentation. The simulation program is used in routine form and the results of the simulation are applied to daily operation. Parallel to all these steps, the model itself has to be documented as well as its usage.

4 Classification of Simulation Software

4.1 Simulation Languages Compared to General Programming Languages

Simulation models can be programmed by using conventional programming languages like Fortran, C, or Pascal. Nonetheless, in many cases it is more appropriate to use a general purpose simulation language. The essential advantage is that the formulation of the model is easier, since many building blocks and functions typically used in simulation are provided by simulation programming languages. This results in less effort for creating and changing simulation programs in comparison with ordinary programming languages. Good examples are statistical functions or integration algorithms: these — crucial for simulation models — are provided in simulation languages but not in general purpose programming languages. By using a simulation language, the user is released from many tasks which in case of an ordinary programming language he has to take care of.

On the other hand, the following aspects have not to be neglected: The specialization of simulation languages for specific application areas (e.g., SIMAN for manufacturing) can constitute a limitation in comparison with conventional programming languages. The limits when using a simulation language even in its own application area could rather fast be reached if unusual situations have to be modeled since simulation languages provide only a relatively limited number of elements even if these elements provide rich functionality. In addition, a user of a simulation language has to learn how to use the simulation language and has to learn and (!) understand the concepts on which the simulation language is based.

The following table provides a classification of simulation software according to Feldmann and Noche [7, 13]. The classification also provides a more or less historical view of simulation software.

Table 1. Classification of simulation software

Level	Basis	Example
0	implementation-language	Fortran, Pascal, C
1	basic components	Simula, SmallTalk
2	components for specific modeling task	GPSS, Siman, Slam
3	components specific for the application area	Simple, Dosimis
4	components for specific sub-parts for an application area	Fad, Lasim

Level 0: The implementation language provides no aspects specific to simulation. Every code relevant for simulation has to be provided by the user or in form of libraries. One of the early library-system was GASP [16]

Level 1: The language provides general programming concepts like classes, co-routines, list-processing — concepts which are crucial for simulation programming, these programming concepts, however, are used in programs outside the simulation area as well. A good example is Simula[6]: Simula started as an extension of Algol to provide important concepts useful for simulation programming (e.g., classes, co-routine, processes). The general importance of these concepts was recognized immediately so that Simula is now considered a programming language for general purposes and not a simulation language. Even C++ was originally developed for simulation purposes! Another language which provides many concepts important for simulation is Smalltalk. This is certainly not a simulation language, but simulation programming in Smalltalk is straight forward and many examples in the basic textbook [8] are taken from simulation.

Programming languages with built-in simulation features (like e.g., Simula and Simscript) seem to be ideal: the language has the same power as a general programming language and additionally provides all the features necessary for simulation. Since simulation programming languages are used by a much smaller community than ordinary programming languages, developments in simulation programming languages are slower compared to a general purpose programming language (e.g., quality of compilers, reception of modern language design concepts, etc.). This results in these languages not being able to keep up with general programming languages.

Level 2: These are the so-called general purpose simulation languages, since they can be used in many different application areas. Even if support for specific „world views" is provided — like for continuous simulation, for discrete event, activity scanning, or process orientation — they are so general that they can be used for many different application purposes. Nonetheless there are differences, e.g., Siman as a general purpose simulation language in addition provides specific constructs for the modeling of manufacturing systems.

Level 3: These systems are called simulators. They are designed for specific application areas like production and logistics. They are module-orientated and even though they are designed for specific application areas, they provide a relatively wide application spectrum.

Level 4: These are very specific simulators, like e.g., for modeling AGV[2] systems, of FMS[3], or warehouse systems.

4.2 Simulation Languages Compared to Simulators

In the beginning, simulation models were written in general purpose programming languages. As early as the early sixties, starting with GPSS, general purpose simulation languages have been use. In the early eighties, a new type of

---

[2] AGV: Automated Guided Vehicle

[3] FMS: Flexible Manufacturing Systems

software tool appeared, called simulator. The simulation model is not described in language form, but in form of a question-answering process or by graphically selecting pre-built model components and combining these components into a model thus providing the parameter values for the components. Simulators are specialized for specific application areas.

A typical simulator provides a restricted set of components which the user is able to select from with the model being created by these components. In our days, most of these systems provide a graphical user-interface where the components are displayed in form of graphical icons. The user picks up such an icon from a list of icons and places these icons in a graphical lay-out of his model. The components are parameterized, so that after the placement of such a component, the user fills out a form, providing the required parameters of the component. Most of the simulators currently on the market serve in the area of production and logistics with the typical components as follows:

- static elements like inventories, buffers, or working stations;
- dynamic elements like tools, transporters, or jobs;
- control-elements like shift-plans, strategies (like routing-plans), or break-downs.

The advantages and disadvantages of simulators compared to general purpose simulation languages are:

Advantages of simulators:

- A simulator can be used without or with little programming knowledge; sometimes even without deep understanding of the application area.
- The usage of the simulator can be learned in a relatively short time as long as the „world view" of the simulator is in accordance with the application area.
- The problem can be expressed in a straight forward form.
- The problem has not to be transformed into an abstract simulation language.
- In many cases, the time spent for the creation of the model is shorter compared to using a general purpose simulation language, especially as far as animation is concerned (see Banks [3]).
- Many simulators provide standard interfaces to other information systems.

Disadvantages:

- Simulators are slow concerning run-time behavior.
- A user of a simulator can only use standard components of the simulators. This is sufficient if standard problems have to be solved. However, if a problem does not fit exactly into the application area or if some aspects of the problem are not covered by standard elements of the simulator, the user has to overcome these shortcomings. In this case the big advantage of simulators turn into big disadvantages since now very „creative" solutions have to be found. The situation is analogous to the usage of so-called Forth Generation

Languages. One possibility to overcome these shortcomings is to use program language interfaces which are provided by most of these simulators. At this point, however, a user falls back to an even lower level by using a general purpose programming language in a situation where he would have started with a general purpose simulation language.

Conclusion: A careful analysis has to be done before deciding in favor of simulators or general purpose simulation languages. As long as the problem matches with the simulator's world view and no specific aspects have to be considered, a simulator is probably the right decision. If efficiency or non-ordinary aspects have to be regarded, a general purpose simulation language is probably better. When making a decision for a simulator, the user has to have a clear understanding of the application. With that knowledge, the different simulators on the market have to be compared which means that extensive information has to be collected and compared. If all the necessary elements cannot be provided by a single simulator, a decision in favor of a general purpose simulation language is probably better: only one software tool has to be bought and only one language has to be learned. This leads to cost- and time-savings in the long run.

5 Simulation in Production and Manufacturing Logistics

Simulation has always been one of the Operation Research tools which are applied successfully in practice. Simulation has established itself as a technique of high practicability, especially for solving real world problems found in production and manufacturing logistics. In the last decade there has been a dramatic increase in the use of simulation models in manufacturing. Nowadays, manufacturing systems are often so complex that other Operation Research techniques cannot be applied. Mathematical models solved by analytical methods — despite of their theoretical elegance — could be considered as well. In practice, however, they can seldom be used, since in most cases, they are too complicated to be solved. Models which can be solved analytically are normally so far away from the real system that such solutions have nothing to do with the real problem anymore. Therefore, such a solution can only be considered as a „right solution for a wrong problem". Nonetheless, the right solution for the wrong problem is a wrong solution! On the other hand, these type of systems can be analyzed and experimented with very successfully by such a powerful and flexible tool like simulation.

In a sense, simulation can be described as an „inexpensive insurance policy," allowing to test-drive a new design for usually a quite small percentage of the investment costs. To use simulation for test-driving a new manufacturing system can now be considered to be state-of-the-art. The success of simulation in manufacturing can be attributed to the following reasons: Computing costs which twenty years ago were so high that they were often prohibitive, have now been reduced to such an extent that they are not considered to be relevant as costs in a simulation study any more. Whereas the investing in a simulator or a simulation language (software costs and (!) education) has to be taken into account, cost for simulation runs can be neglected. Simulation languages and simulators improved over the years and led to a more flexible and easy use of simulation. In addition, graphical support like animation resulted in a better understanding and interpretation of simulation.

According to Biethahn, Hummeltenberg, and Schmidt [4], simulation is now the most important tool for the analysis and the design of manufacturing systems. Scharf and Spiess [17] formulate possible questions giving good examples of how simulation can be used, from the strategic planning level to the operation level.

- How many machines are needed in a specific production area?
- What will future sales figures (to be used in strategic planning) be like?
- How many resources are necessary to produce a specific product mix or volume?
- What has to be the size (capacity) of inventory buffers in a production line in order to produce a specific production volume based on experience about breakdowns of resources?
- What has to be the number, type, and physical arrangement of carts, conveyers, and other support equipment, like pallets and fixtures, to ensure all work stations can be supplied with sufficient material?
- What impact has a change of the lot size to the leadtime; what impact has a different priority rule on in-process inventory?

5.1 Who Needs Simulation Support?

In [15], simulation is called the „enabling technology for enterprise integration". The following groups in the enterprise must have access to models of products and production:

- Product and process design teams
- Production system designers
- Production managers

Product and Process Design Teams. These design teams are responsible for developing new products and their associated manufacturing processes. For this purpose, they need models to evaluate product performance versus product specifications, to asses manufacturability, and to estimate costs over the manufacturing life cycle. More and more, dynamic aspects of the product and of the production process have to be taken into account. Different levels of details are needed: system models, subsystem or assembly models, and unit process models.

System level models relate design characteristics to performance. Subsystem or assembly models concentrate on cost aspects, performance, and manufacturability, analyzing the implications of individual design decisions. These models are used to evaluate the ability of existing manufacturing processes to produce the product at an acceptable quality and yield, and to allocate design space, tolerance, and error budgets to avoid expensive, technology limiting configurations.

Unit process models optimize the production process parameters to achieve the required product characteristics, to prevent damage to the product from the processes, and to reduce cost or environmental impact. Frequently, these models are highly detailed.

Production System Designers. These designers are responsible for developing the manufacturing infrastructure. This is a classical area for the use of simulation: to plan capacities (e.g., for machines), to develop control algorithms (e.g., for AGVs), and to insure quality and flexibility. Critical issues are throughput, makespan, location of bottlenecks, utilization of equipment and personnel, costs. Of increased importance is the flexibility of the production system design, becoming now a key design aspect: plants are today designed to produce a large mix of products economically with the lot size becoming smaller and smaller. The time horizon is usually months, seldom only weeks, in some cases even years.

Production Managers These managers are responsible for day-to-day decisions at the operational level: loading and sequencing of machines, inventory control, changing of control strategies (e.g., for FMSs) or queuing disciplines, deciding on tool management policies, etc. Simulation is used for many tasks such as support for master production scheduling, production scheduling, factory reconfiguration, and determination of corrective actions in response to unforeseen events. This is a challenging job since the manufacturing area is a very dynamic environment and requests quick decisions if sudden and unexpected events happen like: machines breaking down, processes going out of control, material not conforming with specifications or not arriving on time, changing of due dates, scrap rates exceeding normal level, or the product mix changing wildly. Some tasks are equivalent to those in production system design (e.g., developing of control strategies), for production managers, however, the level of detail is much greater. On the other hand, the time horizon is shorter, usually a few days or sometimes even only fractions of a day.

In the following sections, various fields in production and manufacturing logistics will be discussed where simulation can be applied. Fig. 3 gives an structured survey on these fields. The diagram follows the traditional Y-model with the left branch of the Y displaying the primary MRP-orientated functions and the right branch displaying the primary technical functions. The upper part of the diagram represents the planning functions. There, the Y-branches are separated, indicating that the integration between MRP and the technical information systems is not very tight. The lower part represents the control functions. There is only one Y-stroke, indicating the MRP and technical information systems have to be tightly coordinated. On top, there is a „hat", representing the strategic planning level where there is not such a strict separation between MRP and technical functions. Even if the diagram indicates a clear separation for the single fields, in practice they overlap a lot.

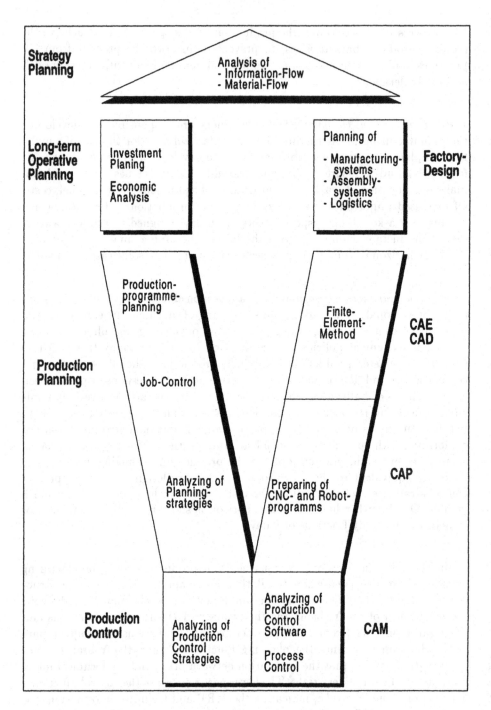

Fig. 3. Application fields for simulation [21]

5.2   Strategic Planning and Long-Term Operative Planning

The strategic planning — including the long range operative planning — comprises all long range decisions. Important decisions like the choice of production sites as well as the introduction of new products have to be based on solid calculations since an error on this level may have disastrous consequences for the firm. Simulation is an important decision support tool since it allows to experiment under different scenarios (varying market developments or financial situations). Simulation can help to reduce the uncertainty of these long range decisions and allows to anticipate cash-flows. Investment planning as well as economic analysis depend on the production system structure. This means that for this long range strategic planning the planning of the production system infrastructure has to be considered as well.

Enterprise Modeling. Simulation in this area deals with broad enterprise management issues. Models on this level tend to be not very detailed. Issues addressed include technology management, strategic information systems planning, and management of corporate culture. Such high level models can be used to drive and control models on lower levels.

5.3   Facility Planning

Facility plans and decisions are taken on several levels. The highest one is that of capacity planning, thus providing enough productive capacity of all kinds to meet the needs of the enterprise. Further down in the planning process, this capacity must relate to specific quantities and conditions of floor space, land, buildings, and equipment. Layout planning in the manufacturing area is concerned with the physical arrangement of production facilities like machines and inventory areas. Criteria for this are economic aspects and technological restrictions. Different levels of detailing can be applied.

Planning of the Logistic System. Part of facility planning is the planning of the overall logistic system. The main task is to provide economically the quantity in demand of the resource in demand in the place in demand at the time in demand in the condition in demand.

Besides the planning of material flows — both within the company as well as shipping and receiving of goods — planning of efficient information flow has also to be performed. Due to the increase in product variability and a shorter product life cycle not only a flexible and efficient manufacturing systems is needed, but also the factor time has to be considered as one or even the most important factor of the competitiveness of the enterprise (see e.g., Yamashina [1]). Simulation is again the appropriate tool to deal with this increased planning complexity, being able to include all dynamic aspect of system behavior.

Layout Planning. Probably „the" classical area of simulation in manufacturing is layout planning. This means placement of machines, inventories, buffers, and tracks on the factory floor. The criteria used reflect on economic and technological restrictions like pre- and post-operations (cooling times), transportation times, safety aspects, etc. Detailed layout planning can be performed on different levels. Probably only by simulation, a user can faithfully adhere to the details of the system no matter what detail is needed, it can be included in the simulation model. All relevant system characteristics can be taken into account. The inclusion of a suitable level of detail provides a realistic setting for experimenting ([18]). If planning on the factory level, details about the different machines are irrelevant; the main concern is information about transportation paths, determining the potential connections between the single production areas. If planning the layout for a manufacturing island, detailed information about, e.g., personal, capacities, processes etc., have to be taken into account.

Material Flow Planning. Today, careful material flow planning is an important task of production system design and has to consider very complex aspects, since the performance of the production is directly depending on the quality of the underlying logistic system. Material flow planning is concerned with providing material to the resources (machines), so that the required operations (processes) can be performed. Due to the separation of operations, the transport volume has increased dramatically. Besides the „classical" material flow system, the conveyer, now several new kinds of transportation systems have been invented like, e.g., AGVs (automated guided vehicles). This development was caused by a wider range in product variability and a shorter product life cycle, making a simple flow-orientated production less feasible from an economic point of view. AGVs depend on precise planning and good general strategies to minimize transport times and to avoid bottlenecks. The number of AGVs should be kept as low as possible, for AGVs are expensive investments, and the more AGVs are on the floor, the higher will be the probability of „traffic-jams". Simulation in this area is routine.

5.4 Simulation support for CAE, CAD, and CAP

Finite Elements. Shorter product life cycles require cost effective and flexible product design methods. From the very beginning in a product design process, it should be possible to make specific statements about future products characteristics. CAE has considerably improved the product design process. Numerical methods, based on exact mathematical methods, can be used to simulate mechanical behavior. This technique is called Finite-Element-Method (FEM). The advantage of this kind of simulation is that even during early stages of the engineering process knowledge about the later part behavior is available and can be taken into account.

The increased use of the FEM requires powerful programming systems which allow to process geometry information fast and at reasonable costs. Those systems are now integrated with CAD systems, allowing to use the geometry information stored in the CAD-model directly. This coupling between CAE and CAD speeds up the FE model building process considerably.

Process Planning. In the area of process planning, simulation can help to reduce the number of necessary experiments, since it allows to perform some of these experiments on computer models. In this form, many process parameters can be fixed in advance. Process planning uses mathematical methods similar to those in product design, namely FEM, Boundry-Element-Method, and Finite-Difference-Method. For a successful use of these methods, CAP (Computer Aided Process Planning) has to be integrated with CAD and CAE.

NC-Programming To use NC-machines effectively on small lot sizes, NC-programs free of errors are of great importance. Studies prove that to run-in an error-free NC-programs needs approximately 150% of normal processing time. This time, however, is doubled or even tripled for erroneous programs ([9]).

One way to check NC-programs is to use a graphical simulation system. This enables to test-run the NC-program without using the NC-machine. Some programming errors like collisions can be recognized and corrected. Currently, technological errors like a wrong feed cannot be recognized, these errors, however, are quite easily corrected by the operator (compared to geometrical errors).

5.5 Production Planning — MRP-Systems

Production planning is another classical application area for simulation. Due to more and more complex production systems, the production planning requires powerful supporting systems for all stages of planning.

The major weakness of current MRP systems is the successive planning philosophy. The separation of material requirement planning and capacity planning causes long lead times, leading eventually to bad results. Although simultaneous planning models have been developed, they are subject to severe restrictions and can therefore not be used in most industries. Simulation can help to overcome some of the deficiencies of current MRP systems. It allows to „play around" with the data at hand to try out alternatives.

Simulation can already be used in the early stages of offer making to ask what-if questions concerning e.g., due dates. In this form, the planner can evaluate potential influences of such new orders on all areas affected in a company, such as inventories, the sales department (additional orders can cause delays for other orders), or purchasing.

Master production scheduling (MPS) can benefit a lot from using simulation. MPS is one of the most important functions of medium term production planning. On this planning level, the actual economical objectives are determined which define the basic conditions for the following planning tasks (as material requirements planning, capacity planning, or job sequencing). A good master production schedule is the basis for the reliable determination of delivery dates of customer orders, the efficient utilization of resources, and the alignment of operative activities with strategic objectives of the enterprise. Although MPS is the basis for many of the following planning decisions, the function is often hardly supported in traditional manufacturing planning and control systems. Simulation can be used to improve this situation since simulation enables the planner to review the effects of each decision. (Notify, that at this stage only a rough-cut capacity planning is being performed; short-range planning with finite capacities is discussed in the next section.)

Another area of application for simulation is the general analysis of planning strategies. In this form, it is possible to test even alternative types of production organizations, like, e.g., changing to a wage-incentive system. The best type of organization had already been determined during facility planning, if the product-mix is changing or new products have to be launched however, another organization of the production might be more effective. Simulation can help to analyze the behavior of the production system in case of different product-mixes or change in production volume.

5.6   Production Control and CAM

Simulation for production control enables the planner to experiment upon the facility or a certain part of it with the actual orders on hand. The result of a simulation run is a feasible Gantt chart. Therefore, simulation can be used as a tool for scheduling. The planner is able to try out different dispatching rules (like FIFO or SPT), lot sizes, or shift plans to evaluate the performance of the production system and the impact of these decisions to all affected areas of the company. Baker's [2] statement from 1974 still holds:

One of the most thoroughly studied and widely applied areas of scheduling research involves the dynamic version of the job shop model. In the dynamic model, jobs arrive at the shop randomly over time, so that the shop itself behaves like a network of queues. In this context, scheduling is generally carried out by means of dispatching decisions: at the time a machine becomes free a decisions must be made regarding what it should do next.

The effects of dispatching procedures in queuing networks are very difficult to describe by means of analytic techniques. Nevertheless the study of scheduling in dynamic job shops has made considerable progress with the use of computer simulation models. The rationale for using simulation methods in job studies is not different from the rationale for simulation in dealing with any other complex system: short of testing alternative policies in the actual system, there is no way to anticipate fully how different operating procedures will effect system behavior. Experimentation with a computer simulation model has made it possible to compare alternative dispatching rules, test broad conjectures about scheduling procedures, and generally develop greater insight into job shop operation.

The main difference between „playing around" in the previous section and in this stage are the data used and therefore the aim of the simulation study. Now, actual orders are used and the focus is on each single job, in contrary to the mid-range planning where only general measures like average machine utilization, makespan, throughput, etc. are observed.

There is a significant difference between simulation for facility planning and simulation for production control: Simulation for production control uses a given model of the real facility to improve the production, whereas simulation for facility planning tries to optimize the design of the facility.

A further application for simulation in production control is the analysis of production control software: The steady development of new production processes also requires new software to control these processes. It is recommended to test such software before using it in real production processes. Simulation enables the planner to run this software under „real" conditions in a model without any risk for the production system.

References

1. *JIT in Japan: What can we learn from their experience.* Springer Verlag, Berlin, 1987.

2. K. R. Baker. *Introduction to sequencing and scheduling.* John Wiley & Sons, 1974.

3. J. Banks, E. Aviles, J. R. McLaughlin, and R. C. Yuan. The simulator — new member of the simulation family. *Interfaces*, 21(2):76–86, March-April 1991.

4. J. Biethahn, W.Hummeltenberg, and B. Schmidt. *Simulation als betriebliche Entscheidungshilfe, Band 2.* Springer-Verlag Berlin, Heidelberg, New York, London, Paris, Tokyo, Hong Kong, Barcelona, Budapest, 1991.

5. W. G. Bulgren. *Discrete system simulation.* Prentice-Hall, Inc., Englewood Cliffs, New Jersey, 1982.

6. O. J. Dahl, B. Myhrhaug, and K. Nygaard. *Simula 67 common base language.* Norwegian Computing Center, 1970.

7. K. Feldmann and B. Schmidt. *Simulation in der Fertigungstechnik.* Springer Verlag Berlin, Heidelberg, New York, Tokyo, 1988.

8. A. Goldberg. *Smalltalk 80.* Addison-Wesley, 1984.

9. Hohwieler, Junghans, Linner, and Neubert. *Anwendungen der Simulation in der NC-Programmierung*, pages 191–216. Vieweg Verlag Braunschweig/Wiesbaden, 1993.

10. W. Kreutzer. Of clouds and clocks — a guided walk through a simulator's toolbox. *NZOR*, 11(1):51–120, January 1983.

11. W. Kreutzer. *System simulation — programming styles and languages.* Addison-Wesely Publishing Company, Sydney, 1986.

12. T. H. Naylor. *Computer simulation experiments with models of economic systems.* John Wiley & Sons, New York, London, Sydney, Toronto, 1971.

13. B. Noche and S. Wenzel. *Marktspiegel Simulationstechnik in Produktion und Logistik.* Köln, 1991.

14. C. D. Pegden, R. E. Shannon, and R. P. Sadowski. *Introduction to simulation using SIMAN.* McGraw-Hill, USA, 1990.

15. J. Polito, J. Jones, and H. Grant. *Enterprise integration: a tool's perspective.* 1993.

16. A. A. B. Pritsker. *The GASP IV simulation language.* John Wiley & Sons, 1974.

17. P. Scharf and W. Spieß. Computersimulation hilft bei der optimierung von produktionsanlagen. *VDI-Z 134*, page 72 p., 1992.

18. Th. I. Schriber. *The nature and role of simulation in the design of manufacturing systems*, pages 5–18. SCS, 1987.

19. R. E. Shannon. *Systems simulation —- the art and science.* Prentice-Hall, Inc, New Jersey, USA, 1975.
20. B. Zeigler. *Theory of modeling and simulation.* John Wiley & Sons, 1976.
21. M. Zell. *Simulationgestützte Fertigungssteuerung.* R. Oldenbourg Verlag, München, Wien, 1992.

Dynamic Modeling of CIM Systems

Wolfram Conen

Wirtschaftsinformatik der Produktionsunternehmen,
University of Essen, 45117 Essen, Germany

Abstract. *Dynamic Modeling* is not a precise defined concept. Nevertheless, its ambiguity makes it appropriate to use in discussions about requirements for a *modern* modeling methodology. "Real" systems (as CIM systems are) tend to be complex and dynamic. Models of such systems (or better: of parts of such systems) should serve different purposes: They should allow to answer certain questions about the underlying system and it should be possible to integrate/combine/interconnect the models to answer more sophisticated questions or questions on a different level of abstraction (e.g., strategic management questions if the underlying models are models for supporting control decision on an operational level). "Dynamic" modeling might emphasize mainly three aspects: (1) the utilization of the model for representing systems (or parts of systems) with certain dynamic properties, (2) the embedding of the method into an iterated (and dynamic) modeling process, and (3), as CIM systems are "living" systems[1], the system of models must be able to consistently capture changes of the underlying system's organization and consequences of changed modeling purposes/goals. These are the components of the dynamics of modeling, as discussed here. The article presents Petri nets as one possibility to provide answers to inquiries into dynamic systems, to embed models into a coherent modeling process, to react on changes of perspective and organization, and to integrate models over different levels of abstraction to support operational, tactical and strategic decision making.

1 Introduction

Firstly, concepts relevant for the consideration of any complex system, are discussed: models, systems, and the quality of models. Section 2 adds a few qualitative aspects specific to CIM systems. Section 3 introduces the proposed solution to the modeling problems mentioned above: Petri Nets. Section 4 presents certain Petri net types in more detail, with an emphasize on Coloured Petri Nets. Here, relations to the concepts of Sections 1 and 2 are pointed out. Section 5 concludes the article with some remarks on the possible improvement of decision making in a CIM context through the use of Petri nets.[2]

---

[1] Their structure and the goals of the decision makers change dynamically.

[2] Please, allow a general remark: Most of the concepts mentioned in this article are presented in an abstract form and related to abstract concepts, as complex/dynamic systems in general. Nevertheless, since most of them are of elementary and essential nature, it should be possible to fruitfully transfer and apply these concepts to CIM-related considerations. Detailed examples are left out due to required brevity.

1.1 Models and Systems

The purpose of models, as discussed here, is to provide a base for answering questions about the system modeled. The ultimate benefit of modeling is to increase decision making capability [31]. There exists a bunch of reasons why it might be impossible to examine the system of interest directly:

- the system is too complex, only some relevant subparts and details should (could) be examined
- an examination of the system itself is too dangerous or even impossible
- the system is in a design state or should be redesigned and a direct design of the system or direct interventions in the system are too expensive or expose an incalculable risk without experimenting with appropriate models.

Models are abstractions of systems. They serve as a surrogate for the system they model—but, nevertheless, models are systems themselves [18]. A multitude of definitions of the term "System" can be found in the literature. At one end of the spectrum this starts with explanations as "A system is what is distinguished as a system" [5], at the other end, completely abstract and formal definitions in a specific context are given. Fishwick presents in [4] the following system theoretic definition as a base for his reflections on system modeling (from [17]). Later, we will compare this definition with the formalism defining Coloured Petri Nets.

Definition 1. A deterministic system $\langle T, U, Y, Q, \Omega, \Delta, \lambda \rangle$ is defined as follows:

- T the time set. For continuous systems (see [1]), $T = \mathcal{R}$ (reals), and for discrete time systems, $T = \mathcal{Z}$ (integers)
- U the input set containing the possible values of the input to the system
- Y the output set
- Q the state set
- Ω the set of admissible (or acceptable) input functions. This contains a set of input functions that could arise during system operation. Often, due to physical limitations, Ω is a subset of the set of all possible input functions $T \rightarrow U$
- δ the transition function. It is defined as $\delta : Q \times T \times T \times \Omega \rightarrow Q$
- λ the output function, $\lambda : T \times Q \rightarrow Y$

For the moment, it seems to be sufficient to view a system as a collection of related components of (possibly) different types; with a well-defined border to its environment, and a connection to this environment via inputs and outputs of an appropriate form.

Zeigler differentiates between controllable and uncontrollable system parts [31]. If any decision maker wants to change the behavior of the system, the controllable parts are naturally the aim of influence. Nevertheless, it could happen that changing controllable parts of the system will enhance or diminish the extent of the controllable parts itself. Zeigler divides the objectives motivating the modeling of a system (or system part) in three categories, depending on their level of intervention: management, control, and design. Management is characterized by a limited power to intervene. One can set goals and determine broad courses of action, but these policies cannot be spelled out in full detail. Execution must be delegated to, and interpreted at, subordinate levels—so the transition between intention and implementation is ambiguous.

In a control context action is deterministically related to policy. Nevertheless, constraints on the scope of intervention limit the control action to the selection of alternatives within fixed extant domains.

In contrast, design connotes greater scope for choice. The designer augments, or replaces, a part of the existing system. Implementation of a design is relatively expensive and infrequent while control and management are continued "on-line" activities. Management, control and design can be thought of as lying along axes representing degree, cost and time scale of intervention on the controllable part of a system (see also [31]).

The effect of each intervention is unsafe, due to the existence of uncontrollable system parts and their interactions with the controllable parts. Models encode knowledge about these interactions. "Their sine qua non is the ability to integrate sets of disconnected relationships whose joint implications would otherwise be difficult to draw. They are especially useful to the extent that they enable extrapolation beyond the data so far acquired." [31]

Decision makers not only differ from each other due to different competences with respect to the level of intervention. They control the system or parts of the system from decision-dependent perspectives: Operational, tactical or strategic decisions have to be made. The view at the system which is necessary to make a strategic decision on a certain level of intervention might be completely different from the view which allows to make successful decisions on the same level of intervention within every day business. A methodology of modeling has to offer the possibility to structure the models (and thus the underlying system) hierarchically in such a manner that the different levels of abstraction (e.g., by means of aggregation) assemble a top-level view onto the system in a coherent and consistent way allowing the best-possible strategic decision. Here, best-possible means: The loss of information implied by using abstractions has to be as small as possible with respect to the question(s) the model should help to answer, and, on the other hand, the gain of simplified insight into the consequences and frame conditions of the decision in question has to be as large as possible.

1.2 Model Quality

As modeling should serve as a base for every extensive consideration of dynamical, complex systems, the *quality of models* becomes important. A few selected quality aspects are listed below:[3]

---

[3] There is (almost) no relation between the position of the aspect in the sequence and its importance.

- **Verification**: The correctness of the model can be proved. This implies the presence of a sufficiently precise formalism to formulate/analyze the model.
- **Validation**: Meats the model the requirements allowing to serve as a surrogate for the modeled system? Behaves the model as the modeled system would do? The test samples for validation should be chosen well enough to justify the use of the model for prediction of the system's behavior (generalization).
- **Reuse**: A model designed to answer certain question should remain usable under slightly different conditions and with respect to changed questions.
- **Modularity/Integration/Hierarchy** — A multitude of models is necessary to model various parts of complex systems (as a CIM system). These models should fit together to allow the answering of complex questions related to a collection of system parts or of questions on high levels of abstractions (integration and hierarchy). This implies a modular design of models which does allow to connect them as appropriate and necessary.
- **Flexibility**: The model should be flexible enough to allow its adaptation to changed management goals, changed organization of the system, and thus changed questions or modeling goals. The mechanisms provided for the realization of flexibility should allow smooth and, in an ideal case, *correct* transitions between models. Here, correct means that the changes in organization/goals/questions lead in a verifiable way to changes of the model.
- **Execution**: The execution of a model is a prerequisite to its validation. All the arguments given for simulation in the literature (see, for example, the simulation section in this book) could be repeated here. Execution enables the modeler to prototype models, to validate the model, and to do an easier transition from the model to an information system (if this is his modeling goal)[4].
- **Communication**: Modeler[5], model-user[6], modeled person[7], and persons affected by the model[8] have to communicate by means of and about the models. Valid models could emerge in a human organization only if discussions focused on the *How* and *What* of modeling and models are possible. The used modeling method should provide mechanisms to support discussions at and between different levels of modeling-expertise. Only the "understanding" of models and modeling process can lead to high acceptance of the decisions made with help of the models.
- **Usability**: Tools for managing the modeling process and supporting the model life-cycle (modeling, validation(execution), maintainance, prototyp-

---

[4] This list is definitely not complete.

[5] This includes here any person directly involved into organization and course of the modeling process.

[6] For example, an information system development team, or a decision maker at a strategic management level.

[7] For example, experts (and their knowledge), workers in a manufacturing sub-system, etc.

[8] For example, workers in a part of the organization which will be re-structured due to the results of the modeling process (as a consequence of decision).

ing) should be available. The method should not be proprietary to ensure broad discussions about the method and further developments of the method. It should be attractive for applications in practice as well as for research activities.

- **Teaching**: The method used to develop models (and the model itself) should be easy to teach and should capture the interests of students *and* practicians.
- **Embedding**: The model should be embedded into a modeling process (model life cycle), and the modeling process should be embedded into a decision making process in a way supporting decision quality as much as possible.

Some of the quality aspects seem to define mutual exclusive goals (e.g., the model should be verifiable and communicable), some are not orthogonal (e.g., Reusability and Flexibility). It is one of the most important tasks for a strategic management of modeling processes to formulate (ideally measurable!) quality criteria and to weight them according to their importance for the project (see, for example, [3]). The list of quality aspects presented above is (most probable) incomplete and no metrics or even hints for measuring the grade of their fulfillment is given, but even a vague notion of quality can be of substantial use for the selection of an appropriate modeling method. Petri nets satisfy the above mentioned criteria up to a certain level, and this level is high enough to justify it to consider the use of Petri nets as *the* basic instrument for modeling tasks in a complex manufacturing environment.

The next section relates modeling to decision making and modeling in CIM system. Section 3 sums up some possible advantages of Petri nets and presents some of the basic ideas behind Petri nets. Section 4 gives a short history of the development of Net Theory and introduces the most important net types. The text is concluded by some remarks on the application of Petri nets in a CIM context.

2 Modeling of CIM Systems

As we already saw, the role modeling plays for good decision making is quite multi-faceted. Ideally, modeling provides a skeleton consisting of models which represents the structure of the surrounding system's "body" and enables decision makers to initiate controllable moves or changes of the body.

The skeleton metaphor leads to an important detail: connectivity of models. Each bone represents an (hopefully) adequate model for a part of the over-all system (e.g., a certain conveyor), modeled to answer specific (and often very narrow) questions (e.g., "Could we tune the conveyor to improve it's throughput?") under certain assumptions (e.g., "We are not interested in how many transport jobs really arrive, we want to prepare the conveyor for peak-level situations (which might never occur)"). This question might result from a control goal at operational horizon. Nevertheless, the conveyor contributes to the behavior of the autonomous manufacturing area he belongs to. The above mentioned model emphasizes technical aspects of the conveyor—and these technical aspects might also be interesting for strategic decisions on a design level of intervention (e.g., "Should we introduce a new conveyor system?"). But, to answer such a question, the management has to examine the conveyor being embedded into the processes in the autonomous manufacturing area. Many of the assumptions made for solving the throughput-question have to be dissolved and explained through interactions between (modeled) subsystems. And that's the point: To be relevant in a system of models which should provide a base for good decision making on all levels of intervention and for different horizons (i.e. operational, tactical, strategic), the model has to be connectable to models of physically or logically neighbored system parts. If it's possible to use all models under different views, connected or not to other models, the skeleton of models could be *the* (abstract) instrument of integration in a CIM system[9]. (Remark: One consequence of this is that viewing the computer as *the* integrative medium in complex manufacturing systems is a far to technical perspective—using computers do not ensure good decision making, it surely is an instrument, but it is not a solution to the integration problem. Here, and in the hole article, *good* decision making simply means decision making which leads to more effectiveness, a better market position, and, thus, the earning of more money.

To ensure connectivity of models the following should hold:

- The method the modeler should use is problem adequate (e.g., allows the simulation of the conveyor), easy to use, and powerful.
- The method is, in principle, understood and accepted by modelers and model-users of adjacent system parts. This ensures the acceptance of decisions based upon the model.
- It is possible to further abstract models to connect levels of strategic, tactical, and operational decision making in a coherent manner.

The use of Petri nets might be a solution in the context of motivation and communication. That is the reason why we decided to present Petri nets as an example/suggestion for dynamic modeling of CIM systems. Following the presentation of some basic and a few more sophisticated aspects of Petri nets, we will briefly review the above mentioned problems again in Section 5.

3 Modeling with Petri Nets

A model is used if the inquiry into a system can not be answered by examining the system itself. Reasons might be that the system simply doesn't exist (modeling for design), the examination of the system might be too costly or too dangerous, or the system might be too complex. Thus, modeling is a process of abstraction which reflects in the resulting model(s) essential aspects of the system and its behavior. Naturally, it depends on the inquiries which aspects of the system and its behavior are essential (and relevant). The models presented in this paper are based upon Petri nets and Net Theory. Net Theory was developed in the

---

[9] For description of CIM systems, please, refer to appropriate parts of this book.

search for natural, simple and powerful methods to describe and analyze flows of information and control in information processing systems. Petri nets are abstract models of information– and object–flows. They allow to describe/model systems and processes on different levels of abstraction and granularity in a homogeneous language. The need for such a theory emerged when it became obvious that computers can do more than evaluating numerical functions. In those cases where the co-ordination of information– and object–flows between concurrently working system parts are the main tasks the computer has to fulfill, computability becomes unimportant and terms as "synchronization", "conflict", "deadlock", "safeness" and problems related to those concepts receive more and more attention.

Petri nets were introduced by C.A. Petri in the early 1960s as a mathematical tool for modeling distributed systems and, in particular, notions of concurrency, non-determinism, communication and synchronization [20]. Thiagarajan points out that Petri nets model the twin notions of states and changes-of-states (or transitions), within the guiding principles (compare [29]):

- states and transitions are two intertwined but distinct notions that deserve an even-handed treatment
- both states and transitions are distributed entities
- the extent of change caused by a transition is fixed and does not depend on the state at which it occurs
- a transition is enabled to occur at a state if and only if the fixed extent of change associated with the transition is possible at that state.

The complexity of a system and its behavior determine a framework of conditions a modeler has to obey. Complex systems enforce modular and hierarchically ordered model design. The modeler must be able to select a level of abstraction/detail for solving a specific subtask which allows to understand the subsystem and it's co-operation with the other system parts. One of the main advantages of the Net Theory is the homogeneity of its modeling language for different levels of abstraction and the (often) simple possibility of transitions between adjacent levels.

The design of control structures for parallel processes where the parallelism depends on certain resource constraints is often more difficult then the design of control structures for complex sequential task. The reason is that concurrent processes can change their state independently from each other and it might be difficult to keep track of the overall state of the system. An ideal instrument for modeling such systems would allow a descriptive representation of the model which can be interpreted in all application areas and allows a verification of its design. Petri nets are such an instrument (according to [28]).

The modeling of systems in which events can occur causally independent (concurrently), but constrained, is the main application area for Petri nets. Not only information systems are of this kind—every manufacturing environment is characterized by a multitude of concurrent processes. The general applicability of concepts like concurrency allows the utilization of Net Theory in a broad context. Nevertheless, Net Theory is not the "overall theory of everything"—Net models deal with discrete quantities—only discrete or discretizable systems can be modeled.[10]

The integration of informations about structure and causal dependencies between events into one model expose the possibility to develop a theory connecting structure and behavior of the system.

Properties of the models (and the underlying system) which can be formally analyzed, include:

- **Finiteness/Boundedness**: Has the system a finite/infinite number of states (in other words: is it in principle possible to implement the system?). Are there certain bounds on the number of tokens in places[11]?
- **Reachability**: Can certain states occur? This may include "wanted" as well as "unwanted" states.
- **Deadlocks**: Is a state of the system be reachable which disables all possible transitions to subsequent states?
- **Cyclicity/Home Markings**: A state which can be reached from all it's subsequent states is called home state. If the initial state of the system is a home state, the system is cyclic (or resetable).
- **Liveness**: There may exist transitions in the system which can not be reached from certain states of the system. Such transitions are called *dead* with respect to certain states. They represent disabled system parts. If the transition can be reached from all system states (or certain states), the transition is called alive (alive with respect to certain states).
- **Fairness/Starvation**: Even in systems where all system parts are alive (and thus may become enabled again and again), it might happen that certain states do never occur, for example, due to the resolution of conflicts (e.g., while allocating resources). This indicates that the system isn't designed/modeled fair (some system parts might "starve"). There exist various definitions of differently "fair" behavior - Jensen uses *impartial, fair*, and

---

[10] For example, discrete systems can also be modeled through state automata. Automata theory is a state-oriented language, the global state of the system is described. Net Theory views local states (constituting the global state) and events as equally important. The preconditions of events and their consequences can be described explicitly. This opens up the possibility to represent the local and the causal structure of the system in the model. The state graphs of automata theory do not allow the answering of question about the causal independence of events or the occurrence of conflicts (non-determinism of behavior), as Petri nets and their occurrence graphs do.

[11] Petri nets consist of places and transitions, connected via arcs. The distribution of tokens in places constitutes a certain state of the net. States change through removing/adding tokens from/to places. This is the effect of "firing" transitions. Certain conditions have to be fulfilled before transitions are allowed to fire (the transition is enabled). Transitions can be concurrently enabled making it necessary to select a transition for firing. This is done non-deterministically. This short introduction is completed and explained in the next Section.

just[12] (compare [2] for PT-nets). An elaborated and interesting presentation of different notions of Fairness can be found in [16].

Among the analysis instruments applicable to Net Theory models are *occurrence graphs, invariants, reduction rules,* and *performance analysis.* The basic idea behind *occurrence graphs* is to construct a graph containing a node for each reachable state and an arc for each state transition. A few reduction possibilities for the size of the occurrence graph exists, trying to preserve the possibility to analyze, for example, the liveness of certain transitions. Even the size of reduced occurrence graphs leads to problems, as well as the fact that it is necessary to fix all system parameters before an occurrence graph can be constructed. The basic idea behind *invariants* is to construct equations satisfied for all reachable states. It is possible to construct invariants for hierarchical nets by composing invariants stemming from nets of lower level of abstraction. Additionally, invariants can be found without fixing the system parameters and this can be done synchronously to the design of the model and thus can be used to improve the design process quality. Nevertheless, since using invariants requires high mathematical skills, it might be difficult to use them in industrial system development (e.g., due to a lack of communicability)[13]. The use of *reduction rules* aims at simplifying the net structure while preserving the properties of interest (e.g., boundedness). Proofing the soundness of reduction rules might be complex and cumbersome - but performed once (considering all possible consequences for all possible nets), the rules can be used by any user without knowing the proof. While preserving the "interesting" properties, the reduced nets generally do not allow any explanation why certain properties are *not* present, hence the results of reductions are non-constructive. *Performance analysis* can be done for nets extended with a time concept by means of simulation or (if the nets are not too large) Markov chain analysis. This complements the examination of the logical correctness of systems, their dynamic properties and functionality aimed at by the most Petri net applications.

The main advantages of the utilization of Petri nets for modeling/design of complex systems are (according to [28]):

- The graphical representation of the models improves the communication between modeler and expert.
- Net models can be used on different levels of abstraction. The resulting models are connected via simple coarsen/refinement operations.
- An underlying theory allows the verification of the models or of certain properties.
- Software for modeling, analyzing and simulation of the models is available[13].

The next Section will present some net types in detail.

---

[12] impartial $\rightarrow$ fair $\rightarrow$ just

[13] e.g., Design/CPN by Meta Software Corporation, 125 Cambridge Park Drive, Cambridge MA 02120, USA, or *Leu* by LION Gesellschaft für Systementwicklung mbH, Uni Tech Center, Universitätsstr. 140, 44799 Bochum, Germany

4 Types of Petri Nets

We will present some of the many variations of Petri nets. Starting point are
black and white nets which are conceptually simple and straightforward to analyze. After the introduction of basic concepts and ideas behind Condition/Event
Systems and Place/Transition Nets, a brief history of Petri nets will follow.
Then the presentation introduces more complex nets, namely Coloured Petri
Nets which are currently very "popular" and well-supported through modeling
tools[14].

4.1 Black and White Nets

A simple (black and white) Petri net is a bipartite digraph. The graph consist
of nodes and directed edges. Bipartite means that it has two different kinds of
nodes, which are called *places* and *transitions*. Edges connect places to transitions
(known as input arcs, and the corresponding places known as input places) or
transitions to places (known as output arcs, and the corresponding places known
as output places). A Petri net can be *marked* by indicating the em tokens which
are contained in each place at a point in time (drawn as dots). These completes
the syntactical aspects of black and white nets. If the input places of a transition
all contain (at least) one token, then the transition is *enabled* (or armed or eligible
for firing). If the transitions *occurs* (or fires), then one token is removed from each
input place and one token is added to each output place. It should be remarked
that this is not a *move* of tokens—the added tokens are completely new tokens,
the removed tokens are destroyed. This *occurence* of transitions constitutes the
semantics of Petri nets. A Petri net is executed by establishing an initial marking
(normally called M_0) and then, at each subsequent point in "time", one or more
eligible transitions are chosen for firing.

It is worth noticing at this stage that the ability of a transition to fire is
determined solely by local conditions, namely the presence of tokens in adjacent
input places. The only global consideration in the above description concerns
the choice of an enabled transition for occuring. Care must be taken in conflict
situations as in Fig. 2, since both transitions are enabled but the occurence of
one disables the other. If non-conflicting (or independent) transitions are enabled
under a marking of the net, they are concurrently enabled. This means that the
two (or more) transitions may occur "at the same time" or "in parallel". Such
groups of concurrently enabled transitions are called *step*. A single transition
can even occur concurrently to itself (if, for example, there are enough tokens

---

The selection of the different types of nets is mainly guided by their relevance for
the development process which led to more complex Petri net types. This list is not
complete in any respect, there is almost an uncountable number of net types, some
are enhancements of existing types, some of them are generic types on their own.
We will mention a few more concepts later in this article, especially the different
notions of time. By the way: some "enhancements" are the expression of a better,
more sophisticated perspective onto existing nets, like the very interesting temporal
logic for Place/Transition (and higher) Nets introduced by Reisig [25]

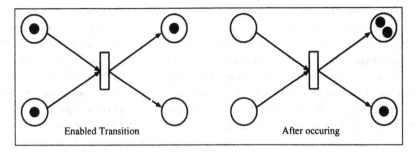

Fig. 1. Firing of a Transition

in the input places of the transition to allow occurence twice or more times). Again, dependency or independency of transitions is determined by viewing at the situation given under a certain marking, i.e., there could be one marking under which t_1 and t_2 are both enabled and independent (simply add one token to the input place in Fig. 2[15] and another marking under which they are enabled and not independent (see Fig. 2). Obviously, there are lots of pairs of transitions in a net which are always independently enabled due to structural reasons (they simply do not share any input places).[16]

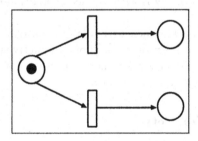

Fig. 2. Conflict!

It should be noted that the complete state of the system is given by the mark-

---

[15] There is still a conflict between the *steps* $\{t_1, t_2\}$, $\{t_1, t_1\}$, and $\{t_2, t_2\}$ - but not between the transitions.

[16] Please, note that terms as *independent* could be differently defined for different net types. In net types with constraints on the token capacity of places (like Condition/Event systems) the sharing of an output place between enabled transition always leads to a conflict. Furthermore, in Condition/Event systems and PT-nets it is possible to determine dependent transitions not only by viewing at single markings (or certain conditions fulfilled by markings) but by analyzing the structure of the net (i.e. place and transitions), leading to statements such as: Whenever t_1 and t_2 are enabled, they are dependently enabled.

ing of the places. The transitions do not hold any state information, but represent changes-of-state[17]. The occurence of an enabled transition or step transforms a marking (e.g., M_0) into another marking (e.g., M_1). Here, M_1 is called *directly reachable* from M_0. If there is another transition or step which occurrence could transform M_1 into, let's say, M_2, then M_2 is called *reachable* from M_0. Thus, an execution of a net can be viewed as a sequence of markings—but, since different transitions/steps can lead to the same sequences of markings, this is not a very precise notion of execution. It is often more appropriate to view an execution as a sequence of occuring transitions/steps (if one is more interested in describing the "functional" aspect of the execution than in describing solely state information, or, in other words, the declarative perspective onto the net [4]).

4.2 Condition/Event Systems

Black and white nets are often *interpreted* as Condition/Event Systems. Here, each place is interpreted as a condition which holds if and only if a token is present at the place. Each transition is interpreted as an event, the occurrence of which leads to a change in the associated conditions (indicated by the adjacent places). This interpretation of a black and white net puts additional constraints on the net: a place can only hold one token at a time (since a condition can only be true or false[18]) and a transition cannot fire if a token is already present in an output place (a consequence of the first constraint). Thus, the net of Fig 1 would not be eligible for firing interpreted as a condition-event system (but Fig. 2 would).

Such interpretations, together with the change of conditions when events occur, can be used to prove properties about the activity of the systems being modeled using propositional logic. Further properties or invariants can be derived using linear algebra techniques (see [23]).

4.3 Place/Transition Nets

Black and white nets can be generalized to Place/Transition Nets, allowing the presence of multiple tokens in a place, the removal and addition of multiple tokens from input and output places (respectively) when a transition fires, and the limitation of the number of tokens in each place. This is still essentially a black and white net, and can be mapped into the basic form already introduced.

4.4 Steps Towards High-level Petri Nets - A Brief History

To overcome the restrictions imposed by the low level of abstraction simple Petri nets offer, a number of high-level Petri net types were introduced in the last decade. High-level Petri nets extend the concept of a net by differentiating

---

[17] This is not true under all circumstances, it is possible that the firing of a transition leads again to the pre-firing state of the net.

[18] In a classical, two-valued logic.

between tokens, i.e. giving them a type of arbitrary complexity and a data value, and make the firing of the net dependent on the availability of appropriately coloured tokens. Since higher-level Petri nets can be mapped to black and white nets they have no greater modeling power. However, the differentiation of tokens gives, in practice, much greater descriptive power [14]. The step from low-level to high-level Petri nets can be compared with the step from assembly languages to modern programming languages with an elaborated type concept. In high-level Petri nets, each token can carry complex data, describing, for example, the entire state of a process or a data base [13]. To clarify the motivation and steps leading to the development of "higher" Petri nets, a brief history is presented before Coloured Petri Nets are explained:

Carl Adam Petri presented Condition/Event Systems in his doctoral thesis. They were widely used (and still are) for theoretical considerations of basic concepts and analysis methods[19]. The development of Place/Transition Nets (PT-nets) is one of the results of the contribution of a large number of people to *new* net models, *new* basic concepts and *new* analysis methods. These net models allow a place to contain several tokens. They were (and again: are) often used for practical applications—but the net model was still to "simple" to cope with real-world application. This realization was motivation enough for countless researchers to develop extensions to PT-nets (e.g.: Time delays, zero testing of places, transitions with firing-priority to solve conflicts, etc.). Most of the net types possessed almost no analytic power [13]. They could be used to describe certain systems but their new features made it difficult to map an analysis method developed for a certain net type to another.

One result of this improvement process had been the presentation of Predicate/Transition Nets (PrT-nets) in [6]. They were introduced without an specific application context in mind and form a generalization of CE- and PT-nets similar to the steps leading from propositional logic to predicate logic. PrT-nets can be formally related to CE- and PT-nets in a way which allows to generalize most of the basic concepts and analysis methods[20]. The generalization of place and transition invariants made some technical problems apparent. To overcome those difficulties, the first version of Coloured Petri Nets were defined in [10]. The main difference between the two net models lies in the way the relation between a binding element and the token colours involved in the occurrence are established: With expressions in PrT-nets and functions in early CP-nets. Since the functions in CP-nets were difficult to read and to understand, the further development led to an improved net model which combined aspects from both strongly related net types, PrT- and CP-nets [13][21]. These nets ("mod-

---

[19] In [27] and [29] Elementary Nets (EN) were introduced to overcome some technical problems induced by the original definition of CE-Nets—nevertheless, the basic ideas of EN are very similar to those underlying CE-Nets.

[20] In [7] an improved version of PrT-nets based on sigma algebra is described.

[21] These nets were originally named High-level Petri Nets (HL-nets), but since this term is widely used as a generic name for CP-nets, PrT-nets, HL-nets and several other kinds of net models, they were called CP-nets again (to complete the confusion).

ern" CP-nets) have two different representations: an expression representation for applicational convenience using arc expressions and guards, and a functional representation for analytical purposes using linear functions between multi-sets. There are formal translations between both representations (in both directions). This form of CP-nets were presented in [11] and [12][22].

4.5 Coloured Petri Nets

In Coloured Petri Nets, the type of the token is called *colour set* and its (date) value is called *colour*. For a given place, all tokens must have colours of a specific colour set. This type is called colour set of the place. This colour set determines the domain of values a token can carry. Their use is analogous to the use of types in typed programming languages. Each net is augmented by the declaration of the used colour sets. Jensen selected an adaptation of standard ML (see, for example, [9]), called CPN ML, as language[23] for the declaration of colour sets and expressions used to inscript arcs, transitions and places. Again, a distribution of tokens on the places is called a *marking*. Each place can be augmented by an initial expression which evaluates to the *initial marking*. The marking of each place is a multi-set over the colour set attached to the place. A multi-set[24] simply allows the existence of more then one token of a specific type in a (multi-)set. It is straightforward to extend the operations on sets to multi-sets (with one exception: the difference operator)[25].

The higher complexity of tokens induces more complex transitions. The colours of the tokens in input places to transitions can be inspected by the transitions, and this information may determine the effect of the transition, namely the colour of the output tokens. More elaborated arc expressions are used to specify a collection of tokens with well-defined colours to be removed from the input places. Each arc expression evaluates to a multi-set over the colour set of the connected place. The arc expressions surrounding a transition are allowed to contain variables. These variables can be bound to different values and this enables a transition to represent a number of closely related but different *executions*. The variables are the key instrument of collapsing a set of related transitions in low-level nets into one transition in CP-nets. The arc expression

---

[22] A full Petri net bibliography (more than 4000 entries) is published in the proceedings of *Advances in Petri Nets*, see, for example, [21].

[23] It is possible to select other languages to augment the nets, an interesting variant using object-oriented features and a pascal-like language is available to public on ftp::ftp.utas.edu.au:departments/computer_science/loopn, it was developed by Charles A. Lakos at the University of Tasmania. The distribution comes complete with some valuable papers describing the specifica of his nets (including simulated time, net modules, inheritance/polymorphism of modules and colour sets, objects as tokens, etc.), some applications, and includes the net-simulator LOOPN (written in C) [14].

[24] Peterson used a similar construct, called bags, in his book [19].

[25] Notational remark: Let c be a colour in the colour set C. If a place P contains two tokens of colour c, the notation 2'c is equivalent to the multi-set {c,c}.

of arcs to output places still evaluates to an appropriate multi-set, but the different possibilities of bounding values to the variable in the expression can lead to a different multi-set each time the transition occurs. If the arc expressions surrounding a transition t contain variables, the transition t is enabled if there exists at least one binding of values to variables such that the evaluation of the arc expressions results in multi-sets currently present in the input places of t. A transition can be enabled under a marking for different bindings. Since it is convenient to view transitions and bindings together in this context, the following is used: A pair (t, b) of transition t and binding b is called a *binding element*. A binding element may be enabled under a marking and transform one marking to another. The terms *Reachability* and *Direct Reachability* are used as above. It is simple to extend the notion of independence and concurrent enabledness to binding elements. Again, independent binding elements can form steps which can occur "in paralle". It is further possible to augment the transition with guards. Guards are boolean expression over the same set of variables as used in the surrounding arc expressions. Guards express constraints on the transition which have to be fulfilled to allow the transition to be enabled[26]. When a variable appears in more than one arc expression surrounding a specific transition (or in the guard and one or more arc expressions) the variable will be bound to the same value in each of its appearances. The binding specifies a colour for the variable and this colour is used for every appearance. It should be noted that the appearance of a variable x around a transition t_1 is completely independent of an appearance of x around a transition t_2. Both appearances can be bound to different values in one single step.

Thus, a CP-net consists of three different parts:

- the net structure (i.e., places, transitions and arcs)
- the declarations of colour sets
- the net inscriptions (i.e., the "text" strings attached to each structural element of the net).

The net inscriptions presented so far describe the basic types of inscriptions used in CP-nets. There are other inscriptions, as time delays and to describe hierarchical relationships, used. The complexity of a CP-net model is distributed among the above mentioned parts—as one can see from almost every motivating example in the literature, especially from the one presented in the first chapter of Jensen's CP-net book ([13]), the more sophisticated elements of high-level nets are mainly used to simplify the structure of the net: Colour sets allow to be more expressive in describing the modeled state of the underlying system, and thereby make it possible to reduce the number of places. Arc expressions and guards with variables allow to be more expressive in describing changes-of-state, thus leading to a reduction of the number of transitions. It is sometimes very tempting to reduce the structure of the net as far as possible—but this might not be very useful! The structure is the most descriptive part of any net type.

---

[26] Guards are normally enclosed in square brackets, e.g., $[x = 5]$, to avoid confusion with arc expressions.

If it looks confusingly, it is surely a good idea to represent more aspects of the system in declarations and inscriptions—but, on the other hand, the structure is the best part of the net to use as a starting point for communication between modelers, between modeler and modeling process supervisor, between modeler and expert, and between modeler and model-user!

In other words: The possibility to coherently represent parts of the model in a graphical way and to use these parts as a base for communication throughout the whole modeling process (including revisions, improvements etc.) is one of the main advantages of Petri nets, and should not be given away by restricting the role of structure in the model too much. What makes Petri nets more attractive with respect to model quality than other graphically oriented modeling methods is the way complexity can be distributed among the components, and the way certain properties of the model can be analysed in either a strictly formal way (invariant analysis, reachability) or a "simple", typically human way: Let's have a look at the net.

To allow a few considerations about the different/common features of Petri nets and system theoretical systems a formal definition of CP-nets is given (from [13]) :

Definition 2. A non-hierarchical CP-net is a 9-tuple
$CPN = (\Sigma, P, T, A, N, C, G, E, I)$ satisfying the requirements below:

1. Σ is a finite set of non-empty types, called colour sets.
2. P is a finite set of places.
3. T is a finite set of transitions.
4. A is a finite set of arcs such that:
   $P \cap T = P \cap A = T \cap A = \emptyset$.
5. N is a node function. It is defined from A into $P \times T \cup T \times P$
6. C is a colour function. It is defined from P into Σ.
7. G is a guard function. It is defined from T into expressions such that:
   $\forall t \in T : [Type(G(t)) = Boolean \wedge Type(Var(G(t))) \subseteq \Sigma]$.
8. E is an arc expression function. It is defined from A into expressions such that:
   $\forall a \in A : [Type(E(a)) = C(p(a))_{MS} \wedge Type(Var(E(a))) \subseteq \Sigma]$
   where $p(a)$ is a place of $N(a)$.
9. I is an initialization function. It is defined from P into closed expressions such that:
   $\forall p \in P : [Type(I(p)) = C(p)_{MS}]$

States and changes-of-state are the core part of both formalisms, CP-nets and deterministic systems. Nevertheless, their representation is much more elaborated in CP-nets. Time is not integrated into the CPN-formalism. There are different approaches available to embed (aspects of) time into Petri nets. Lots of problems can be analysed with a temporal logic defined for CE- and high-level nets without explicitly representing time in the net. Classic operators of temporal logics, as *Next* and *Always*, are available to enhance the descriptive power of propositional resp. predicate logic to formulate and analyse properties of the net (see [24] or [25]). It seems often appropriate to model time with event counters completely by means of the standard net-formalisms [26]. The best-known extensions of Petri nets to explicitly integrate time into the net are Time Petri Nets (see [28]) and Timed Petri Nets ([22]). Time Petri Nets augment the transitions with minimum and maximum time delays defining the interval a transition t has to wait after the tokens in their input places would allow an occurrence of t. In Timed Petri Nets there is an explicit, non-zero duration of transition occurrences. At the beginning of an occurrence of a transition the tokens are removed from the input places and after the duration the tokens become visible in the output places[27]. [28]

The relationship between a CP-net and its environment is not an explicit part of the CPN-definition. The initial marking/expressions can be viewed as a kind of input to the CP-net, but their is no compulsory need to use input information at each execution-step of the net. Since their is no logical fraction between control- and data-flows in Petri nets and places offer an easy way to couple subsystems/sub-nets by using them as input/output places for different nets, it is possible to model large systems, like CIM systems, within the same formalism and to reduce the number of "magical" input/output-relations to a minimum. But, still the problem remains to provide the net with input informations in an integrable way (e.g., interaction with humans). For example, this is solved in the Petri net based tool $\mathcal{L}eu$ (see Footnote 11) by differentiating between automatic transitions and transitions connected with roles users are playing[29]. $\mathcal{L}eu$ is mainly used for modeling and managing business processes integrated with data modeling (EER) and organization modeling (defining roles as sets of permissions to execute activities and, therefore, to change data.

One fundamental concept of Petri nets is their non-determinism. The system-theoretic definition of systems given above defines deterministic systems. This reduces the descriptive power of models build according to the deterministic formalism, but it does not open a real gap between the two formalisms, since non-determinism can be included in system theory and the above definition is kept as simple as possible for descriptive reasons.

The modeling of systems as given in definition 1 by means of CP-nets seems to be adequate. Net type definitions can be viewed as another perspective of systems which defines the core features of state and change-of-state in a coherent manner and emphasizes both aspects equally. This enables a modeler to be as precise as he likes in modeling the inside-capabilities of the underlying system. Furthermore, the introduction of hierarchical elements in Net Theory allows it to

---

[27] One may also augment places or arcs with time (p-timed, a-timed)—this is more or less a matter of perspective.

[28] Transitions with duration somehow contradict the strictly state/event oriented character of Petri nets - in which state is a system while a "long" event occurs? They are also more difficult to analyse—but they allow quantitative simulations of the modeled system, especially if distribution over the occurence duration of transitions are used.

[29] This is further divided in [8] into automatic activities, individual and social interaction activities.

build models on different levels of abstraction, either bottom-up or top-down, to constitute differently detailed representations of the examined system. Since it is always a good idea to gain some experience of what the system to be modeled really is when a modeling process is initiated, the possibility to use Petri nets on almost any level of abstraction encourages a modeler to start a typical modeling process with an attempt to get a grip on the key features of the modeled system and to continue with the modeling of more (top-down) and less (bottom-up) abstract perspectives of the system. Hierarchy helps to keep the parts of the model tractable and communicable.

Hierarchical Coloured Petri nets, as described in [13], allow a formal translation of the hierarchy into an one-level net and ensure the analysability of the net. There is no unbridgeable gap between the levels of abstraction as it sometimes seems to be present in other approaches to the modeling of complex systems (see [4]).

5 Modeling of CIM Systems with Petri Nets

The use of a single method for modeling (on) different levels of abstraction and intervention is still a fiction. Nevertheless, Petri net based methods (e.g.,CPN (with the possibility to integrate IDEF, see [13]), $\mathcal{L}eu$ (utilizing FUN-SOFT nets and integrating the model life cycle) , INCOME (modeling on top of Database Case Tools)) offer a variety of possibilities to model systems problem-adequate, but still based on the same notion of states and changes-of-state introduced by C.A. Petri, and the natural integration of functional (transitions/events) and declarative (places/states) aspects (see [4]). Recent extensions to Petri nets made them more attractive/appropriate for certain applications (as, for example, Fuzzy Petri Nets for operative production control purposes, see [15]). Modelers at every level of intervention and abstraction should be able to chose an appropriate Petri net based method to solve their modeling task—and they should still be able to communicate by means of the models with modelers/decision makers on a physically/logically adjacent level. A senior manager deciding strategic questions as well as an engineer evaluating a specific conveyor systems for control purposes should be able to fulfill his tasks better (with respect to the economic goals of the system)—evaluated from a local and a global point of view.

Petri nets could serve as a base for communication between decision makers. They could be used to harmonize goal systems. The connectivity of models could support the evolution of controllable and understandable CIM systems. Moreover, Petri net models fulfill most of the model-quality criteria given in Section 1. To our opinion, Petri nets are an ideal instrument for integrating models (and therefore systems) by means of *dynamic* modeling.

References

1. Cellier, F. E.: Continous System Modeling. Springer (1991)
2. Chrzastowski-Wachtel, P.: A Proposition for Generalization of Liveness and Fairness Properties. In: Proceedings of the 8th European Workshop on Application and Theory of Petri Nets, Zaragoza, 215-236, (1987)
3. DeMarco, T., Lister, T.: Peopleware: Productive Projects and Teams. Dorset House Publishing, (1987)
4. Fishwick, P. A.: An Integrated Approach to System Modelling. ACM Transactions on Modelling and Computer Simulation, Volume 2, No. 4, (1992)
5. Gaines, Brian: General Systems Research: Quo Vadis. In: General Systems Yearbook, 24, 1-9, (1979)
6. Genrich, H.J., Lautenbach, K.: System Modelling with High-level Petri-Nets. Theoretical Computer Science 13 (1981), North Holland, 109-136
7. Genrich, H.J.: Predicate/Transition Nets. In: W. Bauer, W. Reisig, G. Rozenberg (eds.): Petri-Nets: Central Models and Their Properties, Advances in Petri Nets 1986 Part I, Lecture Notes in Computer Science Vol. 254, Springer (1987), 207-247
8. Gruhn, V.: Communication Support in the Workflow Management Environment LEU. To appear at Workflow-Management Conference, Linz, Austria, Oct. 94
9. Harper, R.: Introduction to Standard ML. Technical Report ECS-LFCS-86-14, University of Edinburgh, Department of Computer Science (1986)
10. Jensen, Kurt: Coloured Petri Nets and the Invariant Method. Theoretical Computer Science 14, North Holland (1981), 317-336
11. Jensen, Kurt: Coloured Petri Nets. In: W. Bauer, W. Reisig, G. Rozenberg (eds.): Petri-Nets: Central Models and Their Properties, Advances in Petri Nets 1986 Part I, Lecture Notes in Computer Science Vol. 254, Springer (1987), 297-329
12. Jensen, Kurt: Coloured Petri Nets: A High-level Language for System Design and Analysis. In: G. Rozenberg (ed.): Advances in Petri Nets 1990, Lecture Notes in Computer Science Vol. 483, Springer (1991), 342-416
13. Jensen, Kurt: Coloured Petri Nets, Volume 1. EATCS Monographs on Theoretical Computer Science, Springer-Verlag (1992)
14. Lakos, C.A.: LOOPN User Manual enclosed in the public ftp-distribution of LOOPN, contact charles@probitas.cs.utas.edu.au
15. Lipp, H.-P.: Einsatz von Fuzzy-Konzepten für das operative Produktionsmanagement. In: atp - Automatisierungstechnische Praxis 34 12, Oldenbourg Verlag (1992), 668-675
16. Manna, Z., Pnueli, A.: The Temporal Logic of Reactive and Concurrent Systems - Specification. Springer (1992)
17. Padulo, L., Arbib, M. A.: Systems Theory: A Unified State Space Approach to Continous and Discrete Systems. W. B. Saunders, Philadelphia, Pa. (1974)
18. Page, Bernd: Diskrete Simulation - Eine Einführung in Modula-2. Springer (1991)
19. Peterson, J.L.: Petri Net Theory and the Modeling of Systems. Prentice Hall (1981)
20. Petri, C.A.: Kommunikation mit Automaten. Schriften des IIM Nr. 2, Institut für Instrumentelle Mathematik, Bonn (1962)
21. Plünnecke, H., Reisig, W.: Bibliography of Petri Nets. In: G. Rozenberg (ed.): Advances in Petri Nets 1991, Lecture Notes in Computer Science Vol. 524, Springer (1991), 317-572
22. Ramchandani, C.: Analysis of Asynchronous Concurrent Systems by Timed Petri Nets. MIT, Project MAC, Technical Report 120, (1974)

23. Reisig, W.: Petri Nets, An Introduction Springer (1985)

24. Reisig, W.: Spezifikation, Modellierung und Korrektheit von Informationssystemen. In: G. Scheschonk, W.Reisig (eds.): Petri-Netze im Einsatz für Entwurf und Entwicklung von Informationssystemen, Reihe Informatik Aktuell, Springer (1993), 19-28

25. Reisig, W.: Elements of a Temporal Logic Coping with Concurrency. Bericht TUM SFB 342/23/92A (1992)

26. Richter, G.: Ereigniszähler. In: G. Scheschonk, W.Reisig (eds.): Petri-Netze im Einsatz für Entwurf und Entwicklung von Informationssystemen, Reihe Informatik Aktuell, Springer (1993), 72-84

27. Rozenberg, G: Behaviour of Elementary Net Systems. In: W. Bauer, W. Reisig, G. Rozenberg (eds.): Petri-Nets: Central Models and Their Properties, Advances in Petri Nets 1986 Part I, Lecture Notes in Computer Science Vol. 254, Springer (1987), 60-94

28. Starke, Peter H.: Analyse von Petri-Netz Modellen. Teubner, Stuttgart (1990)

29. Thiagarajan, P.S.: Elementary Net Systems. In: W. Bauer, W. Reisig, G. Rozenberg (eds.): Petri-Nets: Central Models and Their Properties, Advances in Petri Nets 1986 Part I, Lecture Notes in Computer Science Vol. 254, Springer (1987), 26-59

30. Wedekind, H.: Objektorientierte Schemaentwicklung. Reihe Informatik, Band 85, BI Wissenschaftsverlag (1992)

31. Zeigler, B. P.: Multi-Facetted Modelling and Discrete Event Simulation. Academic Press (1984)

Fuzzy Modeling and Control

Petr Horáček

Czech Technical University of Prague, Faculty of Electrical Engineering,
Department of Control Engineering
Technická 2, 166 27 Prague 6, Czech Republic

Institut National Polytechnique de Grenoble, ENSIEG
Laboratoire d'Automatique de Grenoble
B.P. 46, 38402 Saint-Martin-d'Hères, France

Abstract. The paper presents fuzzy modelling as a tool for design of nonlinear controllers. An architecture and function of a fuzzy logic controller (FLC) is described. Three implementations of FLCs are recognized: look-up table, fuzzy relational and fuzzy logic-neural network based systems. Fuzzy logic - neural networks (FLNN) are analyzed in greater details. Architectures suitable for a FLNN identification for plant and operator modelling are discussed, passive and active regimes for tuning of controller parameters are explained. Unsupervised and supervised identification techniques are described. Principles of direct inverse, model reference, internal model and optimal predictive fuzzy control are explained. Finally, commercial software and hardware tools for design, implementation and evaluation of fuzzy controllers are summarized.

1 Introduction

Automation is an important issue in Computer Integrated Manufacturing. Production and assembly lines, AGVs, robots, etc. have both continuous dynamical and discrete event driven cooperating parts. The CIM systems are large scale, nonlinear and require "intelligent" control. We understand the term "intelligent" as a synonym for "nonlinear", where both continuous time and logical actions are combined. Fuzzy control imitates to some extent the human way of reasoning leading to nonlinear control law. Fuzzy logic controller (FLC) is generally a nonlinear dynamical system. However, the versatility of a FLC is compensated by the design complexity. There is an extensive number of parameters to be adjusted and it is not easy to decide which of them and in what manner they have to be tuned in order to get the desired performance of the control system. This paper presents a systematic approach for fuzzy model identification and controller design.

Fuzzy control or decision system has one or more internal blocks implemented as a fuzzy system processing and generating fuzzy values in the form of fuzzy sets. In most of present days control architectures a fuzzy system is equipped with fuzzification and defuzzification interfaces and the assembly is called a fuzzy controller. As such, a FLC is a nonlinear mapping from an input space X to an output space Y equivalent to the function

$$y(t) = f(x(t))$$ (1)

or a general input-output autoregressive model with external inputs

$$y(t) = f(x(t-1), x(t-2), ..., y(t-1), y(t-2), ...) ,$$ (2)

when a FLC contains buffers for storing past input and output values of x and y and a feedback. The input x is augmented by the vector of the past values of y. In the following text, x will stand for general input and y will denote general output. The function f is not defined analytically but as a mapping between relatively small number of labeled reference points, or rather set of points, and by an interpolation mechanism in between. The input and output reference points are logically linked by if-then rules. The reference points are defined by the fuzzy set kernels. The interpolation between the points depends on the shape of membership functions of the reference fuzzy sets, particular method of input-output implication, the method of inference and the algorithm used for defuzzification. The function f is thus defined by a set of parameters and a set of algorithms operating on them. The advantage of such a rule-based system over the system defined analytically is that the internal structure of a system is transparent to the designer and also to the user as the parameters on which the mapping function is based are expressed linguisticly in human terms and inference algorithms imitate the human way of reasoning.

It is not possible to introduce fuzzy logic and fuzzy set theory in depth here and we refer to the references [3, 9, 12, 20, 21] for the definition of basic terms used throughout the paper. The next chapter describes a fuzzy system as a "brain" of our nonlinear controller.

2 Fuzzy System

A fuzzy system transforms a vector of fuzzy sets as interpreted values of input linguistic variables to a vector of fuzzy sets, interpreted values of output linguistic variables. The transformation process is called approximate reasoning. The internal function of a fuzzy system is defined by reference fuzzy sets and logical relations, described by rules, between them. The alternative is to integrate the reference fuzzy sets and rules into fuzzy relation R between the input and output domain elements. An example of the fuzzy relation for single-input single-output system is graphically represented in Fig.2. When the input fuzzy set is equal to the reference, the fuzzy system generates the reference output pointed by a rule which is fully activated. Reference input implies reference output. The technique used to encode relations between input and output references is called fuzzy implication rule and actual data are passed through a fuzzy system by fuzzy inference.

2.1 Linguistic Variable

The notion of a fuzzy set allows us to define operations with vague and imprecise concepts and we use fuzzy sets for representing linguistic variables. A *linguistic variable* is characterized by a quintuple $(x,T(x),X,G,M)$, where x stands for the name of the variable; $T(x)$ is the term set of x, that is the set of labels used to identify fuzzy sets defined on X; G is a syntactic rule for generating the names of values of x; and M is a semantic function for associating the meaning of a label with the corresponding fuzzy set represented by its membership function.

Consider the following example which is related to Fig.1, where *temperature* is a linguistic variable, $T(temperature)=\{low,comfortable,high,very\_high\}$ and $X=[-10,50]$. We might interpret "*low*" as "*a temperature below about 10°C*", "*comfortable*" as "*a temperature close to 20°C*", "*high*" as "*a temperature roughly between 30 and 35*" and "*very_high*" as "*a temperature above about 40°C*".

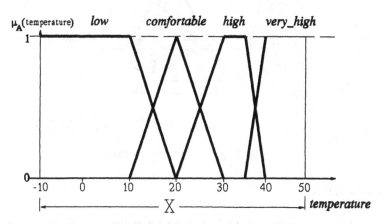

Fig. 1. Fuzzy partition of the universe of discourse

2.2 Fuzzy Implication

Let A and B be the reference input and output fuzzy sets defined on X and Y. Two schemes can be used for coding the relation $R : A$ implies B , a functional scheme

$$R : \quad R(x,y) \quad \forall \ x{\in}X \ , \ y{\in}Y \ , \tag{3}$$

or a linguistic scheme

$$R : \quad if \ A \ then \ B \ . \tag{4}$$

The function R can be either continuous analytical function of its arguments or is defined only in discrete points (x_i,y_j). The discretized version of fuzzy relation $R(x,y)$ is also called fuzzy relation matrix or fuzzy associative memory (FAM). The linguistic encoding of an implication stores definition of reference fuzzy sets A and B separately from binary relation between their labels. Binary relations between reference fuzzy set labels are stored in a table, linguistic associative memory (LAM). For coding a fuzzy relation R between two domain elements x and y belonging to the reference fuzzy sets

A and *B*, defined on *X* and *Y*, we use the formula

$$\mu_R(x,y) = \tau\{\mu_A(x),\mu_B(y)\} \quad , \tag{5}$$

where τ is a T-norm operator symbolizing an operation of conjunction. Frequently used fuzzy implication operators are $\tau=min$ or $\tau=product$. This gives the following fuzzy relations,

$$\mu_{R_c} = min\{\mu_A(x),\mu_B(y)\} \quad , \qquad \mu_{R_p} = \mu_A(x)\cdot\mu_B(y) \quad . \tag{6}$$

Graphical interpretation of the min-rule, introduced by Mamdani, is in Fig.2 .

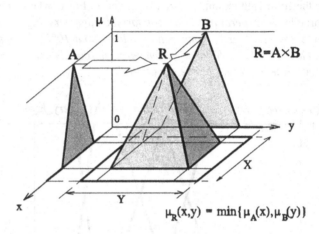

$$\mu_R(x,y) = min\{\mu_A(x),\mu_B(y)\}$$

Fig. 2. Fuzzy implication relation

2.3 Fuzzy Inference

A fuzzy inference, also inference rule or inference engine, is a mechanism of deriving an output from an input which is not necessarily equal to the input reference fuzzy set. There exist two distinct approaches used in fuzzy inferencing, depending on the form in which the knowledge base is stored during inferencing: integrated and distributed knowledge based inferencing.

Integrated inferencing. The technique applies compositional rule of inference on precomputed fuzzy relation *R*, where all the implication information is stored. The output fuzzy value *y* is obtained by projecting the current fuzzy input *x* through the fuzzy relation *R* using selected form of sup-star compositional rule of inference, i.e.

$$y = x \circ R = \{[y, \sup_x(\mu_x(x) \star \mu_R(x,y))], \forall x \in X, y \in Y\} \quad , \tag{7}$$

where $\star$ can be one of the operators from the family of T-norms

$$\star \in \{min, \cdot, \odot, \cap, \ldots\} \quad .$$

Note that in practice max is used for sup and min for $\star$. Let *A'* be a current value of an input linguistic variable *x*. The output value *B'* of *y* corresponding to *A'* is computed as

$$\mu_{B'}(y) = \max_{x} \min\{\mu_{A'}(x),\mu_{R}(x,y)\} \quad , \quad x \in X \ , \ y \in Y \ . \tag{9}$$

The process of reasoning, where a singleton a' is inferenced through the fuzzy relation R, is in Fig.3 .

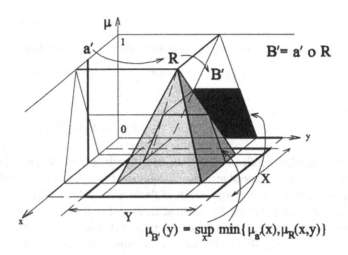

$$\mu_{B'}(y) = \sup_{x} \min\{\mu_{a'}(x),\mu_{R}(x,y)\}$$

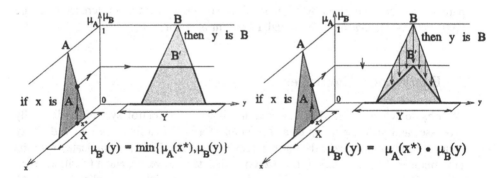

Fig. 4. Reasoning with min implication

Fig. 5. Reasoning with product implication

Distributed inferencing. This inferencing rule does not operate on precomputed implication relation *R* stored in FAM. Both implication and inference are included into the algorithm operating on distributed knowledge stored separately in the form of reference I/O fuzzy sets and a LAM. The method of reasoning for Mamdani's and Larsen's implication is shown in Fig.4 and Fig.5, where a crisp value is supplied as an input. This method requires less memory capacity and is used in most commercially available software tools for design of fuzzy logic controllers.

Note that there is no difference between the result of distributed and integrated inferencing when max-min compositional rule of inference is used.

IF-THEN Rules. Fuzzy IF-THEN rules determine internal logical structure of a fuzzy system relating labels of the input and output reference fuzzy sets. The rules for multiple input cases are typically of the following form. The logical connective "and" in the antecedent part of the rule is interpreted as a fuzzy conjunction in the Cartesian product space of the concerned reference fuzzy sets. The antecedent part of the rule

$$\text{if} \quad x_1 \text{ is } A_1 \text{ and } x_n \text{ is } A_n \quad \text{then} \quad y \text{ is } B \tag{10}$$

$$a \; n \; t \; e \; c \; e \; d \; e \; n \; t \quad | \quad | \quad consequent$$

is evaluated as a fuzzy set with the following membership function

$$\mu_{A_1 \times \cdots \times A_n}(x_1, \ldots x_n) = \min\{\mu_{A_1}(x_1), \ldots \mu_{A_n}(x_n)\} \quad , \text{ or}$$

$$\mu_{A_1 \times \cdots \times A_n}(x_1, \ldots x_n) = \mu_{A_1}(x_1) \cdot \ldots \cdot \mu_{A_n}(x_n) \quad , \tag{11}$$

$$\textit{generally}$$

$$\mu_{A_1 \times \cdots \times A_n}(x_1, \ldots x_n) = \top\{\mu_{A_1}(x_1) , \ldots , \mu_{A_n}(x_n)\} \quad .$$

In fuzzy control, the rules are not chained, like in expert systems, but function in parallel. Sentence connective "also", equivalent to "or" interpreted as a fuzzy disjunction, describes the parallelism. Selection of a fuzzy logic operator for the operation "also" depends on the implication used. For the min or product implication a union operator in the form of max operations is one of the most frequently used. Reference fuzzy sets, IF-THEN rules , implication rule and inference algorithm are design parameters of a fuzzy system. Reference fuzzy sets are often represented by few parameters, rules are given by a relational matrix, implication and inference rules are constructed selecting the T-norm and T-conorm operators.

3   Fuzzy Logic Controller

A fuzzy logic controller (FLC), a building block of a control system, traditionally processes and generates crisp data. The brain of a FLC is a fuzzy system with fuzzy data at the input and fuzzy data at the output ports. The FLC communicates with the environment via a defuzzifier, i.e. fuzzy-to-crisp data converter, and a fuzzifier, crisp-to-fuzzy data converter. Instead of crisp data, fuzzy data might be supplied to the FLC. In this case the fuzzifier is not present.

3.1   Fuzzification Interface

Fuzzification is a straightforward operation. Let x_i is a crisp value provided by a sensor, then the corresponding fuzzy set is a fuzzy singleton

$$A : \mu_A(x) = \begin{cases} 1 \\ 0 \end{cases} , \text{ for } \begin{array}{c} x = x_i \\ x \neq x_i \end{array} , \quad \forall x \in X . \tag{12}$$

3.2   Defuzzification Interface

The defuzzification interface transforms the fuzzy set resulting from the reasoning into a crisp value. Different defuzzification algorithms may be used, among them: mean of maxima - MOM, center of area - COA, center of gravity - COG, center of largest area - COLA, first of maxima - FOM, center of sums - COS and height - HGT. The algorithms operate on the output fuzzy set created after the aggregation of rule conclusions or directly on degree of fulfillment of rules. Let B be the output fuzzy set defined as

$$B = \{(y_j, \mu_B(y_j)) \mid y_j \in Y, j=1,2,...n\} \ . \tag{13}$$

Three often selected defuzzification algorithms , COG, MOM and COS, are described as follows :

$$y_{COG} = \frac{\sum_{j=1}^{n} \mu_B(y_j) \cdot y_j}{\sum_{j=1}^{n} \mu_B(y_j)} \quad , \quad y_{MOM} = \frac{\min_{y \in Y}\{y \mid \mu_B(y)=hgt(B)\} + \max_{y \in Y}\{y \mid \mu_B(y)=hgt(B)\}}{2} \tag{14}$$

The COG method gives smoother output then the MOM which resembles rather a relay characteristic. It is reasonable to combine both methods in some cases. COS automatically includes an aggregation of rule conclusions and applies the COG algorithm afterwards. The aggregation is simply done by an algebraic sum of truncated output reference fuzzy sets which may lead to a function with values greater than one. The result is not a fuzzy set. However, thanks to the defuzzification effects that are not against human sense, the method is tolerated and used in practice. The COS algorithm takes the form

$$y_{COS} = \left(\sum_{j=1}^{n} y_j \cdot \sum_{i=1}^{K} \mu_{B_i}(y_j) \right) \bigg/ \left(\sum_{j=1}^{n} \sum_{i=1}^{K} \mu_{B_i}(y_j) \right) \ , \tag{15}$$

where i is a rule index, K is a total number of rules and B_i is a reference fuzzy set in the consequent part of the i-th rule. The results of three methods are compared in Fig.6.

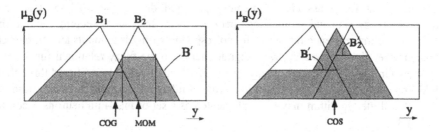

Fig. 6. Comparison of COG, MOM and COS defuzzification algorithms

3.3 Controller Implementation

We distinguish three different implementations of a nonlinear mapping of crisp inputs to crisp outputs used for the function of the central block of the controller: a look-up-table, a fuzzy relation and a fuzzy logic-neural network. All of them are equivalent from the input-output point of view. However, their internal representation is different.

Look-up-Table. The look-up-table is a point to point mapping from the input to the output spaces of crisp values. The look-up-table is prepared off line for every combination of inputs. Nonlinear characteristics are stored as tables and the only task is to read from the table according to the current input an interpolate between the points. No fuzzification, inference and defuzzification is applied on-line. An example of the nonlinear 2 inputs - 1 output controller is in Fig.7. The look-up-table controller has no parameters adjustable on line.

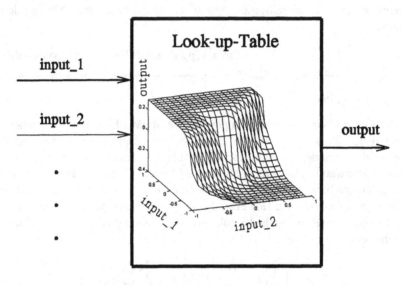

Fig. 7. Look-up-table based controller

Fuzzy Relation. The fuzzy relation based controller has a richer internal representation. It consists of the database and set of data processing algorithms. The database is a fuzzy associative memory prepared during the off-line stage from reference fuzzy sets and implications between them. Instead of keeping membership functions of the reference fuzzy sets for corresponding linguistic inputs and outputs and their interrelations separately, we integrate them into the fuzzy relational function R , as shown in Fig.2. The current fuzzy, or fuzzified, input value A' is filtered through the FAM by applying sup-star compositional rule of inference. In most cases max-min rule is used and the algorithm may be implemented as a set of max-min neurons according to Fig.8.

The fuzzy relation R may be modified on line by unsupervised and supervised learning mechanisms known from the theory of neural networks. The internal structure of the block is in Fig.9, where an output y is generated by the block running in the operation mode and the reference output y^r is supplied to the system in the learning mode. The result from the inference operation is a fuzzy set B' which has to be defuzzified on-line by an explicitly defined defuzzification algorithm. Fig.10 shows how to connect an output from a fuzzy sensor with a relational FLC.

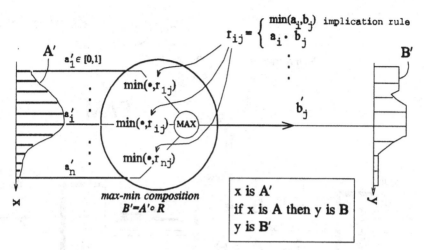

Fig. 8. Fuzzy min-max neuron

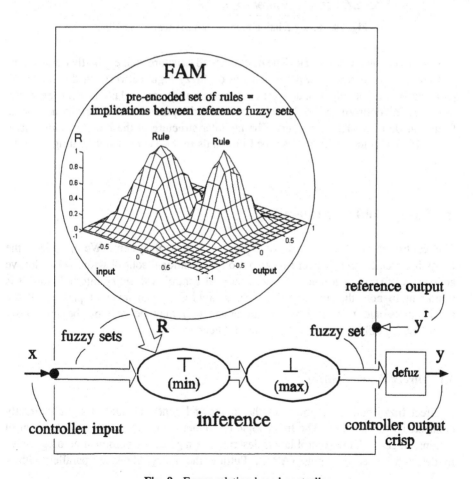

Fig. 9. Fuzzy relation based controller

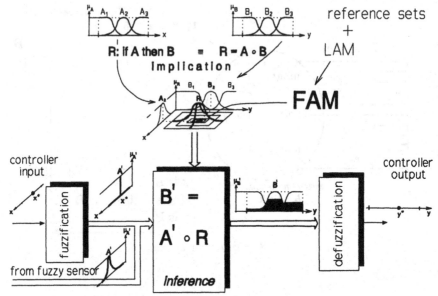

Fig. 10. Fuzzy relation based controller with fuzzy input

Fuzzy Logic-Neural Network. The most detailed structure with explicitly defined data and algorithms operating on them is based on fuzzy logic network. In this model all parameters, i.e. the input and output reference fuzzy sets, relations between them, inference and defuzzification algorithms, are stored separately and are easily accessible from outside as model parameters. The modular structure of the block is shown later in Fig.14. This block will be described in details in the next chapter dealing with the controller design and adaptation.

4 Fuzzy Control System

The controller is always a function part of a control system. We introduce the following architectures: direct fuzzy control, conventional control with fuzzy additive actions and fuzzy supervision and conventional control with fuzzy interpolation. It is important to note that the nonlinear block, a FLC, is an inherent part of all the architectures and that its internal implementation may be one of the previously described regardless the control system architecture.

4.1 Direct Fuzzy Control

A direct fuzzy controller processes the inputs and generates control signals directly connected with the plant. The fuzzy logic reasoning is directly responsible for control actions. In general, the control law is described by a generally nonlinear autoregressive model with external inputs (ARX). Putting the fuzzy block in parallel with a conventional controller may be advantageous in situations when special corrective actions have to be generated by the controller if some special situations happen as indicated by the input data.

4.2 Fuzzy Supervision of Conventional Controllers

In supervisory control, the outputs of a fuzzy system are not used for control actions directly. Instead of acting on a plant they act on adjustable parameters of conventional controllers. One frequently used supervision is shown in Fig.11, where parameters of a conventional PID controller are manipulated by a real-time supervisor, internally implemented as a fuzzy system.

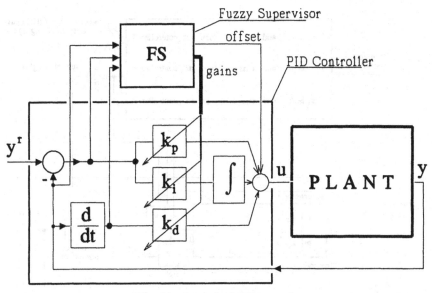

Fig. 11. Supervised PID control

Such a structure extends the functionality of a classical PID controller, still used in most of the control systems in practice, and allows using PID's for control of nonlinear and time-varying plants. Linear state feedback gain manipulation is another example of fuzzy supervision, however the supervision of a PID is more transparent to the designer.

4.3 Fuzzy Interpolation between Analytical Controllers

Instead of supervising parameters of a single controller we can interpolate between multiple analytical controllers running in parallel. We get a fuzzy logic control law selection. The rule consequents have different form from that of (10) and the model is called Sugeno-Takagi model of fuzzy reasoning [19].

There are several forms of rules, depending on controller input variables. The rules might be defined for instance as follows

$$R^j: \quad \textit{if } x_1(k-1) \textit{ is } A^j_{x_1} \textit{ and } x_2(k-1) \textit{ is } A^j_{x_2} \textit{ and } \dots x_n(k-1) \textit{ is } A^j_{x_n} \quad \textit{then}$$

$$u^j(k) = \sum_{i=1}^{n} a_i^j u(k-i) + \sum_{i=1}^{m} b_i^j e(k-i) + c_0^j, \qquad (16)$$

where $x_1, \dots, x_n$ are measured process state variables, e is a regulation error and u is the controller output. The process state variables are compared with the reference fuzzy sets A. a, b and c_0 are parameters of, in this case, linear analytical controllers.

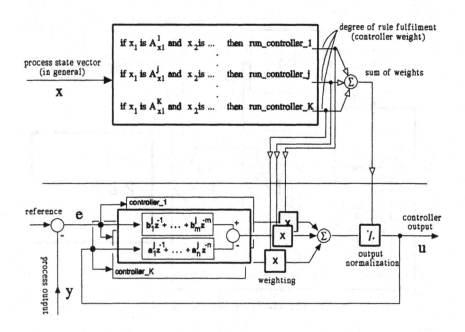

Fig. 12. Fuzzy supervision between analytical controllers

The crisp output is calculated as the weighted average of rule contributions, i.e. by the Fuzzy-Mean method. The reference fuzzy set B is replaced by a crisp output value of an analytical controller activated by the rule. One analytical controller is associated and fired just by a single rule. In another words instead of constant fuzzy value B we use fuzzy singleton whose position moves according to the current state of an analytical model fired by the rule. Fig.12 explains how the system functions. The use of this architecture is desirable when different analytical controllers have been designed for particular setpoints where linearized model of the plant dynamics exists.

5 Fuzzy Logic-Neural Network

A Fuzzy Logic-Neural Network (FLNN), proposed by Lin [13], is a particular implementation of a fuzzy system equipped with fuzzification and defuzzification interfaces. FLNN resembles the multilayer connectionist architecture of artificial neural networks but the function of neurons is layer-dedicated and the connectivity between layers is restricted. The internal structure of a FLNN follows from what was said about a fuzzy system and a fuzzy controller in the chapters 2 and 3. The FLNN is principally built of five layers of neurons.

Layer 1 - input linguistic neurons. Layer 1 is an input buffer for fuzzy values of a linguistic variable. The number of neurons, signal buffers, is equal to the number of inputs. In most FLNNs used for control purposes an input is crisp and its representation is simply a fuzzy singleton.

Layer 2 - input term neurons. Neurons of layer 2 play the role of classifiers. They classify the current value of the corresponding input linguistic variable into the reference fuzzy set by means of fuzzy set operation conjunction for which min operation is normally used. Neurons filter the input signals. The reference fuzzy set related to the particular neuron is encoded into the neuron by means of its membership function. There are two distinct forms of membership function representation: parametrized analytical function and point-by-point representation of discretized MF of a general shape.

Layer 3 - rule neurons. A neuron of the level 3 represents a rule premise and the set of rule neurons, together with the output synaptic links to the level 4, form a rule base. Output synaptic links play the role of the implication. Adding the function of layer 2 and 4 neurons we get an inference engine. Layer 3 neurons determine a logical form of rules. The antecedent part of a rule is formed by the links from the input term neurons. The link to the output term neuron stands for the rule conclusion. Input links play the role of AND and the output link that of THEN.

Layer 4 - output term neurons. These neurons are programmed the same as the layer 2 term neurons but for the output reference fuzzy sets. In addition, they aggregate conclusions of rules having the corresponding reference set in the consequent. The sentence connective ALSO (OR) is partially done. Note that an operation of fuzzy sets disjunction is related to the connective ALSO. Synaptic links to the layer 4 and the layer 4 itself play the role of the implication.

Layer 5 - output linguistic neurons. Two neurons of this layer are related to the output linguistic variable. The first is an input buffer during the learning stage when the output reference signal activates the network in the reversed direction of a signal flow. The second, aggregation-defuzzification neuron, aggregates and subsequently defuzzifies rule conclusions concerning the output linguistic variable represented by the pair. The aggregation completes the ALSO connective partially computed in the preceding layer.

There are natural connectivity constraints between the layers. No flexibility is allowed between the layer 1 and 2 as a particular input linguistic neuron must be connected to all term neurons with labels related to this linguistic variable. A rule can finally have only a single conclusion, that is a rule neuron has links maximally to one term neuron per output linguistic variable.

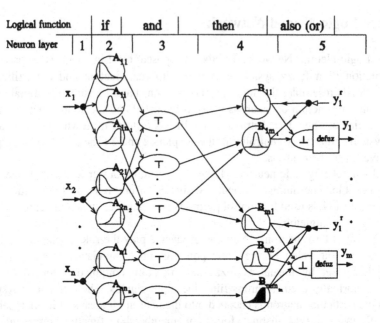

Fig. 13. Fuzzy logic-neural network

The connectionist implementation of a fuzzy system including defuzzification is in Fig.13. Our desire is to keep the function of the FLNN transparent, interpretable and fully parametrized in order to be able to understand what knowledge, where and how it is encoded. The possibility of setting the initial network configuration and the parameters on linguistic level is another important feature. We introduce a general description of a neuron function, similar to that of conventional neural networks. Two functions determine properties of a neuron. The first one aggregates input signals of the neuron transmitted from the p connected neurons of the preceding layer. It has the form of

$$y^{(k)} = f(x_1^{(k)}, x_2^{(k)}, \ldots x_p^{(k)}; w_1^{(k)}, w_2^{(k)}, \ldots, w_p^{(k)}) \quad , \tag{17}$$

where y is an aggregated input, x_i is a particular input, w_i is a parameter, usually a weight, associated with this input and k is the layer index. The second function, the signal function, processes the aggregated input and sends the resulting signal to the neuron(s) of a subsequent layer,

$$output = o_i^{(k)} = a(f) \quad , \tag{18}$$

where a is an activation function and i is an index of the neuron in the k-th layer. We summarize the functional description of layers as follows.

Layer 1 :
$$f = x_i^1 \quad , \qquad a = f \quad . \tag{19}$$

Layer 2 : Parametrized analytical membership functions are suitable for encoding into the input term neuron. Frequently used are bell-shaped, sigmoidal, triangular or trapezoidal functions defined by 2 parameters for symmetric and 3-4 parameters for asymmetric functions. For the case of bell-shape function we get

$$f = \mu_i^j(x_i; m_{ij}, \sigma_{ij}) = -\frac{(x_i - m_{ij})^2}{\sigma_{ij}^2} \qquad , \qquad a = e^f \quad , \qquad (20)$$

where m_{ij} is the center or mean and σ_{ij} is a parameter related to the width of the fuzzy set. A general function can also be considered in the form of a table where i is an index of the linguistic variable and j is an index of its reference fuzzy set

$$A_{ij} = \{x_i(k), \mu_j(x_i(k))\} \quad . \qquad (21)$$

Layer 3 : Rule neurons have no parameters and their logical AND function is described as

$$f = \min(x_1, x_2, ..., x_p) \quad or \quad f = x_1 \cdot x_2 \cdot ... \cdot x_p \qquad , \qquad a = f \quad . \qquad (22)$$

Layer 4 : Switching between two neurons connected to the same output reference term neuron is done according to the regime selected. The function of an operation mode uses T-conorm operator of fuzzy disjunction as

$$f = \max(x_1, x_2, ..., x_p) \qquad , \qquad a = f \quad , \qquad (23)$$

or Lukasiewicz's bounded sum T-conorm operator

$$f = \sum_{i=1}^{p} x_i \qquad , \qquad a = \min(1, f) \quad . \qquad (24)$$

In the reversed mode, used during the network training, the neuron of the layer 4 together with that of the layer 5 functions the same as a couple of level 1 and 2 neurons. A neuron stores an output reference fuzzy set in the form given by (20) or (21).

Layer 5 : Buffers have a trivial function

$$f = y_i \qquad , \qquad a = f \quad . \qquad (25)$$

Aggregation-defuzzification neurons aggregate different rule conclusions concerning the same linguistic variable y_j , coming out of the net and defuzzify. To complete these operations the neuron receives the rule activation levels x_i and parameters of the output reference fuzzy sets m_{ij} , σ_{ij} and compute crisp outputs depending on the defuzzification method and a class of the reference fuzzy set processed. For the modified COG/COA method we get

$$f = \sum_{i=1}^{p} (m_{ij}\sigma_{ij})x_i \qquad , \qquad a = \frac{f}{\sum_{i=1}^{p} \sigma_{ij}x_i} \qquad , \qquad (26)$$

and for the case of singletons

$$f = \sum_{i=1}^{p} m_{ij} x_i \quad , \quad a = \frac{f}{\sum_{i=1}^{p} x_i} \tag{27}$$

The internal representation of a fuzzy system with crisp-fuzzy and fuzzy-crisp converters is highly modular. There is no problem to alter the layer function and to trace how the fuzzy sets, or rather their parameters, are treated passing through the network. The role of the network configuration and parameters stored in the neurons is clearly recognized and this property will be used inversely during the network training and adaptation.

6 Architectures for Fuzzy Model Identification

Design of a control system requires at least one of the following sources of formalized information: (i) model of a plant, (ii) model of an operator controlling the plant manually, (iii) model of the desired behavior of the control system. Our desire is to use the fuzzy logic neural network as a general interpolator in the modelling and subsequent control. The forward and inverse models of a plant are used in the model based control. The forward model will be used for off line experimentation and prediction purposes. The inverse model cascaded with the plant may possibly cancel the plant dynamics so we are able to drive the plant output freely due to the fact that the input-output characteristic of a serial connection tends to be one. Thus the model of the plant is directly suitable for control purposes. The other way of a control system design is to copy a model of a controller, approaching the inverse plant model, which is implicitly used by the human operator. In both cases driving inputs and resulting outputs of an object to be identified are available for controller design. The third source of information, limited to the controller tuning, is to be used in a different way as the model of a controller plays an active control role during the tuning process.

Direct and inverse plant models represent nonlinear system input-output mapping function which can be viewed as nonlinear discrete time difference equation of the form

$$y(k+1) = f\left(y(k),...,y(k-n+1),u(k),...,u(k-m+1)\right) , \tag{28}$$

where y is a vector of plant outputs and u is a vector of control actions of a controller. The given model predicts the value of the future output depending on the n past values of this output and m past values of the system input.

6.1 Forward Modelling

In the forward model, the inputs to the model are the control inputs to the system and the outputs are the measured output signals. The forward model runs in one of two modes as follows

$$y^m(k+1) = \hat{f}\left(y^p(k),...,y^p(k-n+1),u(k),...,u(k-m+1)\right) ,$$

$$y^m(k+1) = \hat{f}\left(y^m(k),...,y^m(k-n+1),u(k),...,u(k-m+1)\right) .$$

(29)

In the first mode the predicted future value of the output is derived from measured past values of the plant, while in the second mode the prediction is derived from the model itself. There is a feedback from the model output to the model input. The second mode may be used when dealing with noisy measurements since it avoids problems of bias caused by noise on the real system output. The nonlinear mapping function f is implemented as a fuzzy relation with fuzzification and defuzzification interface.

6.2 Inverse Modelling

The inverse model, cascaded with the plant it models, cancels the dynamics of the plant so it is possible to govern the plant output arbitrarily. This simple idea is the base for the use of an inverse model as a controller. However to obtain the inverse model is not straightforward. Assume that we implement the mapping f as a FLNN. This computational network has the learning capabilities and interpretable logical structure after learning. It is important to emphasize that the human readable parameters of the network interpreted as fuzzy IF THEN rules enable us to validate whether the parameters learned have logical sense or not. The nonlinear input-output relation on the fuzzy-neural network modelling the plant inverse is

$$u(k) = \hat{f}^{-1}\left(y^*(k+1),y^p(k),...,y^p(k-n+1),u(k-1),...,u(k-m+1)\right) . \quad (30)$$

The inverse model receives the future plant output y^* modelled by the training reference known ahead, current and past plant outputs and the past values of the plant inputs.

6.3 Identification Techniques

We recognize unsupervised and supervised techniques for the FLNN model identification. In the unsupervised case, the FLNN is activated from both sides by correlated reference data and thus the FLNN cannot play an active role of a controller during the identification stage. It is an off-line identification. The block diagram for the identification of an operator is in Fig.14 and the similar scheme may by used for plant identification.

The supervised identification techniques works either off-line or on-line. This is enabled by the fact that the FLNN works in the operation mode and generates outputs during the structure and parameter identification. The output generated by the network is compared with the reference and the error computed is used for the adaptation of the FLNN. The off-line version of the operator identification is shown in Fig.15.

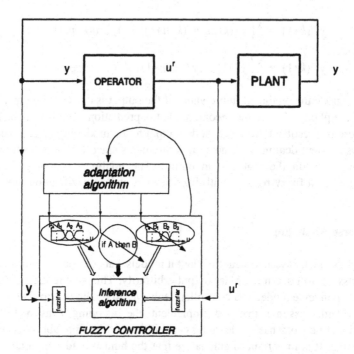

Fig. 14. Unsupervised identification of an operator

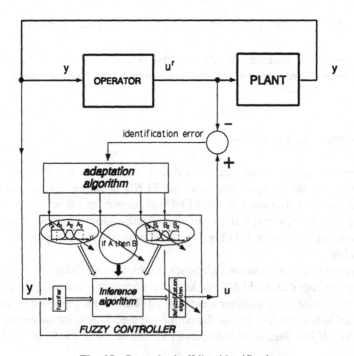

Fig. 15. Supervised off-line identification

Two architectures for inverse modelling of a plant are given in Fig.17 and Fig.18. The on-line supervised identification of a controller is shown in Fig.16. The desired trajectories of e^r and u^r are regarded as target references and the difference between the reference and the reality implicitly influenced by the current setting of FLNN's parameters is used for minimizing the difference during the time.

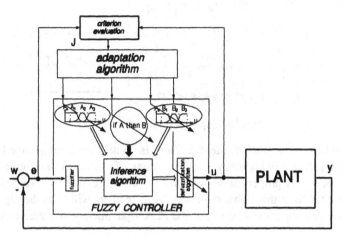

Fig. 16. Supervised on-line identification

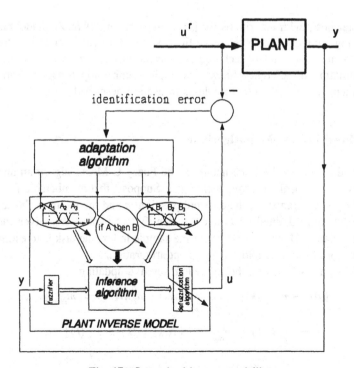

Fig. 17. Supervised inverse modelling

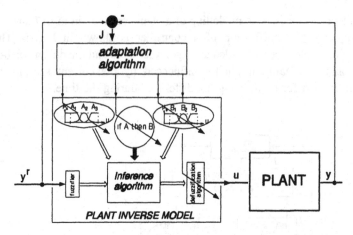

Fig. 18. Supervised inverse modelling

The problem of a FLNN identification is generally a constrained nonlinear optimization problem with many local solutions. The success in finding a global optimum depends on proper initial conditions, the setting of the FLNN parameters in our case, not far from the global optimum. Next chapters will treat the algorithms for unsupervised and supervised tuning of the FLNN parameters in a greater detail.

7 Unsupervised Identification

This chapter deals with algorithms for preliminary tuning of fuzzy model parameters, that is setting input and output reference fuzzy sets and internal logical structure, LAM. Fuzzy C-Means is a standard technique of unsupervised fuzzy clustering suitable for rough estimation of reference fuzzy sets, their membership functions and relations between them [2, 18]. The other methods in use are described here.

7.1 Modified Feature-Map Algorithm

This method [10] is of the same nature as the Fuzzy C-Means algorithm and belongs to a family of statistical clustering techniques. Suppose, that membership functions are described by two parameters, the centroid position m and the width of the set σ. Thus instead of functions defined by a table we identify two parameters. Given the training input/output data $x(k)$, where k states for the sample index, the task is to estimate these parameters for all the sets allowed by the configuration.

The algorithm is based on the following iterative updating

$$\| x(k) - m_{closest}(k) \| = \min_{\forall i \in [1, \varphi]} \| x(k) - m_i(k) \| \quad \Rightarrow \quad m_{closest} \quad , \tag{33}$$

$$m_{closest}(k+1) = m_{closest}(k) + \alpha(k) \ [x(k) - m_{closest}(k)] \quad , \tag{32}$$

$$m_i(k+1) = m_i(k) \quad , \quad m_i \neq m_{closest} \quad , \tag{31}$$

where $\alpha(k)$ is a monotonically decreasing scalar learning rate and p is a total number of fuzzy sets to be placed. This adaptive algorithm is used independently for each input and output linguistic variable. The iterations can be stopped after elapsing the time allowed for the iterations, measured in number of iterations, and the current centroid positions are read.

Once centroids are found the width of membership functions can be determined by the N-nearest-neighbors heuristic where the following objective function is to be minimized with respect to the widths σ_i's :

$$J(\sigma) = \frac{1}{2} \sum_{i=1}^{p} \left[\sum_{\forall j \in p_{nearest}} \left(\frac{m_i - m_j}{\sigma_i} \right)^2 - r \right]^2 \quad , \tag{34}$$

where r is an overlap parameter. An alternative for 2-phase methods, where the second stage serves for fine tuning, is a simplified determination of the width by the formula

$$\sigma_i = \frac{| m_i - m_{closest} |}{r} \quad . \tag{35}$$

7.2 Unsupervised competitive learning

After locating the membership functions, the training signals are passed through the input term nodes in the forward direction and backwards through the output term nodes. The signals filtered by the term nodes are then used for activation of the internal logical structure of the fuzzy model and used for strengthening or weakening the links between the input and output reference fuzzy sets. These links are between the layer 3 and layer 4 neurons. We get the firing strength of each rule node. Based on these rule firing strengths, denoted as $o^3_i(t)$, and the level of activation of term nodes at layer four, denoted as $o^4_j(t)$, we want to connect the correct consequent to the existing rule. This will be done by a competitive learning mechanism [11]. We suppose that the FLNN is fully connected at this time. Each rule has connection links to all the output term nodes. To select a single link we assign a weight to these links denoting w_{ij} a weight for a link between rule node i and output term node j. Various competitive learning laws may be used for an iterative tuning of the link weight. The calculations are done when new training data arrives according to the formula

$$\dot{w}_{ij} = o_j^{(4)} (-w_{ij} + o_i^{(3)}) \quad , \qquad w_{ij} \in [0,1] \quad . \tag{36}$$

The link strength can change only when the receiving neuron is activated. The strength weakens when the transmitting neuron is not activated enough at the same time. When the rule conditions are fully satisfied and the checked conclusion is also highly active, the link increases its weight, i.e. the link strengthens.

After the competitive learning when all the training data have been processed, a link with maximum weight leading from the rule node is chosen and the other links are deleted. If neither from the link weights exceeds a given threshold all are deleted and this tells us that the rule can be deleted since it does not effect the outputs.

The rule base can be further compressed, logically filtered. Compression relies on grouping of rule nodes into a single one when the following conditions are simultaneously satisfied: (i) rule nodes have exactly the same consequences, (ii) some rule conditions are common to all the considered rules, (iii) the union of the remaining conditions of these rule nodes represents the whole term set of some input linguistic variable.

8 Supervised Identification

In the supervised identification we know in what direction and how much the parameters should change. The model quality measure is expressed by an analytical function with model output and reference output as the main arguments. The integral criterion is considered when the past model behavior influences the current model parameters updating. Introducing an explicit criterion, we formulate a nonlinear optimization problem. Two groups of methods are considered, genetic (GA) and backpropagation algorithms (BP). The first one represents global search, the second local search methods.

8.1 Genetic Algorithms

The identification of parameters which defines reference membership functions and an LAM can be carried out by the scheme of a genetic algorithm (GA) [5]. The algorithm is a sort of simulated evolution search algorithm to find the global optimal solution of the unknown function $y=f(x)$. The method belongs to the family of random search algorithms. A typical GA relies on three operations: reproduction, crossover and mutation. The flow of genetic algorithm can be simply explained by the following steps:

1: Determine the string length to represent the variable x in the form of binary code *1,0* and establish an initial string.

2: Construct an initial population using the initial string from the step 1.

3: Transform each string of population into decimal code and then find the fitness value of function $f(x)$.

4: Select the proper strings according to the fitness value to form a gene pool.

5: Obtain a new population through the evolution process of crossover and mutation between the genes of STEP 4.

6: Repeat steps 3-5 until some measure is satisfied.

GAs differ from conventional optimization and search procedures in four aspects : (i) GAs work with a coding of the parameter set rather then with the parameters themselves; (ii) GAs search in a set (population) of potential solutions, not a single solution; (iii) GAs use objective function and not derivatives or other auxiliary information; (iv) GA's use probabilistic transition rules to guide their search. GAs work as blind search algorithms as they exploit only the coding and the objective function value to determine plausible trials in the next generation. GAs ignore the problem-specific information. When this information is available it is advantageous to combine GA with other problem oriented methods.

The only stage where the problem information is involved is the coding of a potential solution into a string and evaluation of a fitness function related to the string. The remaining operations of GA are problem independent.

Coding. There is no general rule of how to encode the fuzzy model parameters into a chromosome. Chromosomes can be strings of bits *(1 1 0 1 0 0 1)* but also lists of real numbers *(0.2 0.7 1.0 0.0)* or list of symbols or labels *(LP SN ZE SP ZE)*. Consider triangular shaped input-output membership functions. They can distort, change the base, and translate. An optional constraint can be considered for the triangles at the extreme limits of the universe of discourse, they have to be the right triangles. Just two sets can overlap and the constraint

$$\mu_{A_i} + \mu_{A_{i+1}} = 1 \tag{37}$$

is satisfied in the overlap region. Under these constraints, the only parameters are the positions of kernels of triangles. One of the useful coding schemes is shown in Fig.19.

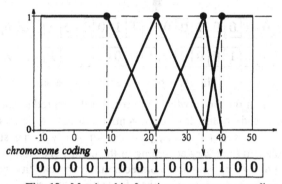

Fig. 19. Membership functions parameters encoding

We can parametrize either the input fuzzy sets, output fuzzy sets or both. To obtain reasonable fit to the reference input/output map, designers use shiftable singletons at the output while the input space is partitioned by fixed fuzzy sets with triangular membership functions. Scaling factors, the input channel gains, can be added to the parameter list as well.

Fitness Function. The fitness function tells us how good is a string. There is no unique form of this function. The fitness function inversely proportional to

$$F(string_i) = \sum_{k=1}^{K} (y(x_k;i) - y''(x_k))^2 + \alpha N_i \quad , \quad \alpha \geq 0 \tag{38}$$

is used in many cases. Here x_k is a k-th vector of model activation data, the inputs, y is a fuzzy model output and y' is the reference output. N_i corresponds to the total number of ones in the i-th string, i.e. the current number of fuzzy sets, α is a weighting factor. The first part of the criterion F_i, which is to be minimized in this particular case, evaluates how well the fuzzy model, determined by the string i, fits to the reference input-output map. The second part penalizes the increasing number of fuzzy sets.

Genetic Operations on Chromosomes. Ones the coding scheme and the fitness function is specified an initial set of potential solutions, i.e. set of strings called a generation, is generated and an iterative procedure of applying standard genetic operations on chromosome strings starts. These operations are selection (reproduction), crossover and mutation.

The selection copies chromosomes with higher fitness value into the gene pole used for further operations. Crossover recombines pairs of chromosomes by swapping parts of them from one or several randomly selected point(s) to create two new chromosomes as shown in Fig. 20.

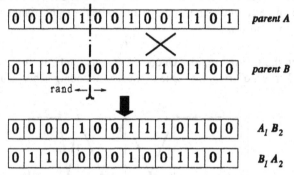

Fig. 20. Crossover genetic operation

Mutation is a random operation which ensures that the reproduction and crossover operations will not lead to the loss of potentially useful genetic material. It is performed on randomly selected items, positions of newly borne strings. Several versions are possible and they vary according to the coding scheme. In binary coding we negate every bit of a string with given biased mutation probability, where relatively higher probability is applied for clearing then for setting the string item. Mutation of a string with real coding is usually done by setting randomly selected item to zero. It corresponds to removing one of the fuzzy set from the active list used in the model. This operation is of course occasional and happens with low mutation probability. Ones removed, it can be optionally activated later but with even lower probability. Similar disable-enable operation is applied on strings of labels. The alternative is to randomly select two items in a string and exchange their content.

Elitest strategy is an optional operation which randomly removes one solution from the newly borne generation and replaces it by the best one from the last generation.

Termination test. To stop the iterations of the algorithm one of the following stopping criteria or their combination can be used : max number of iterations allowed, stabilization criteria, low evolution in the population.

Internal Logical Structure Identification. Suppose that we have n inputs and one output system and $n_1, n_2, ..., n_n$ different fuzzy partitions of a universe of discourse for corresponding input linguistic variable $x_1, x_2...x_n$. Thus the maximal number of distinct conclusions is equal to a max number of rules $n = n_1 n_2 .. n_n$. Suppose that we have identified also the number m , $m \leq n$, of output reference fuzzy sets or just their centroids represented by singletons. The task is to assign labels from the list

$B=\{\emptyset,B_1,B_2,...B_m\}$, to n rules or rather n rule antecedents. Empty label $\emptyset$ indicates that the rule is disabled and will not be evaluated. The internal architecture of the FLC is:

$$i\text{-th rule}: \quad \text{if } x_1 \text{ is } A_{i,1} \text{ and } x_2 \text{ is } A_{i,2} \text{ and } \cdots \text{ and } x_n \text{ is } A_{i,n}$$

$$\text{then } y = b_i \quad , \quad i=1,2,...n_1 \cdot n_2 \cdots n_n \tag{39}$$

Every rule antecedent is described by a particular combination of input reference fuzzy set labels A_{ij} and the complete rule base can be viewed as

rule		antecedent	consequent	
1	:	$(A_{11},A_{12},...,A_{1n})$	b_1	$b_i \in \{\emptyset,B_1,B_2,...,B_m\}$
n_1	:	$(A_{n_1 1},A_{12},...,A_{1n})$	b_{n_1}	$i=1,2,...,n_1 \cdot n_2 \cdot \cdots n_n$
				$m \leq N = n_1 \cdot n_2 \cdot \cdots n_n$
$n_1 \cdot n_2 \cdot \cdots n_n$	:	$(A_{n_1 1},A_{n_2 2},...,A_{n,n})$	$b_{n_1 n_2 \cdots n_n}$	

A string of labels

$$S = (b_1, b_2, ..., b_N) \tag{40}$$

represents potential solution of the assignment problem. The function (38), or rather the value inversely proportional to it, may be regarded as a fitness function. Note that we balance minimal structure solutions, where higher number of rules is disabled, with the fit accuracy. The string S represents an assignment and form potentially valid fuzzy model. The fitness function is given by (38), where the second penalty term minimizes the model complexity.

8.2 Error Backpropagation

Suppose that the FLNN topology has been identified and the input and output spaces partitioned. We assume that the FLNN parameters are set close to the searched optimum with respect to the training patterns available. The task is to complete the fine tuning. Supervised identification based on backpropagation gradient technique is used for the local search. The supervision is governed by the criterion

$$J(\theta(k)) = \frac{1}{2}\left[y'(k) - y(x(k),\theta(k))\right]^2 , \tag{41}$$

where y is an output computed by the net, y' is a reference output and x an activation input processed simultaneously by the identified object and a network. The current net output, and consequently the criterion, depend on the net parameters. These parameters may be parameters of reference membership functions as considered here, but also weights of synaptic links between the layers. The criterion (41) can be more general and penalize bad conditioned logical structure of the net, e.g. number of rules, complexity of particular rules, etc. [1]. $y(x,\theta)$ is a composed function with parameters hidden in different layers of the net. Standard gradient algorithm for minimizing the criterion with respect to the vector of parameters θ is given by the equation

$$\theta(k+1) = \theta(k) - \eta(k)\nabla_\theta J(y(x(k),\theta)) = \theta(k) - \eta(k)\Delta\theta(k) \quad , \qquad (42)$$

where η is a gradient algorithm step size. It follows from the elementary differential calculus that the gradient of a composed function is given by the formula

$$\nabla_\theta J(y(x(k),\theta)) = \frac{\partial J(y(x,\theta)))}{\partial\theta}\Big|_{x(k)} = \frac{\partial J}{\partial y}\frac{\partial y}{\partial\theta}\Big|_{x(k)} \quad . \qquad (43)$$

The formula will be used in network error backpropagation. When the neuron with a parametrized function is reached during the process, the error propagated through the neuron is differentiated according to the neuron's parameters. These partial derivatives are then used for local parameters updating. We derive general formulae for an abstract neuron with a parametrized function as follows

$y = y(x_1,x_2,...,x_n;\theta)$ *neuron function*

$\dfrac{\partial J}{\partial y} = \delta_y$ *neuron error*

                                                                     (44)

$\dfrac{\partial J}{\partial x_i} = \dfrac{\partial J}{\partial y}\dfrac{\partial y}{\partial x_i} = \delta_y\dfrac{\partial y}{\partial x_i} = \delta_{x_i}$ *backpropagated error to i-th predecessor*

$\Delta\theta = \dfrac{\partial J}{\partial\theta} = \delta_y\dfrac{\partial y}{\partial\theta}$ *parameter update*

As all the neurons are algorithmic we can propagate the net error from the end to the neuron where it is born and compute the partial derivative of J with respect to the parameter. The partial derivative is then used for the parameter updating in a straightforward manner. The particular formulas for error propagation are based on analytical function of a layer dedicated neurons introduced in the chapter 5 and can be found in [7, 13].

The supervised identification algorithm has two phases : (i) network state evaluation, i.e. computation of activation levels invoked by the current external input pattern x, in the forward direction; (ii) network error backpropagation in the backword direction. The second phase is initialized with the output error evaluation by the formula

$$\delta_{y_j}^{(5)} = \frac{\partial J}{\partial y_j} = -[y_j^r(k)-y_j(k)] \quad , \quad for \ \ j=1,2,...,m \qquad (45)$$

This error is backpropagated through the network till the layer 2. The neurons carrying parametrized functions, that are the layer 4-5 and layer 2 neurons, are updating their parameters leading to decrease the network error in the next iteration run. The nonparametric neurons only direct the signals propagated through the network according to the current network state computed during the first phase.

Several parameters may be freezed and excluded from the list of parameters to be tuned in order to reduce the dimensionality of the optimization problem and to prevent the changes of parameters we believe they are already set correctly. This is possible in the FLNN as the present state of the network is interpretable in linguistic terms.

9 Model Based Fuzzy Control

A model based control system incorporates a block realizing a model of the plant dynamics. Both direct and inverse models can be used, the first one playing the role of a predictor of the evolution of the plant output and the second for elimination of a plant dynamics when cascaded with the plant. The chapter describes various possibilities of how to include the identified plant model into the control system to meet the control objectives as summarized by Hunt [8] for conventional neural nets.

9.1 Direct Inverse Control - DIC

The functionality of this approach is based on the idea that a serial connection of a system and its inverse leads to an identity mapping between the desired system response, i.e. the input to the inverse model, and the controlled system output. The inverse model acts directly as a controller. This control method is common in robotics applications. However it is difficult to develop an exact inverse model in reality and serious robustness problems occur. The lack of robustness is mainly due to the absence of feedback which can be added to the scheme by means of on-line learning, adjusting of inverse model parameters.

9.2 Model Reference Control - MRC

In the MRC, as shows in [16], the desired performance of the closed-loop system is specified by a stable reference model, defined by an input-output pair $\{r(t)...$reference signal $,y'(t)...$desired response to the reference$\}$.

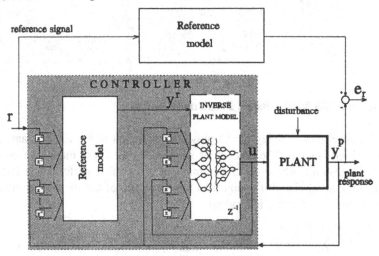

Fig. 21. Model reference control

The control system attempts to make the plant output $y^p(t)$ match the reference model asymptotically, i.e.

$$\lim_{t \to \infty} \| y'(t) - y^P(t) \| \le \varepsilon \qquad (46)$$

for some specified constant $\varepsilon \ge 0$. Fig.21 shows the structure, where the error (46) is used to train the FLNN model working as a controller. In the case where the reference model is an identity mapping the MRC is equivalent to the DIC.

9.3 Internal Model Control - IMC

Here, a system forward and inverse models are used directly as blocks within the feedback loop. In IMC a system model is placed in parallel with the real system. The difference in the output signals is used for feedback and processed by a controller subsystem in the forward path. The architecture of IMC is shown in Fig.22, where an optional filter is designed to introduce desirable robustness and tracking response to the closed-loop system. IMC is limited to open-loop stable systems.

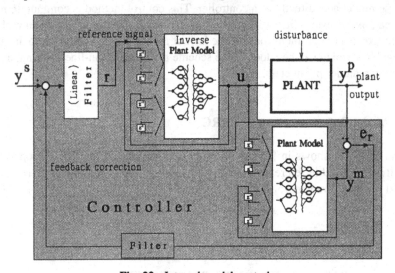

Fig. 22. Internal model control

9.4 Optimal I/O Model Based Predictive Control

In this approach [14] a fuzzy model provides prediction of the future plant response over a specified horizon. The predictions supplied by the fuzzy model are used by an optimization routine which seeks minimum of a specified performance index with respect to a control signal u', subject to the constraint of the dynamical model. The most frequently used index is quadratic

$$J = \sum_{j=N_1}^{N_2} \left(y^*(k+j) - y^m(k+j) \right)^2 + \sum_{j=1}^{N_2} w_j \left(u'(k+j-1) - u'(k+j-2) \right)^2 , \qquad (47)$$

where the constants $N1$, $N2$ define horizons over which the tracking error and subsequent control changes are considered and w_j are the control weights. As a model prediction error increases in time it is not wise to apply the complete trajectory of $u'=(u'(k),u'(k+1),...,u'(k+N_2-1))$ over the specified horizon. Rather the first entry from the sequence of u' is applied. Two versions of this control system architecture are shown in Fig.23 and Fig.24. The second version uses the second FLNN block for identifying the input-output (I/O) behavior of the optimization algorithm in order to replace it during the on-line control phase.

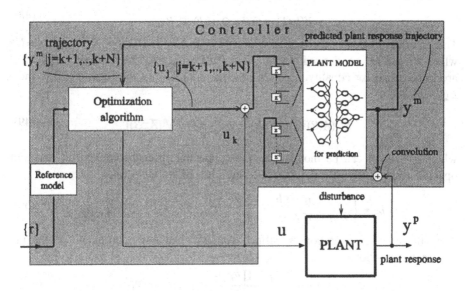

Fig. 23. Model-based optimal predictive control

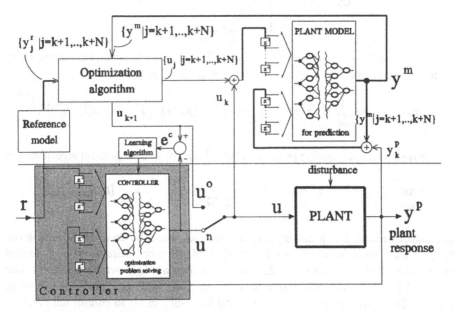

Fig. 24. Model-based optimal predictive control

9.5 Fuzzy Optimal Model Based Predictive Control

Optimal control is aiming at a maximal satisfaction of control goals over constraints specified for the system. The control objective specified by (47) is defined outside the fuzzy model and represents crisp goal evaluation, where all the variables are considered as crisp. The other approach [18] is to express the control objectives as elements in corresponding fuzzy partitions.

Consider the state model of the plant as a single step ahead predictor of the form

$$X(k+1) = X(k) \circ U(k) \circ R \ . \tag{48}$$

The criterium evaluates the degree of satisfaction $X(k+1)(U(k)) = X^*$ and $U(k)=U^*$, where X^* and U^* are desired, referential fuzzy sets. Let A,B be two fuzzy sets which are to be tested for equality. The set-theoretic definition of the fuzzy equality index $EQ(A,B)$ is

$$EQ(A,B)=\frac{1}{2}[(A\subset B)\wedge(A\supset B)+(\overline{A}\subset B)\wedge(\overline{A}\supset B)] \ , \tag{49}$$

where $\subset$ and $\wedge$ can be modeled by the φ-operator (pseudocomplement) and min-operator respectively. Choose the following form of the φ-operator

$$a_i \ \varphi \ b_i = \begin{cases} 1 & if \ a_i \leq b_i \\ b_i/a_i & if \ a_i > b_i \end{cases} \quad , \quad a_i = \mu_A(x_i) \quad , \quad b_i = \mu_B(x_i) \ . \tag{50}$$

Then the equality operation is defined pointwise for each coordinate of the unit hypercube as follows

$$EQ(a_i,b_i) = \begin{cases} \frac{1}{2}\left[\dfrac{b_i}{a_i}+\dfrac{1-a_i}{1-b_i}\right] & if \ a_i > b_i \ , \\ 1 & if \ a_i = b_i \ , \\ \frac{1}{2}\left[\dfrac{a_i}{b_i}+\dfrac{1-b_i}{1-a_i}\right] & if \ a_i < b_i \ . \end{cases} \tag{51}$$

The equality index is used as a fuzzy measure for similarity between the current fuzzy value and a reference. The control objective is given as follows :

$$\max_{U(k)} J \ , \quad J=\frac{1}{nm}\sum_{i=1}^{n}\sum_{j=1}^{m}\gamma_{ij} \ , \quad \gamma_{ij}=\min \ \{EQ(x_i(k+1),x_i^*),EQ(u_j(k),u_j^*)\} \ . \tag{52}$$

The above stated optimization problem is solved iteratively by a gradient algorithm, i.e.

$$\max_{u(k)\in[0,1]^m} J \quad => \quad u^{l+1}(k)=u^l(k) + \frac{\partial J^l}{\partial u^l(k)} \quad , \quad u = [u_1,u_2,.....,u_m]^T \ .$$

The calculations updating the grade of membership might be done every sample on-line, or may serve as a scheme for training of reference statements of a logical processor with if-then internal architecture (*if x then u*). Thus we have a point wise modelled function of a state feedback, mapping $x \rightarrow u$. This is a supervised learning scheme. The tuned logical processor is used in on-line mode to control the system.

10 Conclusion

A reference data driven identification of a nonlinear model, based on a fuzzy system implementation, for plant modelling and controller design has been discussed. However, manual design of FLNN topology and its parameters in the form of a rule base and membership functions dominates in the engineering practice at present. The main reason is that bad reference data may lead to incorrect plant model or controller as is often the case in training of neural networks. To prevent this a designer prefers using rather his experience then data driven algorithms of automatic tuning. It is advantageous to combine the both design approaches and this is a challenge for the software tool developers nowadays. The robustness of identification techniques should also be improved.

Linear input-output mapping functions f , corresponding to the model (2), can be generated when particular membership functions, set of rules and defuzzification technique are used. Thus conventional and modern control theory techniques can be directly implemented in the "fuzzy" environment. For instance the Ziegler-Nichols rules for tuning of a PID like controller could be used for its fuzzy equivalent [6].

The manual design is supported by sophisticated commercial software packages like TILShell® (Togai Infralogic), AB-Flex® (Allen-Bradley), fuzzyTECHTM (INFORM GmbH Aachen), FideTM (Aptronix), CubiCalc® (Hyperlogic), Fuzzy Manifold EditorTM (Fuzzy Systems Engineering), Fuzz-CTM (Byte Craft Ltd.), RT/FuzzyTM (Integrated Systems Inc.). General purpose software packages for scientific calculations, system analysis, simulation and control design like MATLAB® (MathWorks) have Fuzzy toolboxes for fuzzy system design and analysis. The fuzzy software packages offers excellent man-machine graphical interfaces, much better then those available for conventional control system design. The tools generate different codes like C and ADA or hardware dedicated and machine code for industrial PLCs (Allen-Bradley, OMRON, SIEMENS,..), processor chips (MOTOROLA,..) and fuzzy processor chips (NeuraLogic,..). VLSI chips for building fuzzy logic-neural networks are already available [4]. These parallel data processing architectures fit together with the parallel nature of the internal function of a fuzzy system and accelerate the real-time operations.

References

1. d'Alché-Buc, F., V. Andres and J. P. Nadal: Learning Fuzzy Control Rules with a Fuzzy Neural Network, *Artificial Neural Network 2*, Aleksander,T. and Taylor,J. (Editors), Elsevier Science Publishers B.V.,vol.1, 1992, 715-719.
2. Cannon, R. L., J. V. Dave and J. Bezdek: Efficient implementation of Fuzzy C-Means algorithm. *IEEE Trans. on PAMI*, Vol.8, No. 2, 1986, pp.248-255.
3. Dubois, D. and H. Prade: *Fuzzy Sets and Systems*. Academic Press, New York, 1980.
4. Design & Elektronik, Zukunfts Technologien - special issue of Design & Elektronik (D), September 1993.
5. Goldberg, D. E.: *Genetic Algorithms in Search, Optimization and Machine Learning*. Addison Wesley, 1989.

6. Horáček, P. and Z. Binder: An Approach for Design and Implementation of Fuzzy Controllers. In: *Proceedings of European Congress on Fuzzy and Intelligent Technologies EUFIT'93*, Aachen, Germany, vol.2, 1993, pp.163-169.

7. Horáček, P.: Design Techniques for Fuzzy Control. In: J.L.Crowley, A.Borkowski (eds.): *Proc. of International Workshop on Intelligent Robotic Systems IRS'94*, INPG Grenoble, 1994.

8. Hunt, K. J., D. Sbarbaro, R. Zbirkovski and P.J. Gawthrop: Neural Networks for Control Systems - A Survey. *Automatica*, Vol.28, No.6, 1992, pp.1083-1112.

9. Klir, G. J. and T. A. Folger: *Fuzzy Sets, Uncertainty and Information*. Prentice Hall, Englewood Cliffs, N.J., 1988.

10. Kohonen, T.: *Self-Organization and Associative Memory*. Springer-Verlag, Berlin, 1988.

11. Kosko B.: *Neural Networks and Fuzzy Systems* - A Dynamical Systems Approach to Machine Intelligence. Prentice Hall, Englewood Cliffs, N.J., 1992.

12. Lee, C. C.: Fuzzy Logic in Control Systems: Fuzzy Logic Controller, Part I,II. *IEEE TSMC*, Vol.20, No.2, 1990, pp.404-435.

13. Lin Chi-Teng and C. S. G. Lee: Neural-Network-Based Fuzzy Logic Control and Decision System. *IEEE TC*, Vol.40,No.12, 1991, pp.1320-1336.

14. Mayne, D. Q. and H. Michalska: Receding horizon control of nonlinear systems. *Trans. IEEE on Automatic Control*, 35, 1990, 814-824.

15. Miller, W. T., R. S. Sutton and P. J. Werbos: *Neural Networks for Control*. MIT Press, Cambridge, MA, 1990.

16. Narendra, K. S. and K. Parthasarathy: Identification and control for dynamic systems using neural networks. *IEEE Trans. on Neural Networks*, 1, 1990, 4-27.

17. Novák, V.: *Fuzzy Sets and Their Applications*. New York - Adam Hilgert. 1989.

18. Pedrycz, W.: *Fuzzy Control and Fuzzy Systems* - second edition. Research Studies Press Ltd., John Wiley, New York, 1992.

19. Takagi, T. and M. Sugeno: Fuzzy identification of systems and its application to modelling and control. *IEEE Transactions on Systems, Man and Cybernetics* 15 (1), pp.116-132, 1985.

20. Zadeh, L.A.: Fuzzy Logic. *IEEE Trans. on Computers*, 1988, pp.83-93.

21. Zimmermann, H.J.: *Fuzzy Set Theory and its Applications*. Kluwer-Nijhoff Publishing, Boston, 1991.

Foundations of Computer Communications

Václav Matoušek

Department of Informatics and Computer Science
University of West Bohemia, Plzeň (Pilsen)

Abstract. The presented chapter contains the description of main communication problems in several kinds of computer networks, basic computer network structures, fundamentals of communication theory, international standards for computer network design, and a short description of significant properties of typical computer networks and open systems.

1   Introduction

Computers are now found in every walk of life: in the home, in the office, in banks, in schools and colleges, in industry, and so on. Although in some instances the computers carry out their intended function in a standalone mode, in others it is necessary to exchange information with other computers. This means that an essential consideration in the design of most forms of computing equipment installed today is the type of data communication facility that is to be used to allow it to communicate with other computers. In many instances, the necessitates a knowledge not only of the alternative types of data transmission circuits that may be used but also an understanding of the interface requirements to the many different types of computer communication networks available for this purpose. Data communications and the allied subject of computer networks have thus become essential topics in all modern courses on computer systems design and their applications.

In many applications of computer networks, providing a means for two systems to exchange information solves only part of the problem. In an application that involves a distributed community of heterogenous (or dissimilar) computers exchanging files or information over a computer network, for example, such issues as the use of different operating (and hence file) systems and possibly different character sets and word sizes must also be addressed if the systems are to communicate in an unconstrained (open) way. It is thus essential when considering the applications of computer networks also to gain an understanding of the various application-oriented communication protocols that have now been defined to create communication environments in which computers from different manufacturers can exchange information in an open way. The three parts of this problems – data communications, computer networks and open systems – consider each of these issues.

A concise definition of what constitutes a computer network is difficult to produce. An easier approach is to first define what a computer network is. When the term *computer network* was first used, it describes any interconnections between computers. Since that time, three subclasses have emerged that are distinguished primarily by their geographical scope.

The first of these is the *wide area network* (WAN). This network spans a large area – possibly several continents. The second major type is the *local area network* (LAN) and as the name suggests, it is confined to relatively small areas such as a building or a group of buildings, for example a university campus or an enterprise. A third type, which is just emerging at the time of writing, is the *metropolitan area network* (MAN). The scope of this class of network lies between LANs and WANs, i.e. spanning a small city or a town.

In naming of these types of network, the main distinguishing factor would appear to be the size of the area covered. This factor has major effects on the technology used to implement the network, its administration and the type of applications that can be implemented on it.

2 Distributed Computer Systems and Computer Networks

The result of a dramatic fall in the cost of hardware components, combined with similar cost reductions and technological advances in the communications field has been to apply computer networks and interconnected computer systems in many areas of human activities, including production planning and production control. Users have recognized the advantages of interconnecting what were independent computer systems to permit interaction, cooperation and sharing of facilities. The feasibility of cooperating distributed computers opens the door to new and more demanding applications. Distributed computer systems have gained in popularity especially as they can offer many additional advantages over centralized systems. Reduced incremental cost, better reliability, extensibility, better response and performance are just some of the potential advantages.

But *what* exactly are a distributed system and a computer network? What are their attributes? *Why* are they capable of offering such attractive benefits? What are the problems? This small paragraph will attempt to answer these questions, and to give some examples of *where* distributed systems and computer networks have been used. The overall structure of a distributed system is shown in Figure 1.

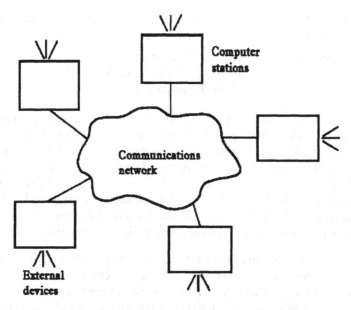

Fig. 1. Overall structure of a distributed system

2.1 What is a distributed system

Distributed processing is a relatively new field and there is as yet no agreed definition. There is rather a spectrum of systems which progress from centralized systems through to a set of diverse, physically dispersed, autonomous computer systems which are interconnected by communications networks to permit the exchange of information. We are mainly concerned with the latter integration of autonomous computer systems to form a cooperative collection of computing components which together combine to achieve some common goal. This is a 'distributed' system. This intuitive description can be refined by taking each of the main components of a computer system and examining the meaning of distribution.

The major components which combine to form a complete system are:

System = hardware
           + system software (control + data)
           + application software (control + data).

However, in many systems the distinction between system software and application software is blurred. For instance in embedded real-time systems the application is often involved in device access and control, and process scheduling (for real-time constraints), both of which are usually the domain of system software. In large data processing systems the data management aspects are often integrated into the system software to provide efficient access and control. The distinction can then be made between control and data rather than between system and application. Distributed systems can then be examined using these three dimensions of *hardware, control* and *data*:

Distributed system = distributed hardware,
and/or distributed control,
and/or distributed data.

1. PROCESSING HARDWARE   A distributed system must contain two or more computers, each with their own local memory and processors. This aspect of physical distribution is the most important factor in defining a distributed system. In order for the distributed computers to cooperate they must be interconnected: hence the need for some form of communication network. The distribution of the computers can reflect the physical distribution of the application or the functional decomposition of the system, where different computers provide different functions in the system.

2. CONTROL   Systems contain both physical resources in the form of processors, terminals, devices etc. and logical resources in the form of processes, files etc. There must be some form of control provided to manage resources and coordinate activities running on the individual processors. The strategy used to manage the resources of the system could be centralized, hierarchical or allow complete autonomy of the individual processors over local resources.

3. DATA   One of the major resources requiring control by both the system and the application software is data. The data being processed can itself be distrsibuted by replication (multiple copies at different locations) or partitioning (parts of the data kept at the diferent locations). Data distribution is often used to enhance tolerance to failures and to provide improved performance by locating the data colse to where it is generated and/or used.

The question is whether *all three* dimensions have to be distributed to classify a system as distributed, or is *one* sufficient?

Closely coupled systems which communicate via shared memory or crossbar switches, and so must be physically close together, are thus not considered as distributed. Loosely coupled collections of totaly autonomous computer systems, which may, for example, be capable of transferring files between each other but do not cooperate to achieve some common goal, are also not considered as true distributed systems. Both classes of system are excluded as they do not raise many of the issues nor possess many of the advantages which we associate with true distributed systems.

The following are the attributes reqiured of a distributed system:

– arbitrary number of system and application processes (logical resources),
– modular physical architecture (physical resources),
– communication by message passing using a shared communications system,
– some system–wide control.

One of the fundamental characteristics of distributed systems is that interprocess message transit is subject to variable delays and failure. There is a non–negligible time between the occurence of an event and its availability for observation at some remote site. It is not possible to build up a consistent and complete view of the state of the system. Rather, the state information is partitioned and distributed. The executive in charge of system–wide control must make decisions using only partial (and possibly inconsistent) information. The possible loss of messages exacerbates this problem. This characteristic of indeterminate message delay and loss identifies the main difficulty in the design and construction of distributed systems.

Finally, based on these conclusions we can formulate the following definition of the distributed system:

A distributed processing system is one in which several autonomous processors and data stores supporting processes and/or databases interact in order to cooperate to achieve an overall goal. The processes coordinate their activities and exchange information by means of information transferred over a communications network.

2.2 Computer network

Communication between distributed communities of computers in industrial applications is required for many reasons. At a very high level, for example, computers located in different enterprises in different parts of the country use public communication services to exchange electronic messages (mail) and to transfer files of e. g. economic information from one computer to another. Similarly, at a local level within a single enterprise or, more often, within a one working place, distributed communities of PCs and workstations use local communication network to access expensive shared resources and to exchange the important production data. The day-to-day transactions and information and data exchanges at all levels of production control are fully dependent on the communication by the reliable computer network.

What is a computer network? Several definitions are accepted in the industry. Perhaps the simplest is: A number of computers interconnected by one or more transmission paths. The transmission path is often the telephone line, due to its convenience and universal presence. The network exist to meet one goal: the transfer and exchange of data between the computers.

Computer networks provide several important advantages to production managing, businesses, and individuals.

1. Modern enterprises today are very dispersed, with offices or production and service shops located in diverse parts of a town, country and the world. Many of the computers located at the sites need to exchange information and data, often on a daily basis or continuously. A network provides the means to exchange data available to the people of the whole enterprise.

2. The networking of computers permits the sharing of resources of the machines. For instance, if a computer becomes saturated with too much work at one site, the work can be loaded through the network path onto another computer in the network. Such load sharing permits a more even and better utilization of resources.

3. Networking also provides the critical function of back-up. In the event that one computer falis, its counterpart can assume its functions and workload. Back-up capability is especially important in systems such as air traffic control. In the event a malfunctions, back-up computers rapidly take over and assume control of operations without endangering air travellers.

4. The use of networking allows a very flexible working environment. Employees can work at home using terminals or PCs tied through networks into the computer at the office, or working place. Many employees now carry portable personal computers on trips and tie into their network through hotel room telephones. Other employees travel to remote offices and use telephones and networks to transmit and receive critical sales, administrative, and research data from computers at company headquarters.

2.3 Communications in the computer network

The computers connected to the computer networks of several types communicate through communications systems having same or very similar properties. Fig. 2 illustrates a simple data communication system. The *application process* (AP) is the end-user application. It usually consists of software such as a computer program, or it could be an end-user terminal/workstation. Typical examples are an accounts receivable program, a payroll program, an inventory control package, or a personnel system.

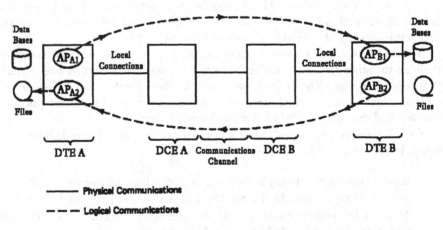

Fig. 2. A communications system

In Figure 2, site A could execute an application process (AP_{A1}) in the form of a software program to access an application process at site B (which is in this case a program (AP_{B1}) and a database). Figure 2 also shows a site B program (AP_{B2}) accessing a file at site A through an application program (AP_{A2}).

The application resides in the *data terminal equipment*, or DTE. DTE is a generic term used to describe the end-user machine, which is usually a personal computer or workstation. The DTE could be a large mainframe computer, or it could be a smaller machine, such as a netbuilder or a personal computer. The DTE takes many forms in the industry, e.g. a sampling device to measure the production process data, a computer used to automate the manufacturing process, an electronic mail computer or terminal, a personal computer in the office etc.

The function of a communication network is to interconnect DTEs together so that they can share resources, exchange data, provide back-up for each other, and allow employees and individuals to perform their work from any location.

Figure 2 shows that a network provides *logical* and *physical* communications for the computers and terminals to be connected. The applications and files use the physical channel to effect logical communications. Logical, in this context, means the DTEs are not concerned with the physical aspects of the communications process. Application $A1$ need only issue a logical *Read* request with an identification of the data. In turn, the communications system is responsible for sending the *Read* request across the physical channels to application $B1$.

Figure 2 also shows the *data circuit-terminating equipment*, or DCE (also called data communications equipment). Its function is to connect the DTEs into the communication line or channel. The DCEs designed in the 1960s and 1970s were strictly communications devices. However, in the last few years the DCEs have incorporated more user functions, and today some DCEs contain a portion of an application process. Notwithstanding, thew primary function of the DCE remains to provide an *interface* of the DTE into the communications network. The familiar modem is an example of a DCE.

The interfaces are specified and estabilished through *protocols*. Protocols are agreementson how communications components and DTEs are to communicate with each other. They may include actual regulations which stipulate a required or recommended convention or technique. Typically, several levels of interfaces and protocols are required to support an end-user application. Today, many organizations are adapting common interfaces and protocols as a result of worldwide efforts to publish recommended *standards* that are vendor and product independent.

DCEs and DTEs can be connected in one of two ways. As illustrated in Fig. 2, they can be connected in a point-to-point configuration in which only two DTE devices are on the line or channel. Illustrated in Fig. 3 is another approach called a multidrop configuration. In this configuration, more than two devices are connected to the same channel.

The DTEs and DCEs send communications traffic to each other in one of three methods:

Simplex: Transmission in one direction only,

Half-Duplex: Transmission in both directions, but only direction at a time is available (also called two-way alternate),

Full-Duplex (or Duplex): Transmission simultaneously in both directions (also called two-way simultaneous).

Simplex transmission is common in television and commercial radio. It is not as common in data communications because of the one way nature of the process, but simplex systems are found in some applications, such as telemetry. Half-duplex transmission is found in many systems, such as inquiry/response applications wherein a DTE sends a query to another DTE and waits for the applications process to access and/or compute the answer and transmit the response back. Terminal-based systems (keyboard + CRT screen terminals) often use half-duplex techniques. Full-duplex (or simply duplex) provides for simultaneous two-way transmission, without the intervening stop-and-wait aspect of half-duplex. Full-duplex is widely used in applications requiring continuous channel usage, high throughput, and fast response time.

The communications networks cannot be realized without the addition of an important component to the system. This component is the *data switching equipment*, or DSE. Figure 4 illustrates the use of the DSE in conjunction with the DTE and DCE. As the name implies, the major function of the DSE is to switch or route traffic (user data) through the network to the final destination. The DSE provides the vital functions of network routing around failed or busy devices and channels. The DSE may also route the data to the final destination through intermediate components, perhaps other switches.

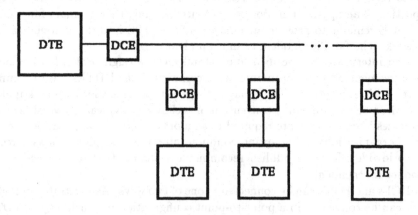

Fig. 3. Multidrop circuits

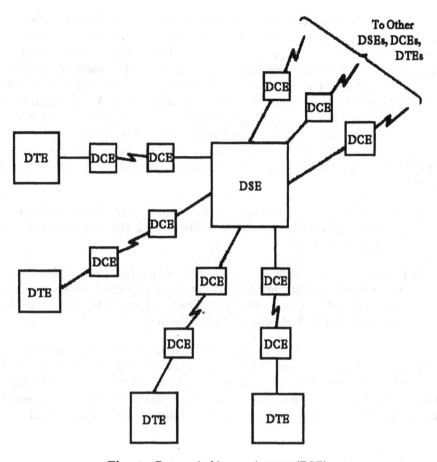

Fig. 4. Data switching equipment (DSE)

3 Network Topologies

A network topology is the shape (or the physical connectivity) of the network. The term *topology* is borrowed from geometry to describe the form of something. The network designer has three major goals when establishing the topology of a network:

- provide maximum possible reliability to assure proper receipt of all treafic (alternate routing),

- route the traffic across the least-cost path within the network between the sending and receiving DTEs (although the least-cost route may not be chosen if other factors, such as reliability, are more important),

- give the end user the best possible response time and throughput.

Network reliability refers to the ability to deliver user data correctly (without errors) from one DTE to another DTE. It entails the ability to recover from errors or lost data in the network, including channel, DTE, DCE, or DSE failure. Reliability also refers to the maintenance of the system, which includes day-to-day testing; preventive maintenance, such as relieving faulty or failing components of their tasks; and fault isolation in the event of problems. When a component creates problems, the network diagnostic system can pinpoint the error readily, isolate the fault, and perhaps isolate the component from the network.

The second major goal in estabilishing a topology for the network is to provide the least-cost path between the application processes residing on the DTEs. This involves:

1. minimizing the actual length of the channel between the components, which usually entails routing the traffic through the fewest number of intermediate components;

2. providing the last expensive channel option for a particular application; for instance, transmitting low-priority data over a relatively inexpensive dial-up, low-speed telephone line, in contrast to transmitting the same data over an expensive high-speed satellite channel.

The third major goal is to provide the best possible response time and throughput. Short response time entails minimizing delay between the transmission and the receipt of the data between the DTEs, and is especially important for interactive sessions between user applications. Throughput entails the transmission of the maximum amount of end-user data in a given period.

3.1 Hierarchical topology

The *hierarchical topology* is one of the most common networks found today The software to control the network is relatively simple and the topology provides a concentration point for control and error resolution. In most cases, the DTE at the highest order of the hierarchy is in control of the network. In Fig. 5(a), traffic flow among and between the DTEs is initiated by DTE A. Many vendors implement a distributed aspect to the hierarchical network by providing methods for the subordinate DTEs to directly control these DTEs below them in hierarchy. This reduces the workload of the central host at the site A.

While the hierarchical topology is attractive from the standpoint of simplicity of control, it presents significant potential bottleneck problems. In some instances, the uppermost DTE, typically a mainframe computer, controls all traffic between DTEs. Not only can this create bottlenecks, but it also presents reliability problems. In the event the upper-level DTE falis, the network capabilities are lost completely if the DTE is not fully backed up by another computer. Nonetheless, hierarchical topologies have been used widely in the past and will continue to be used for many years. They permit a graceful evolution toward

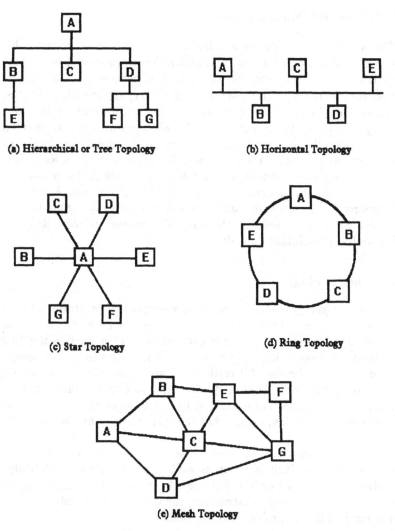

Fig. 5. Network topologies

a more complex network, because subordinate DTEs can be added relatively easy.

The hierarchical topology is also called *a vertical network* or *a tree network*. The word "tree" is derived from the fact that a hierarchical network often resembles a tree with branches stemming from the top of the tree down to the lower level. You might pause at this point and determine if you could draw a tree-type network topology relating to one of your daily activities. One common example is the organizational chart hanging in your office. Indeed, the advantages and disadvantages of a vertical data communications network are much the same of those a hierarchically structured lines of authority with frequent bottlenecks at the upper levels and often insufficient delegation of responsibility.

3.2   Horizontal (Bus) topology

The *horizontal topology* or *bus topology* is illustrated in Figure 5(b). This arrangement is quite popular in local area networks (discussed in the paragraph 8). It is relatively simple to control traffic flow between and among the DTEs because the bus permits all stations to receive every transmission. That is, a single station *broadcasts* to multiple stations. The main drawback of a horizontal topology stems from the fact that usually only one communications channel exists to service all the devices on the network. Consequently, in the event of a failure of the communication channel, the entire network is lost. Some vendors provide for fully redundant channels in the event of the loss of a primary channel. Others provide bypass switches around failed nodes. Another problem with this particular configuration is the difficulty in isolating faults to any one particular component tied into the bus. The absence of concentration points makes problem resolution difficult.

3.3   Star topology

The *star topology* is one of the most widely used structures for data communications systems. One of the major reasons for its continued use is based on historical precedence. The star network was often used in the 1960s and early 1970s because it was easy to control – the software is not complex and the traffic flow is simple. All traffic emanates from the hub of the star, the central site in Fig. 5(c), labelled *A*. Site *A*, typically a special computer, is in complete control of the DTEs attached to it. Consequently, it is quite similar to the hierarchical topology, except that the star topology has limited distributed processing capabilities.

Site *A* is responsible for routing traffic through it to the other components; it is responsible for fault isolation as well. Fault isolation is relatively simple in a star network because the lines can be isolated to identify the problem. However, like the hierarchical structure, the star network is subject to potential bottleneck and failure problems at the central site.

3.4   Ring topology

The *ring topology* is another popular approach to network configuration. As illustrated on Fig. 5(d), the ring topology is so named because of the circular aspect of the data flow. In most instances, data flows in one direction only, with one single station receiving the signal and relaying it to the next station on the ring. The ring topology is attractive because bottlenecks, such as those found in the hierarchical or star systems, are very uncommon. Moreover, the logic to implement a ring network is relatively simple. Each component is tasked with the straightforward job of accepting the data, sending to DTE attached to it, or sending it out on the ring to the next intermediate component. However, like all networks, the ring network does have its deficiencies. The primary problem is the one channel tying in all the components in the ring. If a channel between two nodes fails, then the entire network is lost. Consequently, some vendors have estabilished designs which provide for back-up channels in the event a channel is lost. In other instances, vendors build switches which automatically route the data around the failed node on the ring to prevent the failure from bringing down the entire network.

3.5 Mesh topology

The *mesh topology* has been used in the last few years. Its attraction is its relative immunity to bottleneck and failure problems. Due to the multiplicity of paths from the DTEs and DSEs, traffic can be routed around failed components or busy nodes. Even though this approach is a complex and expensive undertaking (mesh network protocols can be quite involved from the standpoint of the logic to provide these features), some users prefer the reliability of the mesh network to that of the others.

4 Fundametals of Communications Theory

4.1 Channel speed and bit rate

In many present communications systems the data are transmitted from one computer to another in binary images – 1s and 0s. The most elementary method a device uses to send a binary number on a communications path is to switch the signal on and off electrically, or to provide high or low voltages on the line to represent the 1s and 0s. Regardless of how the data is represented on the path – in the form of on/off states, levels of voltage, or directions of current flow – the communication channel is described by its capacity in the number of *bits per second* transmitted. Abbreviations for bits per second are *bit/s, bps,* or *bs*. When one speaks of a 4800 bit/s line, it means a device sends 4800 bits per second through the channel. A bit is simply the representation of the electrical, optical, or electromagnetic state of the line: voltages, current, or some form of radio or optical signal. Seven or eight bits usually comprise a user-coded character, or byte.

A data communications channel utilizing conventional telephone lines is very slow. For purposes of comparison, a channel is classified by categories of *low speed* (0 – 600 bps), *medium speed* (600 – 4800 bps), and *high speed* (4800 – 9600 bps). Only recently, in the last few years, has the industry successfully moved to 9.6 kbits per second (kbit/s) on telephone channels. The typical speeds found beyond 9600 bps are 14400, 19200, 56000, and 64000 bps, and 1.544 megabit/s and 2.048 Mbit/s in Europe. The 1.544 Mbit/s channel is the well-known T1 carrier. This offering is prevalent in transmissions such as high-speed digital channels and digital switches.

The data communications world is fairly slow relative to the computer world. For example, a conventional data processing system with disk files attached to computers operates at about 10 Mbit/s and up. One might reasonably ask, why the slow speed. The answer is that DTEs and DCEs usually communicate through the telephone line. It was the most convenient and readily available path when the industry developed computers and began to interface them with terminals and other computers. The telephone channel is not designed for fast transmission between high-speed computers, but for voice transmission between people, which does not require the speed associated with data transmission.

4.2 Voice communications and analog waveforms

Human voice communications generate acoustical waveforms which propagate through the air. In effect, voice communications are physical energy. When one speaks, oscillating waveforms or high or low, air pressure are created. These waveforms are called *analog* waveforms. They are so named because they exhibit a continuous, repeating occurence and they are nondiscrete, gradually changing from high to low pressure. Of course, one cannot see waveforms in the air because the voice transmissions are air pressure variations.

The telephone handset translates the physical oscillations of the air to electrical energy with same or similar waveform characteristics. The waveform exhibits three primary characteristics that are very important to data communications: *amplitude, frequency,* and *phase.* The amplitude of the signal is a measurement in relation to its voltage, which can be zero or a plus or minus value. Notice the analog aspect of the signal – it gradually increases in positive voltage, then traverses the zero voltage to the negative voltage, and returns to the zero again. This complete oscillation is called a *cycle.* The second major aspect of the signal – *frequency* – describes the number of complete cycles per second or the number of oscillations per second. The number of oscillations per second is also called hertz (Hz).

The term *baud* differs from hertz in that it refers to the rate of signal change on the channel, regardless of the information content of those signals. For example, an 1800 Hz signal can be changed at 1200 times per second. The 1800 Hz describes the "carrier" frequency and the 1200 changes describe the baud. Many people use the term "baud rate", which is actually redundant, since baud implies the rate.

The third major component of the analog signal is the *phase*, which represents the point of the signal reached in the cycle. When the cycle has gone the one-fourth of its phase at the point of maximum positive voltage, it is said to have traversed through 25 percent of its cycle, just as traversing one-fourth the way around the circle represents a distance of 90 degrees.

4.3 Bandwidth and the frequency spectrum

A voice transmission consist of waveforms containing many different frequencies. The particular meld of these frequencies determines the pitch and sound of a person's voice. A human voice occupies the frequencies of approximately 200 Hz to 15 kHz. The human ear can detect frequencies over a broader range, from around 40 Hz to 18 kHz. The range of frequencies (for example, those comprising the human voice spectrum) is called the *bandwidth*. The term bandwidth is also used in computer networks. Bandwidth refers to the range of transmission frequencies that are carried on a communications line. Bandwidth is a critical ingredient in networking because the capacity of a channel is directly related to its bandwidth.

Finally, the Table 1 illustrates the frequency spectrum. The spectrum ranges from the relatively limited bandwidth of the audio frequency, such as voice, through the high frequencies found in coaxial cable, microwave broadcasting, and the very high frequency range where visible light exists.

Table 1. The frequency spectrum

Freq. Range	Design- ation	Typical Uses
10^3	–	Telephone voice frequencies (with low to high speeds)
10^4	VLF	Telephone voice frequencies (with higher speeds)
10^5	LF	Coaxial submarine cables (some high speed batch data transfer)
10^6	MF	Land coaxial cables: AM sound broadcasting (high speed voice & data)
10^7	HF	Land coaxial cables: Shortwave broadcasting (high speed voice & data)
10^8	VHF	Land coaxial cables: VHF sound and TV broadcasting (FM)
10^9	UHF	UHF TV broadcasting
10^{10}	SHF	Short link waveguides: Micro-wave broadcasting (high speed)
10^{11}	EHF	Helical Waveguides (high speed data transfer)
10^{12}	–	Infrared Transmission (local data transmission)
10^{13}	–	Infrared Transmission (local data transmission)
10^{14}	–	Optic fibers: Visible light (very high speed voice and data)
10^{15}	–	Optic fibers: Ultra violet (very high speed voice and data)
10^{19-23}	–	X – rays and Gamma rays

The European telephone bandwidth ranges from 300 Hz to 3400 Hz. If a telephone channel occupies about 3 kHz of band and the bandwidth between the frequency spectrum of 10^3 to 10^4 is 9000 Hz, then this bandwidth is roughly equivalent to three voice channels. However, if one examines the bandwidth between 10^7 and 10^8, the full bandwidth is 90,000,000 Hz and from this bandwidth we can derive (theoretically) 30,000 voice channels (in reality, fewer channels are obtained, due to need to "separate" the channels from each other). Then the bandwidth of coaxial cable and microwave allow the 3 kHz voice-grade bands to occupy different portions of the frequency spectrum.

Since the DTEs' two voice-grade channels now occupy different frequency spectra, the signals can use the same physical media, because they are *linear* in that they behave as if they are independent of each other. The signals are said to occupy or use *subchannels*.

Bandwidth is a limiting factor on transmission capacity within the network. Other limiting factors are the actual signal power of the transmission and the amount of noise in channel. The channel noise is a problem that is inherent to the channel itself and can never be completely eliminated. Noise results from several factors. For instance, atmospheric noise emanates from electrical disturbances in the earth's atmosphere. Space noise can come from the sun and other stars which radiate electromagnetic energy over a very broad frequency spectrum. Noise can also be found on a wire conductor or coaxial conductor because the random movement of electrons within the conductor generates thermal energy.

5 Connections

When DTEs communicate with each other by use of the telephone path, the signal must accomodate itself to a voice oriented analog world. However, DTEs "talk" in *digital* forms. The digital waveform looks considerably different from the analog waveform. It is similar in that it is continuous, repeats itself, and is periodic, but it is *discrete* – it has very abrupt changes in its voltage state. Computers use digital, binary images because semiconductor elements are basically two-state, discrete devices.

Digital transmission is available in many systems today. It has several distinct advantages over the analog channel, and is discussed in more detail in [1], [2]. However, analog channel still dominate the local connections of DTEs into the telephone companies' channels.

5.1 Modems

A method is needed to allow two digital devices to "talk" to each other through the dissimilar analog environment. The modem provides this digital/analog interface. It alters either the amplitude, the frequency, or the phase to represent binary data as an analog signal.

The *modem* is our first example of a DCE. It provides the interface between the digital and analog worlds, as well as the capability to transmit from a digital DTE across the analog channel to a receiving digital DTE. The word "modem" is a shortened term for *modulation/demodulation*. The process modulates the signal at the transmitting modem and demodulates the transmission at the receiving modem.

To be precise, the exact definition of modulation is: the modification of a frequency to carry data. This frequency is called the *carrier* frequency. The data that modulates the carrier (i.e., the data coming from the terminal or computer) is called the *baseband* signal. The term "baseband" usually refers to an unmodulated signal.

The modem modifies the carrier signal (either its amplitude, frequency, or phase) to carry the baseband signal. As illustrated in Figure 6, the AM modem changes the amplitude of its carrier in accordance with the bit stream to be sent.

In this instance, a higher amplitude represents a zero and a lower amplitude represents a one. A more popular modem is called an FM modem (frequency modulation modem) in which the amplitude remains constant and the frequency varies. A binary 1 is represented by one frequency and a binary 0 by another frequency. Another device is a PM modem (phase modulation modem). This modem abruptly alters its phase to represent the change from a 1 to a 0 or a 0 to a 1. The recommended standards used to connect DTEs, DCEs, and DSEs are RS-232-C and V.24/V.28 interfaces (see [1], Appendix C).

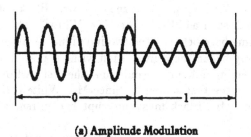

(a) Amplitude Modulation

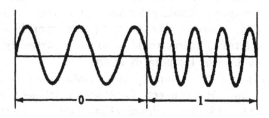

(b) Frequency Modulation

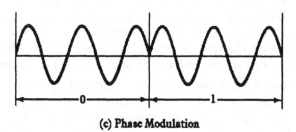

(c) Phase Modulation

Fig. 6. Modulation techniques

5.2 Synchronizing network components

In order for computers to communicate, they first need to notify each other that they are able to communicate. Second, once they are communicating, they must provide a method which keeps both devices aware of the ongoing transmissions. Let us address the first point. A transmitter, such as a computer, must transmit its signal so that the receiving device knows when to search for and recognize the data as it arrives. In essence, the receiver must know the exact time that each binary 1 and 0 is coming across the communication channel. This requirement means that a mutual time base or a "common clock" is necessary between the receiving and transmitting devices.

In this sense, machine communications are analogous to the communications between people. For example, during a party where many people are conversing intermittently among and between themselves, two individuals (say, Mr. Black and Ms. White) must first recognize they are to estabilish communications with each other. If Ms. White' back is turned to Mr. Black, he must first send a preliminary signal, such as "Hello", or "Ms. White". This, in effect, estabilishes a common time base between the two individuals that wish to converse. If an individual simply began the conversation without previous notification, it is quite likely the receiving individual would miss the first part of the transmission; i.e., the first part of the first sentence. Since Ms. White is busy and occupied with other matters, Mr. Black must interrupt her current activity to get her attention.

In the same manner, a transmitting machine must first send to the receiving machine an indication that it wishes to "talk" with it. If the transmitter sends the bit down the channel without prior notice, the receiver will likely not have sufficient time to adjust itself to the incoming bit stream. In such an event, the first few bits of the transmittion would be lost. Consequently, like Ms. White, the receiving DTE must be temporarily interrupted.

This process is part of a communication protocol and is generally referred to as *synchronization*. Short connections between machines often use a separate channel, or line, to provide the synchronization. This line transmits a signal that is turned on and off or varied in accordance with preestabilished conventions. As the clocking signal coming across this line changes, it notifies the receiving device that it is to examine the data line at a specific time. It may also resynchronize the receiver's clock so that the receiver stays very accurately aligned on each incoming data bit. Thus, clocking signal perform two valuable functions: (1) they synchronize the receiver into the trasmission before the data actually arrives, and (2) they keep the receiver synchronized with the incoming data bits.

When there are long distances between computers connected to the network, it makes more sense economically to incorporate the timing into the signal itself instead of using a separate clocking channel. This is known as a self-clocking code. Non-self-clocking codes present a problem in that the clock and the data can be altered as they propagate through separate channels. The clocking signal speeds up or slows down relative to the data signal, which means the receiver has difficulty "locking" onto the data signal.

A self-clocking code is one in which the receiving device can periodically check itself to see that it is sampling the line at the exact time a data bit is propagating into the receiver. This requires (under ideal conditions) the line to change its state very often. The best clocking codes are those in which the state of the line changes frequently, because these state changes (for example, voltage shifts) allow the receiver to continue to readjust itself to the signal.

The "clock" simply provides a reference for the individual binary 1s and 0s. The idea is to have a code with regular and frequent level transitions on the channel. The transitions delineate the binary data cells (1s and 0s) at the receiver, and sampling logic continuously looks for the state transitions in order to delineate the bit streams. Receiver sampling usually occurs at a higher rate then the data rate in order to more precisely define the bit cells.

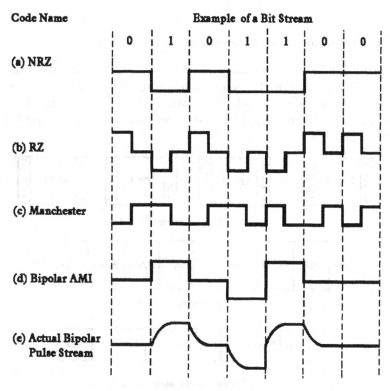

Fig. 7. Digital codes

Figure 7 provides an illustration of several common binary coding schemes used in industry. Each of these has some advantages and disadvantages. They are discussed and briefly derscribed in [1]. Be aware that the signals are not as sharp and square as Fig. 7 suggests. Fig. 7(e) depicts an actual signal. All these signals exhibit one or several of the following four characteristics:

- *Unipolar code.* No signal below zero voltage or no signal above (i.e., algebraic sign does not change: 0 volts for 1, +3 volts for 0).

- *Polar code.* Signal is above and below zero voltage (opposite algebraic sign identify logic states: +3 volts and −3 volts).

- *Bipolar code.* The signal varies between three levels.

- *Alternate mark inversion (AMI) code.* Uses alternate polarity pulses to encode binary 1s.

Figure 7(a) shows the non-return-to-zero code (NRZ), Fig. 7(b) the return-to-zero code (RZ), Fig. 7(c) the very popular *Manchester* code found in many communications systems today, Fig. 7(d) shows one code used by AT&T, and Figure 7(e) shows an actual DC signal as it exists on the communications channel.

5.3 Asynchronous and synchronous transmission

A great number of computers and terminals communicate with each other and with DCEs through the non-return-to zero (NRZ) code. Consequently, clocking becomes a major consideration with these devices. Two data formatting conventions are used to help achieve synchronization. These two methods are illustrated in Figure 8:

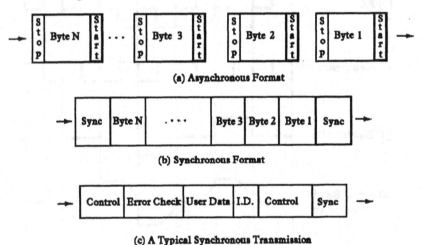

(a) Asynchronous Format

(b) Synchronous Format

(c) A Typical Synchronous Transmission

Fig. 8. Transmission formats

The first approach is called *asynchronous* formatting. With this approach, each data byte (each character) has start and stop signals (i.e., synchronizing signals) placed around it. The purposes of these signals are, first, to alert the receiver that data is arriving, and second, to give the receiver sufficient time to perform certain timing functions before the next byte arrives. The start and stop bits are really nothing more than unique and specific signals which are recognized by the receiving device.

Asynchronous transmission is widely used because the interfaces in the DTEs and DCEs are inexpensive. Since the synchronization occurs between the transmitting and receiving devices on a character-by-character basis, some allowance can be made for inaccuracies between the transmitter and receiver because the inaccuracy can be corrected with the next arriving byte. A "looser" timing tolerance translates to lower component costs.

A more sophisticated process is *synchronous* transmission, which uses separate clocking channels or a self-clocking code. Synchronous formats eliminate the intermittend start/stop signals around each character. The preliminary signals are usually called synchronization or *sync* bytes. The more modern systems call them flags. Their principial function is to alert the receiver that user data is arriving. This process is called *framing*. It can be seen that a long synchronous data message without intermittent start/stop bits could present problems because the receiver could drift from the signal.

We now have enough information on codes and formats to move to other components in the network. However, we will come back to this subject once more. It will be more evident in later paragraphs that the knowledge of codes and signalling states is important to gain an understanding of other aspects of networks and protocols.

5.4 Message formats

A more realistic depiction of synchronous formats is shown in Figure 8(c). Data transported through a computer network usually contain a minimum of five parts:

- sync bytes,

- control field(s) which implement(s) the protocol; that is, manages the movement of the data through the network,

- an identification of the data (at a minimum, the receiver or transmitter identification),

- user data (the application process data),

- an element to check for a transmission error, typically called an error-check field.

Before explaining these five parts in more detail, it should prove useful to describe four terms: *message, block, frame,* and *packet.* These terms have no clear definition in the industry. Indeed, they are often used interchangeably. At this point in our analysis, it is sufficient to describe them all as: a self-contained and independent entity of control and/or user data. Generally, a user packet, message, frame, or block contains at a minimum the five entries shown in Fig. 8(c).

All frames do not contain user data. Network protocols require the exchange of frames among DTEs, DCEs, DSEs, and control centers to effective manage traffic flow, diagnose problems, and perform day-to-day operations. In fact, a substantial amount of network traffic is non-user, overhead frames. The purpose of such overhead is to perform the necessary protocol and interface functions to support user data frames.

The identification (ID) field usually provides a name or number for the receiver as well as for the transmitter. Either the ID or control fields contain sequence numbers, which are used to further identify the specific frames from each sender.

The error-check field is appended by the transmitting site. Its value is derived from a calculation on the contents of the other fields. At the receiving site, an identical process computes another error-check field. The two are then compared; if they are consistent, the chances are very good that the packet was transmitted error-free. This process is called the *cyclic redundancy check* (CRC), and the field is called the *frame check sequence* (FCS).

5.5 The communications port

The illustration in Figure 2 showed the connection of a DTE into a DCE. A DTE can also connect or interface directly into another DTE. A very common approach is a computer connecting directly to a graphics device, a printer, a terminal, or another computer. There are also systems in which DCEs interconnect directly to each other. Regardless of where the connection is made, the input/output communications channel interfaces into the DTE through a communications *port*.

Vendors and technicians have other names for ports. Some commonly used terms are: communications adapter interface, serial port, serial board, board, USART (universal synchronous/asynchronous receiver and transmitter). The intelligence of the port is highly varialble, depending on the type of interface needed. Obviously, the more sophisticated an individual port is, the more expensive it is. The main purpose of the communications port is to interface the communications channel into the DTE and provide for the functions of moving data into and out of the device.

6 Additional network components

The network configurations discussed thus far have consisted of a few channels and devices. In many organizations, the structure is similar to the configuration in Figure 9. We have added several other components to a communication system. First, the DTE (the computer) is connected through various kinds of ports to other computers, terminals, disk files, tape files, and devices such as printers and graphic plotters. One major difference between this picture and previous illustrations is the connection of the computer to a *front-end processor*.

The purpose of the front-end processor (which may or may not be used in your organization) is to offload communications tasks from the mainframe host computer. Many of the communications protocols reside here. The front-end processor is especially designed to do a very limited number of functions but to do them quite efficiently. For instance, it may be responsible for handling errors with the devices attached to it without interrupting the host computer.

311

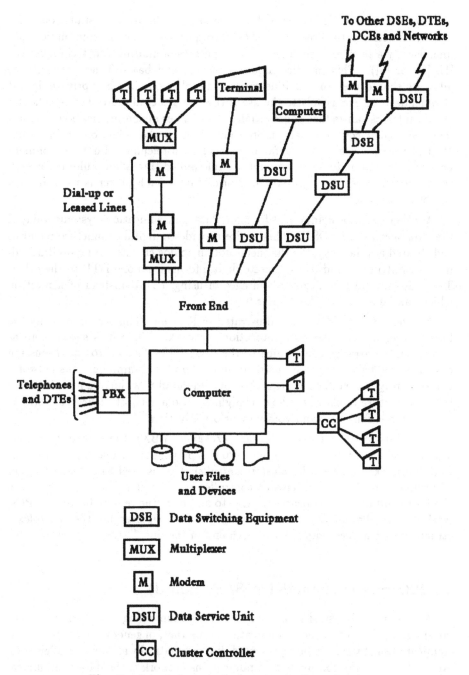

Fig. 9. A typical computer/communications structure

Several of the devices in Figure 9 have been discussed previously. Remember, the *modems* act as an interface between the digital systems and the analog facilities. In addition, several other devices we have not yet discussed are explained below.

The *multiplexer* (MUX) is a device found in allmost all installations. Its primary function is to allow multiple DTEs or ports to share one communications line, usually a telephone channel (for the specifics of multiplexing techniques see [1], [2], or [6]). This practice assumes the channel has sufficient capacity to allow its shared use. In this illustration, the telephone line is a private, leased channel devoted exclusively to the use of the two attached multiplexers. The use of multiplexers can reduce substantially the number of communication channels required. Their cost is usually more than offset by the reduction of line costs. Multiplexers are also very useful in a local (non-telephone line) environment because they can reduce the amount of cable pulled through a building for each connection between the computers. Notice that a multiplexer is used at each end of the channel.

Another common component found in data communications systems today is the *data service unit* (DSU). This device provides a digital channel from end to end. Stated another way, the channel is not an analog channel, but a digital facility transmitting 1s and 0s as discrete digital forms from one DTE to the other. There are several significant advantages in using a DSU instead of a modem, which can be found in [1], Chapter 9.

A *data switch* (DSE) is also illustrated in Figure 9. This switch is really the backbone of a computer communications network. While it is shown here as a very simple arrangement, much of the book [1] is devoted to the discussion of how these switches can be used to build complex communications network. Consequently, we will not address this topic in detail now, but we have included it in this picture to show you how it might fit into a network structure, perhaps similar to one found in your company or organization.

Also illustrated is a *private branch exchange* (PBX), the private telephone switch which is located in many offices and is playing an increasing role in computer networks. Finally, the *cluster controller* (CC) is used to manage a group of terminals or PCs. The cluster controller receives commands or data from the host computer and directs messages to and from the remote terminals, PCs, workstations. This arrangement is similar to the multiplexer, but the multiplexer structure uses a like component at each end of the line.

7 ISO reference model (OSI Standard)

Although the physical separation of the communicating computers may very considerably from one type of application to another, in general a computer communications network can be represented diagrammatically as shown in Figure 10. At the heart of any computer communications network is the data communications facility. However, irrespective of the type of data communications facility, an amount of hardware and software is required within each attached computer to handle the appropriate network-dependent protocols. Typically, they are concerned with the estabilishment of a communication channel across the network and with the control of the flow of messages across this channel. The provision

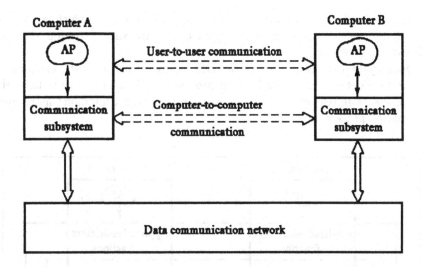

Fig. 10. Computer communications schematic

of such facilities is only part of the network requirements, however, since in many applications the communicating computers may be of different types. This means that they may use different programming languages and, more importantly, different forms of data representation. Also, the computers may use different operating systems; hence, the interface between user (application) programs, normally referred to as *application processes* (APs), and the underlying communications services may be different. For example, one computer may be a small single-user PC, while another may be a large multi-user system.

In the earlier days of computer communications, these issues meant that only closed communities of computers could communicate with each other in a meaningful way. IBM's System Network Architecture and DEC's Digital Network Architecture are just two examples of communication software packages produced by manufacturers to allow their systems to be interconnected together. These proprietary packages, however, of which there are still many in existence, do not address the problem of universal interconnectability, or open systems interconnection. In an attempt to alleviate this problem, the ISO, in the latest 1970s, formulated a reference model to provide a common basis for the coordination of standards developments and to allow existing and evolving standards activities to be placed into perspective with one another. The ultimate aim was to allow an application process in any computer that supported the applicable standards to freely communicate with an application process in any other computer supporting the same standard, irrespective of its origin of manufacture. This model was termed the *ISO Reference Model for Open Systems Interconnection*. It should be stressed, however, that this model is not concerned with specific applications of computer communications networks. Rather, it is concerned with the structuring of the communication software that is needed to provide a reliable, data transparent, communications service (which is independent of any specific manufacturers equipment or conventions) to support a wide range of applications.

Open systems interconnection (or *OSI*) is concerned with the exchange of information between such application processes and its objective is to enable such processes in real systems to cooperate in achieving a common (distributed) information processing goal.

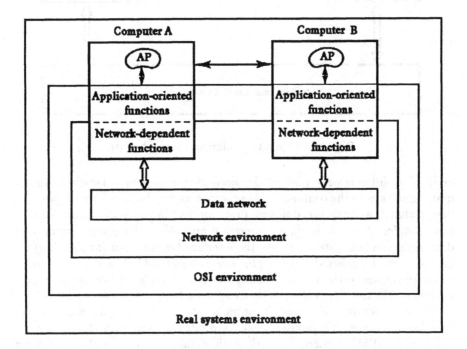

Fig. 11. Operational environments

7.1 OSI layer model

A communications subsystem is a complex piece of software. Early attempts at implementing such software were often based on a single, complex, unstructured program with many interacting components. The resulting comunications subsystem was thus difficult to test and often very difficult to modify.

To overcome this problem, the ISO have adopted a layered approach for the reference model. The complete communications subsystem is broken down into a number of layers each of performs a well-defined function. Conceptually, these layers can be considered as performing one of two overall functions: network-dependent functions or application-oriented functions. This in turn gives rise to three distinct operational environments (see Fig. 11):

1. The *network environment*, which is concerned with the protocols and standards relating to the different types of underlying data communications networks.

2. The *OSI environment*, which embraces the network environment and adds additional application-oriented protocols and standards to allow end systems (computers) to communicate with the one another in an open way.

3. The *real systems environment*, which builds on the OSI environment and is concerned with a manufacturer's own proprietary software and services, which have been produced to meet a particular distributed information processing task.

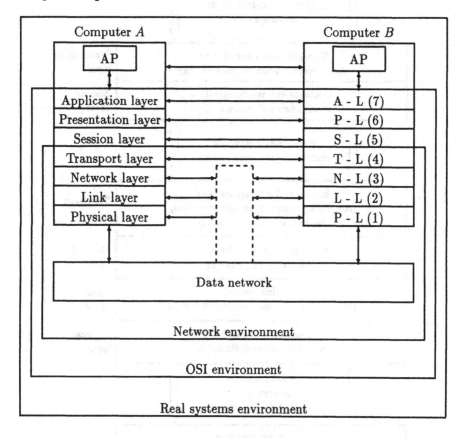

Fig. 12. Overall structure of the ISO reference model

Both the network-dependent and application-oriented (network-independent) components of the OSI model are in turn implemented in the form of a number of *protocol layers*. The boundaries between each protocol layer, and hence the functions performed by each layer, have been selected as a result of the experience gained by earlier standards activity.

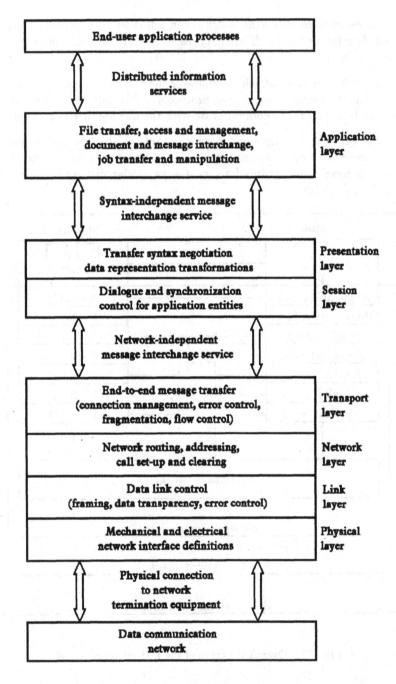

Fig. 13. Protocol layer summary

Each layer performs a well-defined function in the context of the overall communication strategy. It operates according to a defined protocol by exchanging messages, both user data and additional control information, with a corresponding layer in a remote system. Each layer has well-defined interface between itself and the layer immediately above and below it, and as a result the implementation of a particular protocol layer is independent of all other layers.

The logical structure of the ISO reference model is as shown in Figure 12. As can be seen, it is made up of seven protocol layers. The lowest three protocol layers (1 – 3) are network dependent and are concerned with the protocols associated with the data communications network being used to link the two communicating computers together. In contrast, the upper three layers (5 – 7) are application oriented and are concerned with the protocols that alow two end user application processes to interact with each other, normally through a range of services offered by the local operating system. The intermediate transport layers masks the upper application-oriented layers from the detailed operation of the lower network-dependent layers. Essentially, it builds on the services provided by the latter to provide the application-oriented layers with a network-independent message interchange service.

The function of each layer is specified formally in the form of a protocol that defines the set of rules and conventions used by the layer to communicate with a similar (peer) layer in another (remote) system. Each layer provides a defined set of services to the layer immediately above it and, in turn, uses the services provided by the layer immediately below it to transport the message units associated with the protocol to the remote peer layer. Thus, the transport layer, for example, provides a reliable, network-independent message transport service to the session layer above it and uses the sevices provided by the network layer below it to transfer the set of message units associated with the transport protocol to a peer transport layer in another system. Conceptually, therefore, each layer communicates with a similar peer layer in a remote system according to a defined protocol, but in practice the resulting protocol message units of the layer are passed by means of the services provided by the next lower layer. The basic function of each is summarized in Fig. 13.

The application layer provides the user interface to a range of network-wide distributed information services. These include file transfer access and management, and general document and message interchange services such as electronic mail. A number of standard protocols are either available or are being developed for these and other types of service.

The presentation layer is concerned with the representation of the data during transfer between two correspondent application layer protocol entities. To achieve true open systems interconnection, a number of common *abstract data syntax* forms have been defined for use by application entities together with associated *transfer* (or *concrete*) *syntaxes*. The presentation layer thus negotiates and selects the appropriate transfer syntax(es) to be used during a transaction so that the syntax (structure) of the messages being exchanged between two application entities is maintained. Then, if this form of representation is different from the internal abstract form, the presentation entity performs the necessary conversions.

The session layer provides the means necessary for two application layer protocol entities to organize and synchronize their dialogue and to manage their data exchange. It is thus reponsible for setting up (and clearing) a communication (dialogue) channel between two correspondent application layer protocol entities (presentation layer protocol entities in practice) for the duration of the complete network transaction. A number of optional sevices are then provides including interaction management, synchronization, and exception reporting (for the more detailed description see [2]).

The transport layer forms the interface between the higher application-oriented layers and the underlaying network-dependent protocol layers. It provides the session layer with a reliable message transfer facility that is independent of the underlying network type. The transport layer thus masks the detailed operation of the underlying network from the session layer and simply provides the latter with a defined set of message transfer facilities.

The network-dependent layers: As the lowest three layers of the reference model are network dependent, their detailed operation varies from one network type to another. In general, however, the *network layer* is responsible for the estabilishment and clearing of a network-wide connection between two transport layer protocol entities. It includes such facilities as network routing (addressing) and, in some instances, flow control across the computer-to-network interface. The *link layer* builds on the physical connection provided by the particular network to provide the network layer with a reliable information transfer facility. It is thus responsible for such functions as error detection and, in the event of transmission errors, the retransmission of messages. Normally, two types of service are provided: *connectionless*, which treats each information frame as a self-contained entity that is transferred using a best-try approach, and *connection oriented*, which endeavours to provide an error-free information transfer facility. Finally, the *physical layer* is concerned with both the physical and electrical interface between the user equipment and the network terminating equipment. It provides the link layer with the means of transmitting a serial bit stream between two correspondent systems.

The bottom three layers of the OSI model are now well developed. The fourth (transport) layer was approved in 1984. The top three layers are under various stages of development. The more detailed information about all layers of the OSI layered protocol model can be found in [1], Chapters 4 – 9, 13, or [2], Chapter 6, or [3], respectively.

8 Local Area Networks

Local area data networks, normally referred to simply as local area networks or LANs, are concerned with the interconnection of distributed communities of computer-based DTEs whose physical separation is confined to a single building or localized group of buildings. For example, a LAN may be used to interconnect a comunity of computer-based workstations distributed around a block of offices within a single building or a group of buildings such as a university campus. Alternatively, it may be used to interconnect various computer-based pieces of equipment distributed around a factory or hospital complex. Since all the equipment is located within a single estabilishment, however, LANs are normally installed and maintained by the organization and hence they are also referred to as *private data networks*.

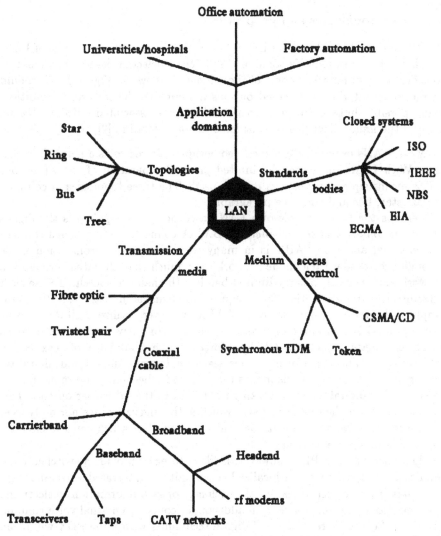

Fig. 14. LAN selection issues

The main difference between the communications path established using a LAN and a connection made through a public data network is that with a LAN, because of the relatively short physical separation, much higher data transmission rates are available. In the context of the ISO Reference Model for OSI, however, this difference manifests itself only at the lower network-dependent layers and hence, as was indicated in the last paragraph, the higher protocol layers in the reference model are, in many instances, the same for both types of network. This paragraph, therefore, is primarily concerned with a brief description of the different types of LAN and the function and operation of the associated network-dependent protocol layers.

8.1 LAN topologies and properties

Before describing the structure and operation of the different types of LAN, it is helpful to first identify some of the different selection issues that must be considered. A summary of some of these issues is shown in Figure 14. It should be stressed that this is intended only as a summary. There are, in addition, many possible links between the tips of branches associated with the figure. Each of the issues identified is considered in some detail in [2], [7].

Most WANs use typically a *mesh* (sometimes referred to as a network) *topology*. With LANs, however, the limited physical separation of the subscriber DTEs means sipmler topologies may be used. The three topologies in common use are star, bus and ring (see paragraph 3).

Perhaps the best example of a LAN based on a *star topology* is the digital *private automatic branch exchange (PABX)*. A connection estabilished through a traditional analogue PABX is in many ways similar to a connection made through the standard telephone network in so much that all paths through the network are designed to carry limited-bandwidth analogue speech. To use such a facility to carry data, therefore, requires modems as discussed in fifth paragraph. The more modern types of PABX, however, utilize digital-switching techniques within the exchange and for this reason they are also referred to as *private digital exchanges (PDX)*. Moreover, the availability of inexpensive integrated circuits to perform the necessary analogue-to-digital and digital-to-analogue conversion functions means that is rapidly becoming common practice to extend the digital mode of working right back to the subscriber outlets. This means that a switched 64 kbps path, which is the digitizing rate normally used for digital voice, is available at each subscriber outlet, which can therefore be used for both voice and data.

The main use of a PDX, however, is likely to be to provide a switched communications path between a localized community of integrated voice and data terminals (PCs, workstations) for the exchange of such information as electronic mail, electronic documents, etc., in addition, of course, to normal voice communications. Furthermore, the use of digital techniques within the PDX will mean that it can also be used to provide such services as voice store-and-forward (that is, a subscriber may leave (store) a voice message for another subscriber for later retrieval (forwarding)) and teleconferencing (multiple subscribers taking part in a single call).

The preferred topologies of LANs designed specifically to function as a data communications subnetwork for the interconnection of local communities of computer-based equipment are *bus* (linear) and *ring*. In practice, bus networks are normally extended into an interconnected set of buses with the resulting topology resembling an unrooted tree structure. Typically, with a *bus topology*, the single network cable is routed through those locations (offices, for example) that have a DTE to be connected to the network and a physical connection (tap) is made to the cable to allow the user DTE to gain access to the network services supported. Appropriate medium access control circuitry and algorithms are then used to share the use of the available transmission bandwidth between the attached community of DTEs.

With a *ring topology*, the network cable passes from one DTE to another until all the DTEs are interconnected in the form of a loop ring. A feature of a ring topology is that there is a direct point-to-point link between each neighbouring DTE which is undirectional in operation. Appropriate medium access control algorithms then ensure the use of the ring is shared between the community of users. The data transmission rates used with both ring and bus topologies (typically from 1 to 10 Mbps) mean that they are best suited for interconnecting local communities of computer-based pieces of equipment; for example, intelligent workstations in an office environment or intelligent controllers around a process plant.

8.2 Transmission media

Shielded twisted pair, coaxial cable and optical fibre are the three main types of transmission medium used for LANs. One of the main advantages of both twisted pair and coaxial cable is that it is straightforward to make a topology, for example, and hence bus networks mainly use these types of transmission media.

Optical fibre cable is best suited to those application demanding either high levels of immunity to electromagnetic interference or very high data rates. Also, it is not easy to make a number of physical taps to a single optical fibre cable. Optical fibre is best suited, therefore, for point-to-point communications as used with a ring network, for example.

Coaxial cable, when used for a LAN, is operated using both baseband and broadband techniques. The latter, instead of transmitting information on to the cable in the form of, say, two voltage levels corresponding to the bit stream being transmitted (baseband), divides the total available bandwidth (frequency range) of the cable into a number of smaller subfrequency bands or channels. Each subfrequency band is then used, with the aid of a pair of special modems, to provide a separate data communications channel. This style of working is known as frequency-division multiplexing and, since the frequencies used in the radio frequency band, the modems are rf modems. This approach is known as broadband working and the same principle is currently in widespread use in the *community antenna television (CATV)* industry to multiplex a number of TV channels on to a single coaxial cable (this kind of TV signal transmission is well known as a *cable-TV*). Each TV channel is allocated a particular frequency band, typically of 6 MHz bandwidth. Each received videosignal is then used to modulate a carrier frequency in the selected frequency band. These modulated carrier signals are transmitted over the cable network and are thus available at each subscriber outlet. The subscriber selects a particular TV channel by tuning to the appropriate frequency band.

In a similar way, it is possible to derive a range of data transmission channels from a single cable by allocating each channel a portion of the total bandwidth, the amount of bandwidth for each channel being determined by the required data rate. For data communications, however, a two-way (duplex) capability is normally required. This may be achieved in one of two ways:

1. *Single-cable system:*   The transmit and receive paths are assigned two different frequency bands on the same cable.

2. *Dual-cable system:*   Two separate cables are used, one for the transmit path and the other for the receive path.

The main difference between the two systems is that a dual-cable system requires twice the amount of cable and cable taps to install. Nevertheless, with a dual-cable system the total cable bandwidth (typically 5 to 450 MHz) is available in each direction. Moreover, the headend equipment is simply an amplifier, whereas with a single-cable system a frequency translator is required to translate the incoming frequency signals associated with the various receive paths to the corresponding outgoing frequencies used for the transmit paths.

8.3   Medium access control methods

When a communications path is established between two DTEs through a star network, the central controlling element (a PDX, for example) ensures that the transmission path between the two DTEs is reserved for the duration of the call. With both a ring and a bus topology, however, there is only a single logical transmission path linking all the pieces of equipment together. Consequently, a discipline must be imposed on all the items connected to the network to ensure that the transmission medium is accessed and used in a fair way. The two techniques that have been adopted for use in the various standards documents are *carrier-sense-multiple-access with collision detection (CSMA/CD)*, for use with bus network topologies, and *control token*, for use with either bus or ring networks. In addition, an access method based on a *slotted ring* is also widespread use with ring networks. For the detailed description of all presented network access methods see please [2], [3], [6], or [7].

8.4   LAN Standards

As LAN evolved in the late 1970s and 1980s, a wide range of different network types were proposed and implemented. However, because of small differences between them, such networks could only be used, in general, to interconnect computers or workstations belonging to the supplier of the LAN. Such networks, of which there are still many in existence, are known, as closed systems.

To alleviate this situation, some major initiatives were launched by various national standards bodies with the aim of formulatting an agreed set of standards for LANs. The major contributor to this activity was the IEEE which formulated the *IEEE 802 series* of standards, as a result of this, these have now been adopted by the ISO as international standards. As can be concluded from the previous paragraphs, however, there is not just a single type of LAN. Rather, there is a range of different types, each with its own topology, medium access control method and intended application domain.

9 Conclusions

The presented chapter has introduced you to the fundamentals of computer communications, wide and local area networks, international communications standards, and computer network access methods. Summarized we can say that each computer network consist of hardware and software means providing the capability to interconnect a variety of data-communicating computers, computer-based equipment and several data-communitating devices within the limited or wide area. All kinds of computer networks have become an integral part of computing and data communications. Today, most computer manufacturers provide some means for connecting their equipment to several computer networks.

Users connected to some computer network (above all to a LAN) can share devices such as processors of several kinds, printers, disk storages etc. The most efficient use of each device can be achieved by allowing access to it by multiple users; the need to obtain individual devices for each user is eliminated. For example, through a LAN, several workstations can access a single disk.

While networks as a whole permit a variety of intelligent devices to exchange data, several important features distinguish a LAN from a wide area networks (WAN):

- Whereas WANs may provide national or international services, LANs provide network services to a limited area.

- Unlike wide area networks, which may be subject to the rules of a regulatory agency such as e.g. the Federal Communications Commission in the USA or the Office of Telecommunications in Great Britain, LANs are privately owned and administered. This means only a LAN can be designed, installed, maintained, and expanded without external involvement or approval.

- Each device connected to a LAN potettialy has equal access to every other network device, each one connected to a WAN has access limited through the system of access rights and priorities.

- LANs are self-contained entities that can be constructed with a relatively small amount of equipment. The geographic coverage that a wide area network provides requires a large array of telecommunications equipment (telephone lines, satellites, and microwave links, to name a few). LANs often connect to a wide area network through a device known as a *gateway*, but the LAN is always a distinct and independent entity.

References

1. *Black U.D.:* Computer Networks. Prentice Hall Inc., Englewood Cliffs, New Jersey, 1987

2. *Halsall F.:* Data Communications, Computer Networks and Open Systems. Addison Wesley Publ., Wokingham, 1992

3. *Miller M.A.:* LAN Protocol Handbook. Prentice Hall Inc., Englewood Cliffs, New Jersey, 1990

4. *Sloman M., Kramer J.:* Distributed Systems and Computer Networks. Prentice Hall Inc., Englewood Cliffs, New Jersey, 1987

5. *Stallings W.:* Local Network Technology. IEEE Computer Series Press, New York, 1985

6. *Tanenbaum A.S.:* Computer Networks. Prentice Hall Inc., Englewood Cliffs, New Jersey, 1988

7. *Tangney B., O'Mahony D.:* Local Area Networks and their Applications. Prentice Hall Int. (UK), Hertfordshire, 1988

Information Management and Information Systems Planning

Roland Traunmüller

Institut für Informatik, Johannes Kepler Universität
Linz

Abstract. Information Management and Information Systems Planning play key roles in the informatization of an enterprise. The following introduction to the field presents definitions, reflects aims and scopes, and discusses levels and aspects of Information Management. Subsequently Information Systems Planning as a pivotal activity connected with Information Management is considered in a similar manner. The planning activities are broken down into components and some examples are sketched. Subsequently attention is given to Information Resources Management. The final sections focus on three developments that have already set forth to modify goals and practice in Information Management and Information Systems Planning: Electronic Document Management and Interchange; Computer Supported Cooperative Work and Business Process Re-Engineering.

Key Words:
Information Management; Information Systems Planning.

1 Reflections on Information Management

1.1 Origins of Information Management

In late Seventies and early Eighties the terms Information Management and Information Resources Management emerged in literature. The fundamental rationale for the coining of these terms was an economic reasoning as formulated by Stonecash in a classic way: " ... fundamentally the argument is that information as a resource - while not identical to the classical resources of people, machines, markets and money - is similar enough that most, if not all, of the traditional resource management principles can be applied without difficulty ..." [32].

The rise of Information Management was provoked by both deficiencies and chances. So Information Management intends to cure failure and inadequacies of existing information systems and to offer prospects at the markets. The then lead of the US-Airlines American Airlines and United Airlines based on their reservation schema was esteemed as actual proof. The promising new idea lead quickly to a prospering of the scientific field. As is often the case, a prosperous scientific field favours the blossoming of various terms and notations. So in the next sections the dimensions of Information Management are considered.

1.2 Dimensions of Information Management

Aims and Scope of Information Management have been extended in course of time with new definitions and terms adding additional aspects:

First, definitions coming from an economics point of view define information as "productivity factor" promoting the economical success of the enterprise [32]. So regarding information as production factor is a prevalent view in Business Science literature [19,22,27,28].

With regard to the importance of resources the expression Information Resources Management was coined [1,16]. The term IRM remained somewhat ambiguous because it was employed as well in similar as well as in different meaning to IM.

A different aspect emphasised was that of design: Information is a "means for shaping and forming the organisation" helping the management to get goals performed effectively. This intention found early supporters in Governmental Informatics and later on became a dominant view [3,25,26,34,35].

Other authors put emphasis on particular aspects: the importance of information exchange between enterprises [31]; the thinking in "systems" and in terms of "cost" and "benefits" [14,15]; the connections with the design phase [21].

Now Business Process Re-Engineering and Computer Supported Cooperative Work dominate the field and will govern the future way of Information Management and Information Systems Planning [10,12].

The literature mentioned above is intended to reflect on different dimensions as a suitable basis serving for further investigation. This intention should be stressed so that above selection is seen as a paradigmatic and subjective choice.

2 Notational Distinctions

2.1 Levels of Information Management

Many authors emphasise the fact that for an adequate discussion it is necessary to discern levels and scopes. A very early definition by Szyperski (1980) states three levels [33]:

Information Management; Information Systems Management; Information Resources Management. Later Mundhenke (1986) makes a similar categorisation [23]: Strategic Information Management; Plain Information Management; Information Techniques Management. Griese (1990) reduces it to a twofold distinction comprising [19]: Strategic Information Management; Operative Information Management.

2.2 Information Management and Information Systems Planning

Nobody argues against a close connection between Information Management and Information Systems Planning:

So from the Business Science side many textbooks contribute major attention to Information Systems Planning (e.g. [13]).

Also the Informatics has been stressed the organisational aspects from the very beginning (e.g. Langefors and his Scandinavian school [20]).

Surely nobody argues against a close connection between Information Management and Information Systems Planning, but the precise definition of the relationship has become a matter of scientific disputation. The methodological questions involved have become even more complicated, because both disciplines, Business Science and Informatics, have developed terminological distinctions of its own. Some authors in the field of informatics are inclined to regard the question of Information Management as a sheer prelude to the Information Systems Planning (e.g. IFIP-8.1 Framework [17]).

Others present stage schema centred on Management Information Systems, so citing Bowman, Gordon and Wetherbe as early examples [2]:
Strategic Management Information Systems Planning; Information Requirements; Resources Allocation. Fortunately, the terminological problems are mainly theoretical in nature. In practice a common sequence of stages has been broadly accepted formulated in the following "alignment process" [36]: Business strategy; IT-strategy; Organisational infrastructure and process; IS infrastructure and process

3. Strategic Information Management

3.1 Primary Questions

Strategic Information Management belongs to the global level of enterprise planning, where all sectors such as production, finances, marketing, personnel, and buildings are considered together. At this level of abstraction information is considered as one resource among others. So the primary aspect is to reflect how in general terms the resource information can be used to the benefit of the enterprise. The major concerns considered can be formulated in the following two primary questions:
Where are weak points in the functions of the enterprise which could be improved?
Where can chances be located for improving the market position of the enterprise?

3.2 Functions in Strategic Information Management

Management functions concern the organisation, planning and control of three factors:
1. Productivity factors such as personnel, finances and IT
2. Production processes such as development and maintenance
3. Management activities such as organising, planning, decision making and control
    Typical questions posed in Strategic Information Management are the following ones:
Analysis of information needs and search for possible solutions
Medium and long-term planning of informatisation
Planning of a global information system architecture
Planning of IT-resources and IT-logistics
Economic analysis of the informatics sector of the enterprise
    It is evident that at the global level of Strategic Information Management questions are usually posed in a general manner. In the same way the answers will be given as global answers. More detailed considerations will be inducted at the subsequent level of Strategic Information System Planning.

4 Information Resources Management

4.1 IM and IRM

The terms IM and IRM were coined at the same time in the United States. From the very beginning no common understanding concerning their usage and distinction has evolved :

Some authors put emphasis on the use of "information as a resource" so reducing the distinction between IRM and IM almost to naught.

Other authors put the accent on "management of resources" so building up IRM as a contrariety to IM.

During the course of time additional interpretations came forth:
In addition, the term IRM is used differently by managers and by data processing professionals.

Also the "I" in IRM is interpreted in different ways by various authors: "Informatics", "Information", and "Information Systems".

4.2 Institutions for IRM in the Public Sector

In the United States IRM was heavily discussed in the Public Sector. The debate was triggered by growing discomfort caused by various annoyances such as [1]:
Evident inefficiencies in the Public Sector
Overflowing information needs of Public Agencies
Severe backlogs in planning, acquiring, and installing equipment
Surreptitious information demands embarrassing private enterprises

In 1980 the idea to introduce IRM to Public Agencies got momentum. So with the "Paperwork Reduction Act" a big leap for institutionalisation was achieved. The act was confirmed in 1986 with the "Paperwork Reduction Reauthorization Act". Through above mentioned acts major activities were set forth and some powerful new Agencies were created [1]:

The Office of Information and Regulation Affairs (OIRA) was established providing central guidance and coordination.

The General Service Administration is aimed at IT logistics and IT counselling.

The Federal Information Locator System (FILS) is a central database disseminating information about the various dispersed databases in the Public Sector in order to prevent unnecessary repeat of data capture.

The "Go-For-12"-Programme is intended to cut down the timely planning and logistics from an average of one and a halve year to twelve months.

4.3 IRM in Manufacturing Enterprises

General enterprise models and consistent data bases are valuable assets of the company. Due to the fact that inconsistencies of the data base generate severe obstacles in integrating applications, efforts put in the planning stage render themselves profitable in the course of subsequent actions. So enterprise models for manufacturing companies have a long tradition starting with KIM (Kölner Integrationsmodell) published by Grochla et al. as a forerunner. Now an elaborate model is provided by Scheer with UDM (Unternehmensdatenmodell).

In KIM a complete functional description of an enterprise was given comprising 350 functions and more than 1500 channels. (One should be well aware that these are the original numbers reflecting the enterprise of the Seventies.) UDM is a complete description of a manufacturing enterprise comprising more than 300 objects [28]. There are also graphical representations given showing objects and their connections as Entity-Relationship-Diagrams.

5 Information Systems Planning

5.1 Need for Strategic Planning of Information Systems

Ideas raised at higher planning level have a chance to become more concrete at subsequent levels. So at the level of Strategic Information System-Planning many question raised at the antecedent level of Strategic Information Management are treated in more detail. Here the term "Information Systems Planning" is used as a short form for "Strategic Information Systems Planning" as is often the case in the literature (e.g., Olle et al.). For the increasing urgency of such planning some reasons can be cited:

Dynamics of the technology market; Broadening scope of the applications; Ongoing improvements of IT-products; Complexity of Information System; Prolonged duration of application development processes

5.2 Benefits from Information System Planning

There are a lot of benefits that can be derived from Information System Planning:
Enterprise Models; Blueprint for an enterprise wide database; Master plan for information systems; Definition of specific DP-projects; Allocation of resources.

Finally specific strategies can be defined to achieve particular goals:
Consolidating of the enterprise's databases; Formulating IT-policies (Operating systems, network infrastructure, mixed hardware, IT-logistics, etc.); Extending or reforming the application spectrum;Creating organisational structures (Information centre, education and tutoring etc.).

5.3 Scope of the Planning Process

Another fact one has to be aware of is the scope of planning. In that way the boundaries of a specific project can be clarified. As a consequence estimations of time and resources will be more reliable.

In principal any procedure in the enterprise would fit in as an object of consideration. Such an undiscriminating attempt surely would soon exhaust the planners' time frame. So reasonable assumptions are to be made from the beginning. A list of topics might include the following ones:

Business plans will typically cover issues such as objectives, goals and critical success factors.

Finance will deal with investments in various parts of the enterprise and with the returns expected. Also current investments in information technology are included.

Mental models present a particular view of the enterprise such as data models, functional models etc.

Competition deals with markets, competitors, products, services, channels and sources of added value.

Organisation would deal with units, interest groups as well as with problems and chances within the enterprise.

Environment covers a large area of factors including government, regulatory bodies, political situation as well as geographical location and particular environmental problems.

Information technology deals with the technological substratum.

5.4 Possible Approaches to the Planning Process

Strategic Information Systems Planning has to provide a global picture of the future system. As a consequence, it should avoid bias and imbalance and it should imply different standpoints. So it might be a reasonable guideline to start with different views and consolidate them afterwards to a single picture. In that way the next sections will deal with the following approaches to the Strategic Planning of Information System:
1. At first a simple model is given representing an organisation driven planning process.
2. Then a business plan and mental model driven approach is given. For the matter of clarity at this point only the descriptive aspects of the process are covered.
3. At last the business plan and mental model driven approach is extended also covering the prescriptive aspects of the process.

6 Organisation Centred Planning Process

6.1 General Remarks

Approaches coming from Informatics have a different focus than approaches from Business and Administrative Sciences. So in an Information Systems point of view the Strategic Planning of the Information System is stressed, whereas questions of Strategic Information Management are considered as a "prelude" and only touched in a general way. So in sections 6 to 8 approaches of Information Systems Planning according to the IFIP-8.1 Framework [17] and the CRIS-literature [6-9] are presented. To these points some remarks should to be added. Within IFIP the Technical Committee 8 on Information Systems has established a Working Group on Design (WG 8.1) which is aimed at Information Systems Design Methodologies. So a series of Review Conferences (CRIS '82, '83, '86, '88) have taken place, to which the author has been appointed a member of the Programme Committee.

From the results of the CRIS-Conferences a Task Group headed by Bill Olle derived commonalties and developed a framework that could be used as well as a means for understanding as well as a body of reference. In 1988 the first edition of "Information Systems Methodologies - A Framework of Understanding" was published with the approval of IFIP Working Group 8.1 and soon has spurred a lot of discussion. In the IFIP-8.1 Framework the stages Design, Analysis and Planning are covered in an elaborated way.

6.2 Approach

The organisation centred approach is aimed at solving imminent business problems of the enterprise and comprises the following steps:
1. Analyse business problems.
2. Identify need for change.
3. Identify information systems.
4. Propose change alternatives and new information systems.
5. Analyse benefits and costs.
6. Evaluate change alternatives.
7. Prepare Plan.

6.3 Components

Components of planning and their interconnections can be derived from steps and are sketched in the attached diagrams which are published in the IFIP 8.1 Framework. The course of derivation is stated below with components given in Italics:

Step 1 introduces *Business Problems, Information System, Business Activities,* and *Interest Groups* within *Organisations.*

Step 2 establishes *Objectives* of the Enterprise and identifies *Needs for Chance.*

Step 3 identifies the existing and planned *Information System* relating it to *Business Activities.* Also cross-references to *Information/Material Sets* are given.

Step 4 uses the information now available to propose *Change Alternatives.*

Step 5 views *Information System* and *Change Alternatives* for deriving from them *Benefits.* In a similar way estimations are made for *Costs.*

Step 6 evaluates *Change Alternatives* to show the preferable ones. Each *Change Alternative* is related to the anticipated support of *Information Systems* and of *Business Activities.*

7 Descriptive Planning Process

7.1 Approach

The descriptive approach is an analytical one and concerned with the principal information needs of the enterprise. Therefore business plans, organisation and mental models are the pivotal basis of the consideration. The descriptive approach will comprise the following steps:
1. Identify business plan components and information needs.
2. Develop business model (from a data perspective).
3. Identify information systems and business areas.
4. Analyse priorities.
5. Prepare plan.

7.2 Deriving Components from Steps

Relevant components can be derived from the above steps in the following way:

Step 1 starts the analysis by identifying the *Objectives* of an enterprise, its major *Organisation Units* and *Business Activities.* Attention is laid on ways of supporting the *Objectives* through *Information Systems.* Measurable *Goals* for each *Objective* are determined and *Critical Success Factors* are stressed. The information to support all these is presented as *Information/Material Set.*

Step 2 develops from *Information/Material Set* a datamodel described with *Entities* and *Relationships.*

Step 3 illustrates how each *Business Activity* uses the data and establishes cross-reference. Resulting clusters of data may represent potential *Information Systems.*

Step 4 derives logical sequences for developing *Information Systems* based on their dependencies for other systems to gather information. This sequence is modified by *Business Priorities,* by a *Cost/Benefit* Evaluation and by estimates for the transition effort.

Step 5 groups Business Areas which provide a scope for subsequent business analysis projects. The Information Systems Plan is then constructed as a sequenced collection of *Plan Projects*.

7.3 Components

In the following some components are explained:

Business Problem is a difficulty being encountered with some aspect of the functioning of the enterprise or its information system.

Organisation Unit refers to all kind of department, section, group, bureau etc.

Unit in Organisation describes the reporting structure among Organisation Units.

Interest Group is a category of people affected by a Business Problem.

Organisation Unit Involved With Interest Group is a cross-reference component.

Objective is a broad result that an enterprise wishes to achieve.

Problem for Objective shows those cases where an identified Business Problem may restrain the achievement of Objectives.

Goal is a significant defined point leading towards accomplishment of Objectives.

Critical Success Factor is on of the determining factors for the achievement of Goals and Objectives.

Information/Material Set may be some kind of information which is either input to a Business Activity or generated by it. Alternatively it may be a collection of goods, parts, supplies, etc.

Business Activity describes what has to be done within an enterprise.

Objective Targets Business Activity highlights the result of the Objective.

Activity Uses Set indicates which Information/Material Set are input or generated.

Entity Type is a class of objects perceived as being

Business Activity Involves Entity Type is a cross-reference which is important for checking, because every entity must be involved at least in one activity and v.v.

Involvement Type categorises the way in which one component uses the other one.

Information From Entity Type is a cross-reference component.

Relationship designates how the entities are interrelated and also may give the cardinalty (1:1; 1:n; m:n).

Entity Type in Relationship discerns a certain relationship.

Need for Change is a statement about something that must be corrected or ameliorated to resolve or entangle a particular Business Problem.

Change Alternative is a suggested new way of doing things or performing activities as a response to some Business Problem.

7.4 Examples of Components

In the following some examples of components are given:

Business Problem: incorrect invoices; late arrival of distribution instructions; inconsistent material quality; no loading bay.

Organisation Unit: enterprise; purchasing department; warehouse and distribution; goods receiving unit.

Interest Group: main board director; stock controllers; head office clerical staff; warehouse foremen; auditors; branch sales manager;

Objective: economic maintenance of high service levels; achieve market leader status.

Goal: achieve 20 percent market share; improve sales by 4 percent each quarter; dispatch goods within 24 hours.

Critical Success Factor: exact forecasts for demand; reliability of stock records.

Information/Material Set: list of suppliers; list of items stocked; stocks of goods; purchase orders; missed supply by dates.

Business Activity: inventory control; ordering of required stock;

Entity Type: supplier; stock item; purchase order; stock on hand.

Involvement Type: "R" input to activity; "U" output from activity.

Relationship: supplier receives purchase order; stock item supplied from supplier.

Need for Change: better order processing; more accurate invoicing; better stock controller moral.

Change Alternative: acceptance of request by van driver; telephone order soliciting; direct ordering by customers.

8 Prescriptive Planning Process

8.1 Approach

The prescriptive approach extends the descriptive one by introducing additional normative elements. Normative elements that are added may be divers planning variables. They indicate different constituents such as name and type of a specific project, the status of the project, allocation of particular resources, priorities to be considered, prioritizing criteria etc.

8.2 Components

Additional components represent the planning variables. In the following some components are outlined:

Information System means in the broad sense used in this chapter a means of recording and communicating to satisfy the requirements of all its users.

Development Status indicates whether the Information System is extant or envisaged

Activity Allocation to System is a cross-reference term.

Entity Type Allocation to System links the Information Systems with the database.

Business Area is a collection of Information Systems that are envisaged.

Plan Project is a project to which resources may be allocated.

Information Systems Plan is a collection of Plan Projects and shows the interdependencies of projects.

Project Status means new development or enhancement.

Project Precedence/Succedence captures the intended sequence.

Information System in Plan Project is a cross-reference component.

Information System Support for Objectives has to ensure that the requirements stated in the strategic planning are indeed met.

Problems Involves Information System identifies the link between problems to be alleviated and information systems.

Cost is a depletion of resources mostly measured on a monetary basis.

Benefit is an improved or a new contribution to the enterprise.

Priority is a ranking measure indicating a desired relative order of development.

Priorization Criterion is a basis for determining Priority (rules, policies, algorithms, etc.)

8.3 Examples of Components

For matter of explanation some examples will be given in the following:
Information System: ICP inventory control and purchasing; rder Processing and DistributionMRP material requirements planning;
Development Status: extant; envisaged.
Business Area: marketing; sales and inventory; finance.
Plan Project: marketing and customers analysis; finance analysis.
Information Systems Plan: '93-'98 plan.
Project Status: new development and enhancement.
Cost: 5 man year development project; terminals for 10 branch offices; 5 week training.
Benefit: better risk management; 30 percent reduction in maintenance costs; 9 percent increase in production line throughput.
Priority: ICP 1; MRP 2.
Priorization Criterion: precedence of planning system; an entity type is used by one third of the activities.

9 Electronic Document Management and Interchange

9.1 Organisational Issues and Enabling Technologies

In summarizing four decades of IT and organisations Bjoern-Andersen states that it is specific for the Nineties that organisational issues have become the driving force. So organisational re-engineering, group support and inter organisational systems are cited [4]. All in all this is a paradigmatic shift in the relation between IT and organisations. This new order is also well mirrored in the arising term of "enabling technologies".

These changing goals and means will be mirrored in Information Management and Information Systems Planning. So the following sections focus on three leading developments:
Electronic Document Management and Interchange
Computer Supported Cooperative Work
Business Process Re-Engineering

9.2 Electronic Document Management

In Computer Integrated Manufacturing documents comprise a variety of symbols and media to represent the ideas and concepts in the documents: text, tables, graphics, images, etc. As a consequence Electronic Document Management (EDM) has become a key technology for CIM and EDM is blossoming in manufacturing enterprises.
Ralph Sprague counts four major reasons for introducing EDM [31]:
1. EDM provides substantial benefits.
2. It automates the paper part of system.
3. Documents are the centre of many applications.
4. Publishing is pervasive.

9.3 Electronic Document Interchange

Electronic Document Interchange (EDI) is the computer-to-computer exchange of standard Business transactions. According to Keen [18] electronic business partnerships will be the standard and interorganizational linkage become a major concern of Information Management. Applications are considered candidates for EDI in case they have the following characteristics: They are applications with many standard transactions; Reporting has to be careful and accurate; There is a gain in making the purchase easier.

10 Computer Supported Cooperative Work

10.1 The Emerging of CSCW

The term "Computer Support for Cooperative Work" was coined by Irene Greif and Paul Cashman in 1984 as a prelude to the first CSCW conference held in Austin, Texas. CSCW was thought of as arising from a particular kind of problem - the needs which some people had to cooperate in groups in doing their work - and so to give rise to the need for particular kinds of systems, to which the term "groupware" came to be applied. Of course systems have always been designed to serve many people. The distinctive feature now identified was that of people cooperating on a task by interacting with each other in or through the machine. Rather than using timesharing precisely for the purpose of sustaining the illusion that users had their own virtual machine entirely to themselves, systems were now conceived directly to support users in their inter-relations. Again, it is not claimed that no previous systems ever did this, rather than explicit recognition of the need would now enter into the design philosophy.

Driving forces for the spreading interest in CSCW shall be cited in the following [30]:
1. dissatisfaction of the users with existing systems
2. increasing expectations of the users
3. "easy" things have already been done
4. concentration on highly formal, bureaucratic procedures
5. the quest for productivity

10.2 Broadening the Support for Cooperation

In this context CSCW is gaining prominence, as holding out the prospect of a different approach to systems design which might address at least some part of these problems. So the challenge is to move out from this base into the effective integration of separate systems, and into the support of "higher-level" organisational processes involving decision-making, negotiation and collaboration - areas characterised by flexibility and rapid change rather than constancy.

A closer inspection of cooperative work situations in enterprises reveals that coordination is not "the only game". Coordinated work represents only one kind of the three major forms of cooperative work that comprises likewise collaboration and group decisions:
1. Coordination - the case considered in the widespreadly used workflow management systems - unifies different activities for the accomplishment of a common

goal. Each activity is in an intrinsic relation to preceding and succeeding ones so making synchronisation a major issue.

2. Collaboration is the case of persons working together without external coordination as it is the case in co-editing and shared drawing. It is necessary to have a common information space and to point at a collective goal.

3. Group decisions need cooperation for the accomplishment of a collective decision. Although diverse opinions and interests may prevail a minimum of mutual trust is required.

There are additional features to be considered in cooperative work: So cooperative ensembles may vary over time. In addition they are marked by an ample local distribution and lack of central control. Likewise work might be as well synchronous as asynchronous.

10.3 CSCW as a Perspective for Manufacturing and Office

Now CSCW has already reached the enterprises. Just to give one example: Under the heading "The Changing Office" the July 94 edition of an acknowledged computing journal with worldwide circulation [24] is dedicated to CSCW concepts and products. They list around hundred commercial products entering the world of business.

The main CSCW application scenarios described are the following ones:
improvements to electronic mail functions (e.g. for forwarding, selecting and storing messages)
shared workspaces for providing common views on a particular subject
group authoring enabling collaborative writing with additions, revisions, comments, and annotations
group decision support systems for argumentation, negotiation, and decision making
meeting rooms, desktop conferences and computer conferences

Some authors have discussed the mission of IT with relevance for generic classes of business processes [5]:
Mechanistic processes: Traditional manufacturing and office systems
Purposive rational processes: IS for accounting, finance, DSS
Normative regulated processes: LEGOL, The Coordinator
Communicative processes: Groupware, meeting rooms

There is a revolution going on in the enterprises. Gradually higher order activities in manufacturing and office will get assistance. In that long run CIM and CIB will open perspectives for entirely new ways of Information Management and Information Systems Planning for manufacturing and services industries as well as the office environment.

11 Business Process Re-Engineering

11.1 BPR: A New Attitude Toward the Organisation

In the late Eighties and early Nineties the term Business Process Re-Engineering (BPR) emerged [10,11] and became a shooting star. Whole conferences such as TC8 '94 Conference [11] were devoted to this topic. Despite different views there a common view on BPR as an integrated perspective and programme based on organisational change: core processes have to be found and redefined with respect to effectiveness and efficiency. BPR brought a frank and bold attitude and this was part of its success. As an example Hammer claimed " ... Forget the 10% and 20% improvements. Go for doubling productivity with only half your resources ... ".

11.2 BPR: A Methodology for Organisational Redesign

Professional practice transformed concepts and guidelines in methods. Although the methods applied are varying in terms they all have two stages that are indispensable:
1. Assessment investigates different business activities, identifies the core processes and selects candidates for change.
2. Re-Engineering gives a vision of future processes and seeks enabling factors. Such enablers may be technical or organisational in character. Subsequently a prototype of the process is specified in detail.

A very elaborate model of participative BPR at the University of Arizona [37] comprises the following seven stages: Preparation, Vision Establishment, Model development, Model Analysis, Action Proposal Generation, Implementation Planning, Project Evaluation.

11.3 BPR in Manufacturing Enterprises

Thompson states that typical candidates for BRR are three types of enterprises [36]:
1. Manufacturing and high technology industries facing domestic and foreign competition.
2. Companies working in recently deregulated industries.
3. Firms recently involved in mergers and acquisitions.
4. Companies in highly competitive service industries.

According to this categories manufacturing enterprises are presumably candidates for BPR.

11.4 BPR: Pouring New Wine in New Bags?

Using information technology as a "change agent" is an idea stemming back from the late Sixties. So the question after the "genuine new in BPR" is a legitimate one. Without doubt there are authentic new features in BPR:
There is a radical questioning of each activity involved.
One is forced to start from the beginning
No "legacies", "taboos", "sacred cows", etc. are spared.

Last but not least, the paramount role of information technology as an enabling factor has to be stressed. It is the state and maturity of technology that makes possible to attain such an high grade of organisational reshaping.

Acknowledgement: The author is indebted to his old friend Joe Aschl. Mr. Aschl who had spent a long time of his carrier in Canada has kindly checked the language of the contribution.

References

1. BAYER, L.: Informationsmanagement und öffentliche Verwaltung, DUV, Wiesbaden, 1992.
2. BOWMAN, B.J., DAVIS, G.B., WETHERBE, J.C.: "Three Stages Models of MIS Planning", Information and Mangement, Vol.6, 1963.

3.   BOTS, P., SOL, H., TRAUNMÜLLER, R. (eds.): Decision Support in Public Administration. North-Holland, Amsterdam, 1993.

4.   BJOERN-ANDERSEN, N., CAVAYE, A.: Reengineering the role of the IS professionals, in [11].

5.   CECEZ-KECZMANOVIC, D.: Business Process Redesign as the Reconstruction of a Communicative Space, in [11].

6.   CRIS82: OLLE, T.W., SOL, H.G., VERRIJN-STUART, A.A. (eds): Information Systems Design Methologies: A Comparative Review. North-Holland, Amsterdam, 1982.

7.   CRIS83: OLLE, T.W., SOL, H.G., TULLY, C.J. (eds.): Information Systems Design Methodologies: A Feature Analysis, North-Holland, Amsterdam, 1983.

8.   CRIS86: OLLE, T.W., SOL, H.G., VERRIJN-STUART A.A.(eds.). Information Systems Design Methodologies: Improving the Practice, North-Holland, Amsterdam, 1986.

9.   CRIS88: OLLE, T.W., VERRIJN-STUART, A.A., BHABUTA, L. (eds.) Computerized Assistance During The Information Systems Life Cycle, North Holland, Amsterdam, 1988.

10. DAVENPORT, T., SHORT, J.: The New Industrial Engineering: Information Technology and Business Process Redesign, in: Sloan Management Review; Sommer 1990.

11. GLASSON, B., HAWRYSZKIEWYCZ, I., UNDERWOOD, A., WEBER, R. (ed.): Proceedings of the IFIP TC8 Conference on Business Process Re-Engineering, Bond University, May 1994, North-Holland, Amsterdam, in press.

12. HAMMER, M.: Reengineering Work: Dont Automate, Obliterate, in: Havard Business Review, Vol. 68/4, July 1990.

13. HANSEN, H.R.: Wirtschaftsinformatik I, 5. Auflage, UTB, Stuttgart, 1986.

14. HÜBNER, H.: Informationsmanagement, Oldenbourg-Verlag, Wien, 1984.

15. HEINRICH, L., BURGHOLZER, P.: Informationsmanagement. Oldenburg, München, 1988.

16. HORTON, F., MARCHAND, D. (ED.): Information Management in Public Administration, Information Management Press, Washington, D.C., 1982.

17. IFIP-8.1 FRAMEWORK: OLLE, T.W., HAGELSTEIN, J., MACDONALD, I., ROLLAND, C., SOL, H.G., VAN ASSCHE, F., VERRIJN-STUART, A.A.: Information Systems Methodologies. A Framework for Understanding, Addison Wesley, Wokingham, 1988 (2nd Edition 1991).

18. KEEN, P.: Shaping the Future: Business through Information Technology, Havard Business School Press, Boston, 1991.

19. KURBEL, K., STRUNZ, H. (Hrsg.): Handbuch der Wirtschaftsinformatik, Poeschel, Stuttgart, 1990.

20. LANGEFORS, B: Theoretical Analysis of Information Systems, Studentenliertatur, Lund, 1966.

21. MARTIN, J.: Information Engineering, Book I, II, III, Prentice Hall, Englewood Cliffs, 1989.

22. MERTENS, P. (Hrsg.) : Lexikon der Wirtschaftsinformatik, Springer, Berlin-Heidelberg-New York-Tokyo, 1987.

23. MUNDHENKE, E.: Informationsmanagement - in den USA bereits Wirklichkeit. DVP 1/1986.

24. PC MAGAZINE: Special edition "The Changing Office", July 1994.

25. REINERMANN, H., FIEDLER, H., GRIMMER, K., LENK, K., TRAUNMÜLLER, R. (Hrsg): Öffentliche Verwaltung und Informationstechnik, Neue Möglichkeiten, neue Probleme, Springer, Berlin-Heidelberg-New York-Tokyo, 1985.

26. REINERMANN, H., FIEDLER, H., GRIMMER, K., LENK, K., TRAUNMÜLLER, R. (Hrsg): Neue Informationstechnologien, neue Verwaltungstrukturen, Decker-Müller, Heidelberg, 1988.

27. SCHEER, A.W.: EDV-orientierte Betriebswirtschaftslehre, Springer, Berlin-Heidelberg-New York-Tokyo, 1991.

28. SCHEER, A.W.: Wirtschaftsinformatik, 3. Auflage, Springer, Berlin-Heidelberg-New York-Tokyo, 1990.

29. SCOTT MORTON, M., S. (ed.): The Corporation of the 1990s, Oxford University Press, New York, 1991.

30. Shapiro, D., Traunmüller, R.: CSCW and Public Administration, in: Bonin, H. (ed.): Proceedings of the IFIP WG 8.5 Workshop "Systems Engineering in Public Administration", Lüneburg, March 1993, North-Holland, Amsterdam, 1993.
31. Sprague, R., McNurlin, B.: Information Systems Management Practice, Prentice-Hall, Englewood-Cliffs, 1993.
32. Stonecash, J.: The IRM Show Down, in: Infosystems 10/1981.
33. Szyperski, N.: Strategisches Informationsmanagement im technologischen Wandel. Angewandte Informatik 4/1980.
34. Traunmüller, R. (ed.): Governmental and Municipal Information Systems, Proceedings of the 2nd IFIP TC8/WG8.5 Conference, Balatonfüred (Hungary), June 1991, North-Holland, Amsterdam, 1992.
35. Traunmüller, R., Fiedler, H., Grimmer, K., Reinermann, H. (Hrsg.): Neue Informationstechnologien und Verwaltung, Springer, Berlin-Heidelberg-New York-Tokyo, 1984.
36. Thompson, D.: Reorganizing MIS: the evolution of business computing in the 1990s. SAMS Publishing, Carmel, 1992.
37. Vogel, D., Orwig, R., Dean, D., Lee, J. Arthur, C.: Re-Engineering with Enterprise Analyzer, Proceedings of the 26th Annual HICSS, Hawai, January 1993.

Overview of Function modelling - IDEF0

Finn Jørgensen

Institute for Product Development
Technical University of Denmark
&
Sant + Bendix A/S, Denmark, member of
Coopers & Lybrand Consulting Group

Abstract: The paper gives a brief overview over the background and contents of the function modelling methodology IDEF0. The purpose is to inform the reader about the origin of the methodology, the methods related to it, why and how to use it, and the terminology and semantics in the methodology.

1 Introduction

In the United States the IDEF0 modelling technique (Integrated DEFinition method 0) is about to become part of the Federal Information Processing Standard (FIPS) complex. This means that all federal institutions e.g. Department of Defence and suppliers to federal institutions are encouraged to use the method when documenting e.g. manufacturing systems. The methodology will hereby in the future be the "standard" methodology to use in the industry.

2 Background for the development of the IDEF-methodologies

As a result of the considerable rationalisation efforts in the production environment, especially on the shopfloor level, which have characterised this century, it is only a minor part of the total costs of manufacturing a product that refers to the hands-on costs. Today, these costs represent 30 - 40% of the total costs of manufacturing a product. (Not to be mistaken with the time that it takes to manufacture a product - or the period of contact, which based on experience is between 5 - 10% of the total time that it takes to manufacture the product.)

Today, non-touch costs related to manufacturing the product represent 60 - 70% of the total manufacturing costs. Non-touch costs do not only include the indirect costs on the administrative level in a company. They also include costs resulting from the entire information work taking place on the shopfloor (e.g., when the shop foreman instructs a co-worker), intermediate stock costs, control costs, quality costs, etc.

When the above is recognised, it is clear that productivity improvements will have the largest positive effect, if found in the non-touch area. Integrated productivity improvements are also found in this area, as these to a high extent involve information.

The IDEF0 method was developed to support the analysis work within both the non-touch and the hands-on areas. It was developed with a view to creating the basis for a complete description of a company's functional structure, data requirements, and data creation.

The work on the development of the method was ordered by the U. S. Airforce, which in 1975 commenced the development programme ICAM (Integrated Computer Aided Manufacturing). The programme lasted until 1985 and is responsible for much of the interest in CIM (Computer Integrated Manufacturing) in the U. S. In connection with the programme, a number of methods were developed. When combined, these methods represent a complete description of a company seen from an information point of view. The methods are:

- Function modelling - IDEF0

- Information modelling - IDEF1 (and IDEF1-X)

- Dynamic modelling - IDEF2

As mentioned, the IDEF0 method is suitable for describing a company's functions and their interrelation - data requirements and creation. This can be done in a gradual detailing with a determined syntax as to how the model could be broken down in an unambiguous way. With this determined structure and syntax, it can also be used for system documentation in connection with e.g., quality control, where individual or departmental responsibilities in connection with company activities play an important role. The background for the method will be described further in the below.

The IDEF1 method was built on the ideas of Entity-Relationship modelling developed by Chen. In connection with the ICAM programme, Chen was asked - together with a number of other experts within data modelling - to find the most suited method. IDEF1 is thus a method for modelling data and the structure of data. An IDEF1 model can be applied as a basis for the creation of a database on a computer. The method is especially suitable for structuring relational databases. If the reader requires more information on the method, please refer to literature on the subject. A few facilities left out from the original Entity-Relationship method have later been added to the IDEF1 method in order to simplify the use of the method. This enhanced version of IDEF1, IDEF1-extended, is called IDEF1-X.

The IDEF0 and IDEF1-X methods supplement each other excellently when the purpose is to map the data and function structure of a company's data. However, the methods do not show the sequence or the duration of activities. The IDEF2 method was, therefore, developed in order to model the dynamic behaviour in a company, e.g., duration of process time on a machine, size of a buffer in front of a machine, and waiting time until a machine is available. The aim was that the model should form the basis for computer simulation (evaluation of consequence) of a certain system configuration. The participants in the development of IDEF2 had, as far as the author has been informed, special interests in their own products that were similar to IDEF2. These modelling tools had already been integrated in marketed products for computer simulation and had, therefore, the largest distribution. The result is that today there exists no agreed recommendation for a standard as to how the dynamic behaviour of a system should be analysed or documented.

The IDEF0 and IDEF1-X methods are today used all over the world. Today, the American Department of Defence demands from any company wanting to be a supplier that their activities and data structure have been documented using IDEF0 and IDEF1. The method is also used by the American National Institute for Standards and Technology (NIST). As mentioned before, the methods are expected to be nominated Federal Information Processing Standards, meaning that an even larger part of the American industry will apply these methods.

In Denmark, the methods were imported in connection with the CIM/GEMS project (CIM/General MEthods for specific Solutions), which lasted from 1986 to 1989. Through a close co-operation with Danish industry, the Institute of Production Management and Industrial Engineering and the Institute for Production Development at the Technical University of Denmark transferred these methods to Danish companies and applied the gained experience to adjust the methods to Danish industry and traditions.

3 The origin of IDEF0

The IDEF0 method is based on the Structured Analysis and Design Technique (SADT) from Softech. The basis of SADT was formed during a project initiated by ITT, where Softech, including "the inventor" Douglas T. Ross, together with MIT (Massachusetts Institute of Technology) formalised the method. Together with Softech, the U. S. Airforce developed the method and demanded through the financing of the development work that it was published so that all companies could use it. On the IDEF Users Group Meeting in Salt Lake City in October 1993 the author of this paper had several talks with Douglas T. Ross, among other things about the difficult negotiations concerning rights to the methods.

In connection with the development work the U. S Airforce had the following requirements to the project:

- Since the architecture is to depict manufacturing, it must be able to express manufacturing operations in a natural and straightforward way.

- Since the subject is so vast and complex, it must be concise and provide a straightforward means of locating details of interest easily and quickly.

- Since it must be used by a wide audience, it must be able to communicate to a wide variety of Aerospace Industry manufacturing personnel as well as to Air Force ICAM Program Office Personnel.

- Since it must serve as a baseline for generic subsystem planning, development, and implementation, it must permit sufficient rigor and precision to insure orderly and correct results.

- Since the baseline must be developed through the co-operative effort of a large segment of the Aerospace industry, it must include a methodology (rules and procedures) for its use that permit many diverse groups to develop architecture pieces and that permit wide-spread review, critique, and approval.

- Since the baseline must represent the entire Aerospace industry rather than any one company or industry segment, the method must include a means of separating "organisation" from function; that is, a common agreement cannot be achieved unless the individual company organisational differences are separated out and only the common functional thread is captured.

These requirements were fulfilled to a very high degree. The degree will not be discussed here, as the below description of the method will supply the reader with the basis for an evaluation hereof.

4 What can IDEF0 be used for in design and implementation of integrated systems?

IDEF0 has several purposes in connection with the building of integrated systems. A recognised procedure when building integrated systems is that the first step is a general "cleaning up" of the company's existing business procedures (also called "broom and bucket"). People working with productivity improvements cannot be expected to be able to assess all the activities or functions are which are taking place in the company. The IDEF0-method could, therefore, help to clarify a number of aspects (all are, of course, not mentioned):

- What functions are being performed in the company, by whom and how long do they take?

- What information do these functions require in order to be able to perform their work and where do they get this information from?

- What information sets the goals, controls, starts, or stops the execution of the function and where does the information come from?

- What information is being created in connection with the execution of the function and to where is it sent.

A proper application of the IDEF0 method is to reach an agreement between analysts (preferably company employees) as to how daily routines are being performed. The method can thus be applied for clarifying and reaching an agreement as to how the information apparatus of the company actually functions.

The work performed on the elaboration of the model is, thereby, to a high degree part of the result. The documentation can be used as a basis for a discussion for the future work and as a basis for future system changes. However, an agreement between analysts as to how the system functions is clearly one of the most important results of an IDEF0 analysis.

In Denmark, the IDEF0 method has been used with good results in connection with for example TBM projects (Time-Based Management), where the focus is on the time consumed in each individual manufacturing function, including administrative functions. Furthermore, the new Business Process Re-engineering (BPR) concept is very suited for IDEF0, as IDEF0 is independent of the organisation structure and only focuses on the functions and processes in the organisation. Activity-Based Costing (ABC) is also a method supported by an IDEF0 analysis in many applications seen in the U. S.

4.1 Top-down Planning - Bottom-up Implementation

The building of a company-wide CIM system cannot take place all at once. You have to start with limited elements of the system in areas where the largest economic effect - or largest demonstrative effect (influencing on attitudes) - can be found. The main principle when building CIM systems is to "think big" but "start small". If you want to work based on this principle, IDEF0 is an excellent tool, as it supports the selection of areas where the first CIM applications should be built and at the same time ensures that these will be integrable with future applications.

5 Limitations to method

The IDEF0 method was originally built to describe the relation and nature of activities. Data represented in the method are not shown in relation. I.e., based on an IDEF0 model it is not possible to build a database or a new paper system (you need an IDEF1-model to cover these aspects). Equally, the method does not describe quantifiable sizes (e.g., size of stock). The method was also not designed to describe detailed process operations (material flow, time on stock, etc.)

6 IDEF0 - terminology and concepts

The below shortly describes the IDEF0 terminology. Based on this chapter, the reader should be able to read and understand the main principles of an IDEF0-diagram.

6.1 Boxes = functions

The above describes the main idea of IDEF0. The method is particularly powerful due to the graphics that are applied to a very high degree. The graphical symbols have a very determined meaning in an IDEF0-model. The graphical symbols making up a model are simply arrows and boxes.

The company's functions, e.g., manufacturing of the product, production control, storage, informing of staff, etc., are described in IDEF0 using boxes. As the method should differentiate between the organisationally limited functions and the "actually performed" functions, the functions are in IDEF0 described using active verbs, i.e., "control production", "manufacture part", "mount gasket", "develop product", etc., and not "production control, "manufacturing of part", "mounting of gasket" or "product development".

The boxes, therefore, also describe a limit to the scope of the function. This is very useful when the function is to be detailed in another diagram.

6.2 Arrows = data or materials

Within the IDEF0 method there are four main types of information, i.e., input, output, constraints, and resource mechanisms. As compared to former methods, the two last categories are new making the method especially suited for describing manufacturing activities. A description of the four types of information follows. In IDEF0, information or materials are described using arrows. Placing of the arrows on the sides of the box is as shown in Figure 1. Input means input to the co-worker.

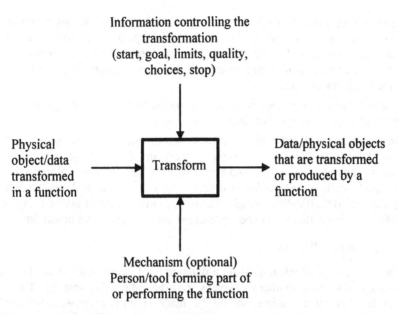

Information controlling the
transformation
(start, goal, limits, quality,
choices, stop)

Physical
object/data
transformed
in a function

Transform

Data/physical objects
that are transformed
or produced by a
function

Mechanism (optional)
Person/tool forming part of
or performing the function

Figure 1. The main elements of an IDEF0 diagram - the box, the arrows and text.

Input: In connection with the manufacturing of products, the co-workers (the functions) require a certain amount of information in order to be able to perform the planned work. This information could be priority lists, drawings, parts lists, quality specifications, etc. The method is especially suitable for describing information, but the arrows can also contain materials and thus help describe material flows. In an IDEF0-diagram, input information is shown using an arrow entering the box on the left side.

Output: In the execution of the work, the co-worker (the function) transforms or creates an amount of information to be applied in connection with the product (times, amount, quality, etc.). This information could also be used for post-calculations or as default times in the future planning. As mentioned, an arrow can also contain materials and not just information about materials. Output is always shown using an arrow leaving the box on the right side.

Constraints: In a production environment all performed activities are subject to a number of limitations determined by functions taking place earlier in the production course. The co-worker's activities are started based on constraints. Constraints can also be verbal instructions from the shop foreman, the arrival of an order card, a routing list, or the item which is to be processed. Constraints can also be information as to when a given activity should terminate or a message about change in choice of tool or material. Constraints means limitations, formulation of goals, control data, etc. Constraints enter the box from the top.

Resource mechanisms: Functions cannot be performed without the presence of a resource mechanism. Examples of resource mechanisms are: machines, calculators, co-workers, electricity, fixtures, tools, etc. At the top level of a model it makes no sense to state all resource mechanisms used by a company. Therefore, a diagram only shows the resource mechanisms where it is necessary to underline that the mechanism must be present or where it is a question of a special type of mechanism. In connection with the documentation of quality control systems, responsibilities as to the execution of activities can be shown using resource mechanisms. Resource mechanisms always enter the box from the bottom.

The content of the arrows is always described with a noun, e.g., drawing, parts list, production plan, flange, instruction, form, etc.

In connection with the elaboration of a diagram there might be some doubt as to whether the information that you want to show should be categorised as input or constraint. The questions to be asked are whether the information in any way initiates, defines goals, defines the function or describes how much time it should take. If so, it is a constraint. IDEF0 defines a main rule to be applied when in doubt. It says that if in doubt, the information should be represented in the diagram as a constraint.

6.3 Drawing of a diagram

The procedure applied when drawing a diagram is of course individual. The IDEF0 method requires that a number of choices are made when diagramming. There is no rule as to how these choices should be made. Figure 2 shows how a typical manufacturing process is diagrammed using the method. The diagrammer should, before diagramming, consider how the entering activities should be grouped. Three functions were chosen in the example, i.e.,

- Plan production
- Procure materials and resources
- Perform manufacturing

In the diagram the boxes are numbered from the left to the right (not shown in Figure 2). Arabic numerals are applied and located in the lower right corner of the box.

6.4 Purpose, viewpoint, and context

In order to guide the focusing of an IDEF0 modelling there are a number of requirements that have to be observed before starting the actual modelling. The previously mentioned procedure can be applied at a very early stage of the modelling. When starting the work with the model to be published (applied), an agreement must have been reached as to: purpose, viewpoint, and context.

Purpose, viewpoint, and context describe the limits, surroundings, and stop criterias of the model. They are characterised by the fact that they are all established before the actual modelling starts and that they are attached to the final model.

Purpose

The reason for describing the purpose of the model is to guide the modelling work in the correct direction and to ensure that an agreement is reached as to when the model has been elaborated sufficiently in order to fulfil the purpose. IDEF0 models consist of a hierarchy of models, gradually detailing one another. It is, therefore, important to establish a criterion for when the modelling work should be finished in order to control the time consumed for this work.

1. Draw the boxes

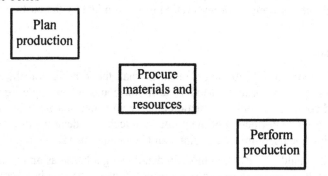

2. Add major inputs and outputs

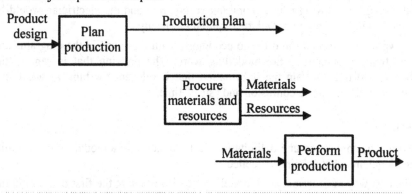

3. Connect input and output and add constraints

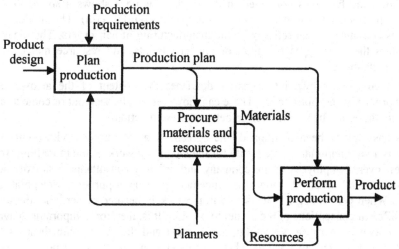

Figure 2. Methodology for drawing diagrams.

An example could be that the purpose of the model is to describe the course of an order through the company from reservation of the product by the customer until it is delivered.

6.5 Viewpoint

The viewpoint of an IDEF0 model ensures that the criteria for the model are established (similar to purpose). Model viewpoint could for example be production planning and control, i.e., only production planing and control aspects are included in the model. Detailed information of the production technical department is for example perhaps not included. Another viewpoint could be quality management.

Viewpoint could be described more in detail using a house as an example. If you want to "model" a house, there are several viewpoints. The family interested in the house will most probably look at the architect's perspectives first. The plumber will look at a diagram showing where plumbing is required, and the electrician would be interested in seeing where to install the required electricity.

Viewpoint is important in order to economise with the space in the diagrams and with the time consumed by the modelling work. The drawing that is sent to the plumber should only contain the information that is relevant for him, as additional information will only cause confusion and "steal" time.

Context

Context means connection and delimitation. Determination of a model's context limits the functional area that is being modelled.

The best way of describing the modelled area is to look at the first diagram in the model. The first diagram in a model only consists of one box, which on an overall level shows the functions contained in the model. Figure 3 shows a so-called A-0 diagram (pronounced A minus 0, which is a specific IDEF0 term). The model only describes activities in connection with the manufacturing of hair dryers. The diagram also shows the main types of input, output, constraints, and resource mechanisms applied in the model.

As mentioned, viewpoint primarily describes the content of the arrows, and context primarily describes boxes. These can, however, not be seen out of context, and there will, therefore, bee some overlap between these functions.

As has already been mentioned, the agreement as to purpose, viewpoint, and context is a very important prerequisite for the modelling work. If the modelling work involves several employees in the company and possibly consultants, it is extremely necessary that all involved "modellers" together agree upon purpose, viewpoint and context. When performing this work, it will most likely become evident that there are many different conceptions of the model to be built. It is, therefore, important to agree upon a drawn up purpose, viewpoint, and context and then work with these as the basis. It is not difficult to imagine what the model would look like, if no agreement has been reached before starting the modelling work, much less the wasted effort contributed by the participants.

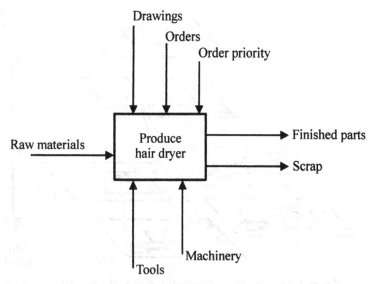

Figure 3. Example of a so-called A-0 diagram, which is limiting the contents of functions and data of the model.

7 Hierarchical decomposition of an IDEF0 model

One of the characteristics of an IDEF-model is the possibility to represent a very complex system in a gradual degree of detailing. It also ensures that details can be found and that connections to other elements in the model can be seen immediately. This is made possible by allowing a box to be "opened", thereby seeing what it contains in a very detailed form.

The hierarchy in an IDEF0 model is based on the so-called node-numbering convention (cf. figure 4). According to this convention, the first diagram, which only contains one box, is called A-0, i.e., A minus 0. The A-0 diagram could be detailed further in an A0 diagram, i.e., A zero. The A0 diagram describes the first subgrouping of the function mentioned in the A-0 diagram.

The advantage of node-numbering now becomes apparent, as the breakdown of the first box in the A0 diagram is called A1, the second box in the A0 diagram is called A2, etc. The above drawing illustrates yet another level, i.e., A12, which contains the details of box 2 in diagram A1 and A32, which is the details of box 2 in diagram A3. It is evident, that all arrows from the top level diagram has to be migrated to the lower level of the model.

Only 6 boxes are allowed in one diagram. This is partly due to the fact that psychologists have proved that we are only capable of containing 5 - 7 things in our head at the same time and that this limit will imply that the hierarchical structure is applied. The lower limit of 3 boxes has been included to ensure that the diagram will contain sufficient details to justify a decomposition hereof.

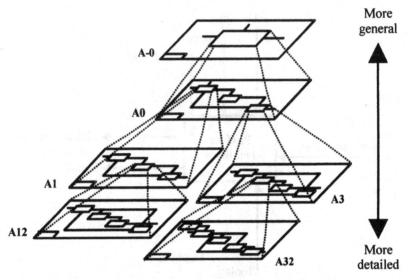

Figure 4. The hierarchical decomposition of an IDEF0 model.
The diagrams can be detailed by decomposition of the individual boxes. The Node number shows the level of the decomposition. Eg. diagram A32 is the detailed description of box 2 on diagram A3.

8 Diagramming rules

The IDEF0 method differs from similar methods in that IDEF0 applies a fixed set of requirements as to how to build a diagram. The aim of these strict requiremnets is to attain a very reader friendly representation as it will be possible for a person familiar with the method to very quickly gain an overview of the contents of a diagram. Another aim is to build a standard representation, which can be understood by all. These requirements should be observed. The below shortly describes the most important rules.

Input, output and, constraint arrows represent "data classes" or "data categories".

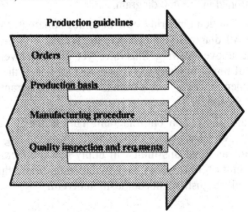

Figure 5. Example of an information stream with the common name "Production guidelines"

Output from a box should always originate 100% from input and resource mechanisms together. Control information is normally not transformed.

Constraint, input, and resource mechanisms can branch out. Data on each branch can be used at the same time (it is for instance copied or retrieved from a data base). When arrows are divided into components, the content is available on each arrow, if otherwise has not been stated (see figure 6).

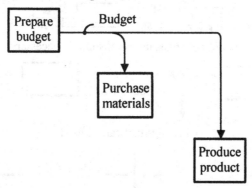

Figure 6. Data on each branch can be used at the same time.

9 Diagramming rules

Below some of the fundamental diagramming rules are mentioned. The main purpose of the (very strict) rules are the aim of reaching a standard representation and high level of readability.

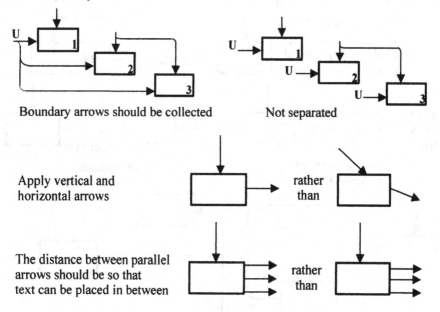

352

Arrows with the same source and destination are joined, if the content is related.

rather than

rather than

Arrows should not cross, unless absolutely necessary

rather than

Crossings that can be used.

Crossings that cannot be used.

When an arrow contains input as well as control, it should be diagrammed as control.

rather than

The boxes should always show control and output. Input is optional.

or

rather than

Control feedback is "up and over".

Input feedback is "down and under".

10 Example of an IDEF0 diagram

An example of an IDEF0 diagram can be found at the end of this contribution.

11 References:

Integrated Computer-Aided Manufacturing (ICAM), Architecture Part II, Volume IV - Function Modeling Manual (IDEF0), *Softech Inc., Wright-Patterson Air Force Base, Ohio 45433.*, AFWAL-TR-81-4023, Volume IV.

A short Course in Systems Development Methodology (SDM), ICAM Technology Transfer course, *Ric Mayer,* Texas A&M University, Dept. of Industrial Engineering, USA.

IDEF0 - Functioning, ICAM Technology Transfer course, *Ric Mayer,* Texas A&M University, Dept. of Industrial Engineering, USA.

IDEF0 - Function Modelling, Course material, *Peter Yeomans,* MicroMatch Limited, 10 Salamanca Crowthorne, England, RG11 6AP, February 1987.

Introduktion til CIM, Funktionsmodellering - IDEF0, course material, *Finn Jørgensen,* Institute for Product Development, Technical University of Denmark, 1989.

Funktionsmodellering - IDEF0, course material, *Finn Jørgensen,* Institute for Product Development, Technical University of Denmark, 1990.

Overblik over Funktionsmodellering - IDEF0. Course handbook in the course Industrial Information Systems, DTU. *Finn Jørgensen,* Institute for Product Development, Technical University of Denmark, 1991.

Funktionsmodellering - IDEF0. Manual for Reviewere. Issued for Asea Brown Boveri. *Finn Jørgensen,* Institute for Product Development, Technical University of Denmark, 1992.

IDEF1-Information Modelling, ICAM Technology Transfer, *Robert E. Young.* Dept. of Industrial Engineering, Texas A&M University, USA.

A Structured Approach to FMS modelling, *S. K. Banerjee; I. Al-Maliki* Int. J. Computer Integrated Manufacturing, 1988,vol. 1, no. 2, page 77-78.

IDEF - Anvendt som driftsteknisk værktøj, *J. J. Ibsen; P. Karsholt,* Institute for Production Management and Industrial Engineering, Technical University of Denmark, publication no. DI.87.10 - B, 1986.

Prototype af et CIM-System, Udviklet og implementeret under anvendelse af IDEF-metoderne, 2. Edition, *Michael Rasmussen & Finn Jørgensen,* CIM/GEMS publication no. CG88/T28, 1988.

IDEF1-Informationsmodellering, course material, *Robert E. Young,* CIM/GEMS project 1988.

354

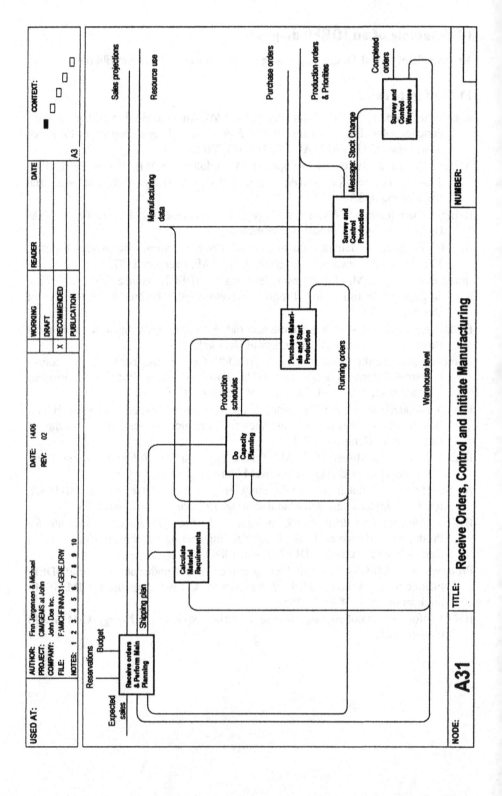

USED AT:

AUTHOR: Finn Jørgensen & Michael
PROJECT: CIM/GEMS at John
COMPANY: John Doe Inc.
FILE: F:\MICHFINNA31-GENE.DRW
NOTES: 1 2 3 4 5 6 7 8 9 10

DATE: 14/06
REV: 02

WORKING
DRAFT
RECOMMENDED
PUBLICATION

x

READER   DATE   CONTEXT:

A3

Reservations
Budget
Expected sales
Shipping plan

Receive orders & Perform Main Planning

Sales projections

Resource use

Calculate Material Requirements

Manufacturing data

Do Capacity Planning

Production schedules

Purchase Materials and Start Production

Purchase orders

Running orders

Survey and Control Production

Message: Stock Change

Production orders & Priorities

Warehouse level

Survey and Control Warehouse

Completed orders

NODE: **A31**   TITLE: **Receive Orders, Control and Initiate Manufacturing**   NUMBER:

Data Modeling with IDEF1X

Heimo H. Adelsberger and Frank Körner

University Essen, D-45117 Essen, Germany

Abstract. IDEF1X is a method for accomplishing logical database design for relational databases. This paper will give the reader an introduction to the methodology and terminology used as well as a summary of the syntax and semantics of IDEF1X.

1 Introduction

In the following, concepts, techniques and procedures for the development of logical models are described, concentrating on the semantic characteristics of data. These semantic information models will support the view of 'data being resources' with a special mark lying on the integration of information systems and the development of databases.

The IDEF[1] method was one of the results of the ICAM[2] program; it comprises several modeling techniques for describing manufacturing systems:

– IDEF0
  A structured representation of the *functional model* describing the functions in the manufacturing system and the information and objects involved
– **IDEF1**
  The *information model,* **representing the structure of the information required to support the functions**
– IDEF2
  The *dynamic model,* representing time-dependent behavioral aspects of the functions as well as information and resources of the manufacturing system

In 1983, the U.S. Air Force founded the IISS[3] project as part of the ICAM program. The objective of the project was the development, management, and use of a consistent semantic definition of data resources. The conceptual scheme was created using the IDEF1 technique.

IDEF1X is an extension of the IDEF1 modeling technique, based on the IISS[4] experiences and on the know-how gained when using IDEF1 for industrial applications. IDEF1X stands out against IDEF1 by an improved graphical representation, an extended semantic content and by simplified development procedures.

The presentation follows closely the original technical report, published by Knowledge Based Systems ([1]).

---

[1] ICAM Definition
[2] Integrated Computer Aided Manufacturing
[3] Integrated Information Support System
[4] Integrated Information Support System; part of the ICAM-program

2 Concepts of Information Modeling

In this section, the importance of information modeling and the overall objectives of IDEF1X are described.

2.1 Data as Resources

During the 80th, many companies began to look on 'data as resources'. This view was stressed by the demand to deal with a fast-growing dynamic (flexibility, growing complexity of data) environment.

An ICAM study showed that — compared to the demands — the data existing when IDEF1X was designed were inconsistent, untimely, inflexible, not in accordance with business needs, and not easy to access. In contrast, the growing 'dynamics' required fast recognition and interpretation of changes, and above all, a fast reaction and adaptation of the data and knowledge throughout the company.

What are data? Data are symbolic representations of facts with a particular meaning. One meaning can be connected to different facts. A fact without meaning is useless, a fact with wrong meaning can cause disasters. Therefore, the main attention is paid to the meaning of data, the *semantics*.

'Information' can be regarded as an aggregation of data for a particular purpose or defined in a special context. Thus, different types of information can be created from the same data. Fig. 1 shows the components of 'information'.

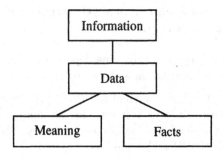

Fig. 1. COMPONENTS OF 'INFORMATION'

2.2 The Three-Level-Architecture

According to the ANSI/X3/SPARC Study Group on Database Management Systems, data management requires three different views on data. Apart from the traditional *External Scheme* and *Internal Scheme*, the group established a third one, the *Conceptual Scheme*, as a neutral view (Fig. 2). The neutral view does not take care of single applications, physical storing, or access of the data.

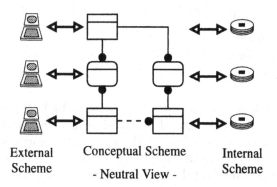

<table>
<tr><td>External
Scheme</td><td>Conceptual Scheme
- Neutral View -</td><td>Internal
Scheme</td></tr>
</table>

Fig. 2. THE THREE-LEVEL-ARCHITECTURE

The conceptual scheme must follow some characteristics in order to guarantee the consistent definition and relationships of data:

- It must be consistent in reference to the infrastructure and correct in regard of all applications.
- It must be extendable in order to define new data without altering data already defined.
- It must be possible to transform the scheme according to the user's demand and according to the necessary storing and access structures.

2.3 Goals of Data Modeling

A DBMS's[5] logical structure does not satisfy the conceptual data definition. In order to define the meaning of data in relation to other data, semantic data modeling techniques must represent a faithful illustration of the real world.

The semantic data model can answer several purposes:

- Planning of data resources
- Development of databases
- Evaluation of standard application software
- Integration of existing databases

2.4 The IDEF1X Approach

The semantic modeling technique IDEF1X was developed to meet the following demands:

- To support the development of conceptual schemes with regard to consistency, extendibility, and the possibility to perform transformations,

---

[5] Data-Base Management System

- easy to understand,
- easy to be learned,
- tested and proven, and
- be automatable.

The IDEF1X method uses an Entity-Relationship (ER) approach for semantic data modeling. The original method — IDEF1 — had already been based on the ER approach (Chen) and the relational theory (Codd). Improvements concerning the graphical representation, the modeling process, and the semantical content (*Categorization Relationship*) characterize the extended method IDEF1X.

The IDEF1X basic elements are:

- **Objects** (persons, places, ideas, events etc.), represented by *boxes*
- **Relationships** between boxes, represented by *arrows*
- **Characteristics** of objects, shown as *Attributes* in the boxes.

3 Syntax and Semantic

In this section, syntactical components, rules, and their meaning will be dealt with.

3.1 Entities

Semantics. An **entity** represents a set of real or abstract objects like persons, places, events, states, ideas etc. For these objects common features, attributes and/or characteristics can be recognized. The single elements are called *entity instances*.

A real existing object may be an entity instance of several different entities in one data model. On the other hand, one entity instance may represent several real existing objects. Entities can be characterized as **identifier-independent** or **identifier-dependent** entities. An entity instance of an identifier-independent entity can be clearly identified without reference to other entities, whereas the identification of identifier-dependent entities depends on the relationships to other entities.

Syntax. In IDEF1X, entities are represented by boxes. Boxes with 'curved corners' represent identifier-dependent entities. Each box gets a unique name, and a consecutive number (Fig. 3) to identify the entity.

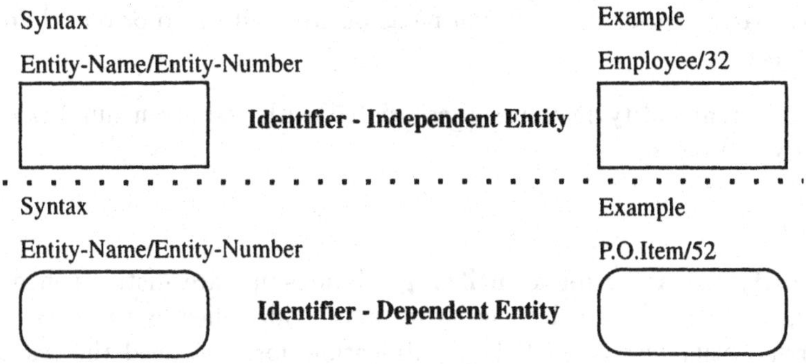

Syntax   Example

Entity-Name/Entity-Number   Employee/32

Identifier - Independent Entity

. . . .

Syntax   Example

Entity-Name/Entity-Number   P.O.Item/52

Identifier - Dependent Entity

Fig. 3. ENTITY SYNTAX

The name of the entity gives a short and meaningful description of the entity in form of a noun phrase (singular subject plus preposition or adjective). The formal definition is given in the glossary. Each entity may appear in several diagrams, yet only once per diagram.

Rules. Entities must follow the following rules:

- Each entity has a unique name which is consistent throughout the entire model.
- Each entity has one or more attributes. An attribute is either *owned, inherited,* or *migrated* (see 3.5).
- Each entity is clearly identified by one ore more attributes (see 3.6).
- Each entity may be related to other entities (via relationships).
- If the entire *foreign key* (see 3.7) is used as *primary key* (see 3.6) or part of it, the entity is identifier-dependent, otherwise identifier-independent.

3.2 Connection Relationships

Semantics. A (specific) connection relationship — called existence dependency relationship — connects a so-called parent entity with a 'child' entity. In this context, an instance of a child entity can only exist if the corresponding parent entity instance exists as well. Each connection relationship is specified by the relationship cardinality. The cardinality defines how many child entity instances exist for each parent entity instance. In IDEF1X, the following cardinalities are considered:

- Each parent entity instance can be associated with zero, one, or many child entity instances.
- Each parent entity instance must be associated with at least one entity instance.

– Each parent entity instance can be associated with zero or one child entity instance.

– Each parent entity instance is associated with the given number of child entity instances.

Identifying- and **non-identifying** relationships are distinguished. In an identifying relationship, a child entity instance is identified by the association to the parent entity (see 3.7). If the identification does not need the association, the relationship is called non-identifying.

To represent multiple relationships (between more than two entities), assertions can be defined, e.g., by using boolean constraints. Another type of constraint is the so-called 'path assertion' which allows to represent dependencies between entities via direct or indirect relationships along several paths.

Syntax. A (specific) connection relationship is represented by a full line between parent and child entity. The line ends with a dot at the child entity. Fig. 4 shows the syntax.

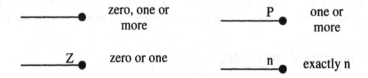

Fig. 4. RELATIONSHIP CARDINALITY SYNTAX

As shown in Fig. 5, a full line represents an identifying relationship between parent and child entity. Child entities in these relationships are always identifier-dependent (represented by a box with curved corners). Furthermore, the parent entity may be identifier-dependent provided it is also a child entity in another relationship. The primary key attributes of the parent entity are inherited as foreign keys to the child entity (see 3.7).

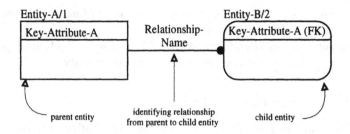

Fig. 5. IDENTIFYING RELATIONSHIP SYNTAX

The dotted line (Fig. 6) represents a non-identifying relationship. Both entities are identifier-independent unless one is also a child entity in an identifying relationship.

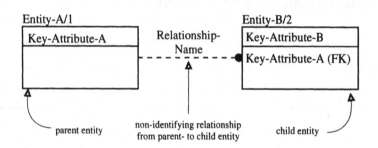

Fig. 6. NON-IDENTIFYING RELATIONSHIP SYNTAX

Each relationship is labeled by a verb phrase (with optional preposition or adverbs). Thus sentences like „'parent entity' – 'relationship' – 'child entity'" can be read. It must be possible to distinguish two different relationships between the same entities, otherwise, the same relationship name can be used several times in a model.

Rules.

- A specific connection relationship always exists between exactly two entities, a parent and a child entity.
- Each child entity instance must be related to exact one parent entity instance.

– Each parent entity instance is (depending on the cardinality) related to zero, one, or more child entity instances.
– A child entity in an identifying relationship is always identifier-dependent.
– Each entity is always related to at least one other entity.

3.3 Categorization Relationships

Semantics. Some real world objects are categories of other real world objects. This case is represented in IDEF1X by a so-called **categorization relationship**. The *generic entity* is the superclass of the *category entities* which are the subclasses. Categorization relationships establish mutual exclusive entities. A generic entity instance can be related to only one category entity instance. The *discriminator* is an attribute of a generic entity, deciding the assignment to the category entity.

The **complete categorization relationship** characterizes a relationship between two or more entities in which each generic entity instance is related to exactly one category entity.

The **incomplete categorization relationship** allows instances of generic entities without a relation to a category entity.

Syntax.

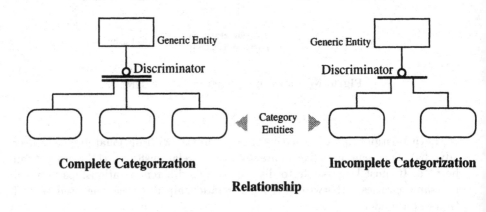

Complete Categorization　　　　　　**Incomplete Categorization**

Relationship

Fig. 7. Categorization relationship syntax

Fig. 7 shows the different syntax of complete and incomplete categorization relationships. The discriminator, the 'circle', of the (in)complete categorization relationship is underlined (once) twice. The cardinality of the categorization relationship is not displayed since it is either zero or one. Each category entity is identifier-dependent, unless the identifier is inherited via another relationship. Catetegorization relationships are not labeled explicitly. The generic entity → category entity direction can be interpreted as 'can be' (or 'must be' in complete categorization relationships). The other direction can be read as 'is a'. Generic entities and the related category entities must have the same key attribute. In addition, category entities can be labeled by role names (see 3.7).

Rules.

- Each category entity can be related to at most one generic entity (at most one categorization relationship).
- Each category entity can be a generic entity of another categorization relationship, thus hierarchies can be created.
- One entity can be a generic entity, distinguished by the discriminator, in different categorization relationships.
- A category entity cannot be child entity in an identifying relationship.
- Primary key attributes of category and generic entities must be identical.
- All instances of a category entity have the same discriminator value.

3.4 Non-Specific Relationships

Semantics. 'Specific' relationships (connection and categorization relationships) define the relationship between two entities (see 3.2, 3.3). In contrast to these relationships, the so-called non-specific relationships allow the cardinality 'many-to-many'. The use of this non-specific-relationship is only allowed at the early stages of development. Later, in Phase 3 (see 4.4), these relationships are decomposed by a third one, the so-called intersection or associative entity, to form specific relationships.

Syntax. Non-specific relationships are represented by a line with dots at both ends. The cardinality can, but need not to be given explicitly. The usual symbols are used ('P' for positive, 'Z' for zero or one, etc.).

Non-specific relationships are labeled in both directions separated by a '/'. Respectively, the 'top-to-bottom' or the 'left-to-right' direction is labeled first, then, after the slash, the 'bottom-to-top' or 'right-to-left' direction. This enables reading the labels in both directions.

Rules.

- Each non-specific relationship connects exactly two entities.
- Each entity instance is related, depending on the cardinality, to none, one or more instances of the other entity.
- Each non-specific relationship must be replaced later in the model building process by specific relationships.

3.5 Attributes

Semantics. Attributes in an IDEF1X model are assigned to an entity in order to represent its characteristics. An attribute instance is a specific characteristic, i.e., a value, the so-called *attribute value*. E.g., the entity 'Employee' has attributes like 'Age', 'Sex', etc.; an attribute instance of 'Age' could be '47'. Each entity instance must have single values for its attributes. Furthermore, each entity is identified by one or more attributes, the so-called *primary key* (see 3.6). So-called *owned attributes* describe basic characteristics of entities. These attributes are not inherited. In addition, entities can have so-called *inherited attributes*. These attributes are inherited from other entities via connection or categorization relationships.

Syntax. Attribute names are noun phrases (like entities). The attribute names are listed in the entity box which is divided by a horizontal line. Above, the primary key attributes are listed, below, the non-key-attributes can be found.

Rules.

- Each attribute must have a proper name whose meaning is consistent all over the model.
- Each entity can own any number of attributes. Each attribute can be owned by exactly one entity.
- Each entity can have any number of inherited attributes which must be part of the related entitiy's primary key.
- **No-null rule**: Each entity instance has values for each attribute.
- **No-repeat rule**: No entity instance is allowed to have more than one value per attribute.

3.6 Primary and Alternate Keys

Semantics. Each entity needs at least one candidate key which identifies each entity instance using one or more attributes. If more then one attribute is needed, the key is called a *composite key*. The single attributes are called key members. If there are more attributes or attribute combinations fulfilling these conditions, all of them are called *candidate keys*. One candidate key is chosen as *primary key* (PK), the others are called *alternate keys* (AK).

Syntax. Primary key attributes are listed in the entity boxes (see 3.5). Alternate key attributes are listed below the 'line' and each one is followed by a number.

Rules.

- Each entity must have a primary key.
- Each entity may have several alternate keys.

- Each primary or alternate key consists of one or more attributes.
- Each attribute may be part of one or more keys (primary or alternate).
- Primary or alternate key attributes may be owned or inherited.
- **Smallest-Key rule**: Primary and alternate keys are formed by the smallest combination of identifying attributes.
- **Full-Functional-Dependency rule**: Using composite keys, each non-key attribute must depend on the whole key.
- **No-transitive-dependency rule**: Each non-key attribute must depend only on the primary or alternate key.

3.7 Foreign Keys

Semantics. Foreign keys are inherited via connection or categorization relationships. The primary key attributes of the parent or generic entities are inherited to the child or to the category entity where they are used as key or non-key attributes.

If the inherited attributes are used in the primary key, there are to cases to be distinguished:

- **Identifying relationship**: The whole primary key of a parent entity is inherited and becomes part of the primary key in the child entity.
- **Non-identifying relationship**: Some of the inherited attributes do not become part of the primary key of the child entity.

Generic and category entities (categorization relationship) represent the same real world object. Thus, the primary key is inherited to all category entities as a primary key.

Sometimes attributes are inherited several times, one child entity is related to one parent entity in several relationships. In these 'multiple relationships', the primary key of the parent entity becomes an inherited attribute in the child entity. For a given child entity instance, the attribute values may be different in different relationships, therefore there are relationships to different parent entity instances. In this case, *Role names* are used for the clear identification.

Syntax. Foreign keys are listed in the entity box as key or non-key attributes, they are indicated by an '(FK)' (Fig. 8).

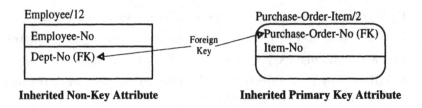

Fig. 8. FOREIGN KEY SYNTAX

As far as the naming is considered, role names must meet the same naming conventions valid for forming attributes. A role name precedes the name of the inherited attributes name, separated by a dot (role name.attribute name (FK)), as shown in Fig. 9.

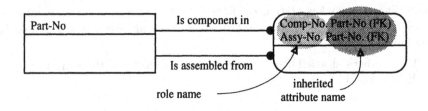

Fig. 9. ROLE NAME SYNTAX

Rules.

- Each entity must have a foreign key for each relationship in which it is child or category entity.
- A primary key of an generic entity must be inherited as primary key to each category entity.
- For each child or category entity instance, an entity must not contain two complete foreign keys, identifying the same parent or generic instance
- Each inherited attribute of a child or generic entity must represent a primary key attribute of the related parent or generic entity. Conversely, each primary key attribute of a parent or generic entity must be an inherited attribute in the related child or category entity.
- Each role name must be unique.
- Each single inherited attribute may be part of several foreign keys as long as it has the same attribute value in the foreign keys of each instance.

4 IDEF1X Modeling

The IDEF1X model building process is divided into two steps, the first step consisting of two, the second one of three phases.

step	phase	name	tasks
Business Model	0	project initiation	create plan, organize team, collect source material, finish first analysis
	1	entity definition	identify and define entities according to source material
	2	relationship definition	identify and define relationships according to Phase 1
Database Design Model	3	primary key definition	identify primary keys for each Phase 1 entity
	4	attribute definition	identify and define non-key attributes, assign non-key attributes to entities

Table 1. PHASES OF IDEF1X - MODELING

4.1 Phase 0 – Project Initiation

An IDEF1X data model is described by its boundaries and contents. The plan for the Phase 0 target development is created by the modeler and the project manager.

Targets of Phase 0.

– Project definition: The questions 'What is to be done?', 'Why is it to be done?' and 'How is it to be done?' must be answered.
– Source material: Plan for source material acquisition (Data Collection Plan).
– Work organization: Definition of project organization, i.e., optional methods for modeling and administration.

Creation of Modeling Objectives.

– Purpose: Definition of the content of the model and its contextual boundaries.
– Scope: Specification of the funcionality of the model.

An important decision is whether an 'AS-IS' (existing system) or 'TO-BE' (planned system) model is created, or both. The formal definition of the problem includes the checking, construction, modification, and/or decomposition of one or more IDEF0 models.

Statements concerning the purpose and scope of modeling process define the modeling objective, like:

'The purpose of this model is to define the current (AS-IS) data used by a manufacturing cell supervisor to manufacture and test composite aircraft parts.'

Creation of the Modeling Plan. The modeling plan outlines the modeling tasks and their sequence:

- Project planning
- Data collection
- Entity definition
- Relationship definition
- Key attribute definition
- Non-key attribute definition
- Model validation
- Acceptance review

The modeling plan serves as a basis for task assignment, milestone planning, and estimation of costs.

Team Organization. The value of a model depends on the acceptance of experts and users. To achieve this in IDEF1X, two mechanisms are available:

- A permanent check by experts to guarantee validity.
- A periodic check by a committee of experts and laymen to create common acceptance.

The model builders are responsible for the exactness of the model. There should be no room or need for the user's interpretations.

A project team is organized according to the following organizational structure: (Fig. 10):

- **Project Manager**
  The project manager is the supervisor of the project, his main tasks are selection of the modelers, definition of sources, selection of experts, formation of the Acceptance Review Committee.
- **Modeler**
  The modeler creates the modeling plan and the model. The source material and modeling techniques prescribed by the project manager serve as a basis for his main tasks. The modeler is responsible for the following tasks: training and education, collecting source material, documentation of the modeling, and checking of the model.
- **Sources** of Information
  Persons or documents providing the required information.
- Subject Matter **Experts**
  Experts have profound knowledge of the manufacturing area to be modeled. Their criticism in the IDEF-Kit-Cycle (Chapter 5.2 and Fig. 27) is very important for the development of the model. Experts must validate the model.

– **Acceptance Review Committee**
  The committee must settle different opinions in the modeling process. Ulti-
  mately, the committee will judge and accept the modeling result.

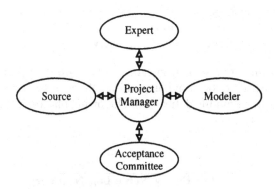

Fig. 10. TEAM ORGANIZATION

Collection of Source Material. The collection of source material is one of the
first tasks. The scope and context of a model is very often based on an IDEF0
model. The modeler must identify the so-called target functions in the function
model. Depending on these target functions, the modeler chooses the sources
(persons or documents).

   The source material may contain the following:

– Results of interviews
– Results of observations
– Principles and procedures
– Outputs of existing systems
– Inputs of existing systems
– Database and file specifications of existing systems

4.2 Phase 1 – Entity Definition

The goal in this phase is the identification and definition of entities concerning
the area to be modeled.

Identification. The entities are identified by using the source material:

– The modeler searches for nouns in the source material.
– The modeler searches for expressions containing 'code', 'number' etc.

Finally, the modeler must decide whether the expressions represent objects with information available or information about an object. In the first case, it is a potential entity, otherwise it probably is an attribute. The following questions support the modeler in deciding the question 'entity or non-entity':

- Is it possible to describe the object?
- Do several instances of the object exist?
- Is it possible to distinguish several instances?
- Does the object describe anything?
  If 'Yes', it is an attribute, else it is an entity.

The result of this analysis is a so-called **entity pool**, containing the names of the entities, identification numbers, and a reference to the source material. The entity pool will be modified during the following phases. Tab. 2 gives a short example for an entity pool.

Number	Entity Name	Log Number
E-1	Backorder	2
E-2	Bill of Lading	2
E-3	Carrier	2
E-4	Clock Card	3
E-5	Commodity	2
E-6	Contractor	4

Table 2. SAMPLE ENTITY POOL

Definition. In this step the entity glossary is established; it contains three components:

- Entity name
- Entity definition
- Entity synonyms
  Synonyms are different names for the same entity.

4.3 Phase 2 – Relationship Definition

In this phase, the basic relationships are identified and defined. At first, non-specific-relationships (see 3.4) are allowed which must be replaced during the following phases. The results of Phase 2 are:

- Relationship Matrix
- Relationship Definitions
- Entity-Level Diagram

Identification of Related Entities. A 'binary relationship' is a connection between two entities. A relationship instance is the meaningful connection of two entities instances. Each entity may be member of several relationships.

It is not the goal to represent all possible relationships, but to define the connection between entities in form of existent dependency (parent-child) or categorization relationships. In Phase 2, however, non-specific relationships are allowed. The first step in Phase 2 is the identification of the relationship by using the entity/relationship matrix (Fig. 11) and the entity-level diagram (see Fig. 12 and Paragraph 4.3).

	Buyer	Requester	Approver	Purchase Requisition	Purchase Req. Item
Buyer		X		X	
Requester	X			X	
Approver				X	
Purchase Requisition	X	X	X		X
Purchase Req. Item				X	

Fig. 11. ENTITY/RELATIONSHIP MATRIX

Relationship Definition. Relationships are defined by the following information:

- Indication of dependencies
- Relationship name
- Textual information

The dependency definition demands for a view in both directions to define the cardinality. The relationship name (verb and optional conjunction) shall create a 'sentence' of the pattern „'first entity' – 'relationship' – 'second entity'". Specific relationship names are interpreted in the direction „'parent entity' - 'relationship name' - 'child entity'", the other direction is not labeled explicitly. The cardinality of a child entity in a categorization relationship is always zero or one. Such relationships imply a 'MAY-BE' relationship and are not labeled. Non-specific relationships are labeled twice, once in each direction. The names are separated by a slash '/' like *name1 / name2*. *Name1* names the directions 'top-to-bottom' or 'left-to-right', *name2* vice versa.

Relationship names must be meaningful. In addition, the exact definition can be given in textual form. The following rules apply to the relationship definition:

- It must be precise.
- It must be concise.
- It must be meaningful and useful.

Entity-Level Diagram Creation. Relationships are represented as Entity-Level diagrams (Fig. 12). The number and scope in the Entity-Level diagram depends on the size of the model and the reviewer's view. If several diagrams are created, the consistency of the diagrams and entity and relationship definitions must be guaranteed.

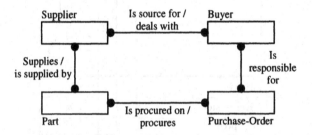

Fig. 12. ENTITY-LEVEL DIAGRAM

Optional diagrams representing the relationships of a single entity are called Entity Diagrams (Fig. 13).

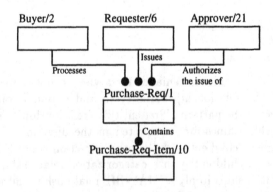

Fig. 13. ENTITY DIAGRAM

This phase demands the following information for each entity:

- Entity definition
- Relationship names and optional definitions
- Representation of at least one entity-level diagram

So-called reference diagrams can extend the information regarding an entity. Reference diagrams are often called *FEO - For Exposition Only* since they may not meet the entity-level diagrams syntax. They serve as a basis for discussion between modelers and reviewers. Fig. 14 shows a reference diagram with alternative representations for one situation.

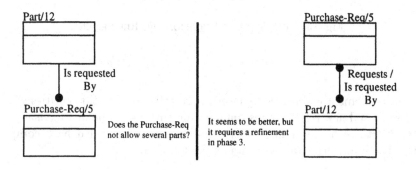

Fig. 14. REFERENCE DIAGRAM (FEO)

At the end of Phase 2, the modeler has collected sufficient information to perform the formal validation of the model using Kits (see 5.2) and Walk-Through (see 5.4).

4.4 Phase 3 – Primary Key Definition.

- Refinement of non-specific relationships
- Definition of key attributes for each entity
- Migration of primary keys to create foreign keys
- Validation of relationships and keys

The results are represented in one or more Phase 3 (key level) diagrams. In Phase 3, the key attribute definitions are created and entity and relationship definitions are refined.

Non-Specific Relationship Refinement. The first step is the transformation of non-specific relationships into specific relationships. This task is supported by the so-called *refinement diagrams*. Refinement diagrams are divided into two sections: in the left side where the non-specific relationship is represented, and the right side, showing the refined specific relationship. Fig. 15 gives a simple example.

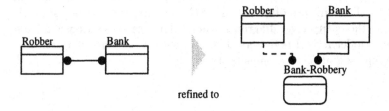

Fig. 15. Non-Specific Relationship Refinement

The refinement is achieved by inserting so-called intersection entities ('Bank-Robbery') and new relationships. Intersection entities can be distinguished from natural entities by the name. Intersection entities are labeled by a composed noun. The refinement process contains the following steps:

- Development of several refinement alternatives.
- Selection of one alternative.
- Insertion of the new intersection entities into the Phase 1 results.
- Insertion of the new relationships into the Phase 2 results.

Representation of Function Views. The complexity of the model may be too high in this phase to reflect its meaning without additional material. Function views are used to support the modeler in the evaluation and validation of the data model. He can use two different methods:

- Selection of source material for one function view.
- Assignment of the function view to certain jobs, processes, or resources which are identified as source material in Phase 0.

Fig. 16 shows a small example.

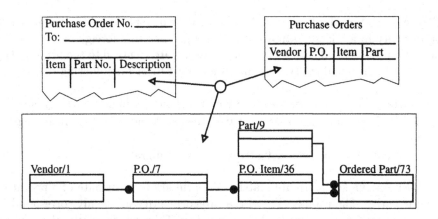

Fig. 16. SCOPE OF A FUNCTION

Key Attributes Identification. In this step, the key attributes, i.e., candidate keys, primary keys, alternate keys, and foreign keys (see 3.6 and 3.7) are identified and defined. Fig. 17 shows the different notations of the key forms.

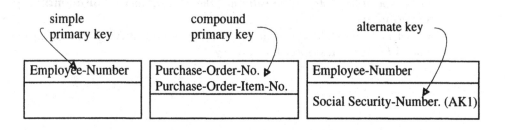

Fig. 17. KEY FORMS

The key identification is done by the following steps:

– Identification of candidate key(s)
– Selection of **one** primary key

Since some candidate keys may be results of key migration (see 4.4), the primary key selection starts with entities being neither child nor category entities.

Key Migration. Key migration is the process of replicating the primary key of an entity in a related entity; the primary key becomes foreign key. After the migration, the related entity's foreign key values are identical to the primary key values of the original entity. Key migration must meet the following rules:

- Migration is made from parent or generic entities to child or category entities.
- For each relationship of two entities, the whole primary key must migrate once.
- Alternate keys and non-key attributes never migrate.

Each foreign key attribute matches the primary key attribute of a parent or generic entity. The primary key of the category entity must only correspond in categorization relationships to the primary key of the generic entity. Using other relationships, the foreign key is allowed to be part of the primary key. Each attribute is either an owned attribute or a foreign key attribute.

- If a primary key of a child entity contains all foreign key attributes, the child entity is called an **identifier-dependent entity** and the relationship is an **identifying relationship**.
- An entity being child entity in several identifying relationships is called **identifier-dependent entity**.
- If an child entity's primary key does not contain all foreign key attributes, it is called *identifier-independent entity* and the relationship is a **non-identifying relationship**.
- An entity being child entity in at least one non-identifying relationship is called **identifier-independent entity**.

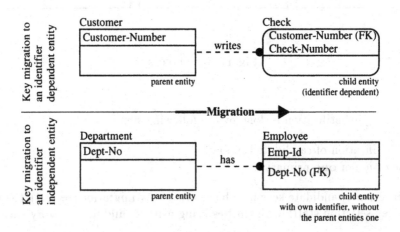

Fig. 18. KEY MIGRATION

Fig. 18 gives two examples of key migration. Identifier-independent entities are represented as boxes, identifier-dependent ones as boxes with curved corners. An identifying relationship is represented as a full line whereas a non-identifying relationship is represented as a dotted line.

An attribute may create several foreign keys in the very same child entity, if the attribute migrates via several relationships to the child entity. In some cases, each instance of a child entity must have the same value in this attribute. This attribute is simply a foreign key. In other cases, these attributes must be distinguished in one instance by using role names (Fig. 9, see 3.7).

Validation of Keys and Relationships. Rules for identification and migration of keys:

- Non-specific relationships are not allowed.

- Key migration only from parent to child entity or from generic to category entity.

- **No-Repeat Rule:** Attributes having more than one value for one entity instance are not allowed (Fig. 19).

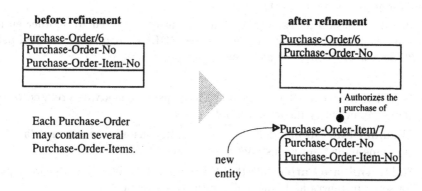

Fig. 19. No-Repeat Rule Refinement

- **No-Null Rule:** Attributes with no value are not allowed (Fig. 20).

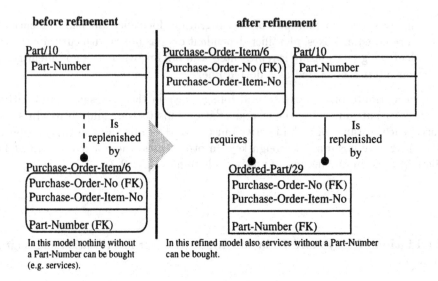

before refinement   **after refinement**

In this model nothing without a Part-Number can be bought (e.g. services).

In this refined model also services without a Part-Number can be bought.

Fig. 20. NO-NULL RULE REFINEMENT

- **Smallest-Key Rule:** Entities with composite keys cannot be split into several entities with less key attributes without loosing information.
- **'Dual Relationships Paths'** must be commented by path assertions.

During the validation phase, new entities may be created (because of no-null and no-repeat rule refinement).

Dual relationship paths only exist if one entity is related to another entity via several relationship paths. In these cases, IDEF1X demands so-called path assertions which classify the paths as:

- **Equal Path** means that each child entity instance is **always** related to the same parent entity instance via both paths.
- **Unequal Path** means that each child entity instance is **never** related to the same parent entity instance via both paths.
- **Indeterminate Path** means that the relationship via both paths are 'equal' for some instances and 'unequal' for the other instances.

If one of the paths is a single relationship and both paths are equal, the single relationship is redundant and must be deleted. This constellation is called 'Triade' (Fig. 21). Path assertions are added as notes to the Phase 3 diagrams and are part of the respective child entity definition.

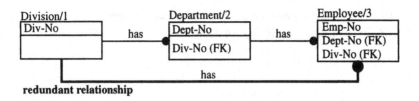

Fig. 21. REDUNDANT RELATIONSHIP

A further result of Phase 3 is the so-called entity/attribute matrix (Fig. 22), containing the attributes used in the model. The items have the following meanings:

- **O** = Owner
- **K** = Key
- **I** = Inherited

Entity		Purchase Req. No.	Buyer Code	Order Code	Vendor Code	Change Number	Ship to Location	Vendor Name	Vendor Adress	Confirmation Code	Confirmation Name	Extra Copy Code	Requester Name	Department Code	Ship Via	Buyer Name	Purchase Order No.	Purchase Req Issue Date	Q.C. Att. Code	Taxable Code	Resale Code	Pattern Number	Payment Type
		1	2	3	4	5	6	7	8	9	10	11	12	13	14	15	16	17	18	19	20	21	22
Purchase Requisition	1	OK																					
Buyer	2		OK																				
Vendor	3			OK																			
Purchase Order	4		I	I													OK						
Requester	6											OK											
Part	9																						
Purchase Req. Item	10	IK																					
Purchase Req. Line	12	IK																					
Approver	21			IK																			
Part Source	22																						

Fig. 22. ENTITY/ATTRIBUTE MATRIX

Representation of Phase 3 Results. After key identification and migration, the function view diagrams are actualized and refined. In Phase 3, these diagrams (Fig. 23) contain the following information:

- Primary, alternate, and foreign key attributes

- Identifier-independent and identifier-dependent entities
- Identifying and non-identifying relationships

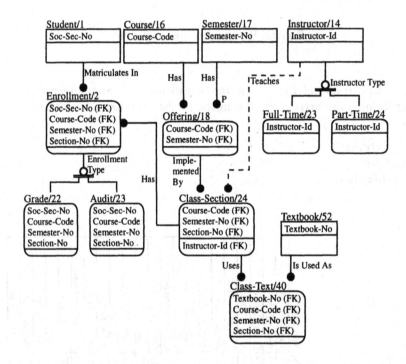

Fig. 23. PHASE 3 FUNCTION VIEW DIAGRAM

Each entity documentation in Phase 3 contains the following information:

- Entity definition
- A list of primary, alternate, and foreign key attributes
- Definition of owned key attributes
- A list of relationships where the entity is a generic entity
- A list of relationships where the entity is a category entity
- A list of identifying relationships where the entity is a parent entity
- A list of identifying relationships where the entity is a child entity
- A list of non-identifying relationships where the entity is a parent entity
- A list of non-identifying relationships where the entity is a child entity
- Definition of dual path assertions

4.5   Phase 4 – Attribute Definition.

Phase 4 is the last modeling phase with the following goals:

- Development of the attribute pool
- Definition of attribute ownerships
- Definition of non-key attributes
- Validation and refinement of the data structure

These results are documented in one or more Phase 4 (attribute level) diagrams. The resulting model corresponds to the fifth normal form, it contains complete definitions and cross references for all entities, relationships, and attributes.

Identification of Non-Key Attributes. In addition to the Phase 3 key attribute identification, the non-key attribute identification is done. Like the key attributes, the non-key attributes get an unique identification number. The attributes pool development is similar to the entity pool development (see 4.2). The modeler searches in the source material for potential attributes often appearing as 'descriptive nouns'. Table 3 shows an example of an attribute pool.

Source Data Number	Attribute Name	Number
1	Purchase Requisition Number	1
2	Buyer Code	2
3	Vendor Name	3
4	Order Code	4
5	Change Number	5
6	Ship to Location	6
7	Vendor Name	8
8	Vendor Address	8
9	Configuration Code	9

Table 3. SAMPLE ATTRIBUTE POOL

Definition of Attribute Ownership. In most cases, the owner entity of an attribute will be obvious. Otherwise, the modeler must browse the source material for the context, the attribute is used in, to decide on the owner entity.

Definition of Attributes. Each Phase 4 attribute must be defined following the principles of Phase 3. Thus the attribute definition contains the following information:

- Attribute name
- Attribute definition
- Attribute synonym(s)/aliase(s)

For the attributes, the modeler can define domains like alphanumeric, text, currency, date, list, interval, etc. Furthermore, he can define constraints or restrictions between several attributes.

Model Refinement. The Phase 4 model refinement must meet the Phase 3 rules (see 4.4). In contrast, Phase 3 non-key attributes are checked and validated. If errors appear, the modeler can insert new entities or mark the appearance. Then he uses an 'N' to mark a no-null rule error or an 'R' to point to a no-repeat rule error. Fig. 24 and 25 show the use of both rules. The entities are inserted later. All new entities and relationships must meet the rules of former phases, thus a new validation is necessary.

All attributes of the Phase 4 model must meet the normal forms (1st to 5th). There is a close connection of attribute ownership and the second normal form. A simple formulation of the second and third normal form is:

„A non-key attribute must be dependent upon the key, the whole key, and nothing but the key."

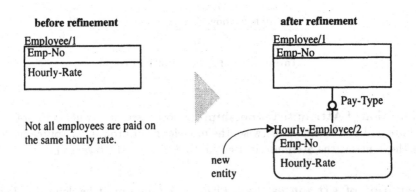

Fig. 24. PHASE 4 - USING NO-NULL RULE

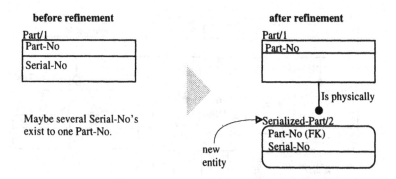

Fig. 25. PHASE 4 - USING NO-REPEAT RULE

Representation of Phase 4 Results. Finally in Phase 4, the function view diagrams (Fig. 26) are enlarged by non-key attributes. Furthermore, the documentation is extended, containing the following information:

- A definition of each entity

- A list of all primary, alternate and foreign key attributes

- A definition of each owned attribute (key and non-key)

- A list of the following relationships with the entity being a parent entity:

  - Generic entity in a categorization

  - Identifying parent relationships

  - Non-identifying parent relationships

- A list of the following relationships with the entity being a child entity:

  - Category entity in a categorization

  - Identifying child relationships

  - Non-identifying child relationships

- A definition of all dual path assertions

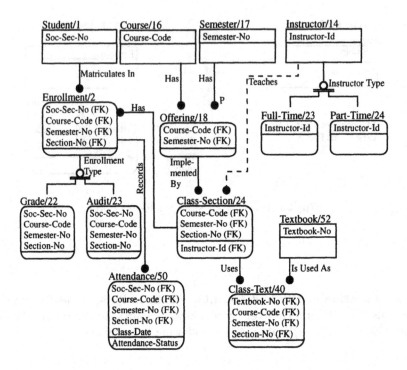

Fig. 26. PHASE 4 FUNCTION VIEW DIAGRAM

5 Documentation and Validation

5.1 Introduction

The goal of IDEF1X is the creation of a consistent and integrated documentation of the semantic characteristics of data. A large part of the documentation and the configuration management can be simplified by software tools. On a very simple level, word processing software can support the definition of entities, attributes, and relationships; interactive graphic tools can support the development of diagrams. These tools, however, are not able to take the content of the model into account, and therefore they are limited in their benefit. Special tools relieve the model builder's task, if they support the following functions:

- Automated creation and layout of diagrams
- Aggregation of data models
- Consistency check and automated refinement of models using modeling rules

– Report generation
– Support of configurations management

As for IDEF1X, such an automated support is desirable but not necessary.

5.2 IDEF1X Kits

A kit is a document containing diagrams, text, glossaries, decision summaries, and background information. The modeler must create at least one kit for each IDEF1X phase. This kit must be checked by experts who comment on the kit. Fig. 27 shows the resulting kit cycle which is used to create commonly accepted models.

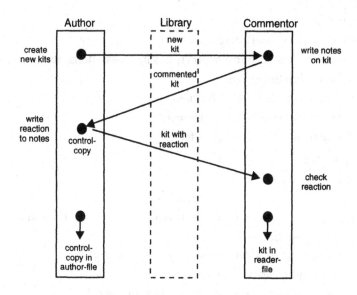

Fig. 27. KIT CYCLE

The model walk-through process (see 5.4) is an alternative to the kits. In the following, the contents of the kits are listed:

– Phase 0 - Kit
  - Kit Cover Sheet
  - Purpose and viewpoint
  - Plan for development and check up of the model
  - Team members and their roles
  - Source material (optional)
  - Author conventions (optional)

- Phase 1 - Kit

  - Kit Cover Sheet
  - Entity pool
  - Entity definitions

- Phase 2 - Kit

  - Kit Cover Sheet
  - Relationship matrix (optional)
  - Phase 2 (Entity level) diagrams
  - Entity reports (definitions and relationships)
  - Relationship definitions
  - Relationship/entity cross-references

- Phase 3 - Kit

  - Kit Cover Sheet
  - Phase 3 (Key level) diagrams
  - Entity reports (definitions, relationships, assertions, and keys)
  - Relationship definitions
  - Key attribute list and definitions
  - Relationship (entity cross reference)
  - Key attribute/entity cross-reference

- Phase 4 - Kit

  - Kit Cover Sheet
  - Phase 4 (attribute-level) diagrams
  - Entity reports (definitions, relationships, assertions, keys, and attributes)
  - Relationship definitions
  - Attribute list and definitions (key and non-key)
  - Relationship/entity cross-reference
  - Attribute/entity cross-references (key and non-key)

5.3   Standard Forms

Each kit starts with a so-called cover sheet (Fig. 28), containing the following information:

- Working information Ⓐ
- Reviewer information Ⓑ
- Content information Ⓒ
- Identification information Ⓓ

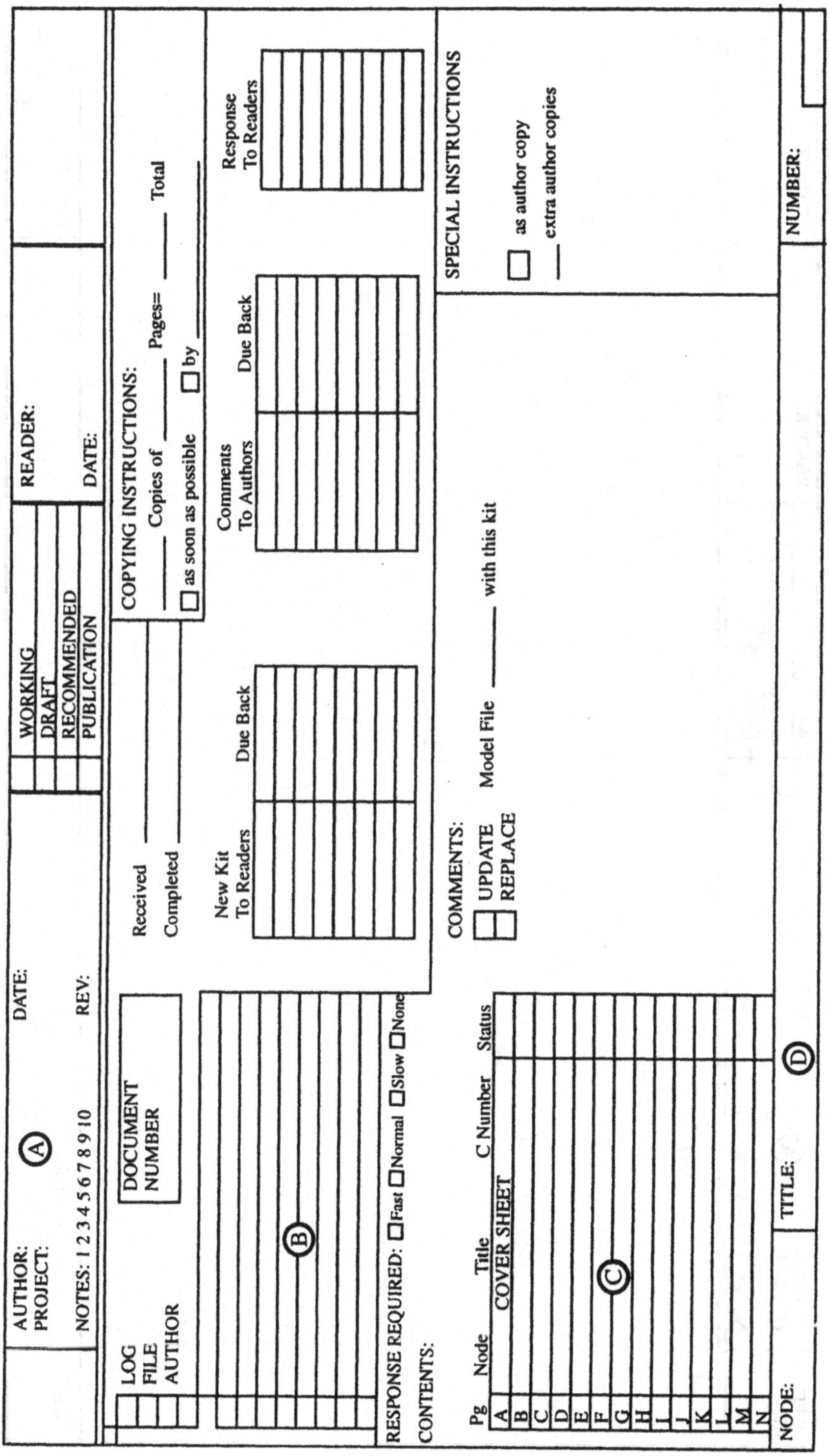

Fig. 28. Kit Cover Sheet

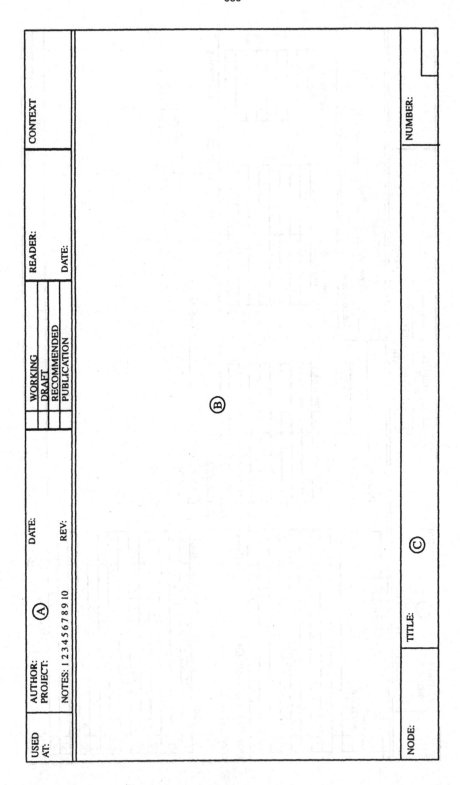

Fig. 29. STANDARD DIAGRAM FORM

389

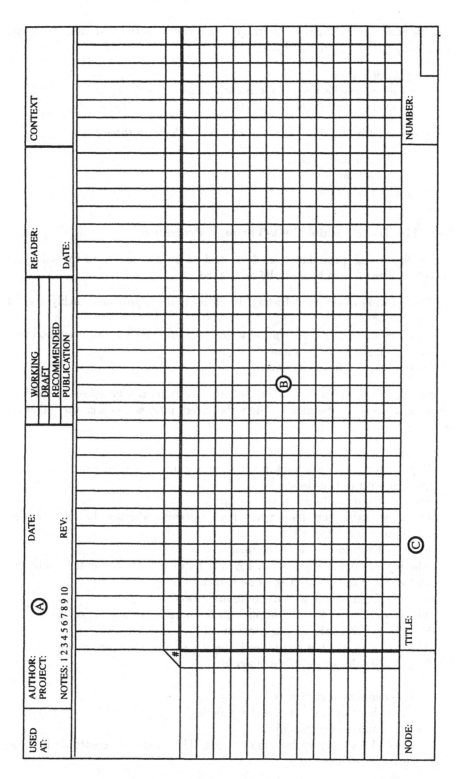

Fig. 30. MATRIX FORM

The **Standard Diagram Form** (Fig. 29) supports the following functions, supporting the structured analysis:

- Context definition
- Cross references between diagrams and other material
- Notes to the single sheet's contents

Therefore, the standard diagram form is divided into three parts:

- Working information Ⓐ
- Message field Ⓑ
- Identification field Ⓒ

5.4 The IDEF Model Walk-Through Procedure

The walk-through procedure was developed as an alternative to the kit cycle provided all team members can be assembled:

- Presentation of the model using the entity pool. It provides quickly a good general view.
- The presentation of the technical terms used provides a vocabulary modelers and readers have in common.
- Presentation of function views.

During the function view walk-through, questions can be asked in order to reveal weaknesses of the model. The walk-trough process contains the following six steps:

- Step 1 - Scan the entity pool
  In this step, the reader obtains a general impression of the model, getting familiar with the entities.
- Step 2 - Read the function view diagram
  Once the entities are known, the reader decides whether the relationships are represented adequately.
- Step 3 - Examine the key attributes
  The reader checks whether all key attributes identify the entity instances and whether the smallest key rule is met.
- Step 4 - Examine the key attribute migration
  The reader checks the migration from the parent to the child entities.
- Step 5 - Examine the non-key attributes
  The reader checks the attributes being not part of the primary key.
- Step 6 - Set the status of the diagram

  - *Recommended as it stands*
    The check up did not reveal any faults.
  - *Recommended as modified*
    The check up revealed small errors. The model is acceptable if these errors are corrected.

- *Draft*
  The check up revealed many errors. A model modification and repeated validation is necessary.
- *Not Accepted*
  The check up revealed so many errors that a new complete analysis is necessary.

References

1. IDEF1X Data Modeling: A Reconstruction of the Original Air Force Wright Aeronautical Laboratory Technical Report Developed Under Air Force Contract No. AF33615-80-C-5155. Mayer, R. J., editor. Knowledge Based Systems Inc. (1992), College Station (Texas)

Automatic programming in CIM based on intelligent tools

Witold Jacak,
Berndt Kapsammer

Institute of Systems Sciences,
Department of Systems Engineering and Automation,
J. K. University of Linz, Austria

Abstract. This paper describes a *realized concept for designing and automatic programming of flexible workcells based on intelligent tools*. A design tool, a task planning tool, a process planning tool and different simulation tools are integrated in the Intelligent Control of Autonomous Robotic System (ICARS). In this system design, programming, execution and testing of an IRb-6 ASEA robots application in flexible machining centres can be performed. ICARS uses non conventional models and algorithms for robot's motion planning. This system provides a broad basis for design and developing an autonomous workcell.

1   Introduction

Robots and robotic workcells are necessary components for an high flexible and automated production system. This high flexibility can be reached only if we demand a comfortable system for design workcells, programming the robots and devices and test the designed components by simulation. For this demand there are many efforts to develop a complex system at the Department of Systems Engineering and Automation at the University in Linz, Austria, in co-operation with the Institute of Technical Cybernetics at the Technical University of Wroclaw in Poland and University of Arizona, Tucson. This system for computer aided planning and programming of autonomous workcells is provided within a research project called "CAD for CIM"[1].

In the recent years many software systems for the graphical designing and off-line programming of robots were built [1 - 8]. But in all these systems you cannot find such features like automatic collision-free motion planning with optimization of dynamical trajectories for robots. These tools and some more features are all integrated in the software system ICARS. It consists of the following components:

- Module (GRIM) for modelling of the virtual workcell
- Technological task specification module
- Process planning module

---

[1] The project "CAD for CIM" is supported by the austrian funds for support of scientific research (FWF).

⚑ Motion planning module
⚑ Simulation module

Some of these modules do also have further submodules which will be described below. A schematic structure of ICARS is given in figure 1.

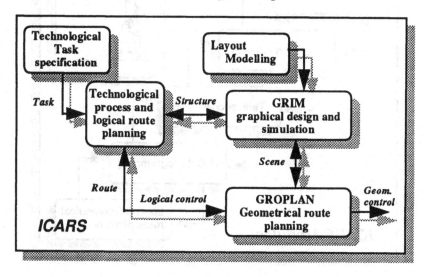

Fig. 1: Structure of ICARS [9]

Global modules are the *task planning module*, the *process and logical route planning module* and the *geometric route planning module*. The *motion planning module* does have single submodules which provides the exact computation for the motion of the robots movements. Thereby the user can choose which type of movement the robot should perform. The most powerful feature in the *motion planning module* is the submodule for collision-free path planning. Another submodule for instance performs the optimization depending on the decision of the user either if it should relate to the shortest time or to the shortest length of the path or to both. The *graphical module* is divided in two main modules. One of it supports the 3 dimensional graphical simulation and the output screen. The second submodule provides the design of simple objects for a further design of the whole workcell layout. A part of the simulation module uses the input and given data and presents the computed simulation results by 2 dimensional graphic charts.

We will now consider the typical design process. The user has to design the necessary components of the cell with all objects like stores, machines and robots of a workcell before using the simulation tools. The design process is a logical order of phases which can be entered iterative. If there arises unfitting results the user can perform new decisions and changes in each design step for a better performance of the future workcell [29].

In figure 2 the logical phases of the design process for programming robotic autonomous workcells are shown.

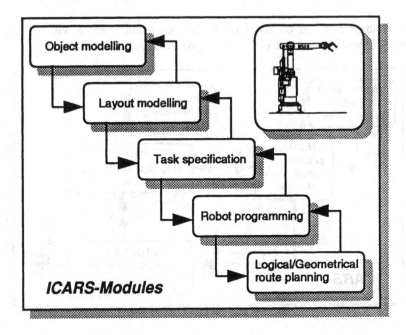

Fig. 2: The phases of the design process with each design step in ICARS

For the design process [30] we consider now each design step and begin with the design of the *virtual robotic cell* with basic geometric design tools and the specification of the *technological task tool*.

2 Virtual Robotic Cell and Technological Task

A *real cell* is a fixed, physical group of machines D (or stores M), and robots R. A *virtual cell* is a formal representation (computer model) of a workcell. In the following chapter we discuss the object and layout modelling for the virtual robotic workcell and afterwards the task planning module.

2.1 Virtual Workcell Modelling

First of all the logical and geometric model of the workcell has to be designed. For the design of a virtual workcell we consider now the definitions for a logical structure and the geometric model of a workcell and the graphical tools: *object designer* and *layout planner*.

2.1.1 Logical Structure of Workcell

In the first phase of the design process, the workcell entity structure is designed; i.e. a logical structure of the cell must be created. The group of machines is divided into subgroups, called machining centres, which are serviced by separate robots.

A robot $r \in R$ can service only machines that are within its service space $Serv\_Sp(r) \subset E_0$ (E_0 - Cartesian base frame). The set of devices which lie in the r-th robot service space is denoted by $Group(r) \subset D \cup M$. More specifically, the device belongs to group serviced by robot r (i.e., $d \in Group(r)$) if all positions of its buffer lie in the service space of robot r. Consequently, the logical model of a workcell is represented by:

$$Cell_{Logic} = \left(D \cup M, R, \{Group(r) | r \in R\}\right) \qquad (1)$$

Based on the sets $Group(r)$ and the description of the task, we can define the relation β which describes the transfer of parts after each technological operation of the task.

$$\beta \subset (O \times O) \times R \qquad (2)$$

where: $\left((o_i, o_j) \beta\, r\right) \Leftrightarrow \left(\{d_i, d_j\} \subset Group(r) \vee \{m_i, d_j\} \subset Group(r)\right)$

and $\alpha\ (o_i) = \{d_i, m_i\}$ and $\alpha\ (o_j) = \{d_j, m_j\}$. ($\alpha$ *is a relation of device or store assignment*, see section 2.2 Technological Task)

To support the logical model construction process, ICARS has a *Taskplan* editor which allows the designer to specify the logical structure of workcell.

For example, we decide that the workcell will be serviced by two IR-b ASEA robots, r_{01}, r_{02}. The machining centre serviced by robot r_{01} consists of devices

$$Group(r_{01}) = \{d_{01}, d_{02}, m_{01}, m_{02}\}.$$

Machining centre of robot r_{02} has following equipment

$$Group(r_{02}) = \{d_{03}, d_{04}, d_{05}, m_{02}, m_{03}\}.$$

Such a division of machines reflects the structure of the technological task. The logical structure of the manufacturing workcell is shown in Figure 3.

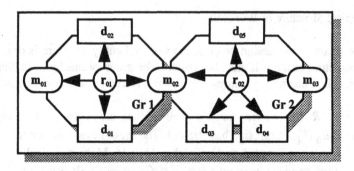

Fig. 3: Logical structure of virtual cell

2.1.2 Geometric Model of Workcell

In the second phase of the modelling process the geometry of a virtual cell must be created. Formally the geometry of the cell is defined as follows [31]:

$$Cell_{Geometry} = (G, H)$$

The first component of the cell geometry description

$$G = \left\{ G_d = (E_d, V_d) \middle| d \in D \cup M \right\}$$

represents the set of geometric models of the cell's objects. E_d is the coordinate frame of object (device) d and V_i is the polyhedral approximation of the d-th object geometry in E_i.

The second component of geometric model

$$H = \left\{ H_d : E_d \to E_0 \middle| d \in D \cup M \right\}$$

represents the cell layout as the set of transformations between an object's coordinate frames E_d and the base coordinate frame E_0 [31].

Consequently, a geometry modelling process proceeds in two stages:

- workcell objects modelling and
- workcell layout modelling phase.

2.1.3 The *Object Designer*

With the object the geometry model of each object is created by using solid modelling [32]. Solid modelling incorporate the design and analysis of virtual objects created from primitives of solids stored in a geometric database. The primitives can be synthetisized by the additionally module called it. The complex virtual objects of a workcell (such as technological devices, robots, static obstacles or parts of any kind) are composed of solid primitives. These primitives can be selected, moved, rotated in Cartesian base frame or in local coordinate frame, coloured and be grouped to more complex objects.

2.1.4 The *Layout Planning Tool*

The workcell's objects can be selected from a list and be placed manually in a robot's work scene at any position and in any orientation. The virtual complex objects (devices, stores, robots and obstacles) are loaded from a graphic library into the Cartesian base frame. With the built in translation and rotation operation tools each object can be moved anywhere in the base coordinate frame. The translation and rotation can be performed in two different modes. The objects can be moved in the base frame or in the local frame.

Actually there is no tool implemented to support an automatic layout planning. In our state of work only one feature supports the user at the layout planning design step. It proofs the distance of each object to a robot and tells if the defined work place of a device or a store is inside the workspace of the robot. The manual optimization of the layout can be performed after analyzing of the results of the simulation.

There exist many theoretical approaches for an automatic layout planning according to the part-machine grouping problems [10-14]. A very interesting and useful approach is given in [15] with the neural network approach.

To obtain fast and fully computerized methods for collision detection, ICARS uses additional geometric representation of each cell's objects. Each volume representation decomposes objects into primitive volumes such as cylinders, boxes, and cubes. We introduce the ellipsoidal representation of 3D objects, which uses ellipsoids for filling the volume. Instead of planes that are in polyhedral approximation, the surface of the object is modelled by parts of ellipsoids. The accuracy of representation depends on the number of ellipsoids and their distribution in the object. The ellipsoid-packing algorithm transforms each primitive solid into the union of ellipsoids, and consequently, the geometry of the virtual object is represented by:

$$V_d = \bigcup_i \varepsilon_i^d \quad \text{where} \quad \varepsilon_i^d: \quad \left(D_i^T(x-r)\right)^T A_i \left(D_i^T(x-r_i)\right) - 1 \leq 0$$

and r_i represents the centre of ellipsoid, D_i is the matrix of major axes, and A_i is the matrix of axes lengths [33].

For example, a parallelepiped is packed by a bigger central ellipsoid and eight smaller spheres in the corners. The packed ellipsoids are a hierarchical representation in this sense that for gross representation only the biggest ellipsoids are used.

The ellipsoidal representation of a virtual object is convenient to test feasibility of robot configurations. Checking for the collision-freedom of the robot configuration can be reduced to the "broken-line ellipsoid" intersection detection problem, which in this case has easy analytical solution [19].

The hierarchy of workcell models using in ICARS is shown in Figure 4.

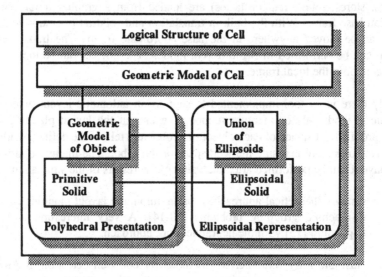

Fig. 4: Hierarchy of virtual cell models

After graphical designing of the more complex objects like workstations, stores, robots and other manufacturing units it must be specified the task.

In figure 5 an example of object modelling and layout planning is shown. Initially the primitive graphical objects are used to build robots, workstations or other objects which are then used to plan the layout.

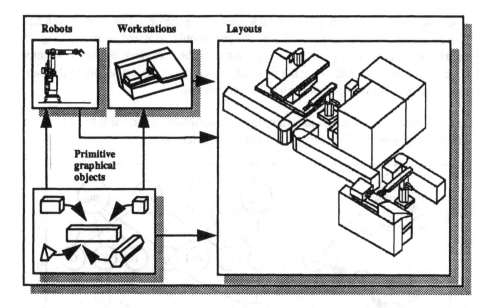

Fig. 5: Complex object modelling and layout planning.

2.2 Technological Task

At the beginning of the design steps it is very important to specify the tasks for each workcell. The other modules use these task specification for further computation and optimization of its realization. The specification of the task will be determined as follows.

The *technological task* realized by the robotic cell is represented by a triple:

$$Task = (O, \prec, a) \tag{3}$$

where: O is a finite set of technological operations (machine, test, etc.) required to process the parts,
$\prec \subset O \times O$ is the *partial order* (precedence relation) on the set O, and
$a \subset O \times (D \cup M)$ is an *assignment relation* for given devices or stores.

The *partial order* represents an operational precedence i.e.: $q \prec o$ means that the operation q is to be completely performed before the operation o can begin.

All defined precedence relations build the production graph of the task. The graph should not have more than one end and may not contain cycles. An example for one production graph can be described as the following sequence: $\{(a\prec b), (b\prec c), (d\prec c), (c\prec e), (c\prec f), (e\prec g), (f\prec g)\}$. That means that the operation b follows the operation a and so on.

The *assignment relation* defines at which workstation the operation of the given task should be performed or at which device the part should be stored after finishing the operation. $(o, d) \in \alpha$ means that the operation o can be performed on the workstation d, and if $(o, m) \in \alpha$, then m is the production store from the set M where the parts can be stored after the operation o has been completed.

In fig. 6 there it is presented an example of a precedence relation and an assignment graph which will be defined by the assignment relation α.

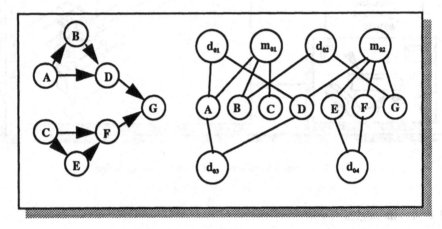

Fig. 6: Example of machining task graph and of an assignment graph.

The technological task described above can be realized in a virtual cell. The virtual robotic cell has an hierarchical structure which contains the models. These represent the dynamical and static data about a real production environment The task specification module is an essential step in the design process for the following other phases. After the specification of the task it is necessary to define the processes of the workcell. Therefore the next chapter will deal with the process planning module.

3   Process Planning Module

3.1   Process Planning Problem

According to the process planning there arises always the known problems of deadlock situations. Above, it was told that it is necessary to find an ordered sequence of operations of a task with a minimum probability of deadlocks (circular-wait-deadlocks) for the process planning. In the opposite the waiting period should be minimal to avoid decreasing of production rates. There exists four situations in which a deadlock can arise if all happens simultaneously (Coffman's conditions[34]):

- ➔ *Mutual exclusion*: Only one operation can use a special resource. The resource is not dividable. If one operation wants to use one hold resource then it must wait till this resource is free again.
- ➔ *Hold and Wait*: In this situation the operation holds its last resource and waits till the next resource is getting free from another operation.
- ➔ *No Pre-emption*: The resources cannot be preempted by another operation. It has to be freed by the holding operation.
- ➔ *Circular Wait*: (e.g. operation o_1 holds resource r_1 and waits now for resource r_2, but o_2 holds r_2 and is waiting for resource r_3 which is used by o_3 and o_3 waits for r_1).

Three algorithms deal with these deadlock situations:

- ➔ *deadlock prevention*-algorithm,
- ➔ *deadlock avoidance*-algorithm,
- ➔ *deadlock detection and recovery*-algorithm.

In the process planning module of ICARS the *deadlock avoidance*-algorithm is implemented. The small difference between *deadlock prevention*- and *deadlock avoidance*-algorithm consists in the moment of happening. While *deadlock prevention*-algorithm prevents a deadlock situation before runtime the *deadlock avoidance*-algorithm tries to avoid a deadlock situation during runtime. One result with *deadlock prevention*-algorithm is that resources are not utilized optimal. The *deadlock detection and recovery*-algorithm recognises a deadlock and tries to recover after a deadlock has happened. The *deadlock avoidance*-algorithm is realized by special backtracking-algorithms and will be discussed in the next subsection.

3.2 The Process and the Production Route

The sequential execution of m single operations is called sequential working process. We define

$$Process = (o_1, o_2, \ldots , o_L), \qquad (4)$$

such, if for 2 operations o_i and o_j of a task the precedence relation $o_i \prec o_j$ is defined then $i < j$ and for all $i = 1, \ldots , L\text{-}1$ there exists a robot r_k which can transport single production parts from a workstation d_i (or m_i) with the operation o_i to the workstation d_{i+1} (or m_{i+1}) with the operation o_{i+1}.

ICARS provides some simple useful features to define a process. If the devices, stores and other objects are defined their identification name like $m_{01}, m_{02}, \ldots$ or $d_{01}, d_{02}, \ldots$ has to be selected in a logical sequence.

A production route is a mapping of one process to the resources of the workcell. Each process can generate more different routes. It is now necessary to define criteria for the selection of the optimal routes.

A process can be realized by different sequences of technological devices (called *resources*) required by successive operations from the list *Process* at the time of their execution. This set of such sequences, denoted by P, is called production routes set. A production route $p \in P$ is an ordered list of resources which has at most $2L+1$ stages, where L denotes the length of the list *Process*. The route is modelled by:

$$p = \left(m_f, res(o_1), res(o_2), \ldots, res(o_L), m_0\right) \qquad (5)$$

where $res(o_i) = \alpha(o_i) = d_i$ if there exists a robot r which can transfer parts directly from d_i to d_{i+1} and $res(o_i) = (d_i, m_i)$ if the robot r can transfer parts only from the production store m_i to d_{i+1}. We assume that there always exists a robot transferring parts from d_i to m_i. By m_f and m_0 we denote the feeder and the output conveyor, respectively.

The production rate is directly related to the order of operations in *Process*. Moreover, a *circular wait* deadlock may occur between pipeline processes [35]. Techniques used to avoid deadlocks result in increased job waiting time, and consequently decrease the production rate. Thus the problem of finding the most efficient route is very important in planning the control of a flexible manufacturing cell.

One criteria is to minimize the throughput-time of products related to the operations and the robot movements. This throughput-time consists of three parts (the machining-time, the waiting-time and the transfer-time of the product itself). The waiting-time and the transfer-time depends strongly from the order of operations of a task and from the robot's movements. Each production route demands another robot's movement, contains different throughput-times and waiting-times and provides different solutions for deadlock-avoidance conditions.

P is the set of all production routes and we build a production route $p \in P$ out of an ordered list of resources. This list contains maximal $2L+1$ elements, where L is the length of the process (number of operations).

The route planning algorithm should take into account the conditions for deadlock avoidance. In order to formulate a quality criterion of process planning, we first describe the procedure of deadlock avoidance presented in [35].

To avoid blocking, the production route p is partitioned into a unique set of Z sublists called *zones*.

$$p = (z_k | k = 1, \ldots, Z),\tag{6}$$

where every zone has the form $z_k = (s_k, u_k)$ and

$u_k = (u_k^i | 1, \ldots, I(k))$ is the sublist of resources which appear only once in a production route. Such resources are called unshared resources.

$s_k = (s_k^j | j = 1, \ldots, J(k))$ is the sublist of resources which are used more than once in a production route and are called shared resources.

Let $C(d)$ denote the capacity of a machine; i.e., the maximum number of jobs to which the machine (or store) can be allocated, and $h(d)$ denote the number of jobs which are currently allocated to the machine d. Assume that the machine v required by currently processed job J belongs to the zone z_k; i.e., $v \in z_k = (s_k, u_k)$. The restricted allocation strategy is defined by the following rules:

- Rule A: If the current capacity of the required resource v is $h(v) < C(v)-1$, than the resource can be allocated to job J,
- Rule B: If the capacity of the required resource v is $h(v) = C(v)-1$, than the resource can be allocated to job J only if there does not exist another fully allocated resource in the zone.

In both cases the entire unshared zone is treated as one resource. The circular wait deadlock can never occur under this restricted resource allocation strategy. The complete formal explanation of the restriction allocation strategy is presented in [35], [37].

Task planner package of ICARS uses the above conditions to formulate the planning *quality criterion*. The goal is to optimize the production rate; i.e., minimize the job waiting times. It is easy to prove that the more elements (machines) belong to a zone, the higher is the probability that a job will wait for resource allocation. Thus, the production routes should contain zones with a minimum number of elements. Based on these remarks, we introduce the measure of route quality defined as:

$$v(p) = \max_k \{ n(k) | k = 1, \ldots, Z \}\tag{7}$$

where $n(k)$ denotes the number of elements in the zone z_k (the subzone u_k is treated as one element) and Z is the number of zones.

The function $v(p)$ is used to evaluate the technological process being planned. The problem solved by ICARS's task planner is to find an ordered sequence of operations from the technological task which is feasible and which minimizes the function $v(p)$.

This is a permutation problem which can have more than one solution. A technological task can be represented by a directed acyclic graph. To solve the planning problem under consideration, the *Taskplan* module is equipped with a backtracking graph search algorithm [37]. The backtracking procedure expands the OR-graph of task operations depth wise, one at a time, along one path (route) with different possible choices of the return point in case of failure.

4 Geometric Route Planner

The production route p for a machining task determines the parameters of the robot's movements and manipulations (such as initial and final positions), to carry out this task. The set of all of robot's motions between devices and stores needed to perform a given process is called a *geometric route* or *geometric control*.

Consequently, the automatic design of a geometric route must determine, for each robot r servicing the process, a set of cell-state dependent time trajectories of the robot's motions.

The fundamental function of the third module of ICARS; i.e. *Groplan* module, is to synthesize the robot's motion trajectories for a given route p. To generate the trajectories, we must have available the geometric model of a virtual cell as well as models of the robot's kinematics and dynamics.

Based on the sequence of operations *Process* and its production route p, the positions table (Frames_Table) for all motions of each robot is created at first. The Frames_Table determines the geometric initial and final positions and orientations of the robot's effector for each robot movement. For movements which realize the transfer of parts, the initial and final positions result directly from the sequence of machines in production route p. For each robot position the collision test is automatically performed.

The robot's motion trajectory planning process is performed by *Groplan* in two stages:

- planning of the geometric track of motion, and
- planning of the motion dynamics along a computed track.

4.1. Path Planner

A robot's path can be described by a sequence of very near lying configurations. These configurations are points in the robot's coordinate system. This coordinate system is the space built by the robot's joint axis

$$\overline{q} = (q_1, q_2, ..., q_n) \tag{8}$$

The variables q_i are the angles of the joints and n is the number of degrees of freedom. All possible values of the joint coordinates $\overline{q}$ forms the configuration space (Q-space) [16], [17]. The vector $\overline{q}$ is also called the joint configuration.

The motion planning is performed in two steps:

> ➣ In the first phase the user has to define a start- and finish-position for the effector. Both positions are going to be mapped directly to the geometrical model of the work scene for each movement. Also we should define the type of the effector's movement. There exists two groups of types. First the user defines a special type of line or curve which the robot has to follow at his path or the user decides that the shortest collision-free path should be computed and performed automatically.
> ➣ In the second phase the type of dynamic of the movement of the previous defined path is going to be defined. The optimization of velocity and acceleration along the given path will also be performed.

For the first group of motion types it is necessary to select a path type of the given types of curves in three dimensional space. The planning module performs an interpolation of the start- and finish-point with the given curve and then there will be computed the joint coordinates q in discrete points. This mathematical operations are going to be performed with the fast-inverse kinematic model [38]. There exists four path types for the movement of the gripper from point A to point B in the three dimensional space (LINE, LINEAR, CYLINDER and CIRCLE).

> ➣ LINE means that the gripper is moving straight forward along a straight line. That is also the shortest distance between start and finish point. It is a linear movement in the Cartesian space.
> ➣ LINEAR means that the gripper is moving with round movements. In the opposite to LINE it moves linear in robot coordinate space. With the LINEAR-movement the angles of all joints will be changed steadily.
> ➣ CYLINDER means that the path of the gripper follows a less bent curve than with LINEAR. The gripper is to be moved linear in the cylindrical base coordinate system (r, ϕ, h).

⤷ CIRCLE means that with defining a third middle-point in the coordinate system the gripper will be moved along a computed circle-path. The path of the gripper moves from start point through the middle point to the finish point.

A special type of path type is the collision-free path type in ICARS. It is a very powerful feature in this robot simulation system for the necessary collision-free paths of robot movements.

Collision-free Path Planning:

The problem of planning collision free and shortest paths for robot's movements exists since the beginning of the existence of robots. For the text oriented programming which implicitly describes the tasks without geometrical path definitions the automatic planning of collision-free and shortest paths is a precondition. To this topic of automatical collision-free path planning there exists only a few papers and publications with useful methods [16], [17], [18], [20].

The essential task in the collision-free path planning is to reduce the complexity for the computation of the path. Therefore it has to be considered the constraints of the robot kinematics. This chapter describes the collision-free path planning based on the discrete robot kinematic model and on the graph search method.

Model of the robot's kinematic: The geometric model of the robot will be represented by a skeleton-arm. We define the vector z as follows:

$$z = (P_1, P_2, ..., P_{n+1}) \tag{9}$$

where $P_i \in E$ is the point of the Cartesian base coordinate system E which describes the position of the joint i. The point P_{n+1} describes the effector's end position. The vector z is called Cartesian configuration or shorter only configuration. The geometric model of the robot is represented through such a skeleton-model.

For the automatic generating of paths and the automatic interpretation of the movement commands it is needed a similar model of the robot's kinematic. Such a model has to provide a direct analyse of the robot's position related to different objects in the work scene. Also this model of the robot's kinematic should be simple applicable for three dimensional graphical simulation of the robot's movements in the Cartesian base coordinate system. For this reason a discrete dynamical system formalism for the robot's kinematic is useful. The discrete model is represented as a dynamical system as follows:

$$M_{robot} = (Z, U, f) \tag{10}$$

where Z is the Cartesian set of configurations (set of states) of the robot's manipulator and U is the set of inputs. The function f is the state transition function and is defined as:

$$z(k+1) = f(z(k),u(k)).$$

The state of the robot's kinematic model is described by the vector z with the points in the base coordinate system. For a possible solution of the kinematic's equations [18] a discretisation of the robot's coordinate space will be performed. Each angle of the joints is partitioned in δq parts and single changes of these angles can only be performed by $\mp \delta q$ or 0. Such statement about the change of the robot's angles leads to the description of the input set of each joint i as a set of a sequence:

$$U_i = \{+1,0,-1\}.$$

where ∓ 1 means a change $\mp \delta q$ and 0 means no change of the angle.

Now it will be synthetisized the model of kinematics in three steps:

- First of all it will be modelled a single joint.
- The second step is to generate the model of the manipulation arm in the two dimensional plane.
- And the third step is to describe the model of the whole manipulator in the three dimensional space.

For each joint the state set is the set of valid angle positions, which will be described as follows:

$$W_i = \{0,1,...,N_i\}, \quad q_i^{max} = q_i^{min} + N_i \delta q \tag{11}$$

The joint state transition will be performed by the rotation δq of a part l_i of the manipulator in the positive or negative direction which is defined by the input u_i. Obviously the output of the joint model is a transformation. This transformation defines the rotation of the point p_{i+1} in the arm-plane where p_{i+1} destinies the position of the points P_{i+1} of the three dimensional space also in the arm-plane.

A new position p_{i+1} after a performed rotation by the angle δq in the defined direction u_i will be computed with the following formula:

$$p'_{i+1} = p_i + R(u_i)[p_{i+1} - p_i]. \tag{12}$$

where $R(u_i)$ is a 2x2 constant matrix.
The model of the i-joint kinematic can be represented as an final state machine:

$$M_{jointi} = (W_i, U_i, Y_i, \lambda_i, \gamma_i).$$ (13)

where $\lambda_i : W_i \times U_i \to W_i$ is the state transition function such that:

$$\lambda_i(w,u) = \begin{cases} \max(0, w+u) & f\ddot{u}r \quad u = -1 \\ \min(N_i, w+u) & f\ddot{u}r \quad u = +1 \\ w & f\ddot{u}r \quad u = 0 \end{cases}$$

and output function $\gamma_i : W_i \times U_i \to Y_i$ is defined as:

$$\gamma_i(w,u) = \begin{cases} R(0) & f\ddot{u}r \quad w = 0 \lor w = N_i \\ R(u_i) & sonst \end{cases}$$

The positions of each joint in the arm-plane is defined by the vector $z_{arm} = (p_n, \ldots, p_{n+1})$. This leads to the representation of the kinematics of the robot's arm as a final state machine in the following definition:

$$M_{arm} = (Z_{arm}, U_{arm}, \lambda_{arm}).$$ (14)

where $U_{arm} = X\{U_i | i = 2, \ldots, n\}$. The state transition is defined recursively:

$$z'_{arm} = \lambda_{arm}(z_{arm}, \overline{u}).$$

$$p'_i = p'_{i-1} + \prod_{k=2}^{i-1} \gamma_{i-k}(w_{i-k}, u_{i-k})[p_i - p_{i-1}].$$

for $i = 2, \ldots, n$ and $p'_1 = p_1$. The variable γ_k is the output function of the FSM model for the joint k.

Because of the discretising of the robot's coordinate space the set of configurations Z is finite. The number of elements of the set Z is equal to $N_1 \cdot \overline{Z}_{arm}$. So the robot's kinematic model can be represented as a final state machine (FSM). The robot's kinematic will be modelled by the following algebraic system:

$$M_{robot} = (Z, U, f).$$ (15)

where $U = X\{U_i | i = 1, \ldots, n\}$ and $f : Z \times U \to Z$. If $z = (P_1, \ldots, P_{n+1})$ and $z' = (P'_1, \ldots, P'_{n+1})$ Then the position of the joint i in the three dimensional space will be described by the following equation:

$$P'_i = \begin{bmatrix} \gamma_1(w_1, u_1) & \overline{0} \\ 0 & 1 \end{bmatrix} \cdot crd^i T^{-1}\left(\lambda_{arm}(T(z), \overline{u})\right)$$

where $T: Z \times Z \to Z_{arm}$ is the transition from Cartesian space to the arm-plane.

The structure of the robot's kinematic model is shown in figure 7.

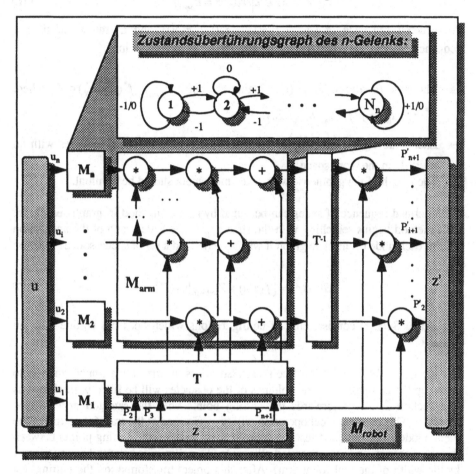

Fig. 7: Structure of the automaton M_{robot}

This previously defined model of the robot's kinematic, also called production system, is a good base for performing simple planning for collision-free paths. Therefor we have to formulate the problem of collision-free path planning in the theory of final state machines.

Method for planning collision-free paths: The problem to find a collision-free path is similar to the problem of generating of an ordered sequence of the states of the automaton $(z(0), z(1),..., z(N))$ and its input which leads from the initial state z_{init} of the automaton M to any state of the final state set. The final state set $Z_f \subset Z$ is defined as:

$$Z_f = \left\{ z \in Z \middle| crd^{n+1} z = P_{final} \right\}. \tag{16}$$

An ordered sequence of states $z^* = (z_{init}, z(1), ..., z(N))$ and at the same time a sequence of inputs u have to be found with the following conditions:

➤ The state (configuration) $z(N)$ obtains the point P_{final}: $f^*(z_{init}, u_N^*) \in Z_f$ where

$f^*(z_{init}, u_N^*) = f(f^*(z_{init}, u_{N-1}^*), \bar{u}_N)$.

➤ Each state $z(j) = f^*\left(z_{init}, u_j^*\right)$ is valid, i.e. the configuration $z(j)$ collides with no obstacle in the work space.

➤ The length of the geometric way of the movement should be minimal.

Such defined sequence of states can be found by the A^*-method in graph-searching. This method begins searching with the state z_{init} in the state graph of the automaton M. For each node of the state graph it will be generated a set of successors as follows:

$$Succ(z) = \left\{ f(z, u) \in Z_{coll.-fr.} \middle| u \in U \right\}. \tag{17}$$

The set of $Succ(z)$ consists only of configurations which does not collide with an obstacle.

For the test of being collision-free there also arises an idea for a transformation to speed up the computation. The volumes of the obstacles will be filled with ellipsoids (see section 2.1.4). Afterwards the cutting points with the skeleton arm will be searched. This mathematical operation will be performed at the transformation of the scene model. The collision-test examines if there exists some cutting points between the skeleton arm and the generated ellipsoids (The ellipsoids have to be made larger by the width of the real robot arm). After this object-transformation the cutting-test between the skeleton axis and each ellipsoids will be reduced to an analyse of the analytical inequality of second order. This means a essential velocity of the collision-test for each state of $Succ(z)$ [18].

The development of the search-graph is initiated in the node z_{init} and the ways of each next node depends on the evaluation function. The evaluation function is the sum of the distance function $k(z)$ and the heuristical function $h(z)$:

$$e(z) = k(z) + h(z) \tag{18}$$

where $k(z)$ is the sum of distances between the configurations of z_{init} and z and $h(z)$ is the distance of the configuration z to the point P_{final}. This function fulfils the monotone conditions where the bi-directional A^*-algorithm finds the shortest path [20], [21]. With the collision-free path planning we retrieve an ordered discrete sequence of robot states:

$$\tilde{q} = (q_{init}, q(1), ..., q(N)).$$

The discrete sequences from q_{init} till $q(N)$ will be approximated to polynomials with third order. The following elements of the sequence $\tilde{q}$ will be connected with cubic-splines. The condition for the spline-polynomials is that the whole curves have to be smooth. So we get a full analytical description of the path:

$$\overline{q} = \overline{q}(s) \quad s \in [0, s_{fin}]$$

where s is the parameter with the length of the path of the effector.

4.2 Trajectory Planner

The motion path is given by a sequence of points $q(i) \in Q$ in the robot's coordinate system as a output from the motion path planning. Thereby it has to perform the smooth operation with the splines-interpolation-method.

Now we consider the following question: How looks the speed distribution along the path like, so on the one hand the physical constraints like motorspeed and boarders for the momentum in the motors cannot be passed over and on the other hand the capacities of the velocity can be used optimal? I.e. how can we find out the velocity-profile for a fixed given path in the Q-space? Thereby it has to be cared for the constraints in the robot's dynamic. This task will be performed in the dynamical optimization module in ICARS.

4.2.1 The Model of the Robot's Dynamic

The motion equations of a robot with n-degrees of freedom can always be represented in the following form [22], [23]:

$$M_{ij}(\overline{q})\ddot{q}_j + N_{ijk}(\overline{q})\dot{q}_j\dot{q}_k + R_{ij}\dot{q}_j + G_i(\overline{q}) = u_i \tag{19}$$

where: M_{ij} are elements of the mass-matrix, N_{ijk} contains the centrifugal- and Coriolis-terms, R_{ij} are the friction forces while g_i is being realized by the weight force, u_i are the joint momentum and forces. The planning of the motion path generates an interpolated and with cubic-splines smoothed function of the parameter s.

$$\overline{q}(s) = (q_i(s) \big| i = 1,...,n)^T \quad s \in [0, s_{fin}] \tag{20}$$

For the computation of the robot's dynamic it will be needed the first and the second derivation of the joint coordinates related to the time [16], [22]. The insertion of the form (20) into the dynamic-equation supplies an equation for the joint momentums of the robot during a motion of a given path [22], [23], [24].

$$u_i = M_i(s)\dot{\mu} + N_i(s)\mu^2 + R_i(s)\mu + G_i(s) \tag{21}$$

where $\mu = \dot{s}$ is the pseudo-speed of the robot's manipulator. This mathematical operation reduces extremely the complexity of the optimization problems.

4.2.2 The Optimization Problem

The destination of an optimal speed profile along the path (20) from $s = 0$ till $s = s_{fin}$ can be formulated as the following one dimensional optimization problem with some constraints:

Define $\mu(s)$ such that

$$J = \int_0^{s_{fin}} L'(s,\mu,\overline{u}) ds \to Min \tag{22}$$

under the following two conditions:

▲ The joint momentums are constrained: $u_i^{min}(s,\mu) \le u_i(s) \le u_i^{min}$ and

▲ $\mu(0) = \mu_0$ *and* $\mu(s_{fin}) = \mu_{fin}$.

In ICARS the quality-criterions have two forms:

(i) for the time optimization:

$$J = \int_0^{s_{fin}} \frac{1}{\mu(s)} ds \tag{23}$$

(ii) for the energy optimization (This form also includes a time-optimization):

$$J = \int_0^{s_{fin}} \frac{1}{\mu(s)} \left(\lambda_1 + \lambda_2 \sum_{i=1}^{n} |u_i| \right) dt \quad and \quad \lambda_1 + \lambda_2 = 1. \tag{24}$$

4.2.3 The Optimization Method

Based on the one dimensional type the general optimization problem with any form of the quality criterion can be solved by known methods of numerical optimization. One special method of the dynamical programming of Bellman suits very good. Because of using much time for this method there we give a new method which is based on Bellman's methods. This new method is also based on the bi-directional graph searching method. Similar to the Bellman's method first the (μ, s)-phase-plane will be supplied with a net $(\Delta\mu, \Delta s)$. The columns of the net define the discrete values of the parameter s and the rows the discrete values of the pseudo-velocity μ. If the pseudo-velocity $\dot{\mu}$ between every pair of net nodes (s_k, μ_k), (s_{k+1}, μ_{k+1}) is constant, then the pseudo-speed μ in the interval $[s_k, s_{k+1}]$ can be presented as a function of the parameter s:

$$\mu(s) = \sqrt{\frac{(s_{k+1} - s) \cdot \mu_k^2 + (s - s_k) \cdot \mu_j^2}{(s_{k+1} - s_k)}} \qquad (25)$$

where $\mu_k = \mu(s_k)$ and $\mu_j = \mu(s_{k+1})$.
If we insert $\mu(s)$ into the criterion (22) then we retrieve the formula that describes the motion expense along of the interval $[s_k, s_{k+1}]$:

$$J(s_k, \mu_k, s_{k+1}, \mu_j) = \int_{s_k}^{s_{k+1}} L'(s, \mu(s), \overline{u}(s)) ds \qquad (26)$$

and where

$$\overline{u}(s) = N\mu^2(s) + R\mu(s) + G + M \cdot \frac{\mu_j^2 - \mu_k^2}{2(s_{k+1} - s_k)} \quad for \quad s \in [s_k, s_{k+1}]. \qquad (27)$$

For each node point of the net (s_k, μ_k) it can be computed the set of succeeding nodes:

$$Succ(s_k, \mu_k) = \left\{ (s_{k+1}, \mu_j) \middle| (s_k, \mu_k) \xrightarrow{\overline{u}(s)} (s_{k+1}, \mu_j) \right\} \qquad (28)$$

where $\overline{u}(s)$ (27) only works together with the condition (i).
The definition (28) allows the production of new nodes by local development in the partial search-graph. So it is obviously possible to use the method with the bi-directional searching of the net [20], [24], where the first searching direction (forward) starts out of the node $(0, \mu_0)$ and the second searching direction (backwards) out of the node (s_{fin}, μ_{fin}). The bi-directional graph searching-algorithm uses the following form as the evaluation function:

$$f(v) = g(v) + h(v) \tag{29}$$

where $g(v)$ describes the expense (26) of the path from node $v_0 = (0, \mu_0)$ (forward) $((s_{fin}, \mu_{fin})$ backward) till the node $v = (s_k, \mu_k)$.

$$g(s_k, \mu_k) = \sum_j^k J(s_{j-1}, \mu_{j-1}, s_j, \mu_j) \tag{30}$$

The heuristical function is defined by the following expression:

$$
\begin{aligned}
h(s_k, \mu_k) &= \sum_{i=k}^{N-1} JB_i^* \quad forward \\
h(s_k, \mu_k) &= \sum_{i=0}^{k-1} JF_i^* \quad backward
\end{aligned}
\tag{31}
$$

where JF_k^* (and JB_k^*) defines the actual and minimal expenses of the transition from the s_k- till the s_{k+1}-column for forward (backward) direction. It is easy to test that this heuristical function is the bottom limit of the optimal criterion-expiration [24]. At the beginning of the computations in both functions $h = 0$. In each step of the algorithm the values increases and their influence to the searching direction is also growing. An exact description of the specified method you can find in [24]. This method reduces the number of analysed nodes of the net. Above that this method is simple for using together with the optimization problems with given time slices for the motion.

For the time optimal problem there exists a fast solution which is also implemented in ICARS [22], [25].

5 Simulation System

The last module in the chain of modules is the simulation system. The computed and optimized paths for robot motion could be shown on the screen. On the one hand the whole workcell is presented in three dimensional representation on the other hand the user can follow the two dimensional outputs and results of each simulation by watching the used time, energy and distances. So it is possible to find at every point of time a better solution for the necessary motions of the robot. Upon these results the redesigning of the workcell can begin for a more optimized production route, for less waste of energy. There is also the possibility provided in ICARS to control directly the real motions of robots in a workcell.

In figure 8 we can see a typical workcell with two robots, three stores and three workstations. There are also the logical and geometric routes from the given task visualized.

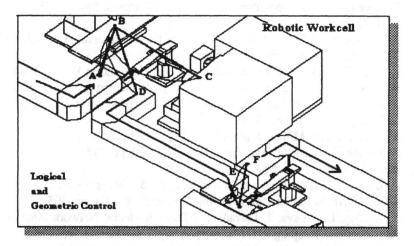

Fig. 8: The simulation of logical and geometric routes generated by ICARS.

6 Conclusion

ICARS is actually used for scientific research only in research laboratories. It is a flexible workcell-simulation system that generates fast simulations based on the intelligent algorithms mentioned above. In the moment ICARS is running only in DOS-mode but the Windows 3.x version with more graphical features will be prepared.

ICARS is available at the Department of Systems Engineering & Automation, J.K. University, Altenbergerstr. 69, 4040 Linz, Austria.

References

1. K. Breitenbach: Simulation und Offline-Programmierung von Industrierobotern mit ROBCAD. In: Robotersimulation ed. D.W. Wloka. Berlin: Springer 1991, pp. 65-88.
2. H. Mayr, H. Öllinger: SMART Simulation of Manufacturing and Robot Task. In Robotersimulation ed. D.W. Wloka. Berlin: Springer 1991, pp. 153-186.
3. D.W. Wloka: ROBSIM - Graphische Simulation von Robotersystemen. In Robotersimulationen ed. D.W. Wloka. Berlin: Springer 1991, pp. 289-326.
4. B. Faverjon: Object Level Programming of Industrial Robots. IEEE Int. Conf. on Robotics and Automation, 2, 1986, pp. 1406-1411.

5. H. J. Warnecke, G. Jordan: Wissensbasierte Entscheidungsunterstützung bei der Einsatzplanung von Industrierobotern. In Robotersimulationen ed. D.W. Wloka. Berlin: Springer 1991, pp. 227-260.

6. R. Speed. Off-line Programming of Industrial Robots. Proc. of ISIR 87, pp. 2110-2123.

7. R. Dillmann, M. Huck: A Software System for the Simulation of Robot based Manufacturing Processes. In Robotics vol. 2, 1986, pp. 3-18.

8. G. Spur, F. Krause: CAD - Technik. München: Carl Hanser Verlag, 1984.

9. A. Wancura: Simulationsgestützte, intelligente Steuerung des Produktionsablaufes in flexiblen Fertigungszellen, handed in by Andreas Wancura in 1993 at the University of Linz, Dipl.-Arb.

10. H.M. Chan, D.A. Milner: J. Manuf. Sys., 1, 1985, pp. 65.

11. A. Kusiak, C.H. Cheng: Ann. Oper. Res., 26, 1990, pp. 415.

12. J.F.K. Purcheck: Prod. Eng., 54,1975, pp. 35.

13. A. Ballakur, H.J. Steudel: Int. J. Prod. Res., 25, 5, 1987, pp. 639.

14. C.O. Malavé, S. Ramachandran: J. Intell. Manuf., 2, 1991, pp. 305.

15. S. Lozano, L. Onieva, J. Larrañeta, J.Teba: A Neural Network Approach to Part-Machine Grouping in GT Manufacturing, IMACS 1993, pp. 619.

16. M. Brady:Robot Motion, MIT Press, Cambridge, 1983.

17. T.Lozano-Perez: Task-Level Planning of Pick-and-Place Robot Motions, IEEE Trans. on Computers 38(3), pp. 21 - 29, 1989.

18. W. Jacak: Strategies for Searching Collision-Free Manipulator Motions: Automata Theory Approach, Robotica 7, pp. 129-138, 1989.

19. W. Jacak: Discrete Kinematic Modelling Techniques in Cartesian Space for Robotic System, in: Advances in Control and Dynamics Systems, ed. C.T. Leondes, Academic Press, 1991.

20. J. Perl: Heuristics, Reading, MA: Addison-Wesley, 1984.

21. J. Perl: Probabilistic Reasoning in Intelligent Systems, San Francisco, Morgan Kaufmann, 1988.

22. K. Shin, N. McKay: A Dynamic Programming Approach to Trajectory Planning of Robotic Manipulators, IEEE Trans. on Automatic Control, 31(6), 491-500, 1986.

23. K. Shin, N. McKay: Minimum Time Control of Robotic Manipulator with Geometric Path Constraints. IEEE Trans. on Automatic Control, 30(6), 531-541, 1985.

24. W. Jacak, I. Duleba, P. Rogalinski: A Graph-Searching Approach to Trajectory Planning of Robot's Manipulator, Robotica, 1992.

25. J.E. Bobrow, S. Dubowsky, J.S. Gibson: On the optimal control of robotic manipulators with actuator constraints, American Contr. Conference, pp. 782-787, June 1983.

26. G. Chroust, W. Jacak: Simulation in Process Engineering - Rozenblit J. (ed.): AIS-93, 4th Conf. on AI, Simulation, and Planning in High Autonomy Systems, Tuscon Sept. 1993, IEEE C/S Press pp. 232-237.

27. G. Chroust: Development Environments: a Multi-level Enactment Scenario.-
Rozenblit J. (ed.): AIS-93, 4th Conf. on AI, Simulation, and Planning in High
Autonomy Systems, Tuscon Sept 1993, IEEE C/S Press pp. 2-13.

28. G. Jahn: Simulation-based Monitoring and Prediagnosis in Intelligent
Robotized Cells. - Crowley J., Dubrawski A. (eds.): IRS-93, Intelligent Robotic
Systems, Zakopane, Poland, July 1993, pp. 91-100.

29. W. Jacak, J. Rozenblit, I. Sierocki: Simulation Based Intelligent Control Design
in Manufacturing Automation - Crowley J., Dubrawski A. (eds.): IRS-93,
Intelligent Robotic Systems, Zakopane, Poland, July 1993, pp. 81-90.

30. G. Chroust, W. Jacak: Simulation Based Process Engineering for Intelligent
Manufacturing - International Conference on Machine Automation ICMA '94
TAMPERE, Feb. 94, pp. 16-18.

31. P.G. Ranky, C.Y.Ho: Robot Modeling. Control and Applications with Software,
Springer Verlag, 1985.

32. W. Jacak: GRIM CAD/CAM system for design and off-line programming of
robotic workcell. Reference manual. Technical Report, University of Linz,
1992.

33. W. Jacak: Strategies for Searching Collision-Free Manipulator Motions,
Automata Theory Approach, Robotica,7, 1989.

34. G. Coffman, P.J. Denning: Operating Systems Theory, Prentice Hall,
Englewood Cliffs, New Jersey, 1973

35. B.H. Krogh, Z. Banaszak: Deadlock Avoidance in Pipeline Concurrent
Processes, Proc. of Workshop on Real-Time Programming IFAC/IFIP, 1989.

36. W. Jacak, J. Rozenblit: Model-based Workcell Planning and Control, IEEE
Trans. on Robotics and Automation (in press).

37. W. Jacak, J. Rozenblit: Automatic Robot Programming - Lecture Notes in
Computer Science 585, Springer Verlag, 1992.

38. W. Jacak: A Discrete Kinematic Model of Robot in the Cartesian Space, IEEE
Trans. on Robotics and Automation, 5(4), 435-446, 1989.

Model based Decision Support Systems

P.J. Drazan

Research Institute for Knowledge Systems (RIKS)
P.O. Box 463, 6200 AL Maastricht, NL

1. Introduction

There is a growing need to support complex decisions made by managers in order to achieve a good control of manufacturing processes. Managers have to reach a good functional integration of the departments, whether production, sales, finance, or else.

Computer based Decision Support Systems (DSS) can provide a consistent and structured support to assist the decisions to be made by providing access to the relevant information (data), by representing the functional relationships between the variables concerned and by the modelling of the represented system to make predictions.

The models simulating the performance of the systems, e.g. flow of production, have been used for a long time. More recently, new major components have been added to form a decision support tool. First, emphasis is put on a good man - machine communication, i.e. on a friendly user interface, second, on the availability of knowledge required to understand the process under consideration and, third, on the support provided to make correct decisions. In this way new developments have been under way, resulting in the concept of DSS. The knowledge provided by the system, often in the form of an explanation and help facility, also provides an important training function for the staff and can thus also be used within its own right as a training support tool.

Decision making means choosing among alternative courses of action in order to attain the required goals. The essence of computer support of decision making is to emulate alternative courses of action by using a model of the process and seeing to which results they lead. The alternative course of action is sometimes called a scenario. The user is playing different scenarios by using different sets of input data. As a result he can see the consequences of his choices. The process of testing different scenarios is called 'what if' analysis. Evaluation of the different scenarios leads to the right choice of decision.

2. Decision Support Systems

2.1. Architecture of a DSS

A production process can be seen as shown in fig. 1. There is a set of inputs entering the process, there is a number of process activities and eventually there is a set of outputs reflecting the results and also monitoring the quality of the process.

Inputs	---	Processes	---	Outputs

	Activities	
Resources	Procedures	Products
Raw materials	Programmes	Services
Costs	Tools	Performances
	Decisions	Consequences

Fig. 1

In order to model a process and to support decision making the DSS has to be configured in a way that fulfills all functions necessary to provide the expected support. It has to attain an architecture which contains the expected functionalities.

The block diagram of a DSS is shown in fig. 2.

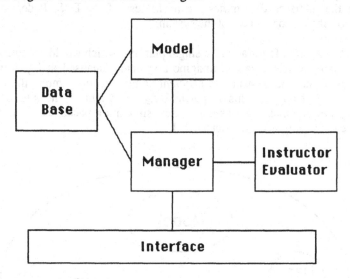

Fig. 2

The system consists of the data base, the model, the manager, the interface and the explanation/instruction component and the evaluator.

The data base, as the name implies, provides all the input information which is needed to run the model. Software packages such as Excel or Lotus, originally developed as spreadsheets, have reached the level of sophistication adequate to serve the purpose. As they are equipped with extensive mathematical libraries they can themselves serve as static models, particularly for financial applications. They are also equipped with 'what if' analysis allowing the user to analyse the different scenarios run by the model.

The model incorporates all representations and functionalities (tables, equations, etc.) necessary to emulate the processes under consideration. The model will be dealt with in more detail in the next section.

The explanation/instruction module provides all help and advice to the user on how to handle the DSS, explains the process, the results obtained and can be used to train the user by way of process simulation.

The manager of the DSS enables the user to control the DSS in the way he desires but, at the same time, protects him from the unintentional misuse of the system and from the insertion of inconsistent data. It controls the flow of information among the individual parts of the DSS, checks the consistency of the data transmission, controls the simulation and allows the user authorised access to the modules of the DSS.

The user interface positioned in front of the manager serves the purpose of easy and user friendly communication with the system. Very often the quality of the cognitive design of the interface determines the usefulness of the DSS. It determines the efficiency of the interactiveness of the system.

It is desirable to provide for a DSS a single platform which would accommodate the individual components. One can imagine a system visualised in fig. 3 where the common platform would ensure an efficient communication among the individual components of the DSS. This is particularly important in the situation where proprietory programmes, such as data bases, simulation packages, data evaluation packages, etc. are imported from outside [2].

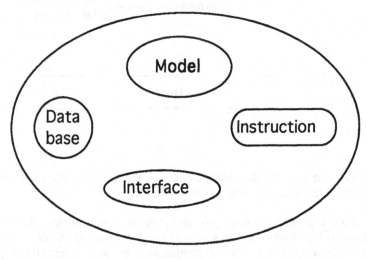

Fig. 3 DSS platform

2.2. DSS design process

In order to attain desired results when designing a DSS one has to go through a number of stages. These stages are:

a) Problem analysis - identification of the process, its boundaries, variables and their functional relationships

b) Design - representation of the process in the computer model by using the identified functional relationships

- design of the evaluator

c) Verification and testing

3. Models

The advantage of using models and the construction of a DSS is that it is cheaper, quicker and safer to experiment with a model and with the DSS than to experiment in the real world. This applies to any field, whether technological, socio-economical or ecological. The added benefit is that the model, by forcing the designer to analyse the real system in great depth, improves his understanding of the problem.

The types of models used to simulate various scenarios vary and so does their complexity.

The simple equation

$$P = \frac{F}{(1 + i)^n}$$

where   P .... is the present value
             F .... is the future value
             i .... is the annual interest rate
and   n .... is the number of years

can be seen as a model representing the process of investment growth. When specific data are entered in the equation it will provide a corresponding output. A different set of data will produce a different output set. Thus different 'scenarios' will be offered by this simple model. The same principle applies to a model of any degree of complexity.

Another type of model is a linear model where the process is represented by linear relationships and subjected to various constraints which have to be satisfied in order to fulfill the task. The following example illustrates this approach:

A company makes two types of personal computers, PC1 and PC2. PC1 requires 300 hour labour and $ 10000 in material costs, PC2 requires 500 hour and $ 15000 in material costs. The profit on PC1 is $ 8000 and on PC2 is $ 12000. The plant has a total capacity of 200 000 days/month and a material budget of $ 8 million.

A minimum of 100 units of PC1 have to be produced but otherwise the market will absorb unlimited number of units. The company wishes to maximise their profits working within these constraints.

The question is how many units of each type are to be made to achieve the goal?

The process is modelled by the following equations:

The profit equation is $Z = 8000 \times PC1 + 12000 \times PC2$
The constraints are:

Labour	$300 \times PC1 + 500 \times PC2 < 200000$
Material	$10000 \times PC1 + 15000 \times PC2 < 8000000$
and	$PC1 > 100$

The model can be expressed in graphical form as shown in fig. 4.

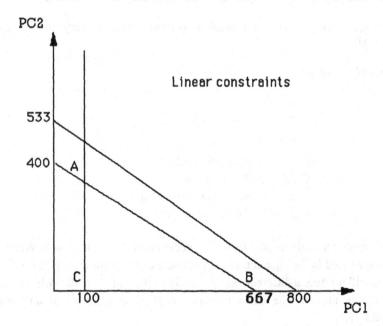

Fig. 4

The following can be easily seen from both the inequalities and from the graph. The labour capacity constrains the production more than the material budget. Subsequently the production regime must be restrained within the triangle ABC. Looking at the profit equation it is apparent that maximum profit is obtained when only PC1s are manufactured. The profit made by this regime of production at the level of 667 units of PC1 is

$$Z = \$\ 8000 \times 667 \text{ units} = \$\ 5336000.$$

3.1. Graphical modelling tools

In this section we shall elaborate on the modelling tools which are particularly suitable for DSS applications.
One such software tool which can be used in this manner is called 'I think'.

There are several important features which modern software should possess. The first feature is that it communicates with the user in his 'language', eliminating any need for him to use a programming language. The programming is done graphically, allowing the user to design his model directly by using graphical representations. The second important feature is that it generates a programming code in the background and the user can check it at any time. By linking the graphical entities into the graph the syntax is automatically checked maintaining the consistency of the representations. The primitives represented by icons, are set at a high level of abstraction allowing thus a wide variety of applications.

The whole concept is based on the presumption that in any system there is a 'flow' (of mass, products, people, money) and there is a 'container' - reservoir, into which this flow is directed, or, alternatively, extracted from. There are two main primitives used in this modelling tool: a 'pipe' which guides the flow and a 'container/reservoir' which contains it. The pipe is provided with a valve which controls the rate of the flow. It is best to demonstrate the modelling tool by providing a simple example.

A petrol station gets regular deliveries of petrol from a tanker which arrives at weekly intervals. The station has a range of customers to whom it supplies the petrol. The manager of the station wishes to achieve a good balance between the deliveries from the tanker and the demand by his customers.

Fig. 5 shows how this process is represented in 'I think'.

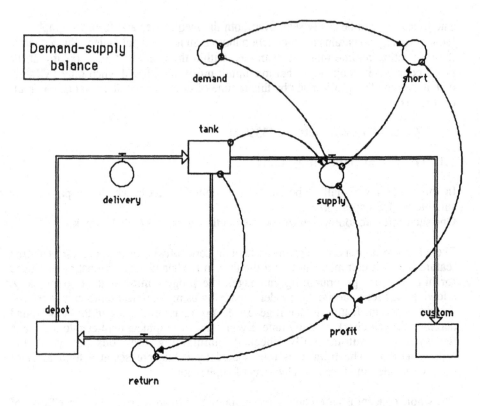

Fig. 5

The tank receives the deliveries via the pipe connecting it with the depot. The pipe represents the tanker which delivers the fuel to the tank. The valve called 'delivery' determines the amount of fuel pumped at each delivery to re-fill the tank. On the other side, the customers empty the tank by filling their vehicles. This process is represented by the pipe connecting the tank with the custom. The rate at which the tank is emptied is determined by the valve called 'supply'. If the demand during the previous period was lower than expected the delivery may exceed the capacity of the tank. In such case the remainder of the fuel has to be returned via the 'return' pipe.

The purpose of the simulation is to calculate the profit/loss of the station. The custom demand is generated randomly within certain limits. The profit is determined by the sold fuel, reduced by the shortage of fuel which loses custom and also reduced by the penalty paid for any returned fuel.

'I think' can represent not only the real, physical flows but also the flow of information. Thus the functional relationship between the variables - profit, supply, demand, shortage and return are represented in the graph by single arrowed arcs.

To create the model the user switches 'I think' into the modelling mode which allows him to pull the primitives in directly by the mouse from the available menu. The primitives are connected by direct manipulation automatically ensuring consistency.

After the model is drawn the mode is switched into 'calculate', enabling the user to specify all functional relationships. An extensive menu of functions is available to accomplish this stage.

To give an example, there is a logical function controlling the valve 'supply' which states that supply equals demand providing there is enough fuel in the tank. If not, then supply can only reach the volume available in the tank:

supply = if (tank>demand) then (demand) else (tank)

When the mouse clicks on the 'supply' valve icon the programme offers a window which allows the user to input the relevant function such as the function shown above. The window also shows the library from which the function can be constructed.

Below is shown the programme listing which describes the model.

```
custom(t) = custom(t - dt) + (supply) * dt
INIT custom = 0

INFLOWS:
supply = if(tank>demand)THEN(demand)ELSE(tank)
tank(t) = tank(t - dt) + (delivery - supply - return) * dt
INIT tank = 0

INFLOWS:
delivery = 4000
OUTFLOWS:
supply = if(tank>demand)THEN(demand)ELSE(tank)
return = IF(tank>5500)THEN(tank-5500)ELSE(0)
tanker(t) = tanker(t - dt) + (return - delivery) * dt
INIT tanker = 3500*30

INFLOWS:
return = IF(tank>5500)THEN(tank-5500)ELSE(0)
OUTFLOWS:
delivery = 4000
demand = random(2,6)*1000
profit = 0.1*supply-0.12*short-0.05*return
short = demand-supply
```

After the model is constructed it is switched into the run mode to obtain the results corresponding to the input data set before. The results may be represented either by graphs or shown in tables.

A table of a short sample of results is shown next.

Months	tank	demand	supply	return	short	profit
Initial	0.00	4743.74			4743.74	-569.25
1	4000.00	3973.05	0.00	0.00	0.00	397.31
2	4026.95	3083.80	3973.05	0.00	0.00	308.38
3	4943.15	5432.82	3083.80	0.00	489.68	435.55
4	4000.00	5488.91	4943.15	0.00	1488.91	221.33
5	4000.00	4111.82	4000.00	0.00	111.82	386.58
6	4000.00	3342.18	4000.00	0.00	0.00	334.22
7	4657.82	4087.15	3342.18	0.00	0.00	408.72
8	4570.66	4749.66	4087.15	0.00	179.00	435.59
9	4000.00	3605.61	4570.66	0.00	0.00	360.56

4. Conclusions

More and more powerful tools are available to construct a DSS. Although the situation is improving for the professional domain user continuously, there is still a problem of one tool not providing all the functionality needed for the application in question. For example, the databases have extensive facilities for the handling of data but usually have limited modelling facilities. Subsequently it is important that the tools are of an 'open design' type so that they can be connected with each other as desired by the user.

As already stated it is usually possible to find suitable software to build the data base and the model. However, the evaluator is the heart of the DSS system in containing the real decision making part. Such a decision may be made by comparing two or more different scenarios run by the model. The requirements of the evaluator are usually application specific and need to be constructed by the domain expert.

5. References

1. E. Turban: Decision Support and Expert Systems. Maxwell Macmillan int. Edtions, 1990.
2. SAM (Simulation And Multimedia), version report; EC Delta programme (1993)
3. 'I think' Tutorial and Technical Reference. High Performance Systems Inc.(1992)

Man - Machine Interface for CIM

Zdeněk Kouba, Tomáš Vlček
Czech Technical University, Faculty of Electrical Engineering
Technická 2, CZ - 166 27 Prague, Czech Republic

Abstract. The aim of this paper is to introduce the progress made by the authors in the development of a methodology for modelling the dependencies among data entries in a dialog box and design of a software tool supporting the user interface programming. An object oriented dialog box model and an event driven user interface architecture are analysed. The principal features of the user interface constructor consisting of a graphical user interface model editor and source code generator are introduced. The formal analysis of dependence among dialog entries [3] is utilised.

1 Introduction

As contrast to well developed data modelling theory (relational algebra, object-oriented data models, ...) the attempts of formal user interface modelling are rare [3,4]. The experience says that the quality of the user interface is one of the most important conditions for success of the particular software application. User interface programming as a part of software developer's activities represents very complicated and tedious work. Those are reasons for development of means which would support the process of user interface design and programming.

Many commercially available programming environments equipped with an interactive user interface design facility already exist. These facilities usually make it possible to design the geometric layout of the data entries on the screen and to specify the validity conditions for particular data entries (e.g. FoxPro's Screen Builder). However, they do not usually take into account the mutual dependencies of several entries. In more complicated user interfaces, the mutual dependencies among entries represent the most difficult part of the programmer's work. Our aim is to describe these mutual dependencies by a model or, more precisely, to design a formalism capable to express mutual dependencies among entries of a general user interface.

The model specifies also events which can arise in the course of the interactive entering data into the dialog box by the user.

Having such a model and the model based user interface constructor the user will interactively define code snippets (chunks of code) for handling particular events to achieve the required behaviour of the user interface. In the course of code generation these code snippets will be bind by the MBUIC into resulting source code.

2 Motivation

Fig. 1 represents schematically an example dialog box of a hypothetical information system on telecommunication lines. The dialog consist of a push button *ACCEPT* and five dialog entries *Line Ident, Transmission Means, Quality* and *Transmission Speed*.

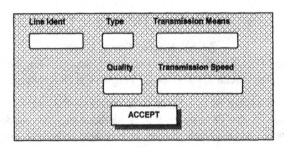

Figure 1. Dialog box schematic diagram

The *Line Ident* entry identifies uniquely the particular telecommunication line. The other entries define the parameters of the line. These four lines are mutually dependent in the sense that not all combinations of their values are allowed.

Let the *Type* entry can have one of the values **RP** (line for transmission of radio programme) or **DP** (data transmission). The *Transmission Means* entry can be either **TF** (telephone equipment) or **TG** (telegraph equipment). The *Quality* entry can have the value of **T**, **A** or **Q** and defines the width of the frequency band. The meaning of the *Transmission Speed* entry is clear.

The possible combination of values is introduced in the following table. The notice **disabled** means that the given entry must be disabled for the given combination of values.

TYPE	TRANSMISSION MEANS	QUALITY	TRANSMISSION SPEED
RP	TF	T, A, Q	disabled
DP	TF	disabled	1200, 2400, 4800, 9600, 19200
DP	TG	disabled	50, 100, 200

The program handling the dialog must ensure:

- the user is not enabled to fulfil incorrect combination of values of mutually dependent entries

- in case that some of the entries (e.g. *Transmission speed*) is realised as a popup, list etc. (i.e. it offers appropriate set of values for a choice), it must react on the change of another dependent entry (e.g. *Transmission means*) by resetting the set of offered values.

To ensure such a behaviour of the dialog the programmer should implement appropriate programming constructions which are repeated with minor modifications many times in the whole application. This fact motivated the authors to analyse a general behaviour of a user interface with the aim to develop a tool supporting easy and comfortable design and maintenance of the user interfaces.

At the very beginning the authors tried to utilise the well known methods of structured analysis. The analysis of the problem has shown that the classical means of the structural analysis like De Marco's Data Flow Diagrams are not suitable for the purposes of the user interface modelling because even for simple screen dialogs the models are too complex due to its generality. It is the reason why a new, more specialised, modelling methodology is needed.

3 Formal analysis

This section provides the formal analysis of the dependencies among the dialog entries in an user interface and the problems of consistence checking. The stuff introduced in this section is an overview of the recent results published in [3].

3.1 The Dialog

Under the term *dialog* we will understand a collection of data items. Each item represents a variable (of simple or structured type) of our computational process. We will call the item the (dialog) *entry*.

The variable corresponding to an entry can be of various types, it can be an integer or real number, a string or a structured type. Nevertheless, the domain of its values is always defined. In the case of a structured type one value of an entry is considered to be a tuple of simple type values.

3.2 Dependent entries

The formalism used in this section is very similar to that of relational database theory [1,2]. The main difference is the terminology (entry corresponds to attribute, etc.).

Let us denote e_1 some entry of a dialog and let the symbol $\text{Dom}(e_1)$ denote the domain of all possible values of the entry (or variable represented by the entry) e_1.

Let the symbol $\text{Val}(e_1)$ denote a particular value of the entry e_1. It means $\text{Val}(e_1) \in \text{Dom}(e_1)$.

Let $e_1, e_2, \ldots, e_n$ be some entries of a considered application. The symbol $\text{Dom}(e_1, e_2, \ldots, e_n)$ denotes a relation specifying all allowed tuples

$$\langle \text{Val}(e_1), \text{Val}(e_2), \ldots, \text{Val}(e_n) \rangle \in \text{Dom}(e_1) \times \text{Dom}(e_2) \times \ldots \times \text{Dom}(e_n).$$

The entries $e_1, \ldots, e_n$ are **mutually independent** iff

$$\text{Dom}(e_1, e_2, \ldots, e_n) = \text{Dom}(e_1) \times \text{Dom}(e_2) \times \ldots \times \text{Dom}(e_n).$$

An important property of this notion is that it is downward closed w.r. to subsets. In other words if all the entries from a set $E = \{e_1, \ldots, e_n\}$ are mutually independent, then all the entries of any subset $E_1 \subset E$ are mutually independent, too.

The reverse statement does not hold. Consider the following negative example: Let $< x, y, z >$ be triplet of real number entries, let us define

$$\text{Dom}(< x, y, z >) = \{< x, y, z > ; (x = 0) \vee (y = 0) \vee (z = 0)\}.$$

Obviously,

$$\text{Dom}(x) \times \text{Dom}(y) \times \text{Dom}(z) = \mathcal{R} \times \mathcal{R} \times \mathcal{R}.$$

$$\text{Dom}(<x, y, z>) \neq \text{Dom}(x) \times \text{Dom}(y) \times \text{Dom}(z),$$

where $\mathcal{R}$ denotes set of all reals.

However, there holds under the specified conditions

$$\text{Dom}(x,y) = \text{Dom}(y,z) = \text{Dom}(x,z) = \mathcal{R} \times \mathcal{R}.$$

More generally, suppose the entries from a set $E = \{e_1, e_2, \ldots, e_n\}$ are not mutually independent. Nevertheless, there can be a subset $E_1 \subset E$ such that all entries from the set E_1 are mutually independent.

3.3 Data dependencies from the user's point of view

We have mentioned the notion of entry dependencies. It seems to be useful to make a remark on dealing with dependencies from a user's point of view.

Let us take into account that the user can fill in the entries sequentially. He can not change more than one entry at once. The problem of data consistency (given by the dependencies) arises here.

The dependence definition through relation (i.e. subset of a Cartesian product of sets) specifies constraints posted on the final values of entries, which appear in the dependence. This definition is static in some sense. It does not describe how to transit from one point to another, i.e. the feasible sequence of entering the data. The following analysis illustrates the problem in more detail.

Let us have two dependent entries e_1 and e_2.

There are several ways how to solve the dependence of both entries, some of them being introduced here:

Postponed consistency check

The user can fill in both entries freely but the change of their values will be accepted after the global check has been done.

The global check tries to verify that $\text{Val}(e_1, e_2) \in \text{Dom}(e_1, e_2)$. If the check is not successful, the user is informed about inconsistency and the data change will be refused (we do not consider details of refusing the change here).

The postponed consistency check is simple to implement but its disadvantage is that until the consistency check (invoked e.g. by pressing a "push button") the inconsistent values of variables may exist.

Immediate consistency check

The other possibility of implementation of an entry dependence is to ensure consistent contents of both entries at every moment.

Let the user change the value of the entry e_1 and let the entries e_1, e_2 be dependent. After filling in (or changing) the entry e_1, the conditional domain of the entry e_2 denoted as $\text{Dom}(e_2 \mid e_1 = \text{Val}(e_1))$ is evaluated.

If the current value $\text{Val}(e_2)$ belongs to the conditional domain

$$\text{Dom}(e_2 \mid e_1 = \text{Val}(e_1)),$$

both the entries are consistent. In the opposite case the value of e_2 will be automatically changed to a default value $\text{Val}(e_2) = \text{V}_{\text{def},2}(\text{Val}(e_1))$ ensuring

$$\mathrm{Val}(e_2) \in \mathrm{Dom}(e_2 \mid e_1 = \mathrm{Val}(e_1)).$$

The mechanism of immediate consistency check should work in both directions.

If we change the value of e_2 so that $\mathrm{Val}(e_2)$ does not belong to $\mathrm{Dom}(e_2 \mid e_1 = \mathrm{Val}(e_1))$, the mechanism evaluates $\mathrm{Dom}(e_1 \mid e_2 = \mathrm{Val}(e_2))$ and it automatically changes the value of e_1 to satisfy the requirement $\mathrm{Val}(e_1) \in \mathrm{Dom}(e_1 \mid e_2 = \mathrm{Val}(e_2))$. This approach can be generalised to multidimensional dependence, too.

Unidirectional immediate consistency check

Sometimes the semantics of entries e_1, e_2 allows us to choose e_1 to be fully truly independent while e_2 depends on e_1. Let us denote such a situation by the symbol $e_1 \Rightarrow e_2$. A natural way of entering data here is to change e_1 first. In such a case we can implement the dependency of e_1 and e_2 by a mechanism called the unidirectional immediate consistency check.

After the value e_1 has been chosen and the $\mathrm{Dom}(e_2 \mid e_1 = \mathrm{Val}(e_1))$ has been evaluated, the user can choose the value from the set $\mathrm{Dom}(e_2 \mid e_1 = \mathrm{Val}(e_1))$ only. To illustrate this idea, assume that our pair of entries takes the value $\mathrm{Val}(e_1, e_2) = <a, b>$. The user's aim is to change the value to $<c, d>$ which is supposed to be consistent. If $d \in \mathrm{Dom}(e_2 \mid e_1 = a)$, the user is forced to change the entry e_1 to $\mathrm{Val}(e_1) = c$, first.

Very often the entries on the screen have the "tandem dependency structure". It means that for the entries $e_1, e_2, \ldots, e_n$ holds

$$e_{i_1} \Rightarrow e_{i_2} \Rightarrow \ldots \Rightarrow e_{i_n}, \text{ where } i_j = 1, \ldots, n, \, i_j \neq i_k.$$

4   OO - Model

The behaviour of the user interface in its nature is the event driven one. The change of the entry's value invokes an event in the dialog which has to be processed and may issue into the change in other dependent entry/entries. This mechanism of events generating and processing can be understood as message transfer among entries. From this abstraction it is very close to the object oriented model.

This section analyses such an object oriented model of the user interface dialog behaviour. It is assumed that the user designs the interface in a graphical representation having a specialised tool at his disposal. The figure Fig. 2 introduces schematically representation of a model.

The *Model editor* deals with two types of graphical objects representing the interface model elements. The rectangular ones represent the dialog entry objects and define the behaviour of the particular entries of the dialog - e.g. the way of entering the data, the format and transformation function for displaying the data, etc. The circle object is linked with one or more rectangular ones and represents the dependence of corresponding dialog entries. The number of entry objects connected with dependence object denotes the arity of the dependence. In general a dependence object is responsible for the consistency checks introduced above. It evaluates the domain of an entry in the condition given by the values of the remaining entries taking part in the dependence.

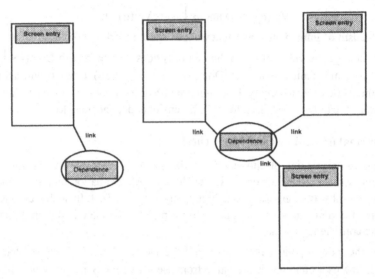

Figure 2. Representation of the model

The user describes the interface entries and their relationships using these two basic graphical objects. The user has at disposal two hierarchies of classes for implementation of the interface model elements - one hierarchy of dialog entry objects and the other one for dependence objects. The roots of both hierarchies are the abstract classes providing the general behaviour. From abstract class Entry the classes for each typical dialog entry like editable field, check box, list etc. are derived. From abstract class Dependence the classes providing the realisation of the various types of dependencies are derived.

It is expected that the user - the designer of the interface - will choose the appropriate class from the corresponding hierarchy. When necessary he derives a new class by overriding and/or adding new methods. He will define the methods in the target programming language. The methods represent code snippets which will be assembled by the interface code generator into the interface source code.

Note: This approach enables the process of the model definition to be of object oriented nature even if the target programming language has no object oriented features.

Internally the user interface program will have the structure shown by Fig. 3. The Entry object has at least the following instance variables:

- Next - the pointer to the next object in the bi-directional list of dialog entry objects
- Previous - the pointer to the previous object in the bi-directional list of dialog entry objects
- Data store - the pointer to the buffer storing the entries' value
- Enabled - variable describing the current state of the entry, i.e. enabled/disabled, visible/invisible, etc.

- Dependence - the pointer to the particular Dependence object or to the Dialog manager in the case of the dialog entry (usually push button terminating the dialog) invoking the postponed consistence check.

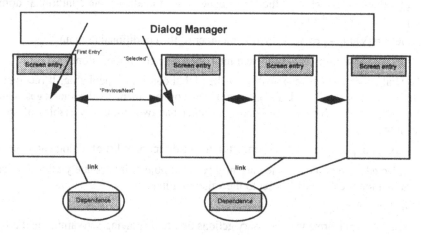

Figure 3. Interface program structure

The Entry object should respond the following messages:
- Select - sets the entry to the "selected" state
- Unselect - sets the entry to the "unselected" state
- Enable - sets the variable Enabled and calls the Display method.
- Display - displays the object in appropriate manner according to its state.
- Transform - transforms the internal representation of the displayed entry into the required form and vice versa, e.g. the string of digits into the number, etc.

From the dependence checking point of view each Entry object responds three important methods:
- When - this method performs the necessary activities when the user attempts to select the given dialog entry. According to the result the selection of the entry is either enabled or disabled.
- Valid - this method performs the necessary activities when the user attempts to leave the selected dialog entry. The main task of this method is to ensure the consistency. This is done by asking corresponding object referred by the instance variable Dependence to perform the consistency check.

The instance variables and methods introduced above are the basic ones, which the abstract class Entry introduces. The derived classes supporting the dialog entries like *List, Combo Box* etc. can have further ones.

Each particular Dependence object provides the services for its related dialog entries. For each object there is a method for consistency checking.

The Dependence object should respond the messages:

- Check - consistency checking

For such objects like lists, pop-ups, combo boxes (i.e. entries which offer a discrete set of values for the choice) the Dependence object evaluates the conditional domain for particular related entries:

- GetFirstValue - provides the first member of the conditional domain.
- GetNextValue - provides the next member of the conditional domain.

The Dialog Manager is responsible for the global control of the dialog. It receives the external events (keyboard, mouse, etc.) and convert them to the messages sent to appropriate entry objects. The Dialog manager has two instance variables of special meaning:

- First - the pointer to the first object in the bi-directional list of dialog entry objects
- Selected - the pointer to the currently selected object. Immediately after the dialog activation the 'Selected' is set to the value of 'First'.

Its methods are:

- Activate - performs the necessary actions before displaying/activating the dialog.
- Deactivate - by analogy - performs deactivation of the dialog.
- Check - performs postponed consistency check by sending the Check message to all Entry objects involved in the list referred by the instance variable First.
- ChangeSelection - attempts to deselect currently selected dialog entry and select the desired one in the following way:
  1. sends the Valid message to the currently selected object. If failed, stops.
  2. sends the When message to the desired object. If failed, stops.
  3. sends the Unselect and Display messages to the currently selected object.
  4. sends the Select and Display messages to the new one.
- NextEntry / PreviousEntry - invokes the ChangeSelection method for the Entry object referred by Next / Previous instance variables of the currently selected Entry object.

5 Object oriented and event driven programming

5.1 The significance of objects

The behaviour of any object on any level of the functional scheme can be expressed by the definition of responses to various events. The events can be generated by the user from the keyboard, mouse etc.

The above mentioned library may contain prototypes of objects. The user may use these prototypes and determine their detailed behaviour by adding new features or by overriding the default ones. The concept of inheritance enables to do that in very uniform, modular and simple way.

5.2 The event driven programming paradigm

What do we understand under the term *event*? Let the computational process be able to recognize a set of changes in the surrounding environment (a mouse click, hitting a key of the keyboard, etc.). These significant changes we call (external) events. They are transformed to software signals. When we speak about events we mean usually just these signals. The events represented by software signals may be generated also internally by the computational process itself, usually as a consequence of a user invoked external event. (We do not consider events caused by hardware interrupts, time slicing, etc.)

The concept of events has its significance in highly modular systems. A system composed of relatively autonomous components (objects in an OOP sense) uses events as the means for communication among those components. That component which recognizes the external event generates the software signal (an event in its metaphorical meaning) thus informing other components about the new situation. It means the system has to be equipped with a signal distribution mechanism. To explain the signal flow we have to define the system components hierarchy.

For clear explanation we can consider the application user interface composed of two dialogs each of them consisting of three entries. This situation may be represented by the hierarchy outlined in Fig. 4.

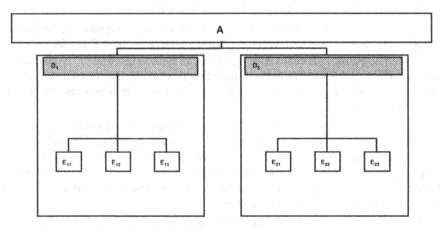

Figure 4. Dialog hierarchy

In Fig.4 the box **A** represents the entire application, D_1 and D_2 are the dialogs and E_{ij} ($j = 1,2,3$) are entries of the dialog D_i ($i = 1, 2$).

Each box in Fig. 4 will be implemented as an (OOP) object. Notice that some objects are collections of other objects (e.g. D_1 is the collection of E_{11}, E_{12}, E_{13}). We can imagine that there is an event processing routine for each object (object's event handler). The root of the hierarchy tree (object **A**) is the only one which is capable to recognize an external event.

Now we are prepared to discuss the event distribution mechanism in an object hierarchy. At least three event types can be distinguished according to the way of their distribution:

- **broadcasts** are events sent by objects which are collections of other objects to all members of that collection in parallel. Each object can either take an action or ignore the event on its arrival.

- **circulars** are events sent by objects which are again collections of other objects. Collection is organized as a circular list of collection members ordered according to actually valid internal linkage. There is always one *current* member within the collection (e.g. selected by the user). In contrast to the broadcast, the circular is sent to a collection of members sequentially starting with the current member. Each member of the collection may react to the event and one of them may even erase the circular event. If an object erases the event, its successors in the collection will not receive that event at all[1]. Notice, that here the ordering of the members of the collection starts to play the role of a priority chain.

- a **message** is an event sent by one object to another explicitly specified one regardless to their positions in the object hierarchy.

Internally the event is a packet consisting of several sections, namely:

- event distribution type (broadcast, circular, or message)
- sender's identification
- receiver's identification (if applicable)
- data section (carrying the detailed description of the event)

At any moment, the user can fill in at most one entry which is represented by the leaf of the hierarchy tree. This entry will be called the **selected entry**[2]. Consider now only those external events connected with the data ntering actions. Thus it is possible to think that the external event (e.g. keyboard hit) is detected by the selected entry. Very often the selected entry itself can execute the external event in which case there is no reason to distribute the event anywhere.

Occasionally, the user presses a special key which should be interpreted as some more global action in the system. The simplest interpretation means that editing of the selected entry has been finished. Usually, this event cannot be processed by the selected entry itself because for instance some consistency checks are to be done. In this case, the selected entry generates an internal event and sends it as a message to the predecessor object in the hierarchy tree.

The addressed object processes the event in an analoguous way. It tries to process the event itself. If the trial is not successful the event is sent higher in the hierarchy, again. If the event arrives at the hierarchy root and the root is not able to process it, the event will be lost (e.g. hitting a special key without any meaning for this application).

When the object recognizes the event, it can be processed in several ways according to the meaning of the event's data section (see above):

---

[1] The parent object can recognize the erasure.

[2] Selected entry is the leaf in the hierarchy tree. There exists a unique path from the root to this leaf. All objects on this path are **selected objects**.

- The event can be processed fully within the object, in which case no more events are generated (e.g. dialog window resizing).

- If the above statement does not hold, the event is sent as a circular to the object's childern. The selected object (i.e. any object on the path from the root to that one which has started the entire event transaction) always ignores the circular without erasing it.

- If the circular comes back (not being erased by any child) to its sender, the event will be passed upwards in the hierarchy as mentioned earlier. This case may be considered as a part of the test whether the object can process the event or not.

Sometimes it may be advantageous to constrain the upward propagation of certain events. It is possible to declare some object in the hierarchy to be *modal* for a set of events. The modality means that the object is in a special mode which ensures that events from the specified set are never passed higher in the tree. This mechanism can be used to ensure the completeness of user's actions. The alert dialogue window offering several choices may serve as an example. The window cannot be left without selecting one of several admissible answers. Thus the alert dialogue window will be modal for all events excluding the event "correct answer selected".

The above described event handling can be considered as a default one. Since each object has its own event handler, it is possible to change the default behaviour and thus achieve the desired event processing.

For instance if some object - a collection member - wants to inform all its peers in the collection, it sends a message to its parent. The parent object converts the message into a broadcast. This is the method how objects on the same hierarchy level can get information about the events within the entire dialog.

6 Conclusion

The concepts of event driven programming and object oriented programming can form a unified and elegant programming style.

The concept of data consistency checks is in agreement with the event driven programming approach. The immediate consistency check may be implemented as a method of the object representing the entry. After entering new data, this method informs other entries about the change by sending a specific event. The dependent entries react to this event by changing their conditional domain and by testing their actual values. Any discovered inconsistency may result in a new event.

The complete design of the dialog in a user interface consists of four stages:

1. The design of the dialog geometrical layout
2. The specification of the dependencies among dialog entries
3. Definition of the supporting code snippets (methods of the entry objects)
4. Target source code generation

The only stage two is independent on graphical environment and programming environment. This is the reason why it is necessary to develop both the library of various dialog entry objects making a bridge between the MBUIC and the given graphical and programming environment and the target code generator.

References

1. Alagić S.: Object-Oriented Database Programming, Springer Verlag, New York, 1988
2. Gardarin G., Valduriez P.: Relational Databases and Knowledge Bases, Addison-Wesley, 1989
3. Kouba Z., Lažanský J., Mařík Vl., Štěpánková O., Vlček T.: Model Based User Interface Constructor for CIM. In: Knowledge Based Hybrid Systems, IFIP Transactions, Elsevier Science Publishers B.V., North-Holland, Amsterdam, 1993
4. Pree W.: Object-Oriented Versus Conventional Construction of User Interface Prototyping Tools, VWG™, Wien, 1992

Designing for evolutionary systems

Jan Hensgens

Research Institute for Knowledge Systems (RIKS),
P.O.Box 463, 6200 AL Maastricht, NL
Email: hensgens@riks.nl

Abstract. The main objective in human-computer interaction design is the development of user-centred systems. Command-based interfaces did meet this objective as long as the users were programmers. But nowadays the group of users is more diversified, the computer has penetrated other professional fields influencing the jobs concerned. This process puts new requirements to interface techniques and methodologies to regain user-centredness. Traditional methodologies appear to be inadequate to prevent all kinds of user problems. Moreover when user problems occur most systems are already finalised. This makes major changes in interface and/or functionality impracticable. Alternative approaches based on rapid prototyping and iterative design must deliver systems open for future adjustments. Users and tasks change whenever new systems are used for longer periods. This chapter describes in more detail an intelligent front-end to an existing program, an evolutionary help system and flexible and adaptable interface tools.

1 Introduction

During the last two decades information technology has changed heavily. The mainframe with its connected terminals lost its pivotal position. The era of the personal computer started and decentralised computing facilities have become available for a broader public. The group of users increased dramatically and these users started to use computers for their own jobs. A lot of effort has been put in developing new software to be used for their daily work. Often these applications were developed in isolated conditions. The functionality of the system has been the main objective [21] and many interface problems have been solved in a pragmatic way. Important decisions concerning the user system dialogue are taken on a local level. These solutions conflict often with a more global design philosophy. As a result of this a lot of these systems lack real usability.

Since the early eighties major progress has been made in the area of human computer interaction. Direct manipulation [22] techniques and the occurrence of graphical user interfaces enable the development of applications better usable for non-experts. Existing applications often needed a complete reimplementation, because the interface statements appeared to be spread over the whole source code. However in technical domains reimplementation could not always be considered. Some simulation programs developed with FORTRAN took years of development time and proved their effectiveness during their long-standing use. This leads to the paradoxical situation that some still used but old applications combine an outstanding functionality with a very poor man machine interaction. The life cycle of these programs expands into the present days with its user friendly driven system approach. Other solutions for updating their interfaces have to be found to prevent these programs to wait for an uncertain reimplementation in the far future. This updating now can be done by the creation of new front-ends, and/or back-ends to the existing application. These must be able to catch old interface events, to interpret them and to transform them into an updated look and feel.

The examples used in this chapter are taken from the CHEMSIM[1] project [11]. The project concerned an interface updating process for an existing simulation program. The project resulted in a methodology and a toolkit to build (intelligent) front-ends and interfaces. This project was done for an external client and it appeared that we could not get access to real users. Our strategy in this situation was to develop flexible and adaptable interfaces, which can be configured on-line by the users of the toolkit or by the end-users themselves. We explicitly choose adaptable systems because of the acknowledgement that end-users are heterogeneous [3]. Firstly they differ on levels of psycho-motoric skills, capabilities, learning abilities, understanding, expectations, motives, preferences, cognitive strategies and abilities. Secondly these individual differences will change in time and can be influenced by temporal and contextual situations. It is important to recognise that using the system changes users, and as they change, they start to use the system in new ways that are impossible to forecast completely [14]. So we believe that new systems should be designed with evolutionary potentials. We hope that the flexible and adaptable interface tools we have implemented will contribute to extend system's life time, to enlarge the group of users and to satisfy individual needs.

The paragraph *Background* gives a short description of some developments, problems and used methodologies in the field of human computer interaction. The way the intelligent front-end for CHEMSIM is designed by applying common interface techniques can be found in the paragraphs *Analysis of the task, user and environment* and *Detailed task analysis*. In *New tools to build and tune systems* a generalisation step is made to find a more generic approach and to specify reusable tools and interface objects. Some of these are described in more detail. Finally in *Conclusions* some provisory conclusions on our work are drawn.

2 Background

Nowadays many computer users are not experienced programmers as was mostly the case in the past. However often they appear to be real experts in their own domain. They do not want to learn the use of computers, they just want to use computers to do their job in a better way. Unfortunately this is only possible with a minimal knowledge of the computer and/or application. Carroll and Rosson [4] call this the paradox of the active user. Direct manipulation techniques improve this situation for the users. Within a very limited time novice users can start working with such application. An interface using direct manipulation gives users the impression that they are working directly with their domain objects. This bridges the gaps of execution and evaluation for the user [16] and the system will virtually disappear from the user's consciousness so that all energy can be directed to his/her work.

---

[1] The application program name CHEMSIM is fictitious . The real application is kept confidential at the request of our client.

With the introduction of personal computers large groups of new users are entering the computing domain and with them new requirements are emerging. Direct manipulation and graphical interfaces are replacing command based interfaces and new types of applications are being built. The old generation of applications implemented mainly procedural tasks, e.g. banking, administration, or mathematical models e.g. oil winning, weather forecasting. With the computerisation of more complex tasks these new systems go far beyond the stage of simple tools. They are penetrating the more creative and design-like tasks. Their interfaces are designed so that these systems are able to take the role of a supportive assistant to their users. This trend implies that more intelligence must be built into this new generation of systems. The knowledge can also be used to put active comments on the user's work. Their strategies, methods and progress with possibly encountered misconceptions or violations on domain based constraints can be commented as is done in critiquing systems like LISP-CRITIC, FRAMER and JANUS [6].

Most users work with different applications at the same time like word processors, drawing packages, information retrieval systems etc. to do their jobs. This puts a strong emphasis on standardising user interfaces for applications running on the same platform. The idea behind standardisation is the potentially positive transfer of knowledge and competence for experienced users crossing different applications. This is especially important when parts of the user's communication are automated and the declarative knowledge has been transformed into procedural knowledge [1]. However, automated actions imply also that the actors are no longer conscious of what they are doing. When interface actions now appear to be inconsistent these experienced users will run into problems frequently.

Standardisation is only partly a side-effect of the introduction of interface toolkits which support programmers with the creation of windows, menus, widgets etc. An example of such toolkit running on the UNIX platform is OSF/Motif [17]. This OSF/Motif, being an open standard layered on top of X-windows [24] appears to become defacto standard in industry. Experiences have shown that choosing for a standard toolkit still does not guarantee a consistent look and feel for the delivered systems. It is necessary to specify and follow additional guidelines and house styles, which integrate a more global design philosophy. Apple was one of the first companies postulating such guidelines [25]. SUN installed a multi-disciplinary team to develop ergonomic specifications and guidelines. This team integrated a clear design philosophy and high-level building blocks in the toolkit OpenLook [18]. In our examples we will see how such high-level building blocks with an explicit behaviour [19] are implemented and used.

The growing importance of the interface is widely accepted. The introduction of visible objects on the screen, the graphical domain representation, the improvement of help systems and manuals, the use of hypertext, the task oriented development approaches and the automatic consistency checking are good instruments to develop better interfaces. But the price to be paid for these improvements appears to be very high. The quantity of the source code grows explosively whenever a system is ported to a windowing environment. This trend can be shown with the "Hello World" example. It is the first exercise for novices in C-programming and its implementation results in a minimal program [12]. The same exercise in a windowing environment can not be implemented anymore as one-liner, the source code requires pages of coding to instantiate the needed window and its interface objects. The development of state-of-the-art interfaces consumes nowadays substantial parts of the limited resources for programming even with the support of existing toolkits. The creation of more high-levelled and reusable building blocks will enable a drastic reduction of the required resources. However developing new interfaces with recent techniques and tools is not yet a guarantee for an easy-to-use system.

Consistent look and feel does not solve all user problems. It is for example no solution for situations where users do not recognise their own work methods, can not use existing domain knowledge or lack necessary information in the presented interface. In that case the acceptance of a new system will be endangered and at the best a long-lasting and problematic use will be established. Such sub optimal system can be used, because human beings appear to be extremely flexible. Even badly designed interfaces can be learned and their procedures can be automated. Expert users tend to forget this problematic use. This tendency covers up most problems of bad design. However some of these interface problems happen to reappear, when the derived problems, like high rates of mistakes and errors, high rates of illness or a reduction of motivation, are analysed.

Good design starts with the imperative rules: know the users, know their tasks and know their environment. This will often be concretised into a formal task analysis. Two kinds of task analyses are needed to guide system design: the analysis of the existing task and procedures of the user and the analysis of the tasks required for operating the proposed system [2]. It could be important to point out strength and weaknesses of humans and computers to find a reasonable match between application and the user's own capabilities. This is called task allocation. Wrong decisions taken during this process of task allocation result into sub optimal human computer co-operation. This leads to unwanted and sometimes even dangerous situations. Imagine what could happen in a nuclear power plant when the monitoring system goes down and the attention of the operators is affected by automatic monitoring which leaves only the boring tasks to humans.

Good systems will have an evolutionary character during their development and use. Users and tasks usually start to change by an introduction of new systems provoking additional and unforeseen user requirements [14]. The best systems are delivered when the developing team succeeds in organising user feedback from the early start of the system's life cycle e.g. the Olympic Message System developed for the Olympic Games in Los Angeles [9]. An active organisation of feedback or user testing helps designers to take correct design decisions. Testing is especially useful when trade-offs e.g. speed versus memorisation are encountered. This formative evaluation can be done with limited numbers of typical users. Nielsen and Landauer [15] concluded that three to five test users for each round of testing is optimal. The broad expertise needed for a good design can only be found within inter-disciplinary design teams. These teams search for workable and not necessarily the best solutions. Designers are used to work with uncertainty and it is a good practice to test more alternatives of (parts of) the system with typical users. The best policy is to build the interface as a module separated from the real application. Separability gives the designer a technical advantage and offers also a management advantage [13]. It enables parallel development of interface and application and it minimises the effects on the application whenever in the future its interface is adapted to additional user requirements or updated to new interface techniques.

3. Analysis of the task, user and environment

The development of new systems starts with a definition of the context by analysing the task, the users and their environment. In our case users are expert chemical engineers who use the concerning simulation program CHEMSIM on a casual and individual base. They use the simulation to predict effects on the production, whenever big changes in the plant are to be implemented. This happens maybe once in the two or three years. CHEMSIM belongs to a previous generation of technical programs (written in FORTRAN) and it receives input from plain, editable ASCII-files. The set of parameters is rather big ($\pm$ 60), the format of its values varies from alphanumeric to hexadecimal, some parameters are heavily interrelated with other parameters and the connection between value and parameter is made based on value's position in the file. The file is used to initialise the simulation which runs for at least two hours on a different platform namely a Cray. There is no user interaction at all at execution time. It appears that users mostly use several runs. They needed the output of previous runs to check the correctness and validity of the input.

The clumsy input and the absence of interaction with users at execution time limit the usability of CHEMSIM, but it does not affect the excellent performance of CHEMSIM as simulation program. Redesign and reimplementation are not considered because of its proved effectiveness, the small group of highly qualified users, the low frequency of use and the high complexity of its algorithms. Interviews done by our client himself with known users (published in a confidential report) showed additional user requirements to improve its usability. They wanted the system
- to provide more freedom to specify parameter values in any order,
- to support an automatic check on validity,
- to use graphical domain pictures and
- to give direct access to a help-system and a technical glossary.

Within the depicted context the only acceptable and practical solution, is adding a knowledge-based front end to the existing conventional system. In this special case it is not necessary to define the Application Programmer Interface (API) between front-end and application. It suffices to deliver a new stand-alone system that generates the required ASCII-file. This generated file will be suitable to serve as data entry for the batch processing of CHEMSIM. By an electronic transfer the unchanged output is sent from the front end to the simulation program on its executing platform. The input specification of CHEMSIM is still done by using plain text-files and it is kept untouched. This is the best guarantee that the creation of the front end and all its changes in the future will not affect the application. The front end remains completely independent of the application itself.

Be aware of the possible existence of additional constraints. In our example the client could not provide access to real users, they were not able to generate a complete list of interdependency rules for the different parameters and the delivered front-end should be maintainable for them in a C-based programming environment. Before we encountered these additional constraints the first phase was already concluded with two prototypes. These were implemented in object oriented Lisp (CLOS) and Prolog and proved the feasibility of our approach. However the new emerging constraints made it impossible to end up with a finalised front-end. We agreed to adjust the project from the specific application CHEMSIM to more general interface methods and techniques. The project delivers some reusable results, which can be used for an interface update process for other existing programs. The client owns more programs working with such input specification where reimplementation is not realistic for the near future. The research objective became the design and implementation of higher levelled interface objects and the development of techniques to be used for these objects.

4 Detailed task analysis

The first activity of the design process of the interface is a task analysis of the usual activities carried out by users in their own environment [20]. The global task for a CHEMSIM user is to specify the input or more precisely to deliver a complete and useful parameter set. In an internal report of the client is stated that "Even experienced users do not know what the program can and can not do, they use a trial and error approach and they always check the detailed output of the program to see if it did calculate what they wanted." (Internal Report [2], 1989). This illustrates that this user task happens to be very difficult with its large set of parameters, varying formats, positioning in file and the numerous and undocumented interdependencies between the parameters. The present procedure for the users is decomposed into sub-tasks. This decomposition delivers a flat task tree, where the task consists of a *fixed sequence* of sub-tasks. Each sub-task is a filling operation for a specific parameter. The global task is finished as soon as the end of the sequence is reached.

Because the interface with CHEMSIM is minimal (ASCII-file) all complexity to ensure the validity of the specified parameter set etc. is left for the user. Such division of labour between human and computer must be expected to be the root for many problems. In concreto the user must be aware of all interdependency rules between the different parameters, he/she must use these rules and must keep track of all the (side-) effects. When no structured and straightforward procedures support the user, as is the case, the load will be placed on user's working memory with its limited capacity [8]. Overloading this working memory during the execution of a complex task leads to many categories of problems. We analysed the collected user problems (table 1) on a semantical level. Some of these problems appear to be or are made worse by violations of Grice's maxims [10]. These maxims are followed in all successful dialogues between humans and are: (1) Maxim of Relation: Be relevant; (2) Maxim of Quality: Be truthful; (3) Maxim of Quantity: Be informative and (4) Maxim of Manner: Be Clear. We came to the conclusion that the problems could be solved a) by a re-allocation of tasks between user and computer and b) by restructuring the global task.

---

[2] Our client did an investigation about the usability of his program, which we call CHEMSIM. The results are presented in an internal report.

1. the structure of organising input parameters is not documented and lacks an intuitive logic
2. the naming of parameters is inconsistent and no help is provided
3. there are no clues about which input parameters are needed for a specific output option
4. there is no check on the ranges of the parameter value for which the mathematical model is valid
5. there is no information about default values, that are used when a parameter is left empty
6. there is no validation of the parameter set as a whole

Table 1. The semantical user problems found by interviewing the CHEMSIM users.

4.1 Reallocation of tasks

Reallocation of tasks will be done by introducing an intelligent front end supporting the users during their input specification. It provides the user with necessary information like default values of a parameter, its range etc. Domain knowledge in the form of explicit interdependency rules will be used to guide the user through the specification process. The editing process of each parameter will be controlled in the background by changing when necessary the validity range, the default values, the obligatory or optional status etc. dynamically. At the same time the front end watches the progress of the global task and it ensures that the end result forms a valid parameter set. Indeed this supposes that the front end has knowledge about the possible, individual and collective settings of the parameters.

When the domain rules can not be provided in advance it appears to be necessary to offer users an interactive rule editor. Such tool enables the specification of the domain rules in a later stage by end users. An effect of this decision is that we must make a shift in our system conception from delivering finished and closed end products to open ones.

4.2 Restructuring the task

Restructuring the task is done by rebuilding the flat task tree. The new task tree is developed bottom up by grouping parameters in small groups according to their clear meaning in the domain. To take a specific example consider a fluid tank. All parameters in the initial parameter set, that belong to the fluid tank are grouped together in a new subset. These parameters are replaced by a single one labelled with the name of the specific tank. Specifying this subset can be considered as a new well defined sub-task. These sub-tasks are again grouped as long as this can be justified by the domain knowledge - for instance specifying all fluid tanks. The consequence of this method is that different sub-tasks fit quite well within a conceptual model of the domain.

The result of this task restructuring delivers a task tree, where leaves are simple tasks, referring to single parameters and where nodes are complex tasks referring to sets of parameters. This hierarchical tree representation of the task (fig. 1) reflects the simulated domain and structures user actions or sub-tasks. Differences between task tree and parameter tree are fading away. All nodes are mapped directly onto interface objects. Simple tasks are connected with specialised parameter editors. These editors show and update automatically the information of default value, validity range, etc. dynamically determined by the domain rules. Complex tasks are connected to graphical windows showing technical pictures of the domain with buttons giving access to next or previous levels in the tree. The interaction can now be guided by powerful domain models and the graphical representations enable the use of direct manipulation techniques to walk through the domain.

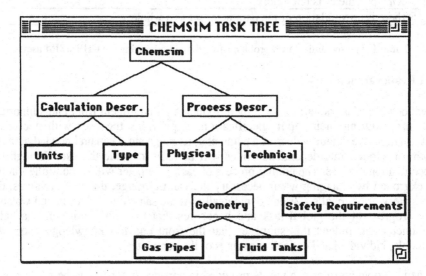

Fig. 1. The top of CHEMSIM's task tree. Complex parameters with hidden nodes are drawn with bold borders.

The internal representation of the parameter tree makes it easy for the system to guide users with their data entry. Domain rules mainly effect the state of the tree and its parameters. The different states of a parameter are forbidden, obligatory and optional. User access to forbidden parameters (simple or complex) is simply denied. However this state can change dynamically when related parameters have been specified forcing the user to follow a certain ordering. Let's look at an example to illustrate this. One rule for CHEMSIM is that calculation description must be completely specified before filling other parameters. The simulation program must know which calculations to produce. At the very start users have only access to parameters belonging to the sub tree Calculation Description. When the calculation description is completed the front end computes the effects on the other parameters using the domain rules. The status of the complex parameter Process Description is changed to obligatory (must be specified) and the status of the process parameters is updated. Those which are not used in the chosen calculation process are set to forbidden and parameters with default values receive the status optional.

Users are allowed to fill in obligatory and optional parameters in any order they want.

The system gives feedback on the status of each parameter in a tree-browser and their linked graphical windows. Updating the state of the parameter tree takes place after each valid entry action. The free order of input makes it difficult for the users to know when the job is finished. An automatic check on completeness is implemented evaluating continuously the progress of the specification process and providing user feedback on this progress.

An early informal user testing showed some important problems with the approach of the levelled graphical windows. Users lost their overview, they disliked the window after window navigation through the tree and there was no logical or intuitive method to provide state feedback. This forces us to look for a complementary graphical representation to the task or domain structure. A text-based tree-browser appears to be able to provide the missing functionality. So the combination of both representations and an easy switch between both was expected to be a workable solution.

The extension with a tree-browser has been evaluated as a real improvement. The tree has several so different viewpoints to its parameters: they can be seen and accessed via the tree browser and the graphical windows. Both representations always refer to the same state of the parameter tree. The next evaluation reveals that mainly the structure of the task tree determines which of the two complementary viewpoints is most usable to provide a clear overview. With a deep structure it is easier in the tree-browser, while in a broad structure the levelled graphical windows are superior. The graphical windows improve user guidance by exploiting existing domain knowledge and give more context information which is useful for the help facility. At the other side the tree browser provides direct access to all parameters at once and integrates the dynamic feedback on the state of the parameters and on the global task progress in a better way.

5. New tools to build and tune systems

We have seen in paragraph 3.2 *restructuring the task* that domain model, task tree and interface objects are becoming intertwined (table 2). The regrouping of parameters in a single tree is done according to the viewpoint of the engineers working with the simulation. Labelling of complex nodes and their connection with technical pictures are taken from the domain. Whenever users access a complex parameter or sub task a graphical window containing the connected picture pops up with its navigation buttons. These nodes are considered to be sub tasks of a global specification process.

The user can choose the wanted abstraction level by walking through the parameter tree which is equivalent to "zooming" actions. In the interface different viewpoints of the tree are supported. Its parameters can be seen and accessed via the different levelled pictures or via the tree browser. Both representations refer to the same state of the parameter tree. The parameter editor for "Gas Inlet" can be called by selecting and clicking on the so-called item in the tree-browser or by pushing the so-labelled button on the technical picture.

Domain Model	Task Model	Interface Objects
CHEMSIM	Parameter Tree (Task Tree)	Overview - Tree Browser - Graphical Windows
Domain Objects	Complex Paras (Nodes)	Technical Pictures
Parameters	Simple Paras (Leaves)	Input Fields - Text Field - Scale - Numeric Field

Table 2. The connection between domain model, parameter tree and interface objects.

5.1 User Interface Generator

The mechanism of building a task or parameter tree and its connection to interface objects is generic. The developed task tree determines completely the interface. Interesting features of this mechanism are: a) generating default interfaces is made possible and b) building more task trees on top of the same parameter set enables an easy adjustment to users with different backgrounds. Generating default interfaces appears to be very useful for our project. It provides a solution for the problem of not having access to real users. At least if a functionality of tuning initial interfaces by end users is provided.

The implemented user interface generator works with rather simple procedures to generate the initial interface. It starts with searching for a parameter tree. If one is found the generator walks through the tree and checks the complex parameters on having their own pictures. A complex parameter without picture receives an empty picture with buttons for all its children. At a later stage users can draw or edit the pictures interactively, they can also select scanned photos and they can drag the navigation buttons to appropriate positions. If there is no parameter tree, users can build one using the provided tree editor and again a default interface can be generated and tuned.

The transition from finished products to evolutionary products (open for changing) has some fundamental consequences. The user group is more diversified. The user can be a real end user or a member of the supporting staff, who must generate and tune the interface within the context of the end users. The imperative rules of knowing the user, the task and the environment are of limited use to us because of the reintroduction of a large distance to the group of real users. At the same time we lack the context needed for the creation of a usable help systems. This forces us to extend our system. An evolutionary help system which can be filled interactively at later stages is developed. At the same time specialised windows are introduced which have flexible and adaptable interfaces. These windows can also be tuned at later stages to find reasonable matches with the task, user's background and/or preferences.

5.2 Evolutionary Help System

One of the most pronounced user requirements for a computer system or application is the availability of a help facility. User observation nevertheless reveals that help systems are rarely used in practice. Most of the help systems appear to be not usable for their intended audience. Reasonable explanations for this paradox are the low priority of help systems in the software design and more important the persons filling it, mostly programmers, are not familiar with the application domain. This makes it problematic to define help needs and to formulate the content. Besides this help needs evolve with the user's knowledge about the system and with the changes in the domain itself. RIKS Help System supports an on-line specification of help messages and an evolutionary approach of the complete help system.

The RIKS help system is a context dependent system. It is recursive and supports different abstraction levels. It can contain textual as well as graphical help and it has hyper text links into the technical glossary. The system can be used as a stand alone application or as a sub system in other applications. The sub system can be connected with applications implemented in C-based environments. This can be done by inserting a few lines of code to instantiate the system, to define context and actual call for the provided help function.

The help system can be used as a tool to build or change its content interactively without any coding. The actual writing of the texts can be left to domain specialists, who are more suited to do the job. Within specific contexts for instance in the graphical window showing the technical picture of a central heating system the help can be called, which, if there is no help available, creates a new link. Whether professional writers, domain experts or end users, they are supported by the presented context as it is shown on the screen. This will heavily reduce the writer's cognitive load. An additional advantage is that filling the help system for an application can be done separately from its implementation. In addition the help can easily be updated with direct feedback from end users which will improve the quality.

However, enabling on-line users to change the content of a help system is not without risks. The distributed ownership will complicate the management of this system. The quality of the help messages requires an additional option to protect (parts) of the help system for unwanted changes. Several methods are available to do this. The ability to change can be restricted to authorised users. The help system is built using the flexible and adaptable RIKS Windows (see next paragraph). The edit facility now can be distributed by using the adaptation mechanism which the help windows inherit from the RIKS Window. All users and user groups receive their own personalised configuration determining their authorisation state. System managers have an additional option namely to set protection levels within the file system (e.g. UNIX system). If a concerned file is write protected, the system recognises this and will not provide edit actions. Write protection enables a temporary freezing of the most evolved parts of the help system.

5.3 Flexible and adaptable windows

Actual end users, their tasks and environmental characteristics can not guide the design of generic interface tools. However each new interface needs a final tuning. This must and can only be done, when required knowledge about user, task and environment becomes available. The same holds for personal preferences whenever these exist for (a group of) users. Postponing these design decisions about communication styles is inevitable. So we want to have a tool that supports this postponing and which supports a flexible interface to be customised interactively. In some cases even end users must be allowed to do this. These objectives are solved and implemented by the creation of RIKS Window as interface object.

RIKS Window has been designed as a specific widget based on top of the OSF/Motif hierarchy. It is implemented within the object oriented paradigm. It can be instantiated (used as it is) and the inheritance mechanisms enable the development of more specialised widgets like a help window, bitmap displayer, glossary window etc. In addition building applications with RIKS Windows supports the consistency. All used windows share the same behaviour. Part of this behaviour is the provided global setting of the communication dialogue. RIKS Windows have *set up* and *help* as standard actions. Programmers can put additional actions in a description list. This list will be used during the instantiation for an automatic generation of the three different menus, e.g. popup menu, menubar and control area. The specific configuration for a window (see below) determines which of these menus will be presented to the user. *Set up* allows the user to change the look and feel of the window and *help* provides context dependent help information.

The settings, e.g. the dialogue style, can be changed interactively in a separate popup window, which presents the items of the setting as a list of toggle buttons. This popup window is called a Property Window and is linked to each particular subclass of RIKS Window. The user can customise the look and feel to a preferred interaction style just by changing the state of these toggle buttons. At this time three different interaction styles are supported: a pulldown menu in a menubar, a popup menu and an always visible control area. By default the menubar is disabled. A possible connection of the class RIKS Window with a separate language processing module (e.g. parser, interpreter, etc.) could provide the user with a natural language input facility which will be propagated to all its different subclasses.

The user can choose any combination of aforementioned menus and this setting or configuration can be saved for the following runs. The choice made depends on the task domain or on personal preferences. However it is recommended to take only one method per window for consistency reasons. If more menus are available at the same time, the user is forced to make an extra choice during execution time: not only which action, but also where this action can be found. The menu system itself deals with the consistency over these different menu styles. The menus can be made context dependent. This will be done by sending hide and show messages to menu items in the description list.

Changing the configuration forces all instances of the same subclass to update their behaviour. There is also an option to specify on the fly the behaviour of all RIKS

Windows belonging to one specific application. This must be done in a specialised Property Window, called the Application Property Window. This window is always accessible through the *global set up* action in the Property Window of the Property Window. Each subclass of RIKS Windows receives by default the values from the application setting.

Storing the configuration is not only useful to keep interaction consistent in time, but enables also an easy reduction of functionality. This selection of functionality can be made by a designer or in some cases even by an end user. For instance disabling the set up switch for a particular subclass of RIKS Window means that the *set up* action is not available anymore for all its instances. Their shared Property Window can no longer be called directly. Via the Application Property Window the *set up* action of all the RIKS Window classes can be determined. Editing windows have an *edit* switch that can be set in the *off* position. Saved in an *off* state means that editing will be impossible in next session. Restoring the edit facilities in the following session via the property window is not possible anymore. The *edit* switch will not be displayed for users without edit authorisation.

6. Conclusions

The help system, the modularity of the interface and the user interface generator prove the feasibility of the open approach to new systems. Both the content of the help system and the tuning of the initial interface can be changed interactively when a system is already in use. This is an important step in the direction of evolutionary systems. It can lengthen their life cycles. Tuning can be done together with or even by the end users themselves. The time consuming and costly knowledge acquisition can be minimised without endangering the links to the specific domain knowledge. Choices made are not definitive anymore, they can always be changed using the same mechanisms. Different interfaces for the same application can be made in a cost effective manner. At the same time the active involvement of the end users improves their commitment, creates a shared ownership and reduces the acceptability problems of the new system.

Adaptation of the communication style to a specific user, user group and/or task can be done manually by using the specialised property windows. The current limited degree of freedom to select the style out of the three different graphical menus, popup menu, menubar and control area, remains open for future extension. This requires only changing of the behaviour of the main RIKS Window. Manual adaptation does not force designers to implement (different) user models. User modelling appears to be a very tedious job. When the adaptation functionality is left for end users it will contribute to their sense of locus of control, not only for doing a task but also for how this task can be done. Locus of control can be vital at the motivational level and it is the belief or perception that an individual has concerning control over 'self' within specific environments [23].

The implementation of the intelligent front end for CHEMSIM, the supporting RIKS Windows and the Help System is done within the object oriented paradigm. This paradigm proves to be very useful for building user interfaces. We succeeded in the creation of reusable components or so-called software-chips [5]. The user interface generator and the help system are actually built on top of the building blocks as they are produced for the intelligent front end. The created reusable blocks [19] cover a broad range from simple widgets (knobs, dials, active pictures etc.), more complex widgets (parameter editors, tree browser, tree editor etc.) to the level of complete sub systems (help system). The inheritance mechanism of object-oriented languages enable a fast and easy specialisation, while the complex interface behaviour linked to the design philosophy is kept intact. The reuse of RIKS Windows via instantiation and specialisation delivers a flexible application with a consistent look and feel enabling a postponed user adaptation.

Future research will investigate the limits of our tools and will define what kind of interfaces can use this approach of user interface generation and adaptation. An extension is to provide on-line support for programmers and designers during interface design. This should supply context dependent and interactive access to psychological principles, guidelines, standards and specific house styles. It would be even better if the design knowledge could be implemented in an active critiquing system [6]: a design assistant.

7. References

1. J.R. Anderson: Skill Acquisition: Compilation of Weak-Method Problem Solutions. *Psychological Review, 94*, 192-210 (1987).
2. C.M. Brown: *Human-Computer Interface Guidelines*. Norwood, NJ: Ablex (1988).
3. D. Browne, M. Norman, D. Riches: Why build adaptive systems? In D. Browne, P. Totterdell, M. Norman (eds.): *Adaptive User Interfaces*. London: Academic Press (1990).
4. J.M. Carroll, M.B. Rosson: Paradox of the active user. In: J.M. Carroll (ed.): *Interfacing Thought*. Cambridge, MA: MIT Press (1987).
5. B.J. Cox: *Object Oriented Programming. An Evolutionary Approach*. Reading, MA: Addisson-Wesley (1987).
6. G. Fisher, K. Nakakoji, J. Ostwald, G. Stahl, T. Summer: Embedding computer-based critics in the contexts of design. *Proceedings of the ACM INTERCHI '93 conference*, Amsterdam, 157-164 (1993).
7. J.D. Foley, V.L. Wallace, P. Chan: The human factors of computer graphics interaction techniques. *IEEE Computer Graphics and Applications, 4*, 13-18 (1984).
8. M.M. Gardiner, B. Christie (eds.): Applying cognitive psychology to user-interface design. Chichester, England: Wiley (1987).
9. J.D. Gould, S.J. Boies, S. Levy, J.T. Richards, Schoonard: The 1984 Olympic Message System: A test of behavioral principles of system design. *Communications of the ACM, 30*, 758-769 (1987).
10. H.P. Grice: Logic and Conversation. In: P. Cole, J.L. Morgan (Eds.): *Studies in syntax. Vol. 3*. New York: Academic Press (1975).
11. J. Hensgens, P. Letanoux: Can complex conventional programs be revitalized? *Knowledge Based Systems, 6*, 17-23 (1993).
12. B.W. Kernighan, D.M. Ritchie: *The C programming language*. London: Prentice-Hall International (1987).
13. W.M. Newman, E.A. Edmonds: The Separable User Interface: a Conversation.

In: E.A. Edmonds (ed.): *The Separable User Interface*. London: Academic Press (1992).

14. J. Nielsen: The Usability Engineering Life Cycle. *Computer, 25 (3)*, 12-22 (1992).

15. J. Nielsen, T.K. Landauer: A mathematical model of the finding of usability problems. *Proceedings of the ACM INTERCHI '93 conference,* Amsterdam, 206-213 (1993).

16. D.A. Norman: Cognitive engineering. In: D.A. Norman, S.W. Draper (eds.): *User centered system design: New perspectives on human-computer design.* Hillsdale, NJ: Erlbaum (1986).

17. OSF/Motif Release 1.1. Open Software Foundation, Cambridge, MA (1990).

18. Open Look graphical usr interface functional specification. Sun Microsystems and AT&T, Reading, MA: Addisson-Wesley (1989).

19. G.R. Overboom, P.R. Maarleveld, P. Letanoux, J. Hensgens: Object-oriented techniques, and what about operation environments? *Conference Proceedings SCOOP Europe 91*, London: SIGS (1991).

20. R. Rubinstein, H.M. Hersh: Design Philosophy. In: *The Human Factor: Designing Computer Systems for people*. Burklington, MA: Digital Press (1984).

21. B. Shackel: IBM makes usability as important as functionality. *The Computer Journal, 29*, 475-476 (1986).

22. B. Shneiderman: The future of interactive systems and the emergency of direct manipulation. *Behavior and Information Technology, 1*, 237-256 (1982).

23. D. Simes, P.A. Sirsky: Human factors: An exploration of the psychology of human-computer dialogues. In: H.R. Hartson (ed.): Advances in human-computer interaction. Norwood, NJ: Ablex (1985).

24. The X-window System, Version 11, Release 4. Massachusetts Institute of Technology (1989).

25. B. Tognazzini: *The Apple II human interface guidelines*. Cupertino, CA: Apple Computer (1985).

Active Subsystems for CIM Environments

Libor Přeučil

Dept. of Control Engineering, AI Division
Czech Technical University, Prague
Technická 2, CZ 166 27, Prague 6, Czech Republic

Abstract. The aim of this section is to introduce concepts and possible approaches which can be used in a build-up of active components for Computer Integrated Manufacturing (CIM) systems. An interesting class of active elements in CIM are the systems capable of autonomous in an undetermined way in partly known environments and/or under not fully defined working conditions. This requires from the system the abilities of autonomous data acquisition and interpretation, decision making, activity planning, and plan execution, all being dependent on the current state of the environment. As the afore-mentioned properties typically appear in intelligent robots, the class of autonomous self-guided robots belongs to possible CIM applications in material transportation, part delivery, etc. The capabilities desired for these systems are obtained by making-use of AI methods, the basic principles of which are sketched below.

1 Role of Autonomous Systems in CIM

To increase industrial productivity, changes in manufacturing technologies have been, and are expected to be, introduced at all levels of industrial production through the use of computers. The development and implementation of Computer Automated Manufacturing (CAM) leads to the realization of the Computer Integrated Manufacturing (CIM) factory as the final entity. This trend is accompanied, however, by a loss of flexibility to change products quickly and maintain cost-effective operations. This is particulary true for small and mid-volume production companies, using traditional job-shop technologies. In contrast to that, high-volume producers employ highly-automated production lines, which become more cost effective in increasing production volume.

The advent of active subsystems which can more or less autonomously participate in production, and effect the production process, improves the situation for the low- and mid-volume producers by increasing production flexibility in the first place.

This is based on the adaptability of such systems to varying working conditions, where the system's actions are driven by a need to reach the desired goal. It is straightforward at this point, that such capabilities are generally provided by robots.

2 Robot Definition

At this point we should perhaps ask what is a robot, being considered for an active subsystem in CIM? What attributes make a machine a robot irrespective of its application?

From this point of view, *robots* are understood to be machines which have the capability of performing independently complex tasks of a physical and mental nature. To overcome problems of changing situations they must interact to some degree in an intelligent way with the environment, and must be able to store experience once it has been gained.

The basic definition considers a robot as a programmable, multifunction manipulator designed to influence the operating environment through variable programmed actions for the performance of a variety of tasks. For industrial environments this would be most specifically the case of materials, parts, and specialized devices movements in the production process.

2.1 Intelligent Robots and Self-Guided Vehicles

More complex robots may also be equipped with mobility units such as wheels, tracks, and legs which provide the capability to move from point to point (mobile robots and vehicles). They may also have sensors to sense unpredictably changing tasks and other external conditions in the environment, thus enabling them to incorporate this information, to store it, and carry out automatic task reprogramming. From this point of view *intelligent robots* are machines which have the ability to solve problems that require, in addition to manipulative capabilities as defined above, autonomous decision-making and task planning capabilities. This "intelligence" resides in the robot's computer, respectively in the computer software making use of artificial intelligence (AI) methods.

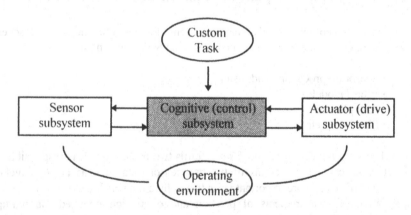

Fig. 1. Common block diagram of an intelligent robot and its links to the environment.

A specific kind of intelligent robot are *self-guided autonomous vehicles* (SGV). Respecting the application field, the common task for this sort of robot is to conduct movements through 2D environments. The robot's goal is typically to *reach a destination* or to *cover a region*. All such activities are closely related to the following functions of the intelligent vehicle.

In the first place there is the navigation task which provides the vehicle's position and orientation at any world location. The self-navigation incorporates data acquisition by various sensor systems and their interpretation with respect to the level of a certain internal knowledge contained within an internal world model (WM). This results in two facts - the vehicle's position and orientation in the 2D world and into hypotheses of the world's properties expressed by the means of obstacle occurrence [1, 3, 4, 6, 11].

Knowing the actual position of the vehicle at any one time, and having previously mentioned knowledge about the environment, the activity planning and plan execution can be resolved. Various approaches, relevant to this sort of problems are introduced in [6, 7, 8, 9, 10].

Generally, both categories of the processes - the world data acquisition and planning - are simultaneously operating functions, where each of them influences the other.

Continuing measurements of the surrounding world and related hypotheses are used to modify the knowledge contained in the world model. This guarantees that the vehicle will always be operating from up-to-date data, and provides the important property of adaptability to changes of the world.

As the WM updating enables the vehicle acting in an partly known and/or dynamic environment, it complicates the processes of activity planning and plan execution. Whenever the vehicle's environment is not completely known in advance before the execution of the planning task, the plan can only be estimated. Afterwards, during the plan execution and using interpreted measurements of the sensor system, the original plan estimate can be (and typically will have to be) modified.

3 Main Functional Components of an Intelligent Mobile Robot

From the basic concept introduced above the system of a global functional block scheme can be derived. One of the possible achitectures of an SGV can consist of:

- Sensor preprocessing and fusion
- World model
- Path planner
- Plan interpreter

The *sensor processing and fusion* fulfils two basic tasks. It is responsible for low level preprocessing operations of the raw sensor data such as noise reduction, dropout detection, etc. More important is it's fusion function which deals with the task of the integration of measurements of position and orientation obtained via multiple measurements and/or sensors. The subsystem concentrates data having the highest reliability and passes it to a local world model. The process of sensor fusion uses - besides knowledge of sensor behaviour - hypotheses on data validity, provided by understanding the content of the local map.

The *world model* as a whole consists of two data-storing parts, the already mentioned *local world map* and the *global world map* and one data-handling system. The related *position location* provides information on the coordinates and orientations of the vehicle via local to global map data matching. As the local map contains temporary data about the vehicle's neighbourhood, the global map contains long-time valid data of the whole work area.

The *scene feature finder* serves the local map data understanding by the means of search for basic elements - features which set up the scene description. This function controls updating of the global and local maps with found scene features and provides adaptability to changes in the environment.

The *path planner* subsystem explores knowledge on the work area structure and topology to solve the task(s) given to the vehicle. The completed path plan is converted by the *plan interpreter* subsystem into a sequence of commands - driving parameters, that feed directly the vehicle drives. The plan interpreter also tightly cooperates with the *collision avoidance* system, so that all the commands during plan execution can be fulfilled. If not, the planner is reactivated and local replanning invoked.

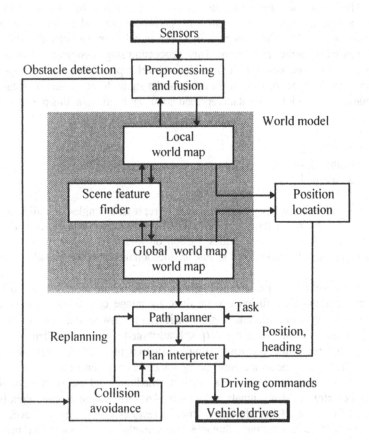

Fig. 2. Principal scheme of the control subsystem in a self-guided vehicle.

3.1 World Data Acquisition and the Sensor Subsystem

The world data acquisition process is the most crucial processing phase in the proposed vehicle structure. As it provides the only links from the real world to the vehicle, it is evident that it can heavily influence the performance of the whole system.

With respect to performance of the used hardware, all the data on the environment can be gathered within certain limits on their quantity and/or rate. The gathering process has to provide sufficient information but must not overload succeeding processing.

The behaviour of the used sensors under various conditions brings another problem. With respect to their physical principles, more or less uncertainty in the measurements can be expected.

These two facts form the main task of the sensor subsystem - *data volume reduction* and *reliability improvement* - by preprocessing and fusion. The incoming information, provided by the sensors, is in a pure signal form, and can be either directly fused, or passed to further processing. This processing stage assigns a semantic context to raw signal data and converts it into higher level representations. The higher level information can later be processed, similarly at the signal level. A useful categorization of possible qualitative levels for data representation can be given at this point:

- Signal level
- Pixel level
- Feature level
- Symbolic level

The *signal level* is a low level one as it usually represents single- or multidimensional signals. Suitable fusion method is signal estimation (filtration) which generally reduces the dispersion of the measured values. Fusion on this level can be used in real-time applications and can be considered as just an additional step in the global processing of signals.

The *pixel level* represents 2D data (images) and therefore is used to fuse this sort of information. As a fusion can be done by image estimation or pixel attribute combination the level of representation is considered for low to medium. An example to this can be a task of fusing edge pixels (pixels attributed with high gradient value), what might serve to create line segments being considered for WM features (see the next example). The fusion process increases performance of image processing tasks.

The *feature level* already represents the medium level and is based on features, which can be extracted from signals and images. An illustrating example might be given as a task of fusion of all straight-line segments (features) with similar direction at one world location (feature attribute). Suitable fusion methods incorporate the methods of geometrical and temporal correspondence and feature attribute combination. In result the fusion reduces processing and increases feature measurement accuracy.

The *symbolic level* is considered for the highest level representation. It can be a symbol with an associated uncertainty measure.

An example of this could be the classification of the result as: "The perceived entity is a box with reliability of 0.85". The fusion making use of logical and statistical inference increases the truth or probability values. Possible combinations of signal-, pixel-, feature-, and symbol-level fusion are shown in Fig. 3.

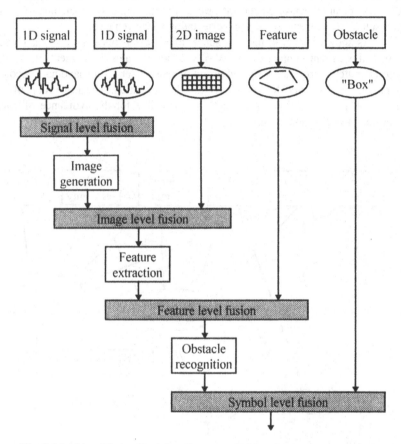

Fig. 3. Multi-level fusion. Each level integrates information of the same nature.

3.2 Data Acquisition Control Strategy

Various input sensors in a multisensor mobile robot are used for different purposes, together setting up a perception task. The perception provides three kinds of outputs. One function is to support the collision avoidance subsystem with the information on region occupancy.

Another function provides the necessary data for the updating of the local map, which contains knowledge of the temporary neighbourhood of the vehicle, accompanied by position and orientation measurements.

The first kind of data - the data on obstacle presence in the vehicle's neighbourhood - is obtained from sensors mounted on the vehicle (range-finder systems, stereo vision, etc.). This sensor system serves two functions of the vehicle: providing data for the world model updating, and supports the location of the vehicle position in the world.

The other kind of data, obtained by the sensor subsystem which uses other kinds of sensors are mostly direct measurements of the vehicle's position and orientation. To obtain this the sensors, such as oedometer, gyro, and landmark beacon systems can be used. Direct position measurements integrated with the range data matching to the global map, result in the current position and orientation of the vehicle.

As shown in Fig. 4, the first subset of these sensors is used to match sensory data with the current content of the WM and then updates the content of the WM to reflect the matching results. A second group of sensors serves for landmark recognition. This process is dedicated for robot position determination. The last group of sensors feeds the obstacle detection part, which is responsible for the avoidance of collisions with obstacles during operation.

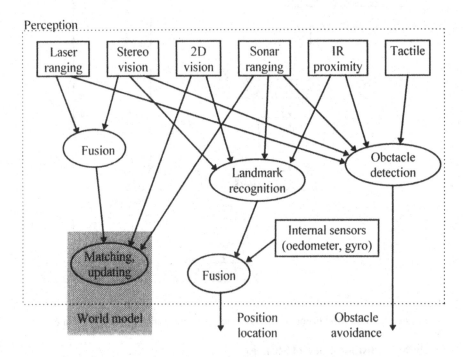

Fig. 4. Typical perception function for navigation purposes of mobile robots.

As the fusion procedure in the given scheme is required at two places, the degree of integration and fusion of sensory data required for each of these tasks can substantially differ.

The simplest approach to fusion for position location is trajectory integration. This calculates the vehicle's position from accumulated rotational and translational motions as determined by internal sensors, such as oedometer and/or gyro. Because of the nature of the always present inaccuracies of any sensors, the total error of location accumulates as the vehicle moves. In order to reduce this cumulative error most mobile robots periodically determine the location of some external landmark. Comparison of the position from the landmark with the one determined by internal sensors therefore keeps the location error within acceptable bounds.

The WM data matching and updating task requires that the sensor information and any associated measure of its uncertainty correspond to the representation used in the world model. Only in that case an integration can take place. Information from different sensors of a similar nature (e.g. laser range finder, stereo vision) can be fused before reaching the matching task, in order to reduce the communication flow, or complexity of the matching process.

3.3 Data Fusion Methods

The methods used for information fusion range from simple statistical approaches to knowledge-based decision making techniques. The choice of a suitable approach depends in the first place on the level of information representation, introduced above, and on a prior knowledge of the sensor response.

Knowing the sensor principle more or less exactly the description of the sensor behaviour - the sensor model - can be built up. Basically, the less accurate the model we have, the more heuristic method has to be used. Unfortunately, in many cases no accurate model is known and therefore sensor behaviour can be described only by typical failure situations. This can be described by heuristic rules, which drive a decision if the measurement has to be refused or accepted.

Signal-level fusion is usually related to processing coming from multiple sensors that are sensing the same entity from the same aspect. The fusion should result in the same form of signals as the original ones but with improved quality. Basically the signals can be seen as random variables corrupted by uncorrelated noise. The fusion can then be considered as an estimation process of the original variable [11]. The most common measure of the signal quality in this case is its variance.

One method of implementing the signal-level fusion is by calculating a *weighted average* of redundant information coming from the sensors. The weights are set up with respect to estimated variances of the signal following the rule: the greater the variance of a certain signal, the lower its weight. This techniques can be also combined with outlying value selection by omitting measurements which are "very far" from the mean value.

If an estimate of signal statistics is known, the *Kalman filter*, instead of the previous method, can be used. Having a linear system with Gaussian error distribution the Kalman filter will provide statistically optimal estimates for the fused data. Besides current estimates of the current values, the Kalman filter also provides a prediction of the future values. Beside the above mentioned main approaches a variety of modifications can be used [12].

Pixel-level fusion is used to improve the information content of 2D data - pixels of an image obtained by a combination of multiple images. A typical example of that is the assignment of depth information to the intensity of images, with respect to their pixels. This can be done, for example, by scene depth reconstruction from two different views (stereo vision) or by utilizing direct range images from a range-finding device. The improvement in data quality via pixel-level fusion can be typically noted through the improvements in performance of the succeeding image processing tasks as segmentation, feature extraction, etc.

The fused image can be created either through pixel-by-pixel fusion or through the fusion of associated local neighbourhoods of corresponding pixels. The former case is mostly used when fusing raw 2D signals. The latter one can already be used for higher-level information representation for fusion of pixel-level feature extraction (e.g. an edge image). It is possible to use many kinds of pixel-level fusion methods, which generally belong to the area of computer vision [12, 15, 16]. Beside these approaches, general methods for data fusion such as the Bayesian approach, Kalman filtering, etc. can be extended to 2D signals as well.

Feature level methods are dedicated for two basic purposes - to increase accuracy and reliability of a feature, and to create additional composite features. A primary feature is created by the assignment of some semantic meaning to the result of processing [13] of a certain portion of sensory data. An example this might again be the case of edge pixel extraction, where each data is assigned its measure of edginess.

Composite features are then created making-use of the previously extracted ones, but bringing a new quality. In our example, as a composite feature can be considered complete edge detection, e.g. every complete edge consists of a set of edge pixels which are connected to neighbours with a similar direction.

Symbol-level methods enable us to put together multiple sensor information on the top level of data abstraction. Basically, the symbol level fusion allows to join information from extremely different sensors, being of very different nature, and the outputs of which can not be easily unified. A model of the world serves for coupling element. The represented data - symbols - originate from two sources:

- From processing of sensory data
- As prior information from a world model

A symbol calculated from the sensory data is matched to the world model content. This can be done by matching features derived from the sensory data to features derived from the world map. The obtained degree of similarity states if the hypothesis on the world state (features derived from sensory data) was a mismatch and/or uncertainty infiltrated from the sensor itself. Frequent successful matches can be used for world model updating.

The other fundamental approaches to data fusion incorporate on this level: Bayesian estimation, Dempster-Shafer evidential reasoning [12], and production rule-based systems with confidence factors [15].

4 Internal World Model

Environment modelling has been pursued through numerous approaches. Generally the task can be viewed from two different perspectives:

- Object recognition
- Navigation

Either way, in both cases the aim is to achieve a useful description of the surrounding world, but the substantial difference is in the level of required abstraction of the model. For navigation purposes it is often sufficient to know whether a specific position is occupied or free, so there is no need to recognize complete objects. Thus, a low level of abstraction may suffice.

One of the simplest representations is the *occupancy (or certainty) grid* which was originally developed especially for use with range sensors (sonars, laser range finders, etc). Occupancy grids represent the space around a robot by an array of cells, in which each cell carries an estimate of confidence that it is occupied. The cells have associated attributes (usually probabilities) labelling them *empty*, *occupied*, and *unknown* [16, 5]. This kind of world representation scheme has, for its simplicity, two basic advantages. It allows very simple fusion of sensory data, because all the sensors mapping the world data to a certain pixel are automatically in spatial correspondence. Besides that, it also permits very simple and efficient path planning methods, based on the potential field principle. On the other hand the representation has high demands on computer memory and therefore it can not efficiently store larger environments.

Mainly due to the mentioned, but crucial disadvantage of the grid-based approach, several higher level abstractions have been suggested. The most common is the *polygonal representation*. The environment is assumed to be represented by a bounded map (a polygon) containing polygonal obstacles. This often considers the world being composed of straight-line segments, which are the primitives approximating the world entities in the WM. It is clear that this kind of information is not on the same level as that which is obtained directly from the sensors. To be able to incorporate some new information about the world into a WM, or else to build it up, the primitives of the world map have to be extracted first.

It has been approved that it is efficient to split the WM into two basic parts - the global and the local maps. Let us describe it for the selected case of the SGV navigation, with a restriction of the further usage of the grid- and polygonal representations of world information.

The *global world model (map)* contains overall data of the vehicle's working area. The global map structure can be derived from a ground plan of the environment, created by relevant line segments (edges), that approximate any obstacles of the real world or their borders, respectively.

If we consider the line segments as graph edges and their joints as nodes, we obtain a planar topological graph representation of the global map, the edges of which can be attributed as:

• Real edge (invisible or visible)
• Virtual edge (always invisible)

and the nodes by their 2D coordinates.

The real edges represent a border, that can not be crossed by the robot (a physical obstacle). The visibility attribute then differs, if the obstacle is detectable by the used sensing technique. Typically, holes in the ground plane are invisible and therefore cannot be used for navigation purposes.

The virtual edges are then dedicated to link the real edges together and to guarantee the connectivity of the graph. Such an environment can always (not necessarily with unique solution!) be partitioned into convex polygonal cells. The virtual edges can be generated to divide the free area of the global map into convex subareas, what might be desired by path planning later on. The visible real edges and consequently the virtual edges are subjects to map a updating function.

If we note that no explicit description of a unique object representation is used here, all the objects in the global map are represented indirectly by sets of their edges (graph loops).

This approach to data representation, without knowing directly "what is an object" in the global map, is found as supporting the updating technique. The greatest advantage is that the updating procedure does not require the recognition of complete obstacles (it does not need to find the whole of the object's border).

The map updating function is then enabled to insert only partial information on the unknown obstacle in the sense of some of it's edges. Such line segments are directly provided by the local map via world sensing and subsequent extraction of the straight-line segments, and can be obtained by making-use of an incomplete set of measurements (aspects) to the object present in the environment.

The *local world map (or sensor-level map)* is used to store temporary information on the nearest environment to the vehicle, gathered by the sensor subsystem. This part of the WM is often assigned to the sensory subsystem. A single system can incorporate multiple local world maps, where each contains certain sensor-specific information serving as an intermediate level for information representation. For the previous example, the local map would contain the hypotheses (short term data representation) on local real edge existence or nonexistence, that serve the global map updating (long term data representation). Besides that, the local map can also support vehicle control functions through direct links to the motor subsystem (e.g. obstacle avoidance, emergency stop, etc.).

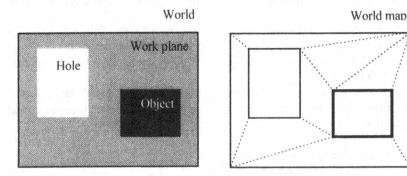

Fig 5. The polygonal world map representation. The thin lines denote inaccessible but transparent borders, thick ones the visible walls, both known as "real edges". The "virtual edges" (dashed) bound the structure together and split the work plane into regions.

The described concept of the WM is medium levelled. Of course, even a higher level WM models for environment representation are possible. The next abstraction level represents symbolic representations. This sort of WM is typically represented by a database containing symbols with certain semantic meanings and the relationships between them.

Besides the high generality of this approach, the symbols are typically highly abstract (e.g. representing certain objects). Therefore the data updating process has to incorporate a recognition stage for the sensed entities. This is a complicated task in itself, and therefore the main disadvantage of the approach.

4.1 Learning Maps

The existence and exact environment location of objects (or better to say primitives that built up a world map) is often not known with absolute certainty. The locations are determined relative to other known locations and not with a globally consistent world map. This happens in the case when sensory measurements are inaccurate or fail.

Similarly, when a mobile robot moves through a partially or completely unknown environment, the robot has to explore the environment and build up the world model as it navigates and stores gathered information. This induces a need to equip mobile robots with a *self-learning ability*. At the level of navigating mobile robots the WM acquisition, and/or the updating process it is under, is understood.

A mobile robot could start operating in an environment with no prior knowledge of its "content". The basic buildup and updating of the global map then has to be done during plan execution, which has been prepared in advance without knowing what all might obstruct the execution.

More frequent is the assumption of a certain initial world model fill-up. This forces the robot to cope only with uncertainty in the environment description, coming from the environment dynamics (its changes). The methods used for WM updating depend strongly on the level of abstraction. For the low-level representation such as occupancy grids are, the fusion process can be considered for an updating technique of the WM.

The map learning process for more sophisticated models on a feature level (see the edge-based concept) can be specified by two steps:

- Extraction of map primitives
- Primitive integration into an existing WM

The extraction of map primitives in a general case represents a search for typical patterns in the input data, or in the sensory local maps, respectively. Such patterns, beside ours straight-line segments, can also be "road crossings" or various landmarks.

Once a feature has been extracted it is expected to be incorporated into the existing WM. As we have assumed for simplification that our WM integrates both sorts of topological and metric knowledge - in other words the WM graph preserves the neighbourhood relation between primitives as well as the distance measure with respect to reality - the WM is a "reprint of the real world". This property makes further abstraction process such as the partitioning of the WM into groups of primitives, possible. A clustering process [5, 18] creates clusters of primitives which are associated with different semantic meanings. In our case of polygonal WM, the clusters of edges being chained in loops representing objects.

The WM learning procedure itself can then be introduced as a sequence of the map primitive extraction and its clustering with the primitives contained in the WM. This actually assigns new primitives to the old ones, creating together a structure with a certain semantic meaning.

If the symbolic level WM representation is used, the WM updating approach can be briefly sketched as follows:

The first step adds the new symbol into the set of existing symbols. The symbol came from the recognition stage not alone, but accompanied by additional descriptors as its position, orientation, etc.

The second step uses the previous one to create new relations with the old symbols, and occasionally omit some old symbols and old relations. The process of new relation build-up must satisfy constraints such as the general assumptions on the WM properties and constraints, that are set up by the already existing relations.

This sets up the rules for the WM creating and update. At this point it should be mentioned that during the process of relation buildup, it might also be found that the current symbol is not consistent with the existing WM and therefore can be omitted from the list.

4.2 Navigation and World Map Updating

The world model provides the robot with the ability of *self-navigation,* but it is also subject to *updating,* which can be seen from the point of the WM as complementary tasks. This induces the question of how to use a certain portion of incoming information, either for robot position acquisition or to make the WM more accurate via updating.

Let us take again our line-segmented WM into account. The self-navigation function serves for the determination of an exact position and heading of the vehicle. The positional information is basically provided by the sensor system, but corrupted by measurement errors. Therefore two kinds of data validation techniques can be used to obtain the exact position.

As the first stage of the position measurement correction is provided directly by the sensor subsystem within the fusion process, the higher-level stage is performed in cooperation with the WM. This can be based on a hypothesis-testing approach. If we interpret the line-segmented range data as edges of obstacles, a hypothesis on the current aspect to the environment can be created. The next supports or denies the hypothesis by comparison of the local map data to global map content. The global position and orientation of the vehicle can then be derived from the matching position of both maps.

As a complementary function to the one mentioned the updating function of the global map can be considered. An update of the global map is indicated if no "suitable" match of a sensor-based hypothesis on obstacle presence to the global map is found, and the hypothesis (the obstacle presence in local map) itself has been acknowledged by multiple measurements.

5 Activity Planning

A standard embedding of the custom tasks given to the robot are two classes of problems to be solved: the *service trip* and *region filling.*

The former task which solves the motion of the robot from the current to a goal position, is considered for the generic case. A slightly more complicated task is the latter problem of region filling. For this specification is not as unambiguous as for the former. The task might be given simply by list of regions to be filled (irrespective of order!), or by the environment area, which is to be filled. The system then has to split the filling area into regions - relatively compact subareas - which will be ordered in a sequence and filled by the vehicle's path, both following some certain optimal criterion.

The area for autonomous activity of the robot can be seen in *how* the given goal can be reached making-use of the given resources [19, 20]. The resources are, in this example, considered to be admissible movements of the vehicle through environment under constrains being set up by:

- Limits on the radii of turns, and vehicle speed (manoeuvring capabilities).

- The planned trajectory should safely avoid all obstacles (obstacle avoidance).

This process is also called *activity (path) planning* .

Moreover for our case, under a certain simplification of the output of the planning process - a path - could be defined as a sequence of of places going from the start point to the goal point along an arc. The start position is given, the sequence of arcs is to be planned.

5.1 Path Planning in Graph-Based Models

Assuming that the end-points of arcs are vehicle states, and the admissible arc of the same length between two states defines a transition between states (an edge), we obtain a *state-space graph*. A planning algorithm provides the sequence of system states which represents an optimal path with minimum cost. This can be based on state-space graph search techniques [1, 2, 6, 7]. Each transition between two states is accompanied by quantitative features, for example: distance of the states, energy consumption, minimal collision risk, or the time elapsed by performing the transition, etc. These serve to evaluate the price of transitions between states, via a chosen *cost function* which enables the employment of *heuristic search* techniques.

Although considerable progress has been made using heuristic approaches [21], these alternative methods do not fully achieve satisfactory performance, and have limitations of their applicability caused by the build-up of heuristic rules. If we omit blind-search methods, let us mention the purely formal approach of *dynamic programming* [7, 22], which can be used for optimal path search. The dynamic programming (backward algorithm) consists of two stages:

The first one generates a tree subgraph of the full state graph which contains all the possible ways from the initial point to the goal.

The second stage then evaluates the cost function step by step from the goal backwards to the initial point following the iterative formula:

$$C(0)=0, \quad C(i)=min_{i-1,i}(C(i-1)+c(i-1,i))$$

Where $C(i)$ denotes the total cost of the path in distance of i steps from the goal. The value $c(i-1.i)$ stand for the price of the transition from $i-1^{th}$ to i^{th} state. At this point it is necessary to show that this sort of optimization method explores in single steps only the information contained in the costs of transitions.

Another classic technique used for finding an optimal path in a graph is the A^* algorithm [1, 23, 24]. In addition to processing the information in the graph itself, as in the previous approach, A^* provides a prescription of how to use additional information on the situation from which the graph was derived. As a result, the algorithm uses far less computational effort than standard algorithms which achieve the same results.

Besides graph nodes, edges and their costs that create a standard graph, A^* uses one more sort of data - a number $h(i)$ denoting an estimate of a lower limit on the cost of moving from the ith node to a goal node. For the most common case of shortest path search the $h(i)$ might be shortest (straight line) distance to the goal. This estimate is a way of incorporating additional information which is usually based on knowledge which is not directly represented within the graph.

The A^* performs search forwards from the start to the goal. The initial path cost is equal to the estimate $h(i)$ between the start and the goal nodes. Let $g(n)$ represent the cost of the selected path from start to successor node n, where $g(start)=0$. Let the current node be the first. Then the algorithm can be sketched:

Choose node N from the set of successors of current node. If the term $g(N)+h(N)$ is not minimal over the whole set of successors, repeat the term evaluation for other successors until the smallest value is found. Consider the node having a minimal term value for a new current node and repeat them all until the goal node is achieved. Total (minimal) path cost is then equal to $g(goal)$.

A less computational effort than A^* can require the IDA^* algorithm [24] which performs depth-first iterative deepening with heuristic cutoffs. Other approaches to path and activity planning in world models represented on symbolic level as GPS, STRIPS, WARPLAN, etc. [1, 20, 23] are not of direct meaning for industrial application in this context.

Search Speedup. There are a couple of ways of how to improve the efficiency of search algorithms. The main idea behind these is that the search can be performed in two directions simultaneously. In the simplest case a search can be performed from the goal (backward chaining) and from the start (forward chaining). Completion of the search is reached, whenever both processes meet each other.

These *bidirectional* algorithms for a single processor [24, 25] are of two kinds:

The *non-wave-shaping* methods aim to find the solution as fast as possible by eliminating the nodes which probably do not lead to a solution. Both searches intersect at a place between the start and the goal.

The *wave-shaping algorithms* always expand the nodes which are found to be the closest ones to the nodes of the opposite search. This ensures that the opposing searches meet in the middle of the path between the start and the goal, this is paid for by the higher computational effort when nodes for expansion are being selected.

As for another way the *island search* can be considered the in which the state space is broken into many smaller (and spatially consistent) subspaces along the estimate of the optimal path from the start to the goal. Each of these subgraph contains a source node (the island) which is assumed to be on the optimal path. The neighbouring source nodes therefore simultaneously serve as a start and goal states for the succeeding search processes.

To illustrate the profit of the previous approach, let us assume having N islands placed at the same distance on the path, the average depth for the search from any island is n/N, where n stands for the length of the solution (number of states which have to be passed from the start to the goal). If the original complexity of the used algorithm without island is, say for A^* case $O(b^n)$, where b denotes the branching factor, the use of N islands requires only $N.O(b^{n^N})$.

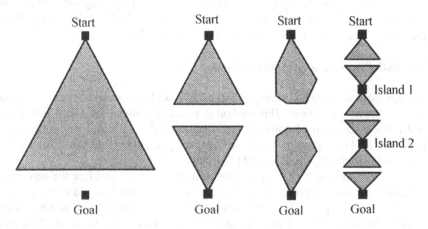

Fig. 6. Various search techniques. From the left: Unidirectional search, bi-directional search, bi-directional search with wave-shaping, and bi-directional search with islands.

The great advantage of the island-approach is that it is suited for implementation on a multiprocessor machine, what can bring an exponential speedup ratio with the number of used processors [24]., although uniprocessor implementation is also profitable.

The only limiting factor is wether the desired island can be determined in advance or not. Their definition can be done either by making use of some prior knowledge of the environment (certain locations which the robot must pass through) or via hierarchical planning , where the islands states are outputs of the prior upper-level planning stage.

Planning over visibility graphs. Another wide-spread technique uses modified representation of the WM. The path planning, respectively the minimum-path search which is conducted not over a standard fixed state-space graph, but a dynamically generated *visibility graph* [28]. The method of creating the graph explores the further knowledge of the physical properties of the world and the vehicle, in order to restrict its complexity. This fact could be expressed in rules guiding the graph build-up:

• Create a node near the object corner only, and only if, the node is visible
from the current node, and no real edge exists between them.

• Do not expand the node to a new position, which is directly unreachable because of the kinematics of the vehicle.

The obtained graph is substantially less complex than the appropriate state-graph, containing all transitions among possible neighboring positions. The important property of the visibility graphs stands in the possibility of dynamic (on-line) generation of the graph during a planning process.

But savings in computational effort when building WM on-line also bring a problem. Having an incomplete WM, there exists no planning algorithm which would guarantee a global optimum of the found solution. A modified A* algorithm performing planning over visibility graphs has been introduced in [28].

5.2 Path Planning in Grid-Based Models

Another method for path planning in grid-based models of the environment is commonly called *distance transform*. This method is basically suitable for global path planning under the assumption of complete information of the presence of obstacles in the environment. The approach of distance transform [26, 28] considers the planning task as backward path finding from the goal to the initial point.

This method covers the environment by a uniform grid (see the certainty grid world model) and then propagates distances from the through free grid elements from the goal to the initial point. The same distance value is assigned to all pixels, which are found in the same distance from the goal. In this manner the distance marking step by step is spread all over the free grid elements unless one of the markings reaches the initial point of the robot.

The shortest path from the initial point to the goal can be then found by tracing downhill the steepest descent path.

12	11	10	9	8	7	6	7	8	9
11	10	11	■	■	6	5	6	7	8
10	9	10	■	6	5	4	5	6	7
9	8	■	■	5	4	3	■	■	6
8	7	6	5	4	3	2	3	■	5
9	8	7	6	■	2	1	2	3	4
10	9	8	7	■	1	0	1	2	3
11	10	9	8	■	2	1	2	3	4
12	11	10	9	■	3	2	3	4	5

Start 10

Goal 0

Fig. 7. Distance marking in the grid-based WM and distance-transform path planning. Minimal paths from the Start to the Goal are: DDDRRRRDRD and DDDRRRRRDD, where D , and R denote the transitions to "down" and "right" pixels.

If there is no downhill path, e.g. the current pixel is on a "distance plateau" (all of the grid cells, including obstacles are initialized to high value), it can be concluded that no path from the initial point to the goal exists.

The optimum criterion as minimization of the path length was intuitively taken in this case. If additional minimization of some other features (e.g. consumed energy, number of turns, maximum velocity, etc.) are required, their influence has to be incorporated into the evaluation of the distance between cells.

As distance transform planning is very easy it also brings disadvantages. The first one is the problem of ambiguous paths. This problem originates from the way of construction the distance markings which do not satisfy the conditions for Euclidean space (in which a robot actually operates). The presented approach does not take into account the distance ratio of two diagonally and rectangular oriented cells. The problems of non Euclidean space is possibly overcome by two methods:

• Scaling of the diagonal distances by square root of 2.

• Assuming 4-connectivity between cells, which allows only "rectangular distances".

Another disadvantage of this approach is its inefficiency in data storing. Representing larger areas by the grid-based approach and evaluation of the distance transform leads to costly computations and high requirements on storage space. This problem can be partially solved by using quadtree data structures for the world model [13], which seem to be a good compromise between network-based, and grid based models. The quadtrees are hierarchical structures, containing grid-like information, but on different resolution levels. However the quadtree overcomes the deficiencies of the grid-based models because it is adaptive to the clutter of the environment. The quadtree fragments space into the finest grid cells only along places where high resolution is needed (along the boundaries of obstacles). The other areas are represented by relatively coarse cells and thus solves the disadvantage of the distance transform which covers the whole environment by the same (and very fine) cells.

Despite high computational overheads, distance transform path planning also offers several advantageous features. The method always finds the shortest path to the goal from all free space cells.

Evaluating the transform for more goals, the method allows us to determine the regions of cells optimal for reaching a certain goal, thus supporting task distribution among multiple robots sharing multiple tasks.

5.3 Partially Unknown Environment

All the previously mentioned methods for path planning are assuming that the knowledge on the robot's environment is complete. Once the path plan has been created, it is expected that it can also be fulfilled.

Whenever, during the plan execution some unexpected event occurs (e.g. an unexpected obstacle is detected by the collision avoidance system), the whole approach fails and the robot stops. It is straightforward that this situation can be expected to appear in real environments. It is possible to overcome this principal insufficiency of the path planning with complete information is possible either/or by:

- Interrupting the plan execution at collision, refining or updating knowledge on the environment and invoking a replanning stage with the current position as the starting position and the original goal.

- Handling the task as the *path planning with incomplete information.*

The latter group of methods [27, 26, 29] assumes that the path planning is limited to the robot's nearest neighborhood - typically a part of the environment which is visible to the sensors from the current position (see planning over visibility graphs). This lets the robot assume having absolutely reliable information on a certain limited area and therefore the problem can then be treated as the one with complete information.

Comparing the incomplete information approach to the previous one brings a question of the global optimum of the planned path. As such algorithms only optimize the path on limited subareas, the global optimum of the solution can not generally be ensured.

Another difficulty with algorithms for incomplete information approach appears in connection with the problem of dimensionality. For an illustration let us assume that an algorithm has to decide where to move next, when being blocked by an obstacle. For the planar case the only admissible decisions are lying in 1D space and are to go either to the left or to the right. For 3D obstacles and if the robot can also move in the 2nd direction it may be easily seen, that an infinity of various decisions "where to move" can be accepted to bypass the obstacle.

This problem can be avoided by (pre)selecting only a limited number of admissible directions in which an obstacle can be passed around. Contrary to the previous disadvantages the class of algorithms working with incomplete environmental models provide the possibility of on-line planning strategies. Such methods actually do dynamic planning over the currently incoming sensory information and always provides updates for the plan for the next motion with the already incorporated changes in the environment structure. This simplified approach substitutes both stages of planning and replanning .

Another important advantage of the algorithms for on-line planning is that there is no need for creating (and storing in a computer memory) a complete state graph, or an occupancy grid, of the whole environment at once.

As can be judged, the nature of the on-line local planning is more compatible to low level knowledge representations as a grid or feature representation, than to the symbolic level one. This enables the coupling of the class of local methods directly to sensory data which might - in an extreme case - result in omitting the global world map from the robot structure.

Many combined methods exist besides these approaches. A frequent case incorporates the previously described approaches into one combined one which uses a global world map in order to ensure a global optimum of the desired solution (up to a certain level) and solves local collisions by on-line planning over visibility graphs.

5.4 Hierarchical World Model

The planning problem becomes very complicated as the environment becomes larger and more complex. The state-space search is typical of exponential computational complexity.

With the rising size and/or desired accuracy of the WM for certain hardware there always exists a threshold over which the implementation of a search algorithm becomes inefficient (slow evaluation, or extremely rising demands on the processor performance and memory)

A successful mission through the environment is based on two major features of the robot: *goal-directed motion* and simultaneous *reaction to changes in environment*. While the first feature is related to the global (coarse) aspects of planning, the second can usually be achieved within the local region surrounding the robot. For example during a transportation task in an indoor environment, which includes long distance motions, the robot is expected to simultaneously avoid local obstacles and move towards the goal as well.

This invokes the idea of splitting the planning process into more stages which are organized in a hierarchical manner.

The upper level provides a coarse plan, which is optimal from the global point of view. Expressed in the terms of our previous example, such a plan might sound like: *move from room1 to room2, move from room2 to room3, etc.*, which says nothing about how a certain movement should be fulfilled. In general the top level planning is suited for higher level knowledge representation (symbolic WMs) which allows it to work directly on the level of entities such as obstacles, space regions, etc. (planning on the symbolic level).

The lower level is dedicated to taking care of the local maneuvering of the robot. Again, for our example it can be the motion from a current position within room1 to room 2, etc., including possible obstacle avoidance in the room area as might be done by nonhierarchical planning.

More planning levels can occur with real tasks, for example planning on the level of places of interest (1st level), passageways (2nd level) which connect the places of interest, and concrete route optimization during motion through passageways (3rd level).

World Models and Hierarchical Planning. Splitting the planning process into more levels requires some modifications in the knowledge representation in WMs. As different planning levels work with different spatial resolution the need of different levels of data resolution appears.

A straightforward modification of grid-based WMs is their representation by quadtrees. The hierarchical structure of a quadtree, beside savings in requirements on storage, allows access to the data on a certain level of resolution.

The case of network-based WMs requires basically the same - different resolution levels, which can be obtained via creating multiple representations of the same WM having the desired resolution. In other words, the top-level graph, respectively it's nodes or edges are expanded through a lower-level graph into subgraphs. This allows an increase in the spatial resolution of the representation with deepening of the current WM level.

As sketched, it can be seen that the hierarchical approach to planning brings on the first place the possibility of on-line (sensor-based) planning, which is efficient in incompletely known environments, and where there is a global optimum guaranteed by the top-level planning (using reliable knowledge from the WM). Another great advantage lies in savings in evaluation time.

6 Plan Execution and Vehicle Control

The method of plan execution, respectively the vehicle control is dependant on the kinematics model of the used vehicle. The simplest case considers a tricycle vehicles, but car-like kinematics also appears very often. A path plan found by a planner is mostly represented by the means of:

- Symbolic command language
- Sequence of succeeding positions

The symbolic language representation describes the plan as a sequence of driving commands. Such macro-commands can define vehicle's actions as in the following example:

(1) *define start position and heading*
(2) *move straightforward for 2 meters, accelerate to speed 0.5 ms^{-1}*
(3) *move 60 degrees turn having diameter of 3 meters, no acceleration.*
(4) *move straightforward for 5 meters, accelerate to speed 2 ms^{-1}*
(5) *move straightforward for 50 meters, no acceleration*
(6) *move straightforward for 5 meters, decelerate to zero speed.*

As the shown commands can not directly control drives of the vehicle, they have to be interpreted first. The interpretation process, done by the driving subsystem, translates the macro-commands into a smooth trajectory. The converted trajectory, typically represented by a dense sequence of desired positions and headings of the vehicle, can then be directly fed to vehicle drive controllers. Positional feedback for the controller is typically closed via gyro and oedometry measurements of the robot's actual position.

The second approach uses sequences of positions (and headings) which the robot has to attain. The robot path between two succeeding points is not defined and is subject to interpolation, which also provides smooth trajectories.

The only difference between the both approaches lies in the level on which the plan has been created. The former one is better suited for symbolic planners or planners working over network-based WMs. The latter one seems to be more practical for use with grid-based WMs.

7 Conclusion

The presented contribution overviews the briefly possible concepts of the structure of an intelligent mobile robot - a self-guided autonomous vehicle. The study is to sketch basic functions of a simple SGVs respecting its application not in laboratory environments, but for usage in real environments. The intent is to follow a concept of cheap and robust intelligence, that would equip SGVs for immediate use in delivery, or cleaning services, in industrial and office environments.

With respect to most of the existing SGVs, the leading idea of the design for such environments is to avoid extraordinary installations in the work area which are normally necessary for the navigation of standard SGVs.

Besides that, by making use of no intensity visual sensor (camera), and designing all the methods with respect to their simplicity, it can expected to make the vehicle inexpensive and therefore suitable for a wide area of practical applications.

References

1. S.C. Shapiro (Ed.): Encyclopedia of AI, John Wiley & Sons Publ., New York, 1990

2. J. Allen, J. Hendler, A. Tate (Eds.): Readings in Planning, Morgan Kaufmann Publishing Inc., Palo Alto, 1990

3. C. Andersen, C. Madsen, J.J. Sorensen, et all: Navigation Using Range Images on a Mobile Robot, journal on Robotics and Autonomous Systems, Elsevier Science Publ., Vol. 10, 1992, pp. 147-160

4. H.A. Vasseur, F.G. Pin, J.R. Taylor: Navigation of a Car-Like Mobile Robots in Obstructed Environments Using Convex Polygonal Callus, journal on Robotics and Autonomous Systems, Elsevier Science Publ., Vol. 10, 1992, pp. 123-146

5. D. Maio, S. Rizzi: Map Learning and Clustering in Autonomous System, IEEE Trans. on Pattern Analysis and Machine Intelligence, Vol.15, No.12, December 1993, pp. 1286-1297

6. T. Lozano-Perez, M.A. Wesley: An Algorithm for Planning Collision Free Paths Among Polyhedral Obstacles, Communication of the ACM, Vol.22, No.10, 1979, pp. 560-570

7. S. Suh, K. Shin: A Variational Dynamic Programming Approach to Robot-Path planning With a Distance-Safety Criterion, IEEE Journal of Robotics and Automation, Vol.4, No.3, June 1988, pp. 334-349

8. Y. Kanayama, S. Yuta: Vehicle Path Specification by a Sequence of Straight Lines, IEEE Journal of Robotics and Automation, Vol.4, No.3, June 1988, pp. 256-276

9. S. Sifrony, M. Shahir: An Efficient Motion Planning Algorithm for a Rod Moving in the Two-Dimensional Polygonal Space, Algorithmica 2, 1987, pp. 367-402

10. E.G. Gilbert, D.W. Johnson: Distance Functions and Their Applications to Robot Path Planning in the Presence of Obstacles, IEEE Journal of Robotics and Automation, Vol. RA-1, No.1, March 1985, pp. 21-30

11. L. Přeučil, R. Mařík, R. Šára, et.al: Vision-Based Robot Navigation in the 3D World, in Proc. of IEEE Conference on Artificial Intelligence, Simulation, and Planning in High Autonomy Systems, Tucson, Arizona, September 1993

12. M.A. Abidi, R.C. Gondzales (Eds.): Data Fusion in Robotics and Machine Intelligence, Academic Press, 1992

13. Bray, V. Hlaváč, M. Šonka: Image Processing and Machine Vision, Chappman-Hall Publ., 1993.

14. J.S. Lee: Multiple Sensor Fusion Based on Morphological Processing, Proc. SPIE, Vol. 1003, Sensor Fusion: Spatial Reasoning and Scene Interpretation, Cambridge Massachussets, 1988, pp. 94-100

15. R. Belknap, E. Riseman, A. Hanson: The Information Fusion Problem and Rule-Based Hypotheses Applied to Aggregation of Image Events, Proc. of the IEEE Conference on Computer Vision and Pattern Recognition, Miami Beach, 1986, pp. 227-234

16. L. Matthies, A. Elfes: Integration of Sonar and Stereo Range Data Using a Grid-Based Representation, Proc. of the IEEE International Conference on Robotics and Automation, 1988, pp. 727-733

17. A. Elfes: Sonar-Based Real World Mapping and Navigation, IEEE Journal on Robotics and Automation, Vol.3, No.3, June 1987

18. B.S. Duran, P.L. Odell: Cluster Analysis, Lecture Notes in Economics and Mathematical Systems, Springer Verlag, 1974

19. L. Kohout: A Perspective on Intelligent Systems, Chappman and Hall Publ., U.K., 1990

20. I. M. Havel: Robotics - an Introduction to Theory of Cognitive Robots, SNTL Publ. Prague, 1980 (in Czech).

21. T. Lozano-Perez: A Simple Motion Planning Algorithm for General Robot Manipulators, IEEE Journal on Robotics and Automation, Vol.3, No.3, 1987, pp. 224-238

22. A.D. Pierre: Optimization Theory with Applications, John Willey and Sons Publ., 1969

23. E. Rich, K. Knight: Artificial Intelligence, Mc Graw-Hill Publ., 1991.

24. P.C. Nelson, A.A. Toptsis: Unidirectional and Bidirectional Search Algorithms, IEEE Software, March 1992, pp. 77-83

25. I. Pohl: Bi-Directional Search, machine Intelligence, March 1971

26. A Zelinski, S. Yuta: Reactive Planning for Mobile Robots Using Numeric Potential Fields, Intelligent Autonomous Systems IAS-3, 1993, pp. 84-93

27. J.V. Lumensky, A.A. Stepanov: Path Planning Strategies for a Point Mobile Automaton Moving Admits Unknown Obstacles of Arbitrary Shape, Algorithmica, 1987.

28. P. Oubrecht: Planning Algorithms for Robotics, Diploma Thesis, Dept. of Control Eng., Czech Technical University, Prague, 1994

29. K Azarm, G. Schmidt: Integrated Mobile Robot Motion Planning and Execution in Changing Indoor Environments. IEEE proceedings of IROS'94 conference, Munich, September 12-16, 1994, pp. 298-305

CAD systems: trends and developments

Pavel Slavík

Department of Computer Science,
Czech Technical University Prague,
Czech Republic

Abstract. This paper discusses some modern aspects in the field of
CAD systems. The reader's attention is drawn to problems concerning
with graphical information processing and their impact on the development of a new generation of CAD systems. These new trends reflect the
changes in the design technology in the last few years. Design technology
changes stem mostly from the growing necessity for intensive communication and collaboration among members of a design team. This requires
to solve problems like storage and transmission of graphical information.
Other important problems to discuss are properties of modern solid modellers. This requires to estabilish criteria for evaluation of solid modellers.
All these topics will be discussed in the following text.

1 Introduction

Our course on CAD systems covers both the introduction and some modern aspects of the field. The following text discusses new trends and developments in
the field of CAD systems. As the scope of CAD systems is fairly wide, we concentrate on the topics of CAD systems where processing of graphical information
is included. In this field, a lot of new techniques have appeared during the last
few years. We will discuss only the most important ones that mirror the mainstream of the development. One of these techniques covers the case when the
growing complexity of products requires employment of new techniques which
allow the cooperative design of products. That means that the design process
is performed in parallel by the members of the design team. This requires new
types of software to support this sort of work and also additional facilities to
allow storage and transport of the design data - graphical data in particular. A
wide variety of standard data formats which are tailored to these specific needs
is required. New developments include integration processes which put together
various fields of the design. This implies the need for new methodology in the
usage of CAD systems. As an example, the mechatronics will be given.

These new techniques have been incorporated in the new properties of the
CAD software. Perhaps the most apparent progress can be seen in the field of
solid modellers. The analysis and the description of properties of such systems
will be given. A very important topic is the evaluation of CAD systems or their
parts. To show how such an evaluation can be performed, and what criteria and
tools exist, a short explanation is given below.

2 New techniques in CAD systems

In the past, many definitions of the term CAD system have been made. These definitions have been subject to the rapid development within the field. There currently exist many new techniques in the CAD field which were not known until a few years ago. The appearance of new techniques is the result of the growing complexity of products designed by means of CAD systems. The kernel of each up-to-date CAD system is the subsystem for solid modelling since it is a source of data for other CAD subsystems. The functionality of this subsystem usually determines the quality of a CAD system as a whole. The following text will be a brief overview of techniques in the CAD field which have appeared recently together with their impact on solid modelling techniques.

2.1 Constraint based modelling

Solid modelling systems have been used since the early 1970s. It is a technology that has always been hard to use as it traditionally brought certain problems to users. The main problem is that a design usually goes through multiple, and often extensive, changes during its design cycle. Traditional CAD systems did not include adequate tools to handle these modifications easily. These systems usually forced the user to keep track of all the relationships between various parts of the design. For example, if the radius of a hole changed, it was the designer's task to remember to change the radius of a shaft that passed through the hole. Even in simple designs, there are many such relationships to be maintained. This is a cumbersome task with which the user must cope. As a result, the designers had to spend a considerable amount of time after every modification to ensure that everything still fit together correctly.

Constraint based modellers [2] are the solution to these problems. They allow the CAD system to automatically track the relationships between parts. These modellers employ a set of rules, or constraints, by means of which it is possible to define how changes to a group of geometric elements should be handled in such a way that the integrity of the design will be protected. The constraint based modellers use three types of constraints: numeric, geometric, and algebraic:

- numeric constraints operate with positions, lengths, diameters, spline parameters, angular values, and other measurable variables.
- geometric constraints handle parallelism, perpendicularity, tangency, and other non-numeric relationships that express relations of one piece of geometry to another.
- algebraic constraints combine numeric and geometric constraints in very simple equations. In more complex cases, the constraints are expressed by a set of equations that can include if- then-else branches, inequalities, etc.

Some modellers use certain type of constraints whereas another type does not. This is usually dependent on the application. The constraint based modellers can define and solve constraints in two ways. This leads to a definition of two types of constraint based modellers: parametric and variational.

In parametric systems, the constraints are solved in the order in which they were specified. This means that all evaluated variables in the currently investigated constraint must have been defined in previously solved constraints. Their values are used to compute a new value of a variable (on the left hand side in the current constraint equation) which is assigned to some object, thus expressing some property of the object. In variational systems, the constraints are solved simultaneously. This means that the order of constraints can be arbitrary.

In case of parametric systems we speak about one-directional constraints. The position of a previously defined point influences the position of the later defined point (new position of the formerly defined point results in a new position of the later defined point).

Example:

$$x_2 := x_1 + \Delta x;$$

$$y_2 := y_1 + \Delta y;$$

(x1, y1 are coordinates of the previously defined point) In variational systems, both points influence their position mutually.

Both approaches are suitable for certain classes of applications. Parametric systems are used for parts and assemblies with exactly specified hierarchies, where each part is specified by a small number of parameters. Variational systems, due to their higher complexity, require more computing power as it is usually necessary to solve a large set of constraints. There are many theoretical problems concerning this solution process. In many cases we can not obtain an exact solution. The solution process often requires approximation and simplifications which should converge to an acceptable result. The variational systems are suitable for work with objects which are defined by a large set of parameters. As the order of constraints can be arbitrary, the user has more freedom during the design process than in the case of parametric system. Commercial CAD constraint based modellers usually use a combination of parametric and variational techniques.

Currently, there exist several commercially available systems which offer tools for constraints definition and handling. Typical examples of such systems include: Bravo (Applicon Inc.), CADDS5 (Computervision), Catia (Dassault Systemes), Unigraphics (EDS Unigraphics), I/EMS (Intergraph), I-deas Master Series (SRDC) and ProEngineer (Parametric TechnologyCorp.).

The constraint based modellers offer the possibility to combine constraints with form features. Here we come to another advanced technique used in current CAD systems - feature modelling [2]. Feature modelling is a relatively new way of storing information about objects in CAD/CAM systems. The traditional solid modelling systems describe shapes of 3D objects by storing sufficient information about the 3D geometry and topology of the object. With feature modelling, beside shape information, other information about an object is also stored in a so-called product model. This model is the extension of the traditional, purely geometric model in that the functional information about the object is added. This can be either information about the function of some part of the object or information about the way some part of the object is manufactured or assembled.

This type of information is usually expressed by means of form features. Form features are a higher level constraint than points, curves and surfaces. Examples of form features can be holes and slots. The designers work with entities which are more natural for them from the point of application. Instead of working with entities in the purely geometrical sense, the designer works with slots, bosses, throughholes, blind-holes, etc. The use of form features has a much wider scope than just solid modelling. The features can identify areas in a product that can be manufactured in one operation with one type of equipment.

In analysis applications, features can identify areas of objects that require, for some reasons, special attention in the analysis and in the assembly process planning. Features can identify connections between subparts. Another use of features is in product model standardization and product data exchange.

The usage of form features allows (at least in theory) complete CAD/CAM integration. When having a central product model, all application modules of a system can have access to it. This relieves the user from entering new information specific for each application module. A very important aspect when using features is that after creation of the initial conceptual model of an object, several engineers can simultaneously work on the product model. This approach is called concurrent or simultaneous engineering. It is based on the fact that each engineer has his own way of looking at the object.

In the case that features are used in concurrent engineering, all applications will have their special set of features. Each set reflects an engineer's way of looking at the object. This requires the existence of automatic conversions between features from different sets. This is currently a field of an intensive research.

2.2 Mechatronics

Nowadays, designers are faced with new problems emerging from new requirements. A good example of this is mechatronics, which has become a buzzword in the last few years. Mechatronics is a solution to the problem where two traditional CAD worlds meet: the world of mechanical design and the world of electronic design. Industries from aerospace to consumer products rely heavily on computer chips, printed circuit boards, and software to perform jobs once handled by larger and heavier mechanical components. The reason is that the use of electronic components ensures higher reliability in comparison with their mechanical equivalents. Last, but not least, the energy consumption is usually several times lower when using electronic components. Electronic components are infiltrating mechanical systems at such a rapid rate that, according to recent surveys, nearly one-third of all products now contain both electronic and mechanical components.

The traditional approach in the design process of such products is as follows: one group does the mechanical design, one handles the electronics, and a third group puts the two systems together. By means of this approach we can reach satisfactory results, but not an optimal one. The reason is that the trade-offs between the mechanics and the electronics can not be explored. One of the main problems is that mechanical and electrical designers speak their own professional language . This results in a lack of mutual understanding. From this point of view, the third group creates a sort of interface between the previously mentioned groups. The skill of the designers in this "interface" group, communication techniques, organization skills and many other factors heavily influence the quality of information which flows between the other two groups. This fact has a significant influence on the quality of the final product designed.

Electro-mechanical systems require some unique design considerations. A good example of this is the thermal characteristics of the system. In the electronic part of the design, it is not complicated to investigate and optimize thermal chracteristics of the printed circuit board. In the case the board is installed with other boards in a sheet-metal container, the situation becomes much more complex. It is necessary to solve new problems like:

- how much heat will be generated by the entire system?
- what will be the best heat distribution in the container?

We can see that these questions take into account global properties of the device designed. This situation requires a new approach to solve these questions. To support the solution of these questions (and similar ones), it is necessary to create an interface that enables mechanical and printed circuit board design systems to share data. Making this statement more general, we can obtain one of the possible definitions of mechatronics (often found in literature): Mechatronics is the integration of mechanical, electronic, and software engineering in product design and manufacturing. Currently a couple of commercial solutions to this problem exist. One of them has been offered by Computervision in the form of an interface between its mechanical design system, Meduse, and its electronic design package,Theda.

It is clear that, in general, it is necessary to create a unified data model for product reflecting both mechanical and electrical properties. To give an idea of what kind of problems have to be solved we should stress that printed circuit board design (PCB) packages work in a 2D environment, whereas the mechanical design packages work in a 3D environment. The solution was to extend PCB design packages in such a way that PCB are represented in the form of a 3D model. This also has advantages from the designer's point of view - a solid model of a PCB and its mechanical surroundings is easier to visualize than a 2D view. Solid models also allow mass properties analysis on a board or in conjunction with its housing. Last, but not least, this complex approach allows the employment of various aesthetic criteria to examine the designed product as a whole from the aesthetic point of view. Mechatronics will also bring new methodologies to the design process as it will require mechanical and electronic designers to communicate to a degree that they have not previously done.

3   Data formats for CAD systems

One very important topic in the field of CAD systems is standardization. In principle, standardization is necessary to ease the portability of data and programs between various CAD systems and computer platforms. This strategy should lead to cost reduction both of software products and of products designed by means of them. The problems of standardization are very numerous and they cover virtually all areas of CAD systems. We will deal further with the data formats used in CAD systems.

3.1 Pictorial data formats for CAD systems

Data in a CAD system has either pictorial or non pictorial form. We will limit ourselves to the pictorial case. The necessity of standard pictorial data formats emerged mainly from the need for their efficient storage and transfer. When speaking about pictorial data formats for CAD systems, we should have in mind that the situation has changed in the last years. Traditionally, the pictorial information for CAD was in vector form only (technical drawings etc.). In the last few years, new techniques are used where the pictures have a raster form (in the following text we will speak about images). It concerns mostly various visualization techniques which are applied for solid models visualization, visualization of simulated processes etc. Thus, we will also mention image data formats and methods for image data handling used for raster images. As pictorial data is characteristic for its great volume, the standard parts of standard formats are usually methods (and corresponding software modules) for pictorial data compression.The compression techniques and their use in the field of pictorial data format will be discussed further.

The first attempt to estabilish a standard format was done in the beginning of 80's when CGM (Computer Graphics Metafile) was defined together with GKS (Graphical Kernel System) standard. The graphical information was stored in the form of simple data structure where the basic primtives were polylines, text, filled areas etc. These primitives were interpreted purely in accordance with their geometric meaning. This scheme served as a guideline for the design and implementation of next formats which have become either national standards or de facto standards (e.g. DXF). Even if the extent of geometric data was very large, compression methods were not used, or they were used only in a very limited extent. Problems with picture information compression emerged when it become necessary to store and transmit pictorial information which had raster form. To store such information in "natural" format would be very complicated task as the volume of data would be very great. Let us imagine just one picture 1k by 1k pixels in true color mode. The memory requirements would be 3 Mbytes. The need for compression methods is very urgent.

The compression methods used are based on various principles. One of them is a quadtree technique where an image is subdivided into a number of distinct areas where each area represents a single color (or grey scale value) [5]. Each area is recursively divided into subareas until the division is completed (areas of interest are represented by a single pixel value) or until a predefined level of division has been achieved. The compression results from the fact that the values associated with the subareas are then stored rather than individual pixel areas.

Other image compression methods are applications of compression methods which have been used for compression of one-dimensional strings. Examples include RLE (Run Length Encoding), Huffman encoding, and LZW compression. The priciple is that the raster image is considered to be a set of lines (e.g. horizontal scan lines) where each line is compressed as one 1-D string. These compression methods belong to the class of lossless methods as after the decompression process we can obtain the complete original image without loss of information.

In some applications, a certain loss of information is acceptable (usually the human eye is not able to recognize that some information is missing) so that lossy compression methods are used.

There are two main methods of this kind:

- DCT (Discrete Cosine Transformation)
- fractal compression.

In the first case, a raster (image) is divided into units (blocks) of 8 by 8 pixels. Each block is transformed by DCT and the result is 64 DCT coefficients. The compression works on the basis that not all 64 coefficients have the same importance when reconstructing the original image block by means of the inverse DCT. On average, it has been shown that only about 5 of the coefficients (out of 64) are actually needed to give good replication of the original image.

Fractal compression stores an image as a collection of affine maps that describe coordinate and color-space transformations from one region of the original image to another. The original image can be reconstructed from this set of maps alone. Unlike DCT compression the fractal compression is not a block-by- block collection of transformed data. With fractal compression, redundancy is reduced at a much higher level. Storage savings come about as the compression recognizes that arbitrary parts of the image "look like" other parts. The compression ratio is very high. Having a ratio on the order 20 to 1, we obtain, after decompression, an image without noticeable degradation in comparison to the original image.

The above given compression methods are widely used in standard image data formats. Next we will give a short outline of the most frequently used raster image data formats:

- TIFF (Tagged Image File Format) was designed to cover existing image file formats for applications, packages, hardware etc. Originally TIFF was designed for information and image interchange between electronic publishing packages. The format is extensible and hardware independent. It supports several image data compression techniques.
- GIF (Graphics Interchange Format) defines a protocol which supports the hardware independent, on-line transmission of raster graphics data. It supports a version of the LZW compression algorithm for its compression.
- PCX format is rather simple in comparison with GIF and TIFF formats. A PCX image file consists of a 127 byte header containing information about resolution, number of colors, color tables etc. followed by an encoded form

of the graphics data. The encoding used is a simple byte oriented RLE technique.

- X-Window define how the raster data is represented within an X- Window system: Pixmaps, Bitmaps and Images. Xlib utilities are available for handling all three types. Due to the wide use of X- Window, the use of these formats ensures wide portability of image data.

- IIF (Image Interchange Facility) is one of the two aspects of the Image Processing and Interchange standard, now under development within ISO. The other part is the Programmer's Imagining Kernel System (PIKS). The IIF should facilitate the interchange of all types of image data between application programs, and provide adequate facilities to the de facto standard currently used in image processing.

- Special formats:
  - scientific visualization:
    * HDF (Hierarchical Data Format) is a multi-object file format for the transfer of graphical and floating-point data between different hardware platforms. It allows storage and retrieval of 8- and 24-bit raster images, color tables and scientific data and accompanying annotations. Related items of information are grouped into sets (raster image sets, scientific data sets etc.).
    * NCDF (Network Common Data Format) is a data abstraction for the storage and retrieval of scientific data. It supports the creation, access and sharing of data in a form that is self-explanatory and network transparent.
  - formats related to multimedia:
    * JPEG for still images
    * MPEG for a sequence of images (e.g. video).

The names of these formats were derived from the names of the expert groups where these formats were developed (Joint Photographic Expert Group and Motion Picture Expert Group). The image compression techniques are based on the DCT and they are a vital component in solving the requirement for storing documents containing still and moving images.

3.2 Formats for storage of 3D information

As 3D applications have become more and more popular in the last years, the problem of storing, retrieving and transmitting 3D information has become a very urgent one [4]. We can see an analogy with the situation in 2D applications a few years ago when the need for data format standardization became very urgent. In this section we will mention some of the data formats which are used in 3D information processing:

- PHIGS Archive File. This format is very common in the field of 3D information processing (and at the same time is one of the oldest ones). This format has been defined as a part of the PHIGS (Programmer's Hierarchical Interactive Graphics Systems) standard that provides tools (in a form of a library of functions) for modelling 2D and 3D graphical objects, displaying them, editing them and interacting with them by means of a display. The objects in the format are defined in terms of PHIGS structures which are composed of a number of other structures and/or graphical elements, thereby forming a hierarchical structure network.

- OFF (Object File Format) is used for a storage of 3D objects for modelling/animation purposes. The format is not very efficient as it allows storage and retrieval of objects only defined in a polygonal form. The format is extensive and there are tools to define new primitives (it can lead to some data compatibility problems by data transfer between two systems). There are no hierarchical data structures available.

- NFF (Neutral File Format) allows the use of the following primitives: polygon, spline and cylinder. It is possible to assign some attributes to surfaces (no textures). As in the case of OFF, no hierarchical data structures are available.

- P3D (Programmable 3D) format is represented by a Lisp program. There are at disposal the following primitives: polygons, spheres, cylinders, cones and Bezier patches. It is possible to create hierarchical data structures. As it is possible to use programming constructs like if and loop statements, the user can easily describe complicated 3D objects. Moreover, it is possible to define characteristics and positions of light sources in the scene.

- Inventor format is defined for the data visualization system Inventor which runs on the Silicon Graphics platform. The user can create, by means of library functions, a formal description of a 3D scene in a form of a graph implementation of a data structure. This data structure (format) can be stored and retrieved in a frame of an Inventor session. These geometric primitives are used: polygons, Bezier patches and NURBS including attributes like textures etc.

- RIB (RenderMan Interface Bytestream) has been created as a part of the RenderMan system. RenderMan was created as a platform - independent rendering standard capable of generating high- quality, photorealistic 3D images. The RIB file forms the basis of communication between RenderMan - compatible programs and among various computer platforms. Programs such as 3D Studio from Autodesk, Microstation from Intergraph Corp., and FastCAD from Evolution Computing all output RIB files. The RIB file contains a detailed listing of each scene element (object, camera, light) and its associated parameters and variables. As the RIB files may be edited in a word processor, the process is similar to editing Postscript files.

3.3 Application oriented data formats

The above given data formats allowed storage, retrieval and transfer of image and geometric data. All are important in CAD/CAM technology. The information was interpreted in purely geometric terms (geometric objects, geometric attributes, geometric operations etc.). Recent trends in data formats show that there is a growing necessity to describe product models in full. This demand stems from integration processes in the CAD/CAM area where the information about a product during its whole life cycle should be stored and retrieved in a compact and consistent way. This information can be transferred from one step in a CIM system to another.

To fulfill the global set of requirements, an international solution is necessary. Product development and maintanance is not limited to a single company or country any more. The first attempt to create such a standard was IGES. IGES grew out of the graphics standard work, seeing the application need for data exchange beyond graphics or picture data. Because of its many inadequacies, many national versions were created (some became national or de facto standards in the field). As new demands occured the need for a new standardization effort emerged. The major step in this direction is STEP (Standard for the Exchange of Product Model Data). STEP's goal is to enable the capture of information comprising a computerized product model in a neutral form without loss of completeness and integrity throughout the whole life cycle of a product.

STEP itself is going to be an ISO standard. European projects (ESPRIT) based on STEP exist. They are mostly targeted for the exchange of geometric data which is semantically handled from the point of applications. Examples of such projects are CAD*I (CAD Interfaces) and CADEx (CAD Geometry Data Exachange). The STEP development is a very complex project. It should not be considered to be a data exchange format, but also as a major conceptual work towards product modelling that will have a major impact on the whole CAD and CIM world in the near future. It will lead to a new generation of CAD/CAM systems or systems components that can be linked via the product data to meet all requirements the STEP standard should fulfill. The product model should include all kinds of product properties, like shape, features, tolerances, material, etc. Moreover the different application areas (mechanical engineering, electrical and architectural engineering, shipbuilding and many others) should be covered considering not only the product itself, but also its whole life cycle.

To specify the exchange functionality, STEP is divided into five major areas [6]:

- reference model and description methods. A formal specification language, EXPRESS, is used to specify the information elements of the resource models.

- information models. E.g. product description, geometry, topology, application dependent models etc.

- application protocols. They describe the specific subsets of entities of the general STEP product model and defines a reference model from the point of an application. item implementation methods (physical file formats, database formats etc.)

- test and verification methods. They are defined along with the standard development.

4 New generation of solid modellers

One of the areas in CAD systems where new techniques were introduced is the area of solid (geometric) modelling. Solid modelling has played an important role throughout the whole history of CAD systems. One of the reasons for this situation is the fact that the solid modelling subsystem (in the frame of a CAD system) was the source of data for other CAD subsystems.

During the years that CAD systems were used, a number of solid modelling techniques were developed, a number of approaches for solid (geometric) model rendering were used etc. This led to the situation which is well known from other fields of computer science: difficulty with data transfer when data was acquired by various techniques. The time has come to develop a unified approach to the creation of geometric models. Such a system which will meet this requirement (besides others) is a candidate for a sort of standard (either an industrial one or de facto standard).

The solid modeller which meets the requirements (ease of data transfer, unified geometric models, use of constraints, etc.) for such a modern modeller is ACIS [1]. ACIS (product of the US based company Spatial Technology Inc.) is a new three dimensional system where the 3D object description is based on the boundary representation model. The system itself has an object oriented architecture that can host a wide variety of component technologies. ACIS facilitates the development of tightly coupled modelling extensions through derived classes, while, at the same time, offering a broad and extensible procedural interface. ACIS is the first solid modeller that can be used for virtually any CAD/CAM application. It offers a unified environment for the modelling of curves and solids. The class of objects the system can manipulate is very wide. Solids can be manifold or non-manifold and can be open or closed (which is not a common feature in commercial solid modellers). Due to the very general nature of the solid modelling environment, the system offers the possibility to intersect curves with curves, surfaces and solids, surfaces with surfaces, surfaces with solids and solids with solids.

ACIS is very fast and reliable. At the same time, the precision of the operations performed by the system is exceptionally high. The system contains tools supporting modern solid modelling techniques. The kernel of the system includes a special database with bidirectional associativity. That means that changes in one model in different stages of the modelling process are projected in changes in data used in other stages of the modelling process. ACIS allows for the use of parametric and feature modelling techniques and also offers tools for parallel work which means that concurrent engineering methods can be employed. All these facilities allow considerable reduction of design time.

Due to the overall flexibility of the system, it is easy to link application oriented modules (e.g. Finite Element Methods Techniques, Simulation Modules, Visualization Modules etc.). The development of new ACIS modules runs under supervision of the ACIS Open Consortium. Currently, ACIS runs on 50 platforms.

ACIS can be considered to be a solid modelling environment for the design and implementation of application systems. In the middle of 1993 there were about 12 systems from 10 companies available on the market [3]:

Company	Software
AGS	Visionael 3D
Aries	Concept Station
Concurrent Technology	Rapidcast
Control Data	ICEM DDN, ICEM PART
Graftek	GMS/Shape
Hewlett-Packard	Solid Designer
Lujuustekniikka Oy	ARGOS
Straessle	KONSYS 2000, RWT 2000
Ford Motor	PDGS

Currently, many companies worldwide work on application modules oriented for various applications. The recent trend uses ACIS applications to support product data exchange - especially based on the STEP format. The following is an attempt to summarize the most important characteristics corresponding to the most frequent requirements:

- open system
- object oriented system
- support for concurrent engineering
- form features & parametric modelling
- unified approach to various solid representations (wireframes, CSG, B-rep,...)
- support for local operations
- support for the NURBS geometry
- communication with other systems (via STEP)
- developments of new methods and tools in the field of CAD systems
- support for link with other modules for CAM, CAE, ...

ACIS meets all these requirements. For this reason, it is considered to be the most successful geometric modeller available in recent years.

5 Criteria for the selection of a CAD system

When purchasing a CAD system, it is necessary to make a precise analysis of CAD systems available on the market so as to choose one which fulfills company needs. A CAD system consists of two main parts: software & hardware. It is necessary to estabilish criteria for both parts. This criteria should be specified very carefully with respect to real needs, as the more general functions the system has, usually, the more expensive it is. In the following, we will list some of characteristics we should concentrate on during the selection process. The list is (of course) not an exhaustive one, but gives a good idea how the criteria should be estabilished for a specific application.

5.1 Software

The software selection criteria of a CAD system is highly dependent on the application and the level of sophistication required. It is also necessary to have in mind that the requirements for a CAD system can be very wide, ranging from a simple technical drawing to animated visualization processes. Next, we will give a short survey of characteristics which should be taken into account when making a decision about the purchase of a CAD system (or its geometric part). The importance of these characteristics varies in accordance with the application and other aspects.

In the following list, each item consists of list of entities that can be represented in the system. Less sophisticated systems include only some of them. This gives us a guideline to evaluate the functionality of the CAD system. Only common entities are listed - very sophisticated systems can include some unlisted entities.

- Modeller part:
  - geometry: polygonal meshes, linear patches, B-spline patches, NURBS patches, particle systems
  - 2D primitives: circle, rectangle, arc, polygon, linear spline, Bezier spline, B-spline, NURBS
  - 3D primitives: face, sphere, cube, cylinder, cone, torus, tube, grid, polyhedron
  - operations: move, rotate, scale, stretch, mirror, copy, extrude, chamfer, filet, blending, cut
  - boolean operations: union, intersection, difference
- Renderer Part:
  - shading: wireframe, Gouraud, Lambert, Phong, Blinn, raytrace, radiosity
  - mapping: reflection, opacity, luminance, shiness, specular, bump, procedural
  - effects: motion, blurr, fog, smoke, water, lens, clouds, rain, depth cue, image compositing
- Lighting Part:
  - types: ambient, spotlights, pointlights, arealights
  - variables: hue, luminance, saturation, cone angle,
- Camera Part:
  - multiple cameras, pan, zoom, track, area zoom, focal, depth of field

In the case of special systems for animation, other criteria are also employed (method of animation, kinematics, deformation etc.). The list of features of single systems informs us about the functionality. Another question is the perfomance and memory requirements. This is highly dependent on the hardware characteristics.

5.2 Hardware

The user who should take decision about the purchase of a graphical system has usually some information from producers. Data from various producers is not compatible, so it is very complicated to make a comparison of graphical performance of two systems. Because the graphical hardware usually creates an extra part in the system it is impossible to derive information about "graphical performance" from figures describing the performance of the basic processor.

The solution is to develop a sort of graphical based benchmark which will be, to a certain extent, hardware independent. The results obtained by means of such a benchmark for various systems will be comparable. Such a benchmark has been developed since 1986 when GPC (Graphics Performance Characterization) Commitee under the auspices of NCGA (National Computer Graphics Association) was founded. All major computer manufacturers took part in the formation of GPC. The results of GPC's activities was PLB (Picture Level Benchmark) which is going to be an industrial standard [7].

The basic idea is that various application classes exist which are characterized by a large volume of typical graphical computations. For this reason, six various tests exist (at present), which examine various aspects of the functionality of the graphical hardware. The possibility to test 3D graphics accelerators also exists. The tests in general are higher level tests, which means that instead of testing performance concerning the processing of sets of polygons within certain time unit, some graphical techniques and their performance are tested. These techniques include: 2D design, 3D wireframe performance and solid modelling.

This strategy corresponds to the experience that e.g. workstations which are designed to work in 3D environment give bad results in performance in 2D environment. The PLB is very flexible and many options exist to modify the PLB implementation in such a way that it fits to the specific hardware. It is possible to specify various configurations of a system. Activities exist to modify the PLB for a specific environment or application. An example is XPC-group which develops X11-Performance Benchmark. The result will be a benchmark which will (similarly to PLB) measure X-systems performance during the run of CAD, desktop or picture processing programs. Another group develops IBT (Imagining Benchmark Tool) which will be applied in the area of picture processing. The performance will be derived from performance of typical picture processing operations.

The development and implementation of benchmarks gives a powerful tool to users to evaluate, in an objective way, the graphical performance of a system within a specific application. This helps to get the best price/performance ratio which has a positive effect on the economic figures of a company which has bought the system.

6 Conclusion

The aim of the above given text was to draw the reader's attention to some up-to-date trends in the field of CAD systems. Our attention has been drawn to **three basic topics:**

- new methods and techniques used in current CAD systems
- support of concurrent engineering methods (data exchange on various level at various stages of the design)
- up-to-date CAD systems and their evaluation.

The acquaintance with these trends should help to open up nontraditional views on CAD systems both in their usage and construction. It was obvious from the given examples that these trends represent directions of the development for the next decade (decades?). It is possible to observe that new trends are based on integration of both new software techniques and new communication techniques between designers in a team. The new generation of CAD systems is (and will be) the place where computer and human factors will further integrated in order to reach optimal use of both computer resources and human resources during the design process.

References

1. ACIS - Technical Overview; Release 1.4., Spatial Technology, Inc. 1992
2. Bronsvoort,W.F., Jansen,F.W.: Feature modelling and conversion - Key concepts to concurrent engineering. In Computers in Industry 2/93, pp.61-86
3. Daemisch,K.-F.: Verflochtene Formen, iX 5/94, pp. 48-58
4. Grau,O.: Raumfahrt, iX 1/94, pp. 130-134
5. Sammet,H.: Application of Spatial Data Structure, Addison Wesley 1990.
6. ISO: STEP - Standard for the Exchange of Product Model Data ISO CD 10303, 1991
7. Utermoehle,M.: Von Aepfel und Birnen, iX 2/94, pp.140-144

Concurrent Engineering

Dr. Steffen M. Fohn   Dr. Arthur Greef
IBM Corporation
Thornwood, New York

Professor Robert E. Young   Professor Peter O'Grady
Department of Industrial Engineering
North Carolina State University
Raleigh, North Carolina

Abstract. In this chapter, the philosophy behind Concurrent Engineering is presented as well as current approaches used to implement Concurrent Engineering. Concurrent Engineering design encourages the simultaneous consideration of all aspects of a product's life-cycle at the design stage. It has been shown to be successful in shortening product development time and costs by avoiding the typical problems associated with sequential design. Companies competing in today's global and volatile marketplace cannot afford long development leadtimes or high costs. Success stories in Concurrent Engineering have primarily relied on the design team approach, a collaboration of people from different departments representing different life-cycle perspectives. However, the design team approach and other approaches suffer in their inability to manage (i.e., store, access, update, etc.) the immense amount of data and information required to perform Concurrent Engineering.

1 Concurrent Engineering

In today's competitive global marketplace manufacturing companies must react quickly to volatile market demands and growing product complexity through efficient design processes that ensure a product's quality, competitive price, and prompt availability to consumers. Success stories [30] have shown that this can be achieved by adopting a manufacturing philosophy termed Concurrent Engineering (CE). This philosophy encourages the simultaneous consideration of all aspects of a product's life cycle, at the design stage. It stresses a parallel approach to design that is different from the more traditional approach where products are designed in isolation and only then considered in terms of their manufacturability, testability, quality, serviceability, etc. [19, 20, 32].

The traditional departmental organization of companies, engaged in product design and manufacture, has resulted in a serial or staged approach to product development. In such an environment a working design advances from one department to another in a predetermined sequence. Each department has responsibility for certain aspects of the product (i.e., manufacturability, testability, quality, serviceability, etc. When design errors are detected the working design must be sent back to the responsible department for redesign. Consequently, serially designed products are experiencing higher development costs and longer lead times from conception to market. Companies have realized that such design iterations are unacceptable in an increasingly competitive marketplace and are transforming their serial design strategy into one that stresses the parallel evolution of a product's design beginning with conception. This new strategy is called *Concurrent Engineering*.

Concurrent Engineering, also referred to as Simultaneous or Parallel Engineering, is a design philosophy which encourages the consideration of all aspects of a product's life-cycle at the design stage. The activity promotes the anticipation of problems and the avoidance of potential delays. Subsequently, design iterations are reduced, development costs are lowered, and lead times in bringing a product to market are shortened.

The simultaneous consideration of different life-cycle aspects in Concurrent Engineering has been attempted through a number of approaches: handbooks, checklists and procedures, design teams, and artificial intelligence systems (i.e., rule-based and constraint-based systems). The suitability of these approaches for Concurrent Engineering, including their advantages and disadvantages, are discussed. There is however, one disadvantage that all the approaches have in common. It is the inability to effectively manage (i.e., access, store, update, etc.) the immense amount of data and information inherent to Concurrent Engineering design. The last section discusses the information access and handling issue.

2 Concurrent Engineering Versus Traditional Engineering

Traditionally product design was performed serially. In the serial approach communication occurs but only after a particular stage in the design process has been completed. Serial design typically adheres to the following steps. First, design engineering takes a product definition and works it into a design, usually without ever consulting manufacturing. Once a design is verified by either simulation and/or hardware prototyping, the design "gets tossed over the wall" to manufacturing where the design is checked for manufacturability, and testability. The process plan is then defined, the production costs are determined, and a production schedule is built. Third, the purchasing department orders the materials, machinery, and tools required. A quality control staff develops a quality control program. After manufacturing, the service department takes responsibility of product service and maintenance. Any errors discovered along the line, results in the design being passed back to the responsible department for rework or redesign. These errors which are typically detected in the later stages are prevalent and

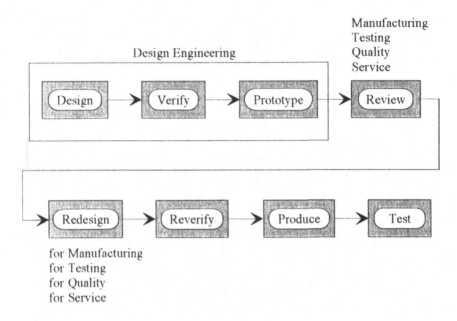

Fig. 1. Traditional Engineering [29]

characteristic of the staged process. These errors are a direct result of the division of the design problem according to departmental responsibility which forces solutions to the individual subproblems but never to whole problem. Quite often many iterations are required to correct these design errors; the end result is a prolonged leadtime from conception to manufacturing and greater total product cost.(see Figure 1)

Concurrent Engineering is also referred to as simultaneous or parallel engineering. It is essentially the collaboration of many people from the different departments representing the different perspectives of a product's life-cycle. The steps of the development process are handled in parallel instead of in serial. This activity results in the anticipation of problems and bottlenecks and helps to eliminate them as early as possibly, avoiding the delays in bringing the product to market. Concurrent Engineering reduces the number of product iterations required in serial design; one reason for this is that first generation prototypes not only meet design requirements but are also within the bounds of the manufacturing constraints (see Figure 2).

In today's market products are experiencing an ever shortening life span. This is primarily due to two main reasons. First, the majority of a product's profits occurs early in its life cycle. Second, the rate of product obsolescence is increasing due to fierce competition and rapid technological advancements. Hence, companies can be more profitable if they keep their products on the market a shorter time and replace

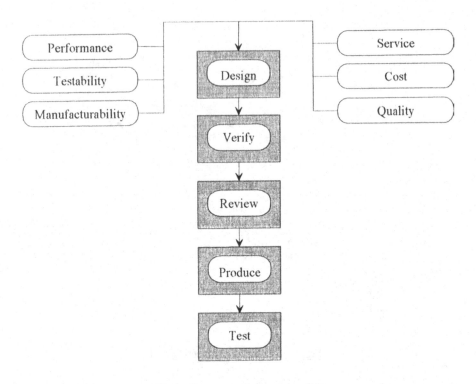

Fig. 2. Concurrent Engineering [29]

those products with ones that are more in line with ever increasing customer demands. As the average product life becomes shorter it is vital that the leadtimes of new products from conception to market shorten as well. Long leadtimes are extremely detrimental to a company's well-being and therefore simply unacceptable.

The effect a design has on the overall product cost is another important issue. Without considering and addressing life-cycle concerns in the design stage it is often too late to make substantial changes to reduce down-the-line expenditures (e.g., the cost of servicing a product is affected by the ease and time in which a product can be serviced). It is the influence of the design itself which establishes the large product cost. Robert Winner of the Institute for Defense Analysis estimates that the development phase for a product spends 1 to 7 per cent of the total project cost, but determines 70 to 85 per cent of the life-cycle costs [34]. For example, Ford Motor Co. estimates that a design's influence on the total product cost is 70 per cent even though the design itself only constitutes 5 per cent of the cost. Concurrent Engineering initiatives attempt to close the gap between the design phase and the life-cycle costs committed.

Concurrent Engineering, then, is not just an alternative way of performing design, but it is a necessary course of action which must be pursued by all companies in order to remain competitive.

3 Success of the Concurrent Engineering Approach

The purpose of this section is to cite some of the successes which have been realized through the use of Concurrent Engineering from industry. They include the following. Emhart Corporation reports a one-third reduction in the leadtime to bring a line of door hardware to the market [8]. John Deere & Co. reports a 30% cut in the cost of developing new construction equipment and 60% cut in the development time. AT&T Co. reported a 50% reduction in the time needed to make a 5ESS electronic switching system [24]. At Mercury Computer Systems the cycle for shipping a new board from design approval to a customer was reduced by over 25 % [26]. At Hewlett Packard Co., the 54600 oscilloscope was developed from idea to finish in one-third of the time it would have without Concurrent Engineering. The oscilloscope was able to be produced for the target price and is composed of only a few modules which can be assembled in 18 minutes. [30] Chrysler's Viper showed the auto industry that a domestic car can go from auto show concept car to full production in under 3 years [27]. Similar experiences from implementing Concurrent Engineering have been noted by other companies: Cisco Systems Inc., Raytheon Inc., and Litton Systems Inc. [30]. L. K. Keys, et. al. [15] state that Concurrent Engineering has been reported to yield not only reductions in development times and costs, but also in inventory, scrap, rework, and defects.

4 Nature of the Concurrent Engineering Problem

Concurrent Engineering like many other design problems is ill-structured. The design need not be bounded by a continuous, convex solution space. Design parameters range not only over the domain of real numbers but also over sets of numbers and symbols. Design objectives are numerous. Currently, there are no algorithms or procedures in existence to handle this type of ill-defined design problem. Furthermore, the Concurrent Engineering problem by its very nature (encompassing design, manufacturing, testing, servicing, etc.) spans an enormous volume and variety of information. Concurrent Engineering necessitates the use of empirical, analytical, and heuristic knowledge [14]. In other words, accompanying the problem's ill-structure is also an immense data and information management problem. Consequently, the design process primarily involves the incremental and concurrent proposal and testing of possible answers subject to the design criteria. In light of this and the fact that design problems have many solutions, designing then becomes a process of finding "a" feasible solution within the design specifications and not one of finding "the" solution [12]. Tools built to support Concurrent Engineering must take these issues into account. The next section discusses some of approaches which have been used to support Concurrent Engineering.

5 Approaches to Concurrent Engineering

There are a variety of techniques or systems that attempt to incorporate product life-cycle concerns in an effort to support the Concurrent Engineering design philosophy.

These include the use of handbooks, checklists and procedures, design teams, and Artificial Intelligence (AI) systems [32]. The techniques are all similar in that they endeavor to integrate and represent different life-cycle concerns into a medium which can be used for product design.

5.1 Handbooks

Handbooks are the most conventional technique available to advise on design. They normally entail compiled life-cycle information in the form standards and design guides. It is the designers responsibility to consider and integrate this information into the design decisions. [3,5]

5.2 Checklists and Procedures

Checklists and procedures have been predominately used for the design review for some of the life-cycle areas, in particular manufacturing. In using this technique designers are usually required to classify the design in order to obtain the appropriate guidelines for the design based on the design feature or problem [2, 9, 21].

5.3 Design Teams

Design teams are interdisciplinary groups composed of members representing all aspects of a products life-cycle (i.e., marketing, sales, research and development, manufacturing, purchasing, testing, quality control, and service). The members transcend traditional departmental boundaries and loyalties to share their ideas and concerns while negotiating conflicts during a product's design phase. The team is responsible for collaborating over to entire life of the product, from conception to obsolescence[8, 23, 31].

5.4 Artificial Intelligence (AI) Systems

There are relatively few computerized systems which have been built to support Concurrent Engineering. The majority of systems which have been built are either rule-based expert systems or constraint-based systems. A constraint-based system refers to any AI system (production rule, frame-based, object-oriented, logic representations) which utilizes the notion of constraints. Constraints are typically used to restrict variable (design parameter) relations and to support inferencing. The majority of rule-based and constraint-based systems, like the handbooks and checklists and procedures, tend to focus on certain pieces of the Concurrent Engineering problem. For example there are several rule-based experts systems which have been developed for Design For Assembly including: DACON (Design for Assembly Consultation) [28], ADAM (Assisted Design for Assembly and Manufacturing) [25], and PACIES (A Part Code Identification Expert System) [4]. Similarly, there are also numerous constraint-based systems which focus on specific portions of the Concurrent Engineering problem, such as computer graphics and design, electrical system design, and mechanical system design. Constraint-based systems, like rule-based systems, are also seen in other domains of design: planning, scheduling, configuration, and general purpose problem solvers. Constraint-based systems to support Concurrent Engineering activities are also beginning to appearing in research literature (see [10], section 3.2, for an extensive review of constraint-based systems).

5.5 Suitability of these Approaches

The handbook approach to Concurrent Engineering forces onto the designer the responsibility of managing (accessing, updating, and storing) an enormous amount of empirical, heuristics and analytical information. This is an overwhelming task for any designer and by its manual nature is prone to error. The approach of using handbooks is further complicated by the fact that handbooks typically focus on a certain perspective of the design (i.e., mechanical design). In this case the designer wishing to perform Concurrent Engineering may require several different handbooks each of which only address certain life-cycle concerns. This problem becomes increasingly severe when the designer must also balance the conflicting objectives between the different handbooks. For example, in Design for Assembly the objective is to simplify assembly by eliminating removable fasteners such as screws while in Design for Maintainability the objective is to provide quick access by using removable fasteners. Since the handbook approach is not wholly integrated, certain life-cycle concerns and their objectives will receive more emphasis possibly resulting in errors only to be discovered later in the product's lifetime.

Checklists and procedures are relatively easy to construct and their use is intuitive. The problem with these lists and guidelines is that they are also "thrown over the wall" but in the opposite direction; hence, they fail to reflect the dynamic change constantly occurring within the design and manufacturing environments [26]. Many of these types of approaches suffer from their dependency on accurate cost information which is often difficult to obtain [16]. The procedures are usually rigidly structured thereby mechanizing the design evaluation into stages possibly biasing the solution towards a particular sub-problem. In addition, checklists and guides tend to focus on one certain area [13]. This characteristic forces the designer into managing and synthesizing the information output resulting from the different checklists and procedures which are required to address the different life-cycle concerns. Again the designer is plagued with voluminous knowledge and data management. As with the handbook approach, this approach is also not wholly integrated, skewing the design in favor of certain life-cycle concerns, possibly resulting in errors to be discovered only later in the product's lifetime.

Design by its nature is collaborative and communal. Teams of engineers of are needed to design most real world complex systems [18]. To engage in Concurrent Engineering, companies usually set up design teams [1, 6, 15, 17, 23, 34] whose members, drawn from all stages of the product's life-cycle, are responsible for ensuring that their design rules are satisfied (or at least considered) at the time of design. Consequently, problems which in the serial approach surface in some later stage are discovered early on in the design team approach to Concurrent Engineering. The approach has been very successful in many sectors of industry; the successes listed in section 3 are based upon design team implementations. Although conceptually, the design team approach to Concurrent Engineering seems plain, its practical execution is not. When considering the life cycle of even the simplest product, the number of aspects to be considered at the design stage, is quite formidable making the design team approach difficult to manage. The management problem can be decomposed into two subproblems. First, there is the management of the team and design process itself (i.e., how and when should what members interact) in order that the design be accomplished in the most efficient and easiest way possible. There are also many other issues that fall into team management but are not directly related to the design process; these include: team member compensation and rewards, etc. [22]. Second, there are massive information management problems; this set of problems is a result of two main conditions in which team members commonly work. First, there is no central medium for storing, accessing, and sharing up to date life-cycle information among team members. Each team member typically uses their own set of computing resources (i.e., databases, spreadsheets, etc.) causing compatibility problems. Second, team members are usually geographically separated. [23] The probability of communication between team members more than 50 yards apart decreases by 80 per cent [22]. In other words, the major problem with the design team approach lies rooted in the inability of team members to efficiently and effectively communicate their concerns to the design effort and in the unavailability of immediate access to life-cycle data from all of the perspectives. Without the assistance of a computerized decision support tool, the design team members must rely on paper communications and face to face meetings, both of which are subject to human delays and errors.

Rule-based expert systems force a designer into making a hierarchy of decisions [11]. Hence, they are not particularly well suited to model a Concurrent Engineering problem because the problem itself is characterized by a wide number of choices and lacks any clear hierarchy of decisions. This makes it nearly impossible to construct all but a very specific system. [14] Consequently, designers are neither able to approach the design from different perspectives of the life-cycle nor design in the absence of complete information. Furthermore, it is difficult to represent the mutually constraining influences that different aspects of a product's life-cycle exert on each other [33]. Constraint-based systems, generally speaking, do not suffer from these problems and seem to offer the flexibility deemed necessary for a Concurrent Engineering decision support tool. The recent growth in the development and emergence of constraint-based systems can be attributed to growing general consensus that design is indeed a constraint-based activity. Although, Artificial Intelligence techniques in general address the problem at a knowledge representation and reasoning level, they are not sufficient to support the data processing for information intensive problems [7].

In summary, the design team approach is effective in alleviating the narrow scopes typically adopted by handbook, and checklists and procedures approaches. This is done by providing an environment of several team members whose responsibility it is to make sure that their life-cycle concerns are addressed during the product's design. Hence, certain aspects of the design's life-cycle cannot be forgotten or ignored resulting in a better overall design in that one is ensured of a design meeting life-cycle requirements. However, the design team approach is still weak in two areas: the management of the team member interaction during design (i.e., the design process and the communication required between the different team members), and the management of the information required to support the design process. Both of these weaknesses must be addressed and resolved in order to efficiently perform Concurrent Engineering. This is especially true for design problems on a large scale. The first weakness, the management of team member interaction (i.e., the communication between the different life-cycle perspectives), can be addressed by AI knowledge representation and reasoning techniques. As mentioned above, rule-based systems are inadequate for modeling Concurrent Engineering problems; this is primarily due to the rigid hierarchy of decisions. Constraint-based systems on the other hand seem to provide the flexibility needed in this type of decision support system to resolve this first weakness. However, the second weakness, management of the voluminous information imperative to the design process, still needs to be resolved (see Figure 3).

Additionally, it is important to point out that there are many tools and classical methods currently used by those designers engaged in any above systems. The significance of these tools and methods dictate their incorporation into the total Concurrent Engineering solution. Some of theses tools are Computer-Aided Design (CAD), Computer-Aided Engineering (CAE), Computer-Aided Manufacturing (CAM), Computer-Aided Process Planning (CAPP), and Computer Simulation. Many powerful methods have been developed for design "optimization", these include: Quality Function Deployment (QFD), the Taguchi Method, Design-For-Assembly (DFA), Design-For-Manufacturabilty (DFM), Statistical Process Control (SPC), Continuous Process Improvement (CPI), Just-In-Time (JIT), Total Quality Management (TQM), etc. [7, 13]. Because of the major role these tools and methods play in hastening product improvement, the Concurrent Engineering design system of the future will also have to interface and integrate these tools and methods.

6 Information Accessing and Handling as a Major Issue

The key ingredient to Concurrent Engineering is teamwork but systems for sharing and managing design information are vital [24]. It is important to reiterate that Concurrent Engineering, is not just a knowledge intensive activity as might be suggested by the previous approaches but is also an information intensive activity. When considering that Concurrent Engineering is meant to consider all aspects of a product's life-cycle from conception to obsolescence, it is obvious that the information requirements are enormous (i.e., associated with each life-cycle concerns are corporate and government standards and specifications, corporate and

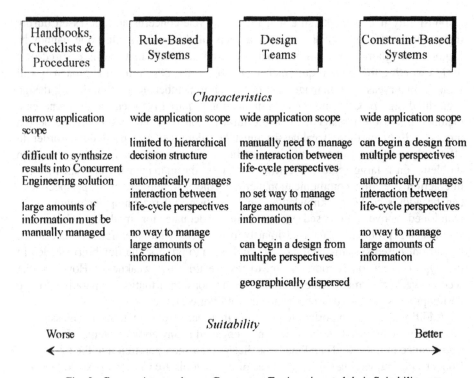

Fig. 3. Current Approaches to Concurrent Engineering and their Suitability

professional society empirical data, etc.). Computer tools built to support Concurrent engineering must take both of these issues into account.

The knowledge issue can be addressed by a computer tool, facilitating the design team approach, by providing a central medium capable of representing individual design team members' concerns and communicating those concerns to downstream product designers. AI techniques (e.g., constraint-based systems) are particularly well suited to handle the knowledge representation and reasoning (i.e., design team member interaction -- the "design process") burden of Concurrent Engineering Problem. However, there are currently no systems which include facilities for the intense information management needed for Concurrent Engineering. Although computer-based methods (Database Management Systems) exist that are good at information management (e.g., data storage, accessing, updating, etc.) they have not been used and integrated into design systems to the extent necessary to support Concurrent Engineering. To successfully resolve the information requirements for Concurrent Engineering, a Computer tool must retrieve and digest information into useful knowledge which not only supports the decision making but also guides it. A Computer tool effective in addressing both of these issues will have taken a significant step towards a more complete decision support tool for Concurrent Engineering.

7 References

1. S. Ashley: The Battle to Build Better Products. Mechanical Engineering 112, 11, 34-38 (1991)

2. G. Boothroyd, P. Dewhurst: Product Design for Manufacture and Assembly. Manufacturing Engineering, April (1988)

3. J. G. Bralla: Handbook of Product Design for Manufacturing. McGraw Hill 1986

4. Y. Chen, R. E. Young: PACIES: A Part Code Identification Expert System. IIE Transactions 20, 2, 132-136 (1988)

5. W. H. Chieng, D. A. Hoeltzel: A Generic Planning Model for Large Scale Engineering Design Optimization with a Power Transmission Case Study. Computers in Engineering 1, (1987)

6. B. Chiesi, T. Parasida, E. Walliser: Concurrent Engineering at Boeing Helicopters. 47th Annual Forum Proceeding of the American Helicopter Society, Part 2, May 6-8, 921-930 (1991)

7. S. N. Dwivedi, R. Lanka: AI and Concurrent Engineering in Factories of the Future. In: J. L. Alty and L. I. Mikulich (eds.): Industrial Applications of Artificial Intelligence. Elsevier Science Publishers 1991, pp. 20-32

8. B. Evans: Simultaneous Engineering. Mechanical Engineering 110, 2, 38-39 (1988)

9. W. Eversheim, W. Muller: Assembly Oriented Design. In: W. B. Heginbotham (ed.): Programmable Assembly, IFS Ltd., UK, 1984

10. S. Fohn: Dynamic Constraint Satisfaction with the Relational Model for Concurrent Engineering. Ph.D. dissertation, Dept. of Industrial Engineering, North Carolina State Univrsity, USA, 1994

11. A. Greef, R. E. Young: Representing Rule-Based Expert Systems as Constraint Networks. GISDEM Technical Report TR-90-22, Department of Industrial Engineering, North Carolina State University, USA, 1990

12. M. Grover, S. Ervin, J. Anderson, A. Fleisher: Designing with Constraints. In: Y. E. Kalay (ed.): Computability of Design. John Wiley & Sons, New York: 1987, pp. 9-36

13. D. Hall: Concurrent Engineering: Defining Terms and Techniques. IEEE Spectrum 28, 7, 24-25 (1991)

14. V. M. Karbharl, D. J. Wilkins: Decision Support Systems for the Concurrent Engineering of Composites. Proceedings of the Seventh Annual ASM/ESD Advanced Composites Conference, Detroit, MI, Sept. 3--Oct. 3, 459-467 (1991)

15. L. K. Keys, R. Rao, K. Balakrishnan: Concurrent Engineering for Consumer, Industrial Products, and Government Systems. IEEE Transactions on Components, Hybrids, and Manufacturing Technology 15, 3, 282-287 (1992)

16. E. Kroll, E. Lenz, J. R. Wolfberg: A Knowledge-Based Solution to the Design for Assembly Problem. Manufacturing Review 1, 2, (1988)

17. S. A. Meyer: Integrated Design Environment-Aircraft (IDEA). An Approach to Concurrent Engineering. 46th Annual Forum Proceedings of the American Helicopter Society, Part 1 (of 2), May 21-23, 509-522 (1990)

18. S. Mittal, A. Araya: A Knowledge-Based Framework for Design. Proceedings of the Fifth International Conference on Artificial Intelligence, Philadelphia (1986)

19. P. O'Grady, R. E. Young, A. Greef, L. Smith: An Advice System for Concurrent Engineering. International Journal of Computer Integrated Manufacturing 4, 2, 63-70 (1991)

20. P. O'Grady, R. E. Young: Issues in Concurrent Engineering Systems. Journal of Design and Manufacturing 1, 27-34 (1991)

21. M. Oakley: Managing Product Design. John Wiley Son, Inc., New York 1984

22. J. V. Owen: Concurrent Engineering. Manufacturing Engineering 105, 5, 69-73 (1992)

23. R. Reddy, R. T. Wood, K. J. Cleetus: The DARPA Initiative: Encouraging New Industrial Practices. IEEE Spectrum 28, 7, 26-30 (1991)

24. A. Rosenblatt, G. F. Watson: Concurrent Engineering. IEEE Spectrum 28, 7, 22 (1991)

25. P. J. Sackette, A. E. Holbrook: DFA as a Primary Process Decreases Design Deficiencies. Assembly Automation 137-140 (1988)

26. S. G. Shina: New Rules for World Class Companies. IEEE Spectrum 28, 7, 23-26 (1991)

27. E. E. Sprow: Chrysler's Concurrent Engineering Challenge. Manufacturing Engineering 108, 2, 35-42 (1992)

28. K. Swift: Knowledge-Based Design for Manufacture. Kogan Page 1987

29. J. Turino: Making it Work Calls for Input from Everyone. IEEE Spectrum 28, 7, 30-32 (1991)

30. R. Wheeler, R. W. Burnett, A. Rosenblatt: Concurrent Engineering: Success Stories in Instrumentation. Communications, IEEE Spectrum 28, 7, 32-37 (1991)

31. R. T. Yeh: Notes on Concurrent Engineering. IEEE Transactions on Knowledge and Data Engineering 4, 5, 407-414 (1992)

32. R. E. Young, A. Greef, P. O'Grady: An Artificial Intelligence Constraint Network System for Concurrent Engineering. International Journal of Production Research 30, 7, 1715-1735 (1992a)

33. R. E. Young, A. Greef, P. O'Grady: Representation Issues for a Constraint-Based Concurrent Engineering System. GISDEM Technical Report, Dept. of Industrial Engineering, North Carolina State University, USA, 1992b

34. W. I. Zangwill: Concurrent Engineering: Concepts and Implementation. IEEE Management Review 20, 4, 40-52 (1992)

Groupwork in the Shop-Floor-Area
Needs Decentral CIM-Structures and Components

Dieter Specht and Frank Fehler

Brandenburgische Technische Universität Cottbus,
Postfach 10 13 44, 03013 Cottbus, Germany.

Abstract. Groupwork in shop-floor-area needs specific information technology concepts. Supporting the complex job spectrum in the production- and assembly sector requires more extensive information and control. Groupwork-oriented data processing structures have to accomplish data- and information-exchange between vertical and horizontal levels of organization. Production controlling, personnel planning, quality management, maintenance controlling and group accounting are the needed modules of an integrated groupwork information system. All availible informations have to be compressed step by step to get high acceptance by the workers. The groupwork controlling dashboard (GCD) represents signific coefficients of global business targets. The introduction of groupwork can be supported by information technology. Client Server Arcitecture of information systems makes a step by step introduction of decentral production- and CIM-structures possible. The decentralization of maintenance shows the concept of an integrated organization and information restructuring as an example. This concept can be transmitted to the decentralization of other indirect tasks.

1 The World-Market Requires Refined Concepts of Production Organisation and Information Management

Today's business success of numerous enterprises has to be based on the effective use of covered potentials. The enterprises have to activate all their available manpower, ressources and materials. Groupwork is an important step to break with traditional organisation and hierarchical structures. It is a way to realize lean production.

The introduction of groupwork is characterized as a process of work organisation, shop-floor layout as well as workers education and development. Common restructuring projects can not be considered to fit the information management. Information management has to supply all target groups with information commensurate with their needs.

Information systems used in modern decentral production should be based on these structures. The urgent need of a specific CIM-concept results from an investigation of shop floor organizations with groupwork. CIM-concepts designed for a hierarchical organisation cannot be used successfully for groupwork [1].

2 Groupwork Changes Work and Work-Conditions

By introducing groupwork, the workers areas of responsibility grow by dispositive, organizational and administrative jobs. The principle and complexity of job sharing as well as the work process of single workers are distinguishing features of groupwork models (Fig. 1).

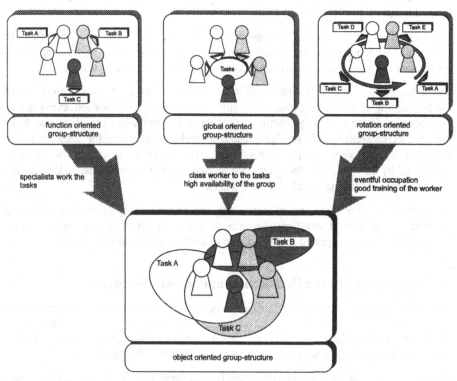

Fig. 1: Groupwork models

Function-oriented groupwork contains a straight assignment of different kinds of jobs to specialized workers or groups. This groupwork model allows to integrate indirect production jobs, for example disposition of materials, quality control and maintenance into the responsibility of the team. It supports the decentralization of production-planning and controlling. Every team member is responsible for a seperate task. The most important advantage of this model is that all jobs are done by a specialist. On the other hand the team members need a defined seperation of labour-tasks. Therefore this model is less flexible. For example the absence of maintenance workers may cut the production process because only maintenance staff know how to perform repairs which are of urgent necessity.

Quite different is a groupwork model which is global-oriented. This structure requires universally trained and qualified team members. All workers do all jobs. Specific requirements decide about the assignment of the workers to jobs. Personnel planning can be made for a short term because the group is highly available. Global-oriented groupwork requires only a limited range of different jobs. The workers have to be qualified and trained for the processing of all coming jobs. That requires investment in human ressources.

A special model of the global-oriented team organization is the rotation-oriented groupwork. The partitition of labour is effected by a rotation principle. The workers are avaible to do different jobs in alternative ways. All group members get a good basic training and practice of all relevant operations. This model is suitable for a limited range of different jobs as well .

The object-oriented type of groupwork is particularly useful for the flexible production of single objects including indirect production jobs. In this model of groupwork the groupmember-structure is deduced from the production object.

The group should consist of employees with different kinds of education. In this way the performing of all existing jobs is possible. According to qualification, every group member should be able to perform all present tasks. But there are also jobs from nonproductive areas beside production and assembly. There are for instance many recurrent or not very complex jobs. They are especially suited to be flexibly performed by all workers of the group. An increasing complexity of the tasks requires a higher level of training for all skilled workers.

The group has to be composed in such a way that the necessary know-how of skilled workers is available at any time. Sometimes this is essential because other skilled workers have to be integrated into the group in order to carry out specialized tasks.

3 Jobspectrum in the Production and Assembly Sector

Describing the structure of groupwork we have to consider the organisation model and the whole spectrum of jobs that have to be done. The responsibility of a team consists of quality management, maintenance and repair, materials management, process control and personnel management as well as production and assembly jobs.

Figure 2 shows as an example the complex job spectrum in the production- and assembly sector. The assignment of every single job has to be individually determined by using specific boundary conditions. In addition to this, an inclusion of upstream and downstream operation of the production process is practicable.

The introduction of team work is not dependent on the used groupwork model and represents an important step towards Lean Management. Different levels of hierarchy can be reduced by transferring important decision competences of the production management. Object-oriented groupwork includes a decentralisation of tasks and competences. The team should be autonomous. Every member of the team is responsible for the success and also for the failure of the whole groupwork. The assignment of orders to single teams, the fixing of quality standards, the redistribution of workers, the material planning is still done by a central production management. The whole fine tuning and the implementation will be done by the team.

Fig. 2: Example of the complex job spectrum in the production- and assembly sector

4 Object Oriented Information Management

Many CIM-activities in the 80's were based on a hierachically organized enterprise. Often this has turned out to be cost-intensive in terms of labour. All applications need high-quality hardware-systems. Many enterprises used host computers and individually designed software. In the last years powerful microprocessors have become cheaper. Therefor it is possible to distribute computer performance corresponding to the requirements. On the other hand the evolution of standard software modules tend towards individual adaptation. In this way modern decentral production structures can be coordinated by a software- and hardware supported integration. Computer integrated manufacturing -CIM- gets a new dimension with object oriented groupwork.

The more complex tasks handled by the groupmembers, the more extensive is the need for information and control. Groupwork requires a more complex access to data than normal manufacturing methods. Groupmembers dealing with the success and failure of the group need information about economical characteristics of the group, as well as information about the enterprise.

Groupwork-oriented data processing structures do not only take data-exchange and information-exchange between vertical levels of organization into account, but also horizontally between individual production units. Direct connection of information is necessary for an optimal communication performance. Data processing units that slow down the information flow are unsuitable.

Many tasks of the group and existing computer equipment require the connection of data processing systems. This is of special importance for computers and CNC-controls of a production unit. In addition to this, an access to higher systems is needed, e.g. production planning and control, quality management systems or maintenance planning systems. (Fig. 3)

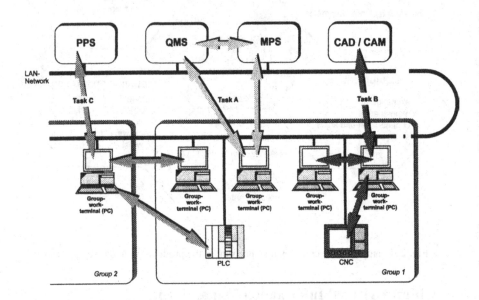

Fig.3: Object oriented information management for groupwork

Complex tasks of production planning and control call for information-technology to cooperate with upstream and downstream groups, e.g. data- and capacity planning. Horizontal coordination becomes more important under the aspect of an extensiv group autonomy. By means of data processing connections with Kanban-Systems and Just-in-time-controll are practicable efficiently.

5 Informationsystems for Groupwork

The logical data processing structure of a Tayloristic production organisation is equivalent to the shape of a tree. The information system structure of a groupwork-oriented production area needs a matrix structure. A modern Client-Server concept makes it possible to integrate all sorts of hard- and software equipment into a data processing stucture [4].

Every integrated system should be both Client and Server. An open concept offers advantages for the introduction and extension of data processing components. Thus, a complex information system structure based on the organisation of production can be developed.

Modern decentral stuctures of production require an integration of data processing. The information management has to get access to make all informations availible in the production department and in the enterprise, if needed.

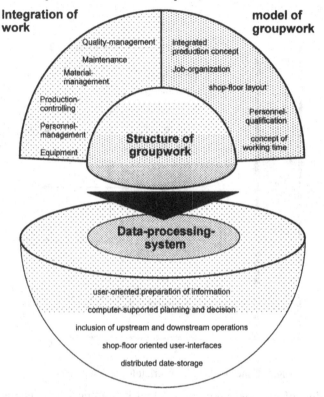

Fig.4: The data processing draft is oriented towards production organisation

Figure 4 shows boundary conditions for the development of groupwork. The extent of the required data processing system is determined by the task spectrum of the team.

Software- and hardware-systems for groupwork, should have to support the worker during planning and control of their tasks. This includes a decision support system for complex organisational and technical problems. Normally these tools are more or less well provided by many systems. These systems are, however, not suitable for use in groupwork. Modern software- and hardware systems for groupwork have to satisfy the requirements on lean business [6]. This means:

- structures of software- and hardware-systems have to be compatible with the decentral organisation of enterprises,
- required information should be available for system users.

The complex job spectrum in the production- and assembly sector contains production and assembly, personnel management, quality management, maintenance as well as material and equipment management. From possible areas of responsibility of a group the following software tools can be deduced:

- production controlling,
- personnel planning,
- quality management,
- maintenance controlling,
- group accounting.

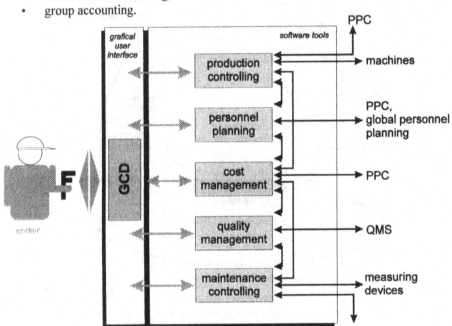

Fig.5: Concept of a groupwork information system

The main window of the production control tool is a grafical user interface for the appointing of production orders. This function supports workers to log and handle job orders. Besides the schedule of machines and facilities will be carried out. The system allows to make capacity plans for all resources depending on the group. Additional functions support the groupmembers for the inventory management to purchase parts and mould blanks. If required the administration of CNC- and industrial roboter programms can be included as well. This tool connects upstream and downstream groups as well as higher order systems to take production controlling datas and hand over answering signals.

Personnel planning tasks can be supported by the personnel controlling tool. It helps the workers to plan labour employment in consideration of required qualifications as well as the shift organization and vacation planning. Another important task of the personnel planning is to find out the qualification profil of the group. Besides the requirement of specialized training courses have to be detected.

The logging of quality-datas: reference dimensions, attributive features and defective components are the basic function of the quality control tool. Reference dimensions logging can be done online when the measuring devices are connected to the groupwork information system by a computer network or off-line by manual data input. To get best results about all steps of quality, tests should to be supported by the computer: The acquisition of test parameters, the evaluation of the test results, the calculation of quality indexes and the output of the test results. The test sheeds for all components of the production programm are stored in the quality control tool. For comfortable groupwork supporting the evaluation and result tracing will be done through histogramms, quality control cards and characteristic process values.

The maintenance control tool includes analysing and planning functions. The basic of efficient maintenance analyses functions is to write down production stops and machine standstills. These datas are available from the BDE-system. Analyses of breakdown periods inform about machines' reliability. This is an important basis for the spreading of the capital expenditure budget. On the other hand the system has to detect and analyse trouble reasons. For this weakpoint analysis will be transacted. The maintenance control tool helps the workers to introduce trouble shooting.

A distinguished planning function is the development of maintenance plans for machines and measuring devices. To get a high capacity utilization the maintenance plans have to be reconciled with the production plans. Production deadlocks can be used for prefered maintenance activities. The administration of decentral spare parts and tool depots belong to the maintenance control as well.

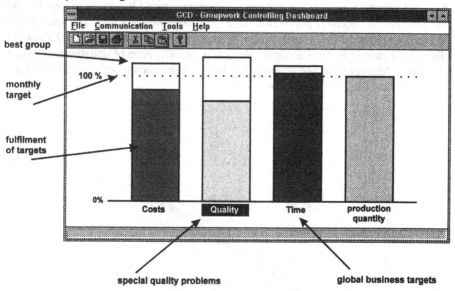

Fig.6: Example of a groupwork controlling dashboard

Basic function of the group accounting tool is the evaluation and tracing of economic coefficients: costs per unit, refuse costs, costs per repair and restoring works, personnel costs, time to pass through, production efficiency. These coefficients have to be represented clear and understandable to identify cost drivers. Among the representation of coefficients per month the user need coefficient trends and evolutions. Benchmarking functions allow the workers to compare the efficiency of their group with other groups. This kind of competition is an effective way to cut costs, inprove production efficiency and reinforce group members identification with their work.

All software tools are supervised by the groupwork controlling dashboard (GCD). It is based on targets which are derived from global business targets. These are costs, quality, time and production quantity targets. Information created by the different controlling tools are compressed for this signific coefficients. The GDC shows one screen to inform the worker about the group performance. Therefore the global business targets are graficaly traced. A bechmarking function is included to compare the target fulfilment of the group with the other groups. If there is an important problem the target will be marked. The users are notified to analyse and repair the problem.

6 Support the Introduction of Groupwork by Information Technology

The introduction of groupwork creates many new tasks for the workers. In order to avoid an overloading of the workers, many indirect tasks should still be done by a central department. The group will take over these new tasks step by step. The Client Server Concept supports this process of rearrangement. It makes a step by step introduction of decentral production- and CIM-structures possible. The introduction will be explained by the decentralization of maintenance activities. Figure 7 shows the decentralization of information technology and the delegation of maintenance tasks to the group dependent on the strategic production organisation and the complexity of the maintenance tasks can be accomplished.

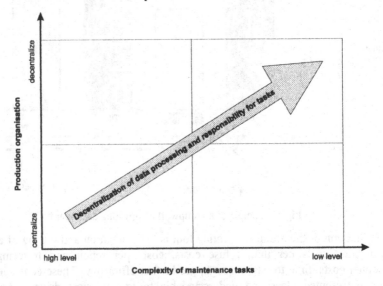

Fig. 7: Portfolio of groupwork oriented maintenance concepts

Depending on the individual conditions of the firm the decentralization of maintenance tasks can happen in four steps. On the basic step the group takes care for simple tasks: keeping clean of work places and tools, greasing machines and carry out easy repair works. After that a stage of job enlargement follows. The groupmembers take over more sophisticated maintenance jobs. With step three starts job enrichment. The group becomes responsible for planning, realization and check up of regular maintenance tasks. The last stage of development is the self-organization of the group. They take over responsibility of all internal tasks and the synchronization of their activies with the other departments.

A maintenance controlling system is used in the maintenance department. It needs to be permanently supplied and extended with new data. Depending on the stage of decentralization the groupmembers need this information for certain maintenance jobs. With Client Server Concepts it does not matter on which PC the data are physically stored, because they are available on all connected systems. The responsibility of data correctness and validity can be adjusted by a simple rule: responsibility of a task means responsibility of the specific data as well. The group members can not take over the responsibility of data before accomplishing the third step of decentralization. After that Data recording and data maintenance will be done at the creation area.

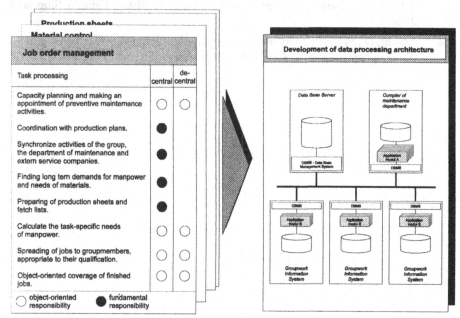

Fig. 8: Concept of a modular maintenance controlling system supporting groupwork

Figure 8 shows the concept of a modular maintenance controlling system based on Client-Server-Architecture. The maintenance department is competent for the coordination of maintenance activities with the production planning. Additionaly the department synchronizes the tasks of the groups, themselfes and external service companies. It finds out the long-term demands for manpower and for needs of material. The preparing of production sheets and fetch lists are tasks of the maintenance department. This data will be stored on a central data base server. All other planning tasks in this example belong to the group. For group members and maintenance specialists there are different applications and user interfaces to support their work.

References

1. Spur, G.; Stöferle, Th.: Handbuch der Fertigungstechnik. Band 6 - Fabrikbetrieb. Carl Hanser Verlag, München Wien, 1994

2. Pocsay, A.; Ripplinger, M.: Controllingorientierte Fertigungssteuerung. CIM-Management (1993) No. 6, pp. 45-48

3. Niefer, H.: Planung, Einführung und Optimierung von Gruppenarbeit in der Teilefertigung. Dissertationsschrift, TU Berlin 1993

4. Testi, F.: Client-Server integrieren die automatische Datenverarbeitung. Werkstatt und Betrieb 127 (1994) 1-2, pp. 25-28

5. Spur, G.; Specht, D.; Herter, P. J.: Job design. In: Salvendy, G.; Karwoski, W.: Design of work and development of personnel in advanced Manufacturing. John Wiley & Sons, Inc., New York, Chichester, Brisbane, Toronto, Singapore, 1994

6. Specht, D.; Fehler, F.: Organisationsorientierte Informationsverarbeitung für die Gruppenarbeit. ZwF 89 (1994) 7-8, pp. 362-365

Does CIM Need AI?

Vladimír Mařík, Jiří Lažanský, Olga Štěpánková

Czech Technical University, Faculty of Electrical Eng.,
Technická 2, 166 27 Prague 6, Czech Republic

Abstract: The parts of AI which seems to be of substantial importance for the CIM tasks are overviewed in this contribution. The stress is aimed mainly at impacts of Knowledge Engineering at the global CIM software architectures, at Distributed AI as a basis for system integration, at AI-centered view on Planning and Scheduling, at the role of Declarative Programming and Machine Learning in CIM tasks. The potential contributions of the stressed topics is discussed.

Two co-authors of this contribution are the co-editors of this volume. They tried to follow the goal of partial linking and integrating other AI-oriented complementary contributions in this book.

1 Introduction

The idea of CIM (Computer Integrated Manufacturing) stresses the global strategic goal: to *integrate* all company activities into a unified management structure exploring a large scale hierarchy of computers. It is expected to embrace corporate product design, manufacturing, marketing, sales management, planning, scheduling, real-time machine control, material handling, assembling, quality control and product dispatching [1].

Although this idea may be considered as a yet unattainable horizon, different software tools for CAD, CAP, CAM, CAQ etc. have been developed and they are successfully applied in solving partial tasks of the entire goal. The ambitious target of CIM emphasizes complete integration of computer-aided activities in the factory as well as the use of the knowledge-based technology within the entire information processing. Unified behavior and intelligent access to the particular subsystems imply the need for integration of the technical, managerial and business activities.

Encyclopedia of Artificial Intelligence [2] characterizes AI as follows: "AI is the study of ways in which computers can be made to perform cognitive tasks, at which, at present, people are better. Examples of problems that fall under aegis of AI include commonsense tasks, such as understanding English, recognizing scenes, finding a way to reach an object that is far overhead, heavy, and fragile, and making sense of the plot of a mystery novel. In addition, AI includes expert tasks, such as diagnosing diseases, designing computer systems, locating mineral deposits, and planning scientific experiments. The techniques that AI applies to solving these problems are representation and inference methods for handling the relevant knowledge and search-based problem-solving methods for exploiting that knowledge. Also the tasks with which AI is concerned may seem to form a very heterogeneous set, they are, in fact, related through their common reliance on techniques for manipulating knowledge and conducting search."

This way, AI is a scientific discipline gathering techniques, approaches and methodologies rather then aiming to develop a unified, consistent theory. Within its frame, the central role of problem specific (and very often heuristic) *knowledge* for problem solving has been recognized.

Besides of the expert systems (as special software *products*) AI provides a wide spectrum of methods and techniques (state space search, space pruning by heuristic knowledge, theorem proving, uncertainty processing etc.) which may be embodied into various software products, thus influencing their structure and enhancing the performance. AI has brought a new philosophy into the software engineering.

The scope of subareas covering the area of AI shifts in time: While new and new topics appear in the AI scope of interest, some older subareas are leaving the AI focus as soon as their background is well-formed and certain theoretical level is reached. This is the case of adaptive algorithms for control that are now considered as a part of control science. Similarly, pattern recognition has evolved into a nearly independent discipline. The substantial part of the current content of AI can be covered by the following list of subproblems:

- Knowledge Representation and Problem Solving
- Pattern Recognition
- Evolutionary Computing and Neural Networks
- Expert Systems (incl. uncertainty processing)
- Machine Learning
- Logic, Declarative Programming and PROLOG
- AI Planning & Scheduling
- Distributed AI
- Qualitative Simulation
- Image and Speech Processing and Understanding
- Robotics
- Knowledge Engineering

Some of these areas have currently a minor impact on CIM (like pattern recognition, speech processing), while some of them play an important role in the further development and integration of complex CIM systems.

In this paper, we would like to point to those important parts of AI that are of substantial importance from the CIM point of view:

- Knowledge Engineering and its impact on the global CIM software architecture;
- Distributed AI providing structures for software integration;
- AI-centered view on Planning and Scheduling;
- Declarative Programming and Constraint Logic Programming in CIM tasks;
- Machine Learning as a tool for building knowledge-intensive software.

Some of these topics are treated in detail in the corresponding contributions in this volume (e.g. Planning & Scheduling, multi-agent approach of Distributed AI, genetic algorithms). Besides, there are included also contributions explaining some new AI techniques promising good potential for CIM applications (Qualitative Simulation or Robotics for CIM).

2 Knowledge Engineering and CIM Software Engineering

CIM-Software Engineering is aimed to the development and maintenance of complex software systems covering the considered CIM-tasks in the factory. Usually, the user company purchases hardware and software components from different vendors. The goal of CIM-software designers is to integrate all these products on different levels of abstraction, starting from the conceptual level up-to the final implementation. This integration should be significantly more sophisticated than simple sequential linkage of the existing systems. It must take all possible information interconnections among subsystems into account, i.e. loops in the so called *product life-cycle model* should be considered. Moreover, the CIM-system specifications are rather performance-based so that constraints resulting from the hardware and manufacturing structures, from the company habits and traditions, from the demands on flexibility of CIM-solutions, etc. form a very important part of these specifications.

The ultimate goal of manufacturing is the final product. In connection with the product, it is often spoken about various models. When studying the product functionality or behavior during the product design, the *product model* is used. For example, when designing an electronic circuit, emulation of the product can be done using a computer. However, it is not our intent to discuss this type of models here.

As we are concerned with manufacturing, we should pay attention to the *product life-cycle*. It consists of all the stages starting at the very first idea of the product purpose, through the design and manufacture, to the maintenance phases, or even to the after-use abolition. The product life-cycle is, in principle, a serial process because it describes the life of one instance of the product. However, the products are usually manufactured in quantities. Thus for manufacturing purposes, it is necessary to consider the fact that the stages of the life-cycles of different individual products run in *parallel*. The parallelism is embedded in the *product life-cycle model* (PLCM) which contains not only the data on the product itself, but also the specification of processes, devices, services, and business options met during the entire product life. The PLCM also mirrors the dynamics of the product development together with the changes in the manufacturing environment.

Traditional *software engineering* (SE) has introduced the notion of *software life-cycle model*. The software life cycle model may be considered as a good analogy to the PLCM. The experience in software engineering can enhance the methods used in PLCM structuring, and vice-versa.

Artificial intelligence is a source of rich experience on software development for solving problems in complex environments. AI has certified the fundamental role of *knowledge*, its handling, and the importance of an adequate problem *representation*. The complexity of some problems has called for entirely new approaches motivated by natural systems exhibiting intelligent behavior, for instance, distributed *multi-agent* systems. Thus, AI provides a good theoretical and philosophical background for solving the problems of integration of autonomous subsystems - a basic task of CIM software engineering.

2.1 Knowledge Engineering and its Influence to Life-Cycle Modeling

Knowledge Engineering is an organic part of software engineering dedicated to the design and development of knowledge-based systems. Thus, the methodology of the AI-systems can find a direct counterpart in system architectures and methodologies used in the CIM area. Is there any difference between the traditional software engineering approaches and the AI methodologies?

The traditional software life-cycle in SE is described by the *waterfall model* presented by Royce in 1970 [3] - see Figure 1. Linear ordering of all successive phases with the sharp borderlines between individual life-cycle steps is the main feature of this model. As a rule, each phase must be completed before the next one is started.

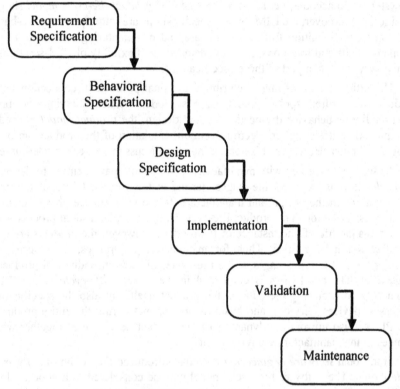

Figure 1: The waterfall model

The problem specifications in classical software engineering tasks are precise. This is not the case of the AI systems that have to cope with ambiguous, ill-defined problems. As an example of such an AI task in the manufacturing area, let us mention the problem of finding the culprit of low quality production. The specification is often not more than a vague performance-based description of the desired system behavior. Such systems can hardly be verified precisely, but their performance may be evaluated and compared to expectations. This is because of the lack of any objective criteria due to the vague requirements. Often, the quality of the AI system behavior can be judged by the end-user subjectively, only. That is why the process of a real-world complex AI system development is never complete and may be recognized as an everlasting *incremental improvement*.

The idea of permanent incremental improvements logically leads to the *fast-prototyping* strategy. It focuses on the rapid development of a *pilot prototype* which is expected to be improved and extended gradually - often many times trough a lot of feedback loops. The fast-prototyping mentioned here is neither prototyping as introductory case-study nor prototyping for experimentation but *prototyping for evolution*. It is not, in fact, prototyping in a strict sense, and it is sometimes referred to *as versioning*.

It is quite clear that the waterfall model does not suit to the Knowledge' Engineering purposes. Some other models have been presented later, e.g. the evolutionary model, spiral model or RUDE (Run - Understand - Debug - Edit) model [4]. But the ideas of a fast prototype which is not built, evaluated and discarded but which is built, evaluated, modified, and re-evaluated, fits well to the RUDE model representing the RUDE cycle as a possible basis for system development.

There is one substantial flaw of the two models mentioned above: They express possible iterative processes within the software development process but they lack any global guidelines encompassing the overall design strategy. They are not life-cycle models (LCM) in a full sense; they may rather be considered as meta-models to them [5]. One of the trials to avoid this flaw was the proposal of the POLITE [6] software LCM. This explores the waterfall model as "a structured and controllable framework" in which a RUDE cycle is invoked *within each stage* of software development.

Much closer to the nature of the practical system development seems to be a modified POLITE model referred to as POLITE-VRC model (VRC - various RUDE cycles) [7] - see Figure 2. In this model, many diverse *iterative cycles not only within, but namely between different LCM phases* are considered. Such feedback cycles, like, for instance, iterative cycles between the phases "Implementation" and "Behavioral Specification", can be carried out on different levels of generality, granularity, etc. The POLITE-VRC model which considers all possible iterative "feed-back loops" in the LCM fits the philosophy of *fast-prototyping* and *versioning* better than the other models. It also covers the philosophy of such processes like *controlled modifications* (the process of generating system modifications by tackling each specific problem at its source [4]) or *structured growth*. In the latter case, the structure of the system is growing into the depth and into details. The already created structure serves as a skeleton of the permanent refinement.

Stressing the role of the RUDE cycles, the POLITE-VRC model may be expressed as a sequence of repeated applications of RUDE cycles embracing different phases of the waterfall model. The role of the waterfall model is to form a skeleton with the goal to rectify the modifications flow within each of the RUDE cycles.

Note that the refinement RUDE cycles can be used to enhance different aspects of the system. They can improve the behavior, structure, knowledge, etc. It is possible to distinguish among, e.g., *knowledge acquisition*, *reasoning*, or *explanation* RUDE cycles [7].

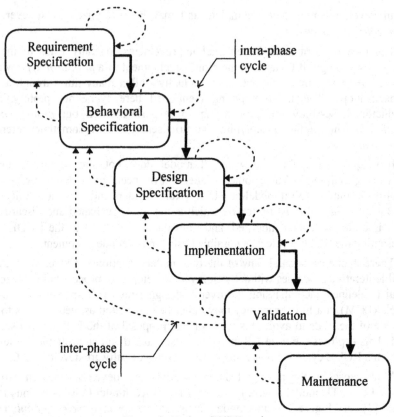

Figure 2: Various RUDE cycles

2.2 Product Life-Cycle Model

A high degree of similarity between the software LCM developed in Knowledge Engineering and the PLCM used in the CIM area is quite remarkable.

The separate phases of the PLCM that were originally organized in a waterfall style at the early stages of CIM methodologies may be recognized as follows [8]: Production Program Planning, Development and Design (CAD), Process Planning (CAP), Production Scheduling (PPS), Manufacturing Control (CAM) and Supervision, Quality Control (CAQ), etc.

Later, it has been recognized that *iterative cycles* among different phases are necessary. For instance, the design phase can be significantly influenced by the knowledge and data concerning the planning and manufacturing phases. It has been reported [1] that up to 70% of the product cost is incurred during the design phase, and 20% during actual production. Very large savings in the total costs are possible if right decisions are done at the early design phase.

In general, studying the PLCM in detail, it may be discovered that there is really an excellent analogy of the POLITE-VRC model, including the terminology used - see Figure 3.

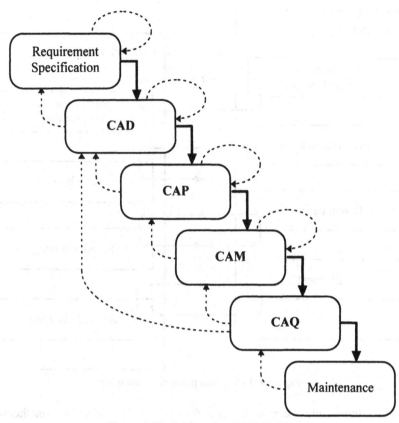

Figure 3: PLCM based on POLITE-VRC

A very tight linkage of the CAD to CAM phases in the frame of the PLCM seems to be crucial for the total efficiency of CIM-software systems. This linkage should be organized through many different iterative cycles. As a matter of fact, a lot of efforts has been dedicated to the development of various *integrated product modelers* for integration of the CAD and CAP phases through iterative cycles (see Figure 4). The "inter-phases" interaction is organized on different levels, starting on the level of rough requirements and finishing on the level of very detailed refinements. Such a structure of the PLCM supports rapid-prototyping (first, rough prototype of product model is considered as the output) with subsequent (maybe, ad infinitum) refinements [9]. It also helps to analyze the individual information-processing functions, and to define the intermediate stages of the global linkage as well as supporting tools for performing the elementary subtasks in detail.

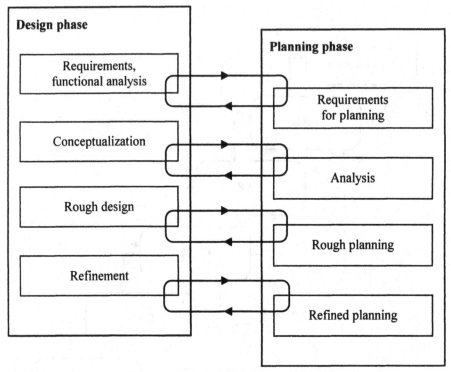

Figure 4: Design and planning relationship

A very important concept is the use of *feature-based techniques* as a methodology for integration of the product design with other stages of the PLCM. The features are defined as form features associated with the semantics of their engineering design. They integrate geometry, technology and function aspects and are treated as objects in CAD/CAM engineering tasks. Tehy serve as entities common to all information processing phases. An important result is a common understanding of features in the design, planning, scheduling and manufacturing phases. A pioneering example of the feature modeling system is the system FEAMOS [10].

The next step towards fully integrated CAD/CAM is the attempt to integrate the planning process with the manufacturing control [11]. In the planning phase the solution space for production plans is rather restricted by static optimization methods. Both the pre-selected plans and the dynamic model of flexible production facilities form the basis for simultaneous creation of the final plan and manufacturing control.

Concurrent Product Engineering. As it can be seen, the detailed *analysis* of the iterative processes within the POLITE-VRC-like PLCM can create a good methodological basis for simultaneous (concurrent) engineering in the CIM-area. This trend is also strongly supported by the AI ideas. Each application of the concurrent engineering requires the exact formulation of the subtasks to be solved in parallel [12]. The AI-inspired constraint-handling approach is usually used for this purpose. The constraints demarcating the local problem solving spaces may be considered as objects in the OO-programming environment and organized in networks which ensure the appropriate constraint-propagation [13]. The individual subsystems having procedural nature and solving separate subtasks may be organized in a modular way. The communication among them is usually based on the blackboard or multi-agent techniques of distributed AI [14]. Multiple hierarchies of different nature (like hierarchies of constraints, procedural subsystems, data models etc.) can cooperate in an OO-programming environment quite efficiently.

3 Multi-Agent Approach to CIM

As mentioned above, the CIM-Software Engineering is mainly aimed at integration of diverse well-developed and well-working software subsystems. These subsystems are usually able to solve partial problems, but the overall CIM-system efficiency is strongly dependent on their successful incorporation into the overall software architecture.

There is another specialty of the CIM-system integration process: The "classical" software development process (used mainly during the partial, elementary subsystems creation) is directed from the conceptual to implementation level. In many cases of CIM-subsystems integration we are facing the situation that a lot of data/procedures/knowledge is already available on the implementation or programming levels. For example, there is a large firm database as well as geometry oriented CAD-system available and it is necessary to create some integration software on the top of them. The initial formulation of the target overall system behavior is not sufficient. There are two ways how to solve the task:

- Either build-up the entire software system by the bottom-up approach (from implementation level incrementally to conceptualization level) and to create different higher-level data/knowledge structures upon that hitherto implemented. This is the way how the system TEPRO [15] was developed.

- Another possibility is to start from both ends and to try to fill the gap between the conceptual and implementation levels by appropriate data/knowledge modeling structures. This approach led to the idea of an *integrated product model* structures (DICAD [8]) or to a similar, so called *integrated reference model* (e.g. described in EXPRESS OO-specification language and used within the IMPPACT Project [10]).

The implementation/programming structures could be considered as constraints for the CIM-system design. Both the approaches mentioned above do not fit the waterfall model as they need iterative incremental model of development. They fit neither the RUDE nor the POLITE models: The RUDE procedure does not reflect properly the hierarchy of data/knowledge, the POLITE model does not consider the between-phases iterative cycles. But the POLITE-VRC model can describe both the bottom-up and bi-directional approaches to the incremental system development.

When studying the history of CIM software integration methodologies, the following four stages can be distinguished:

- *System integration through exchange of data files* (sixties and early seventies): The data file representing the output of one CIM-subsystem was simply considered as the input to other CIM-subsystems. This approach is too clumsy and does not integrate the subsystems at all.

- *System integration based on a huge central database* (late seventies): The intent was to store all the data connected with the product, including the drawing, production plans, marketing data, etc., in one huge (maybe distributed) database shared by all the subsystems. Even the latest database technology does not offer such storage capacity to support this approach.

- *System integration through knowledge-based systems* (eighties): The idea of the knowledge-based linkage was implied by the boom of expert systems. This approach lead to ad-hoc, not general enough, and not re-usable solutions.

- *System integration based on the multi-agent approach* (early nineties): This approach is based on the recent results of both the object-oriented programming and distributed AI.

All the approaches, but the last one, have failed in industrial practice when attepmting to integrate large, complex systems.

The idea of the multi-agent approach considers each CIM-subsystem as an autonomous agent who communicates with the other members of the agents' community. Each agent can be considered as an autonomous computational process which sends and receives some standard messages. The theory of agents originates in the area of *distributed artificial intelligence* (DAI). The DAI is usually split [16] into two primary areas:

Distributed problem solving: The subsystems are viewed as modules that cooperate at the level of dividing and sharing data/knowledge about the current product model

Multi-agent systems: The subsystems are considered as active autonomous intelligent "agents" coordinating their behavior by negotiations or protocols, reflecting their joint goal. A unified system communication channel is usually applied. The more autonomous subsystems, the less global data and strategic knowledge have to be stored and explored on the top control level.

In the case of distributed problem solving the method of shared global memory has been used and implemented in a form of a blackboard control structure. The attempts to organize a global shared memory for CIM tasks, for instance in the form of global product models stored in a centralized CIM database, have failed. This is obvious especially in the case of geographically distributed CIM systems, where an enormous amount of long-distance communication with the central database neccessarily degrades the system performance and becomes costly.

The multi-agent approach is philosophically based on Hewitt's actor model [17]. This technique fits much better the real-life demands on distributed CIM systems, especially if these subsystems are coarse grained and can be considered as loosely coupled. The philosophy of actors (or agents) leads to the possibility of considering agents as structures that combine both procedures and data into a single entity - an *object*. That is why the object-oriented methodology may be regarded as one of favorable schemes applicable in development of distributed CIM systems.

There are four fundamental questions associated with each multi-agent system design:

a) How to decompose the task into subtasks?

b) How to allocate the subtasks to the problem solving units?

c) How and what to communicate?

d) How to control the system globally?

ad a) and *b):* Suppose that the CIM task should be solved at various geographical sites. The decomposition of the entire CIM task is usually already given by the location of the facilities (workshops, design centers, etc.) and specific functions of these facilities. According to the functionality, many typical units can be found (e.g., selling agencies, business coordination unit, manufacturing units, technology development units, etc.). In the real-world tasks also accounting, maintenance of resources, legal affairs, and marketing activities have to be supported.

The presented facts show that the CIM task decomposition and the allocation of subtasks are usually functionally pre-defined and do not create a substantial problem.

ad c): Communication plays dominant role in distributed systems. Every functional unit (module, node) of the distributed system may be considered as a computational process that consists of two parts:

- problem-dependent problem-solving part (*task*) which has no knowledge about the other nodes or about the overall system,
- problem-independent part (*agent*) responsible for communication and coordination activities. The agents contain all the knowledge about the rest of the distributed system. They represent elementary control units inside the distributed system.

From the theoretical point of view, an agent is described by:
- the messages it can receive and send out,
- the model of its own behavior (self-description),
- the model of the agent's community,
- the execution engine.

One of the main problems of practical multi-agent CIM-system implementations is the design of agents. A detailed analysis of many interdependencies, requirements on mutual coordination, and on data and knowledge flow must be done. This work is tedious, time consuming, and expensive. It contains a specific knowledge engineering activity that can incorporate the software life cycle methodologies mentioned above.

In the current systems, the necessary analysis represents the real bottleneck and is usually done "by hand". But some systems (like HITOP-A [12]) already contain modules supporting this analytical process.

A new programming paradigm has appeared in connection with the multi-agent systems. It deals with the *agent-oriented programming* [18] which can be viewed as a specialization of object-oriented programming. An agent is considered as an object with a particular structure which includes some pre-defined elements. The inter-agent communication in the form of message passing is considerably more stylized than in the case of the object-oriented approach. The specific structures of commitment and specialized interaction protocols create an important part of an agent-oriented programming.

Recently, there have appeared environments called *agent factories*. Such environments or development kits provide
- a set of software tools for creating agents and organizing communication among them, these tools should be able to explore the idea of a wide software re-usability;

- an associated design methodology;
- a library of prototype agents (as a support to a "drag-and-drop" philosophy).

The agents can be tailored from the kit on fly using a special developer's agent. This is another typical feature of these environments: All activities connected with the multi-agent system "life", like simulation, monitoring, or interfacing to the user, are organized as specialized agents. Besides the developer's agent, there can be different simulation agents, monitoring agents, and user-interfaces can be considered as autonomous agents, too.

In the area of DAI, very sophisticated methods based on social organization and modeling have been developed. Only a part of them is currently applicable in practice, especially in the area of CIM-software development as there is *no sufficient implementation support available*. The object-based concurrent programming and the agent-oriented programming are very promising approaches, but still very far from providing clear languages with desired agent-structure features. Nearly all the CIM solutions, even based on DAI philosophy, are implemented *ad hoc*. That means that even the theoretically well-founded multi-agent systems are usually implemented from scratch and the CIM-software engineers have to tailor the communication methods and structures. Widely accepted, well-defined methodology and suitable tools are still missing.

From these facts it follows that the job of a CIM-software engineer requires knowledge and skills of a *knowledge engineer*, *software engineer*, and a *system analyst*.

The idea of agent-oriented programming represents a deep elaboration of the object-oriented programming. It is based mainly on encapsulation. The main difference between multi-agent programming compared to OO programming is that the behavior of agents is much more complex than that of objects. Yet another difference is in the inter-agent communication that plays a dominant role in the structural design of the multi-agent system. There has to be defined a very versatile communication protocol among the agents. It has to combine features of the OSI (=Open System Interface) networking model and of the object message passing mechanisms.

A peer-to-peer communication model is necessary for efficient inter-agent cooperation. Local-area networks run according to this model. Every machine in a LAN knows exactly its communication counterpart and the bi-directional communication channel can be readily established. As soon as we start to consider some internetworking concept in a wider area, we have to realize the role of gateways ensuring the LANs' connectivity while simultaneously isolating the intra-LAN traffic. This is the concept of many WANs, especially the concept of Internet. In such an environment, there have to be machines that are ready to accept requests for connections (*servers*) and machines asking for such connections (*clients*). This way the world-wide virtual network is formed.

The problem is how to build an agent-oriented peer-to-peer virtual network over the client-server architecture running the standard TCP/IP [19]. The CIM-software engineer can solve the problem in two possible ways:

The first possibility is to forget the TCP itself and try to build another communication protocol using the lowest IP layer to organize the peer-to-peer communication. This approach has many drawbacks. All of them issue from the fact that the IP itself is an unsafe connection-less best-effort-delivery protocol, only. The data packets traveling over the Internet can get lost, can be duplicated, or can get extremely delayed. These problems must be overcome by some auxiliary algorithms (like acknowledgment - that can get lost, too). All these shortcomings are already solved by the versatile TCP/IP. Thus it is better to avoid reinventing the wheel, and use the standard.

Another possible approach is to organize the communication so that the peer-to-peer behavior is observable from the outside, but the internal implementation is based on the client-server TCP/IP connections. In this case, there is a star-like structure used on the implementation level. It consists of a central node (server or hub) which provides the connection service among the nodes (clients) in tips. The behavioral level resembles a set of peer-to-peer connections. This topology has several advantages. It enables to add new agents during the multi-agent system operation, to control the communications states, to broadcast messages, etc. It was proved that this solution may be efficiently supported in the Internet by the parallel virtual machine (PVM) [20]. The main deficiency of this topology is that when the hub fails, the entire multi-agent system fails, too.

In multi-agent systems, the only way how to accomplish the global control strategy (determining behavior of a set of subsystems) through knowledge "owned" by the individual agents. It is the knowledge that creates a substantial part of the content of an agent. It usually has the form of explicit rules describing the behavior of an agent itself and the behavior of other agents (see e.g., ARCHON [16]). There are general rules (valid for all agents) and very specific rules used to express the control knowledge owned by a single agent. The agent-specific knowledge is very extensive in CIM tasks where it plays a dominant role because of extremely strong functional specialization of CIM-subsystems. That is why, the rules' definition requires a huge amount of knowledge engineering work. This reminds of the knowledge acquisition bottle-neck well-known from the expert systems. Even in CIM systems, this activity can be supported by machine-learning as mentioned later.

4 AI Impacts on Planning and Scheduling in CIM

The AI literature often investigates planning in a very abstract sense. The discussions on planning are sophisticated but the connection to the real-world CIM planning tasks is not very straightforward.

Before proceeding further, let us mention the difference between the concept of planning in AI and common sense of this word. Common-sense planning means some sort of reasoning about a sequence of actions to be taken in future. These actions are usually seen in some pre-determined points in time, or at least some temporal ordering is implicitly considered.

First experiments in AI planning have had this nature. They were most frequently applied to the planning of robot behavior. However, during the development of AI planning methods the scope of AI-planning has broadened substantially. The time as the only variable inducing the ordering of actions has been substituted by other possible quantities. For example, planning is often applied to solve various spatial layout problems where the time no more plays the dominant role. Various kinds of discrete optimization can be considered as planning tasks, too.

There is one common feature of planning in CIM applications. It is nearly impossible to specify the goals uniquely. Usually, in classical AI planning the goal is fully determined by a precise structural description, such as 'tower of three ordered blocks' in the block world - cf. [21]. The goal specification in applied planning often allows a big number of admissible solutions. The desired one can be chosen only within a user interaction or by a set of additional user supplied conditions. Sometimes there exists a utility function that allows to evaluate the goodness of each solution. This situation takes place especially in case of an optimization.

There is another important observation: Any system which is "too automatic" is not well accepted by the end-user. The user often wants to have control over the system behavior, at least to a certain degree. The user appreciates lucid user interface even if the system quality and its scientific background is not very sophisticated. If the system generates a reasonable number of solutions which are nicely displayed on the computer screen and the user can comfortably choose among them, then this system will be judged as better than that one which generates one or two fully optimal outcomes described by an obscure set of logic or algebraic formulae and numbers, only.

The CIM covers (besides others) the entire chain of a product development. The chain starts by the *design* phase where CAD systems are used. The design is, in principle, a very sophisticated planning activity of a creative nature. A very comprehensive explanation of logics and planning in design is given in [22].

Next link of the CIM chain is the task of making a plan how to manufacture the designed product. This phase of activities is often supported by some computer system. The main goal of such systems is to specify *how* to manufacture the newly designed product in the given manufacturing environment. It means to prescribe the *sequence* of desired technological *operations* which are to be carried out to manufacture the final product. This sequencing exactly matches the classical AI planning idea. Only the constraints and circumstances to be taken into account are usually much more complex. Moreover, these systems often provide support to those activities that might make the production more precise, easier, and cheaper. This may involve design of specialized tools and writing and debugging of special programs for numerically controlled machines. Sometimes it can happen that the designed product is too complex for the manufacturing environment available. Then it is often desirable to re-design the product with some new constraints posted.

The next step in manufacturing is the *scheduling* task. Here it is necessary to cope with the problem of parallelism in the manufacturing environment. The main goal is to determine *when* and at which concrete machine to carry out each particular operation on each product. The duration of operations, due dates, machine capabilities and accuracy, and many other constraints are to be taken into account. Another group of demands covers various priorities of orders, overall company profit and other managerial requirements. Tasks of this type lie on the border between AI and *Operations Research* (OR). A typical task here, called *resource allocation* problem, can be considered from the mathematical point of view as discrete (often bivalent) programming.

The main difficulty with the application of the OR methods is in the necessity of explicit specification of all the constraints and the utility function. In real life CIM tasks, the specification represents many thousands of equations and inequalities on hundreds of discrete variables. It is even unrealistic only to explicitly formulate the task. Moreover, in many real-life cases it is very difficult to express explicitly the utility function for the optimization.

From its beginning, AI investigated planning as one of its main tasks. The developed techniques are (as most other AI task) based on state space search. This is no substantial difference to OR. The cardinal difference is in the approach how the state space is being parsed. The traditional OR paradigm is algorithmic while AI tries to exploit knowledge and logic-based reasoning to cope with the complex search problem.

Several more recent results in AI can support planning tasks in industrial practice. These methods are based on *truth maintenance* (purely logic-based methods), *distributed planning* (based on task decomposition) and *incremental planning* [23]. Nevertheless, in all cases the results are very domain-dependent and their generalization is complicated.

However, AI offers other techniques for this purpose. Probably the most promising one is to consider the planning task as a *constraint satisfaction problem* [24]. Various constraint-handling techniques have been developed for efficient pruning of the state space search tree. Their simple use is ensured in combination with *logic programming*.

The basic idea of logic programming methodology is in a strict decomposition of any problem into two separate parts - the declarative and the procedural ones. Within this methodology the programmer is invited to concentrate on the declarative formulation while the implementation of the used logic programming language takes over the responsibility for creation of appropriate solutions (the procedural part). The basic computational paradigm used by logic languages for solving most problems is *generate and test* which is very inefficient for problems of combinatorial nature. That is why, it was suggested to embed consistency checking for constraints into logic programming and restrict this way the search space for the problem solution in advance. This issued to a discipline called *constraint logic programming* (CLP) [25].

During recent years, there have been suggested and implemented several CLP languages differing namely by their ability to express and handle various domains and sets of constraints. CHIP [26] was designed to work with finite domains and rational numbers, PROLOG III [27] with trees, CLP(R) [28] with reals. There have been developed complex full applications using those languages. Among the most impressive ones are the programs for scheduling (ship-unloading in a port, slot allocation for an airport).

The main advantage of CLP methodology is that given a language detailed modeling of the problem can be completed in a very short time. The rapid development makes it possible to experiment with different representations of the problem, thus creating good environment for deep investigation of the application domain. CLP applicability goes beyond the prototyping stage [29]. According to their experience the prototype created by CLP techniques can constitute the kernel of the operational software, which has to be integrated into the full system by providing necessary interfaces. For that purpose there can be utilized some handy tools developed for creation of operational interfaces (the fourth generation languages). Thus CLP methodology reasonably complements the existing set of tools for software development by focusing attention to the hard problems of constraint satisfaction. This is facilitated by a shift from procedural to declarative programming started in seventies.

5 Declarative Programming and PROLOG

We have tried to identify the characteristic principles of the actual approach to CIM Software Engineering. Gradual refinement of concepts and solutions seems to be a common topic of PLCM and POLITE-VRC. This philosophy of software development is strongly supported by utilization of methods for rapid prototyping. Logic programming is one of natural candidates for that purpose.

5.1 Principles of Logic Programming

Theoretical roots of logic programming grow from the idea to utilize the available means for automatic theorem proving (developed in AI research during sixties and seventies) to close the gap between input declarative description of a problem and the target algorithm for its solution. This is depicted by a well known equality [30]

$$algorithm = problem\_description + control.$$

This approach stresses the creative role of the programmer in the problem description phase when decisive steps are to be taken to choose the appropriate part of background knowledge as well as useful problem decomposition etc. This description forms a logic theory, where the actual problem description is to be proven. The used algorithm for theorem proving suggests a sequence of steps applied to reach this goal. Programming languages supporting this feature are often referred to as *declarative programming languages*. Thus, the logic programming paradigm can be interpreted as an attempt to shift responsibility for control from human programmer towards the software environment as much as possible.

Logic programming uses for the problem description a restricted part of the language of the first order logic, limiting its attention to special type of formulas, namely to Horn clauses. *Horn clause* is such a disjunction of atomic formulas or their negations that at most one of them is positive. It is natural to represent Horn clauses with a positive atom in the form of implication: the conclusion corresponds to the single positive atom, while the condition is expressed as a conjunction of all the other atoms present in the Horn clause. A set of formulas characterizing the considered problem (together with its environment) is called a *logic program*, those in the context of classical logic it would be referred to as a theory. The actual problem to be solved is put as a *query* to this logic program - it is described as a formula to be proved within the considered program (theory). As a rule, the query has ever a form of a Horn clause without a positive atom. How the proof is obtained? This is the task of control - it is determined by the choice of theorem proving mechanism. Most often it is one of the resolution strategies. The first implementation of logic programming paradigm has been Prolog [27, 30]. Prolog has gained recently lot of attention. Serious introduction into the field is offered by a number of textbooks, a person with engineering applications in mind would surely enjoy one of the following ones [31, 32].

The logic programming language providing only the basic expressive means mentioned above is referred to as *pure Prolog*. But in order to become of real value for non-elementary examples, the language had to be enriched by further extra-logical and procedural features indispensable in practical applications (arithmetic, input/output predicates, types, editing commands, file handling etc.). All actually available implementations of Prolog provide these types of extensions, moreover they support import-export to classical programming languages.

Originally, mathematical logic was developed as a tool for formalization of mathematical thinking in general. That is why its language is ready to express many complex constructs as well as their properties. Prolog inherits this flexibility due to its roots in logic. All principal ideas, which have appeared within software community during last two decades, have influenced development of logic programming languages. There are many different extensions of Prolog. This language readily incorporates all those principles, which promise to enhance its expressivity or its problem-solving capacity. This is the case of classes of objects and property inheritance refined in Object-Oriented programming, the corresponding implementation of O-O Prolog is described in [33]. It is natural to study parallel processes within logic programming, because sequentiality is forced to Prolog due to the intention to implement it on a sequential machine only. As soon as parallel machines appeared, there were developed parallel Prolog systems, namely Parlog [34] and later even distributed implementations on multi-transputer environments [35]. Moreover, a need to provide tools for expression and handling uncertainty resulted in inclusion of fuzzy techniques [32, 36]. Utilization of fuzzy reasoning significantly extended scope of Prolog applications, ranging from natural language processing to simulation of queuing systems in a way similar to that of Petri-nets.

5.2 What Is Logic Programming Intended for?

Prolog is frequently used for various problems of symbolic processing, e.g. for manipulation with programs, databases or "theories" in general. Such tools are indispensable e.g. when building a system for automatic program debugging or transformations as well as when we intend to apply a more sophisticated problem solving strategy then that of classical Prolog for certain type of problems. What is common to all these types of problems? The characteristic feature of such applications is that programs are treated as data - they represent input to another program. This technique, called *metaprogramming*, is typical for Prolog. It allows to adapt the Prolog system according to the requirements of the actual application, thus to overcome the limitations of Prolog simple strategy of problem-solving.

Program maintenance seems to be one of the most demanding tasks in POLITE-VRC life cycle model, which we have found to be characteristic for CIM software development. This has to be taken into account when choosing the appropriate implementation language: computer support of repeated debugging is badly needed. Efficient debugging demands close interaction between the computer system and its human user. The system has to obtain some decisive information from the user. This way a new problem arises: how to restrict the number of questions to the user to relevant, meaningful, and strictly necessary queries. Declarative programming style facilitates this task. Tools for automatic debugging have been studied using logic programming. The pioneering work in this direction is represented by the system MIS [37]. MIS tries to identify the simplest cause of a difference between intended and real program behavior discovered in specific examples. Its goal is to suggest a debugged version of the original program.

Prolog proved to be a promising object language for the task of "programming by example". This and similar problems have been extensively studied recently within the framework of ILP - Inductive Logic Programming [38], a subfield of empirical machine learning, where the resulting knowledge is to be expressed in the form of a logic program. This target knowledge can be viewed as a "condensed information" allowing to explain/deduce the input empirical data. It can be useful in diverse domains ranging from for automatic evaluation of process-data up to creation of simple means for communication with data-base (very general "query by example"). There have been implemented interesting systems for these purposes, e.g. FOIL [39], GOLEM [40], CLINT [41]. Such systems can help to overcome the bottle-neck of knowledge acquisition ever jeopardizing development of knowledge/expert systems.

Declarative style of logic programming supports clear definition of concepts and their mutual relations. Logic programmer can concentrate on this conceptual task, while the procedural part of the problem can be passed to the interpreter of the logic programming language. However attractive this separation between knowledge and control can seem, it is a source of problems, too. Novices to the field are able to learn basic principles of Prolog rather quickly in order to solve school examples. But then they tend to apply their successful solutions to simple problems, e.g. generate-and-test methodology, in more complex tasks. This results in very inefficient programs, obviously. That is why many people believe that the logic programming approach to computing is inadequate to real-life software needs. The following paragraph will summarize some arguments against this prejudice.

5.3 CIM and Logic Programming

Prolog, one of the best known implementations of logic programming approach is used for various purposes and in very diverse environments, e.g. for communication with databases, knowledge and expert systems, scheduling systems or development of new programming languages. Constraint logic programming CLP represents one of the principal extensions of logic programming leading to impressive increase of speed due to utilization of very effective techniques for scanning the problem space. CLP proves to be useful namely for solutions of problems from the domain of *scheduling and planning*.

In nineties, there can be traced another novel trend in Prolog applications. Logic programming is used as en environment for *software development*. One of the pioneering papers on this application of Prolog is the paper "Use of Prolog for Developing a New Programming Language" [42] published in 1992. It describes development of Erlang - a new concurrent real-time symbolic programming language. Erlang first appeared as a simple Prolog meta-interpreter. Its prototype was used by a user group to test their reactions to the language and it was continually changed/refined in order to meet their demands. The features of Prolog significantly supported this development cycle - Prolog allowed the development team to make required changes easily, often within few hours or at worst within few days. When performance became an issue Erlang was directly implemented using experience of emulators of WAM - Prolog abstract machine used in implementations of Prolog.

In [43], we can find already 10 papers utilizing Prolog as a tool for software development. Let us briefly mention two of them, each concerned with specific type of problems related to CIM.

Robot programming becomes an important issue in real-life applications. Usually, there are distinguished two levels in the programming process: robot-level programming (real time event handling, basic motion control) and task-level programming (planning and reasoning). Actually, there is no generally agreed-on architecture to be used for this purpose. A *classical approach* is based on hierarchical functional decomposition with a strictly vertical control/data flow among components looping through perception, modeling, planning and motor control. On the other hand, the advocates of *behavior-based approach* believe that intelligent behavior can be achieved on a reactive basis by providing quick response to the stimuli of physical world. It is hoped that the best results can be achieved in the *heterogeneous systems*, combining the advantages of both systems. The heterogeneous architecture defines new requirements for robot programming environment. In *A Structured Logic Programming Approach to Robot Programming* [44], the authors summarize them as follows:

1. A programming environment is needed which allows full integration between low-level and high-level robot programming and supports a wide range of robot architectures.

2. Object-oriented programming should be supported, since it induces classification and realization in such a rapidly evolving field. Moreover, it makes it possible to define reusable components and develop applications incrementally. Then object abstractions and mechanisms have to be fully supported although open problems exist, namely definition of real-time object-oriented models and dynamic class creation.

3. Since robot architectures are always intrinsically parallel, concurrency has to be supported. Moreover, a degree of explicit control over parallel computations should be provided in order to deal with the bounded computational resources available and real-time constraints.

4. Many task-level applications have to deal with highly symbolic processing tasks, such as knowledge representation, planning and learning; therefore, high-level programming notions and tools typical of knowledge-based applications are required.

Although these four objectives do not mention any standard feature of Prolog explicitly, the authors claim that Prolog proved very useful to build a system meeting those requirements. Beside well-known Prolog advantages (unification, declarativeness) the authors highly evaluate another yet unexploited feature, namely the chance of giving multiple interpretations to a Prolog data-base (e.g., knowledge-base, object, message mailbox, logic theory). They find this feature to be a powerful unifying factor with respect to programming-in-the-large. It was used for development of a programming system where knowledge-based units can be created as structured objects (contexts) referred to by an explicit name. The symbolic part of the system is implemented in APPEAL logic programming system allowing to create and experiment with different agents. These agents constitute a meta-control level with respect to real-time control. The resulting implementation provides efficient integration of symbolic processing with flexible real-time system explicitly designed in C for robot control. The coordination of interaction of different heterogeneous components and agents is achieved through the contextual platform constituted by several Prolog machines sharing a data-base that acts as an evolving world.

Those advanced CIM sub-systems, which are build from highly automated production units (NC/CNC-machines) linked by a computerized material handling system, are referred to as *flexible manufacturing systems* (FMS). Different products can follow different routs through the FMS system according to the actual state of the system (work load of the machines, input queue etc.). There are two crucial tasks, which significantly influence/characterize behavior of FMS, namely release and machine scheduling with respect to actual data. In general, scheduling is an NP complete problem, the effective approximate solution of which has to apply domain heuristics. This information often represents experience of domain experts or other uncertain data, which cannot be expressed precisely. Fuzzy methodology has been developed as a useful alternative under such conditions. This concept can be naturally imbedded into logic programming. The paper *The Application of Concurrent Fuzzy Prolog in the Field of Modeling Flexible Manufacturing Systems* [45] describes an implementation of a system for release and machine scheduling, where fuzzy approach is successfully combined with inherent parallelism of Concurrent Prolog.

The development of CIM methodology has positively influenced other technical fields, for which coordination of diverse activities is crucial. This is the case of building industry, where cognitive and analytical skills of humans and computers complement each other. Like in CAD, even in this domain there are available different analysis and simulation tools, which have to be integrated into a user-friendly system to simplify everyday tasks of construction professionals. The paper *Computer Facilitated Building Design: Prolog as a General Purpose Language in Cooperating Systems* [46] suggests a CIM-like solution supporting building design including choice of construction material, its cost and thermal performance as well as construction scheduling. Beside KB support, the resulting system allows to by-pass the use of paper documents as a medium of data exchange. The system is a cooperating distributed knowledge based system with the user centered approach. The user plays an integral part of the system - he is provided with the results of the various analysis so that he can make an informed choice when various options are available. Prolog is used to implement a knowledge base controller coordinating a suite of programs and merging the results of the individual programs. One of them is the model of a building coded in Flex [47] - an expert system toolkit implemented in Prolog providing inheritance, demons, and constraints. Combination of object-oriented paradigm with that of logic programming proved to be an elegant solution to many typical problems (version maintenance, isolating platform dependent code, etc.).

The above mentioned projects demonstrate clearly that Prolog is indeed a very useful language for real life applications. Moreover, it is clear now, that efficient Prolog coding is possible, although not trivial. Many projects prove that Prolog programs can be compiled into code which is efficient for the vast majority of needs.

What are the features that make Prolog attractive for applications?

- Prolog is declarative - actual sequencing of the rules applied to solve the considered problem is derived automatically, without intervention of the programmer. That is why new design knowledge can be included without significant change in the whole program. This results in fast iterative cycle during the prototyping period due to flexibility and modular nature Prolog affords.

- Prolog provides programmers with means to express their ideas faster. Moreover the Prolog code is much shorter, thus easier to maintain and debug. This results in increase of programmer's productivity.

- Prolog basic primitives unification and backtracking provide flexible matching and useful automatic searching mechanism. It has powerful data types with automatic allocation and deallocation making it easy to handle complex data which can be accessed without using pointers.

- Prolog is based on the language of first order logic which is very flexible. It provides good means for metaprogramming. Moreover, Prolog can easily incorporate extensions necessary to work with new paradigms like contextual programming, OO paradigm, fuzzy reasoning. We have mentioned some of them already, because they point to directions of further development of the language.

- All these features make it possible to work in small well manageable groups.

Prolog is especially suited for those applications utilizing knowledge intensive tools, symbolic processing ranging from providing interpreters and compilers defining new programming language to straightforward use of infix operators for mirroring the *business language* of the users, metaprograming tasks as writing programs which generate or check, transform or optimize other programs, handling rule-based information with exception handling and niche applications. Moreover, Prolog programs seem to be useful for specification of background knowledge for machine learning applications.

6 Machine Learning

The aim of machine learning (ML) is to equip the computers with an ability to utilize experience to reach better performance. This sort of behavior is vital for all living creatures, because it ensures their adaptation to changing environment. Thanks to learning organisms of various complexities can distinguish dangerous situations or recognize order in a seemingly incoherent set of facts. Of course, learning is an indispensable part of artificial intelligence. But is this domain rape to be applied to industrial problems? Is it not too speculative?

Learning from examples has been one of the central topics in ML community during last two decades. There can be identified two main streams, namely "symbolic" and "evolutionary" learning techniques. The motivation for development of the evolutionary approach comes from natural systems. It is based on modifications of some laws governing development of the biological systems - this is the case of genetic algorithms or neural networks. Genetic algorithms are treated in detail in the corresponding chapter. Let us summarize motivation for development and the main results of the first stream as well as its possible impact on CIM.

6.1 Symbolic Learning

This name is derived from the fact that ML tools process examples in order to extract "knowledge" in a symbolic, declarative, human-understandable form (properties of concepts, hierarchies, if-then rules). Resulting knowledge can be embedded in knowledge and expert systems. It is ready to be shared by the computer and its human user, who will accept explanation referring to this knowledge. This ability to "conceptualize" through symbolic machine learning is a very desirable feature, because it could help to break the knowledge acquisition bottle-neck. Moreover, modern industry recognizes importance of computer support for symbolic knowledge extraction because hardware and software control and monitoring systems produce huge quantities of data, which humans are not able to analyze. What is the intended result of requested analysis? It is supposed to identify one of the following patterns in the provided data (examples):

- the unstructured individuals form classes of related objects (clustering, unsupervised learning),
- particular classes are characterized by well defined concepts (conceptual clustering) or by recognition functions (discrimination),
- atypical individuals are distinguished in the whole population.

Learning tools have been developed to support all the upper mentioned tasks [48, 49, 50, 51]. These tools range from classical clustering techniques based on distance measure between the studied objects up to characterization of the concepts by Horn clause programs in Inductive Logic Programming [52, 38].

One of the best known algorithms of symbolic ML is that one for inductive tree generation [53] from ID3 family. It processes input examples classified to an arbitrary but fixed set of classes. This classification has to be provided by the user (teacher). In general, this type of learning is referred to as supervised learning. On the other hand, unsupervised learning works with unclassified examples. It tries to cluster them or to suggest a pattern or a new structure, which could be appropriate for more concise description of the considered set of unclassified input examples.

Let us describe ID3 algorithm in more detail. All its input examples have to be presented in an analogous way using a fixed language, i.e. a set of attributes, chosen a priori for characterization of the considered task. An attribute is a unary function defined on the studied objects (color, weight, prize, etc.) It can take either discrete or continuous values. Originally, the attributes were supposed to have a finite domain consisting of a discrete set of values only. Later the learning algorithms were invented or modified in such a way that they are ready to handle continuous domains, too.

The input of ID3 algorithm can be visualized in a form of a matrix - each row corresponds to one input example, the columns represent the attributes with one extra column identifying the classification given by the user. From the input examples, ID3 algorithm induces a concise and efficient strategy for identification of a class of a new unclassified example. This strategy is presented in the form of a decision tree. The inner vertices (nodes) of the decision tree are denoted by the attributes of the considered language while the leaves by the names of possible classes. Each of the edges starting in a given node is denoted by a subset of that node attribute's domain. There must hold that the subdomains denoting all edges beginning in the same node are a non-intersecting cover of the full domain (of the considered attribute). Given a new, yet unclassified object, the classification procedure proceeds then as follows:

1. Denote the root as the actual node.
2. For the studied object, find the value of the attribute denoting the actual node. This value will be referred to as the actual attribute value.
3. Identify the edge beginning in the actual mode denoted by the subdomain covering the actual attribute value. Follow this edge up to its end node.
4. If this end node is a leaf, then classify the studied object by the class denoting this leaf, else denote this node as the actual node and GOTO 2.

In a well structured domain, relatively simple decision tree can serve for classification of new objects. It by-passes the necessity to remember all the input examples presented by vast amount of data. The tree can be pruned to obtain even more compact result if the user is ready to accept classification with a small pre-specified possibility of error.

6.2 Machine Learning and Software Maintenance

The "classical" ML algorithms tried to induce knowledge from scratch. Recently, large effort has been devoted to the development of algorithms that exploit background knowledge. The role of background knowledge can range from "heuristics" for efficient structuring of data upto "debugging-like" scenario. In the latter case, the provided examples are used for refinement of background knowledge to improve its descriptive power; it is referred to as *theory revision* [54]. Philosophy of such systems is close to human approach to learning. Such systems cannot be static - they must evolve in time (evolutive systems) and need permanent maintenance. The number of evolutive systems grows continuously calling for attention and development of supporting tools. The PLCM can serve as an example. The evolutive systems are spreading even among knowledge-based industrial applications. MLT Consortium [55] cites following knowledge-based applications, which call for permanent and interactive learning capability utilizing background knowledge:

- Diagnostic expert systems delivered along with new products on the market, for which the diagnostic experience is not completely formalized and fixed. Thus an embedded maintenance functionality is absolutely necessary for these systems. There is a current paradoxical situation in which expert systems can be made for well-known and thus old products or processes.

- Control or monitoring systems in changing environments, either changing with time as a normal aging mechanism or because of possible reconfigurations or extensions of the process (e.g. modification in power supply or telecommunication networks).

- Systems in which knowledge is confidential and is the exclusive property of the customer. Thus maintenance and expansion of the knowledge base is for large part done only by in-house experts, which implies that more support must be given by the system.

- Intelligent design support systems, as design expertise has to continuously evolve to keep up with new technologies and know-how.

6.3 Examples of Industrial Applications

No doubt that the last three problems are relevant to most CIM applications. Industrial companies have identified recently further real problems which can benefit from utilization of the ML techniques. The ML group of Daimler-Benz Concern reports [56] on some topics, which they have successfully handled for their company:

- One is concerned with time series prediction in the domain of annual development of registered cars and trucks. These predictions are used by management in developing short and long term production plans actually. But there is envisioned utilization of the developed algorithms for prediction of exchange rate between $ and DM, too. Correct prediction can influence economy of the company significantly if it accepts payments in different currencies (17% of the total sales of Daimler-Benz is paid in $).

- There has been conducted joint feasibility project with Mercedes-Benz in order to verify that ML techniques can speed-up and facilitate development of knowledge base for a diagnostic expert system. Since obtained results are encouraging (about 3500 rules on 800 diagnoses were induced from about 30 000 protocols), more users intent to follow suggested approach to ES development.

Recently, the ML group of Daimler-Benz Concern has started projects on some dedicated ML applications. Two of them could be readily incorporated into CIM systems:

- A system for intelligent form filling, which applies ML algorithms in order to predict values for the fields at the actual form. The prediction is based on field values from the past forms and on the user supplied field values already entered in the actual form.

- A system for case based reasoning (CBR) in design and planning, which is intended to support reuse of plans and designs. Reuse/modification of good solutions is a strategy widely used by human experts in this domain. Its efficiency depends on access to "related" cases, that is why its automation is highly desirable.

CBR is another active stream within ML community. There have to be developed and studied strategies how to apply CBR - the essential difficulties appear in the choice of case representation, integration of domain knowledge, and intelligent adaptation/repair strategies.

Techniques of machine learning penetrate even into those engineering disciplines where long experience is needed to get a high competence. This is the case of machinery design and analysis of stresses in its components, for example. The finite element method is frequently used for this analysis. The basic problem here is the choice of appropriate resolution for modeling each component part (distance of considered approximation points) - high resolution leads to time-consuming computations, while too coarse a mesh produces intolerable approximation errors. Inductive logic programming has been successfully applied in this domain [57], inducing some meaningful general rules for finite element mesh design from examples of correctly and badly chosen meshes on different components. The resulting rules bind the number of elements on a given edge with the information on geometry of the whole component and on its mechanical properties (e.g. load, freedom of moving), which is provided in the form of background knowledge. This approach represents a knowledge extraction mechanism allowing to "verbalize" knowledge implicitly present in the submitted examples of correct meshes. It is one of the first steps toward the customized CAD systems of the future.

6.4 Learning in Planning and Scheduling

Various systems investigated by physics exhibit behavior resulting in some equilibrium characterized by minimum energy. Learning and adaptation known from biology and evolution theory are processes converging to overall improvement in systems' behavior. It means that these systems provide a very good inspiration for the solution of optimization tasks. Networks of neurons - or their artificial counterparts called neural networks - possess very similar properties. Machine learning has adopted some of these principles known from nature. Many such systems have been investigated and simulated yielding surprisingly good results. This research issued into two classes of algorithms: *simulated annealing* [16] and *genetic algorithms* [58].

However, some other methods observed in human way of thinking have influenced the machine learning, too. Among others, we can mention *case-based reasoning* [59]. This method is based on the idea of empirically verified examples of successful solutions of similar problems and the belief that feasible slight modification of the solution will keep the goodness.

Case-based reasoning is the intrinsic method used in CIM and in technology generally. It is used mainly with design and planning. Roughly speaking, the designer's technical experience is based on his/her knowledge of a number of successful designs and exploitation of this experience is done through combinations and extension of the known solutions. New designs or plans come from partial modifications of existing ones. This idea has been applied especially in production planning.

The following situations are frequent: The factory manufactures standard products (e.g. vehicles). The customers require slight modifications of the standard product. Then it is necessary to modify the production plans to fulfill the customers' demands.

The ground idea exploiting the principle of case-based reasoning is to have some prototype plans for the standard production. The plans for customer modified products are then generated by the admissible modifications of the prototype plan. If we introduce a state space describing the manufacturing procedure together with the production resources we can thus think about some optimal trajectory represented by the prototype plan. The derived plan is then a slightly distorted sub-optimal trajectory which has to fulfill the same constraints as the original plan. As the distortion should be minimum it follows that only pre-specified actions for the plan modification are allowed. Such actions might be substitution, deletion, or addition of several operations. In the state space model we can think about some tube around the optimal trajectory in which the trajectories of modified plans remain.

The genetic algorithms (GA) are used for many purposes. One of the interesting and most successful applications has been used for finding optimal schedules.

Scheduling is a very complicated task of discrete optimization when it is necessary to pair resources and actions from some given sets keeping many constraints imposed on these pairs. The optimization is performed by minimization of some utility function defined for those pairs.

The principle of GA follows the evolution in nature. There are generations of given size and each individual is characterized by some combination of chromosomes, in the computer simulation represented by a bit string. When creating next generation, two main operations are defined: *mating* and *mutation*. Mating takes two individuals and creates other two ones in the next generation randomly exchanging two parts of the chromosome chain. The two individuals for mating are chosen with the probability proportional to their quality. This is an analogy to the natural selection. To prevent from getting stuck in a local minimum, the mutation is introduced. It randomly changes some chromosomes of some randomly chosen individuals in each generation. It has been observed that the GA are surprisingly robust.

GA have been used to solve scheduling problems which are considered to be NP-complete. The main advantage of GA is that they can search for optimum continuously and at every moment we can get the best solution found so far. Another strength of GA is their low sensitivity to probability measures used to pick up the candidates for mating and mutation.

The genetic algorithms have been experimentally used for the task of the layout design (i.e. for a specific type of a planning task) in the system PACEB developed for a Czech factory producing concrete prefabricates tailored to customers' requests [60]. The results achieved by GA-approach proved to be comparable to those obtained by a complete state space search. The GA-algorithms provided proper solutions several times faster and, moreover, the designer is liberated from the need to design domain-specific search-pruning heuristics.

The crucial problem is the task representation. It means that we have to find a suitable mapping of the problem onto the bit strings (chromosomes). It is suspected that the optimal mapping itself is an NP-hard problem.

GA have been successfully applied to the classical production scheduling [61], layout planning, logistic problems, and many other CIM tasks where the more traditional approaches fail.

7 Conclusions

Information processing is the central point of CIM-ideas. It deals with huge and complex problems of data and knowledge handling which cannot be captured by the "classical" software engineering approaches and methodologies.

The knowledge-based approach (applied in isolated CIM-modules) has been recognized as useful and necessary for the entire CIM-area in the early eighties. The CIM-community has realized the decisive role of knowledge for efficiency improvement.

The models and methodologies developed within the frame of Knowledge Engineering correspond perfectly well to the models and methodologies used in CIM-software engineering. They formalize successfully the fast prototyping, iterative software development cycles etc. Moreover, a product life cycle has been recognized as a good analogy to the KE-software life cycle from both the formal as well as practical points of view. It provides, together with the methods of DAI, a suitable starting point for the concurrent CIM-engineering.

Planning and scheduling is an AI domain which is expected to provide solid theoretical backgrounds and efficient methods for all planning/scheduling processes in the CIM-area. Most of the existing results have - unfortunately - only toy-like nature. The AI-planning is very far from a complete success, but the first tools applicable in industrial practice are already provided by the AI-developers to facilitate its task. The CLP languages seem to offer very strong and general tools, which can significantly improve and influence important phases of CIM software development.

The last achievements in the area of declarative programming and machine learning can be explored with success as well. These methods offer good theoretical basis for the real-life applications of the RUDE-cycle-like methodologies.

Actual machine learning techniques offer a number of valuable tools which can play an important role in solving some problems occurring in industrial applications, namely that of customization of the general shells of knowledge-based systems and maintenance of evolving systems. ML from examples can help to find rules which are typical for given environment, CBR supports reuse and modification of successful solutions, while theory revision approach promises applications in the difficult task of evolving system maintenance. On the other hand, the techniques of unsupervised learning can suggest new concepts or language constructs which deepen our understanding to the considered domain. A similar goal is persuaded by data-mining techniques searching for functional dependencies implicitly present in the considered example data [62].

Artificial Intelligence influences the progress in CIM more than expected. It has significantly changed philosophy used by CIM-system designers and programmers [63]. It has introduced new software- and product-life cycle models and methodologies. It helps to overcome the so called "software crisis" in complex software development by introducing object oriented techniques based on certain theoretical results of AI.

The influence of AI on both the Software Engineering as well as CIM concepts proves to be much greater than usually considered and expected. *AI provides not only isolated methods, but completely new insights into the software development methodologies* especially in the case of huge, complex CIM-software systems.

References

1. H. Nicholson: Interconnected Manufacturing Systems, London: Peter Peregrinus Ltd., 1991

2. S. C. Shapiro (Ed.): Encyclopedia of Artificial Intelligence, John Wiley & Sons, Inc., 1990

3. W. W. Royce: Managing the Development of Large Software Systems. In: WESCON Technical Papers, Los Angeles, Vol.14 (1970), pp. 328-338.

4. D. Partridge: Engineering Artificial Intelligence Software. Oxford: Intellect 1992.

544

5. L. Johnson, H. G. Galal, E. N. Johnson: Development Methods and Environments for Knowledge Systems. In: EXPERSYS-92, ITT-International, Gournay sur Marne - France (1992), pp.197-202.

6. J. Bader, and others: Practical Engineering of Knowledge Based Systems. Information and Software Technology 30 (1988), pp. 266-277.

7. V. Mařík, T. Vlček: Some Aspects of Knowledge Engineering. In: V. Mařík, O. Štěpánková, R. Trappl (Eds.): Advanced Topics in Artificial Intelligence. Lecture Notes in Computer Science 617. Heidelberg: Springer 1992, pp. 316-337.

8. H. Grabowski, and others: An Integrated CAD/CAM-System for Product and Process Modelling. In: Advanced Geometric Modelling for Engineering Applications. Amsterdam: North Holland 1990, pp. 403-420

9. J. Milberg: Effizienz- und Qualitatsmanagement fur Simultaneous Engineering. In: Proceedings of PTK'92. Berlin: IPK, 1992, pp. 59-66.

10. O. Bjorke, O. Myklebust: IMPPACT, Trondheim - Norway: Tapir Publ., 1992.

11. C. Altmann: Dynamische Prozessgestaltung in flexiblen Fertigungssystemen durch integrierte Arbeitsvorbereitung. Vienna: Hanser, 1991.

12. R. S. Engelmore, A. Morgan (Eds.): Blackboard Systems, Reading, Mass.: Addison-Wesley, 1987

13. F.-L. Krause: Leistunssteigerung der Produktionsvorbereitung. In: Proceedings of PTK'92. Berlin: IPK, 1992, pp. 166-184.

14. Y. M. Huang, J. Rozenblit: Architectures for Distributed Knowledge Processing. In: Neural and Intelligent Systems Integration, J. Wiley, 1991, pp. 437-455.

15. J. Lažanský, V. Mařík: Knowledge and Data Concepts in AI. In: Cybernetics and Systems Research. Signapore: World Scientific Publ. 1992, pp. 1601-1608.

16. D. G. Bounds: New Optimization Methods from Physics and Biology, Nature 329 (1987), pp. 245-219

17. C. Hewitt: Control Structures as Patterns for Passing Messages. In: Artificial Intelligence, Vol. 8 (1977), pp. 323-363

18. W. J. McClay: A Query Server for Diverse Sources of Data and Knowledge. In: C. Moss, K. Bowen (Eds.): Proceedings of the First Int. Conf. on The Practical Application of PROLOG, London: Association for Logic Programming 1992

19. D. E. Commer: Internetworking with TCP/IP, Volume I., Prentice-Hall, 1991

20. D. Alred, et al.: AGATHA: Applying Prolog to Test and Diagnosis of Printed Circuit Boards. In: C. Moss, K. Bowen (Eds.): Proceedings of the First Int. Conf. on The Practical Application of PROLOG, London: Association for Logic Programming 1992

21. E. Rich, K. Knight K.: Artificial Intelligence, McGraw Hill, 1991

22. R. Coyne: Logic Models of Design. London: Pitman 1988

23. S. Steel: Notes on Current Trends in AI Planning. In: V. Mařík, O. Štěpánková, R. Trappl (Eds.): Advanced Topics in Artificial Intelligence. Lecture Notes in Computer Science 617. Heidelberg: Springer 1992, pp. 316-337.

24. M. S. Fox: Constraint-Directed Search: A Case Study of Job-Shop Scheduling, London: Pitman 1987

25. P. Van Henterick: Contraint Satisfaction in Logic Programming, Cambridge, MA: MIT Press 1989

26. CHIP 2.1 Reference Manual. London: ICL 1990

27. A. Colmerauer: Opening the PROLOG-III Universe, BYTE Magazine 12 (1987), pp 177-182.

28. J. Jaffar, S. Michaylov: Methodology and Implementation of a CLP System. In: Proc. of 4th Int. Conference on Logic Programming, Melbourne (1987)

29. M. Rueher, B. Legeard: Which Role for CLP in Software Engineering? An Investigation on the Basis of First Applications. In: C. Moss, K. Bowen (Eds.): Proceedings of the First Int. Conf. on The Practical Application of PROLOG, London: Association for Logic Programming 1992

30. R. Kowalski: Logic for Problem Solving. New York: North Holland 1979

31. I. Bratko: Prolog Programming for Artificial Intelligence. 2nd ed., Addison Wesley 1990

32. T. Van Le: The Techniques of Prolog Programming. New York: John Willey 1993

33. F. G. McCabe: Logic and Objects. London: Prentice Hall 1992

34. S. Gregory: Parallel Logic Programming in PARLOG: The Language and its Implementation. Addison Wesley 1987

35. U. Glässer, G. Hannsen, M. Kärcher G. Lehrenfeld: A Distributed Implementation of Flat Concurrent Prolog on Multi-transputer Environment. In Proc. of the First Int. Conf. of the Austrian Center for Parallel Computation, Berlin: Springer 1991

36. C. Geiger: ConFuP - Concept of a Parallel Logic Programming Language with Fuzzy Semantics. Diploma Thesis, Univ. of Paderborn, November 1993

37. E. Shapiro: Algorithmic Program Debugging. Cambridge, MA: MIT Press 1983

38. N. Lavrač, S. Džerovski: Inductive Logic Programming, Techniques and Applications. Ellis Horwood 1994

39. R. Quinlan: Learning Logical Definitions from Relations. Machine Learning, 5 (3), 239-266

40. S. Muggleton: Inductive Acquisition of Expert Knowledge. Turing Institute Press in assoc. with Addison Wesley Publishers, 1990

41. L. De Raedt: Interactive Theory Revision: An Inductive Logic Programming Approach. London: Academic Press 1993

42. J. L. Armstrong, S. R. Virding, M. C. Williams: Use of Prolog for Developing a New Programming Language. In: C. Moss, K. Bowen (Eds.): Proceedings of the First Int. Conf. on The Practical Application of PROLOG, London: Association for Logic Programming 1992

43. A. Roth, L. Sterling, C. Spenser (Eds.): Proceedings of the Second Int. Conf. on The Practical Application of PROLOG, London: Royal Society of Arts 1994

44. E. Denti, A. Natali, A. Omicini, F. Zanichelli: A Structured Logic Programming Approach to Robot Programming. In: A. Roth, L. Sterling, C. Spenser (Eds.): Proceedings of the Second Int. Conf. on The Practical Application of PROLOG, London: Royal Society of Arts 1994, pp. 187-206

45. C. Geiger, G. Lehrenfeld: The Application of Concurrent Fuzzy Prolog in the Field of Modelling Flexible Manufacturing Systems. In: A. Roth, L. Sterling, C. Spenser (Eds.): Proceedings of the Second Int. Conf. on The Practical Application of PROLOG, London: Royal Society of Arts 1994, pp. 233-252

46. R. M. Drogemuller, J. D. Smith: Computer Facilitated Building Design. In: A. Roth, L. Sterling, C. Spenser (Eds.): Proceedings of the Second Int. Conf. on The Practical Application of PROLOG, London: Royal Society of Arts 1994, pp. 593-612

47. P. Vasey, D. Westwood: Flex Expert System Toolkit. London: Logic Programming Associates Ltd. 1990

48. R. Michalski, J. Carbonell, T. M. Mitchell (Eds.): Machine Learning: An Artificial Intelligence Approach. Volume I. San Mateo, California: Morgan-Kaufmann 1983

49. R. Michalski, J. Carbonell, T. M. Mitchell (Eds.): Machine Learning: An Artificial Intelligence Approach. Volume II. San Mateo, California: Morgan-Kaufmann 1986

50. Y. Kodratoff: Introduction in Machine Learning. London: Pitman 1988

51. Y. Kodratoff, R. Michalski (Eds.): Machine Learning: An Artificial Intelligence Approach, Volume III. San Mateo, California: Morgan-Kaufmann 1990

52. S. Muggleton (Ed.): Inductive Logic Programming. London: Academic Press 1992

53. J. Quinlan: C4.5: Programs for Machine Learning. San Mateo, California: Morgan-Kaufmann 1993

54. L. De Raedt: Interactive Theory Revision: An Inductive Logic Programming Approach. London: Academic Press 1992

55 MLT: A sythesis report. in Proceedings of MLnet Workshop on Industrial Applications of Machine Learning - ILWS'94. Douran, France: 1994, pp. 113-125

56. G. Nakhaeizadeh: Some Applications of Machine Learning in Daimler-Benz Concern. In: Proceedings of MLnet Workshop on Industrial Applications of Machine Learning - ILWS'94. Douran, France: 1994, pp. 215-223

57. B. Dolšak, A. Jezernik, I. Bratko: A knowledge base for finite element mesh design. In: Proc. Sixth ISSEK Workshop. Ljubljana, Slovenia: Jožef Stefan Institute 1992

58. D. E. Goldberg: Genetic Algorithms in Search, Optimization and Machine Learning. Reading, Mass.: Addison Wesley 1989

59. S. Slade: Case-Based Reasoning: A Research Paradigm. AI Magazine 12(1) (1991), pp. 42-55

60. J. Lažanský, Z. Kouba, V. Mařík, T. Vlček, O. Štěpánková: Optimization and Decision Making in CIM. In: Proc. of "Design to Manufacture in Modern Industry" Conference. Bled, Slovenia: University of Maribor 1993

61. C. Muller, E. H. Magill, D. G. Smith: Distributed Genetic Algorithms for Resource Allocation. In: Proc. Workshop "Schedulling of Production Processes", 10th European Conference on AI. Vienna: 1992, pp. 37-45

62. D. H. Fisher Jr., M. J. Pazzani, P. Langley (Eds.): Concept Formation: Knowledge and Experience in Unsupervised Learning. San Mateo, California: Morgan-Kaufmann 1991

63. G. Spur, H.-J. Germer, C. Lehmann: Impact of Geometric Modelling for CIM. In: Advanced Geometric Modelling for Engineering Applications. Amsterdam: North Holland 1990, pp. 3-24.

Multi-Agent-Systems - A Natural Trend in CIM

Klaus-Peter Keilmann

University of Essen, D-45117, Germany

Abstract. During the last few years Computer Integrated Manufacturing (CIM) became an important aspect in the manufacturing industry. The different components of CIM - as there are the CAx techniques on the engineering side and manufacturing planning and control systems (MPC-systems) on the operational side - now show their influence on organizational structures and the working environment. Limits and problems of the fully automated manufactory - often called the "factory of the future" - have been recognized. Two ongoing trends can be observed: the decentralization of MPC-systems, leading to distributed MPC-systems, and the integration of the different modules of CIM, resulting in new requirements for the underlying information systems as well as for the persons using these integrated systems. The emerging field of distributed artificial intelligence (DAI) seems to offer interesting solutions to these requirements. Especially multi-agent-systems (MAS) have shown their feasibility in the area of CIM. The principles of coordination and communication are central aspects of these systems. Different prototypes have been implemented, for example HBBS: a hierarchical blackboard system for concurrent engineering, also used in the area of distributed manufacturing planning and control.

1 Introduction

Current markets require products that fulfill the specific wishes and needs of the customers. Therefore it forces the manufacturer to strengthen the customer orientation and the flexibility of his manufacturing units to fulfill customer demands fast. This enforces more complex information and manufacturing systems. Current development of computers and sensors supports this direction, but as more complex these systems are the less is their technical reliability [1]. How to deal with these complex systems is one more problem. Humans - the people that have to use and to work with these systems - cannot deal with over complex systems. Work nowadays can be characterized by a change from physical to psychic demands. Cause of the black-box character of most of the computational units the worker does not know what is going on inside. He has to react to instructions of a systems and therefore his behavior is no longer determined by himself. Another point is the lack of qualification to deal with these new systems. On the one side are technical skills - for example the knowledge of the command language, which button to push under which circumstances- on the other side knowledge about how to use the system - for example the use of a pointer device in the area of CAD instead of using a pencil [1].

As an overall postulation there must be the aim that the human has to be the controlling one in such complex systems. This is only possible if he knows the programmed functional logic inside his information system and is able to control it [1]. Segmentation at the manufacturing level and the development of distributed systems for CIM seem to offer possibilities to achieve this goal.

The conceptual idea behind segmentation of production is to simplify or to minimize the connections and relationships between the different manufacturing units. The aim is to gain more flexibility and productivity in the area of manufacturing. Team orientation is one of the important basic ideas. In all manufacturing concepts that follow the idea of team orientation, separated operating units will be defined, which are responsible for the manufacturing of a specific group of parts or a specific service. These units operate, within a specific framework autonomous with respect to the planning of the different necessary working steps. Different kinds of team oriented organizational structures can be seen, as there are manufacturing islands, flexible manufacturing centers, flexible manufacturing cells and flexible manufacturing systems [2]. Distinctions between these different forms of organization can be made in terms of complexity, the amount of available operations, the internal kind of material flow and the flexibility of the underlying transport system [3]. Segmenting the manufacturing is not enough. Responsibilities (e.g. for resources, scheduling, money, human resources) and decision making have to be delegated to lower levels. Due to increasing integration and networking capabilities of computerized manufacturing systems functional, organizational, and social separated units and structures should come to closer relationships [4]. DAI in our opinion offers technological and organizational possibilities to implement this new paradigm of manufacturing.

An overview of DAI, a classification of systems and agents and references to basic literature will be given in chapter two. Chapter three deals with the basic aspects of cooperation and communication. Chapter four contains a description of the structure of the hierarchical blackboard-system HBBS and its usage in the area of distributed manufacturing planning and control systems.

2 Distributed Artificial Intelligence

Distributed Artificial Intelligence (DAI) is a relative new field of research. DAI is aiming to develop methods and techniques for solving complex problems by means of intelligent behavior of a distributed system. The first workshops and conferences (DAI workshops) took place in the United States, later in Europe - the Modeling Autonomous Agents in a Multi-Agent World (MAAMAW) workshops. The first monograph on DAI appeared 1987: Distributed Artificial Intelligence edited by M. Huhns [5].

Research in DAI can be split in Distributed Problem Solving (DPS) and Multi-Agent-Systems (MAS) [6]. In contrast to distributed systems the different modules have to cooperate, therefore DPS and MAS have to be separated from Parallel AI [7].

Parallel AI: These are systems that use parallelism as a possibility for faster computation. The decomposition of the problem is done by defining separate subproblems where no cooperation is necessary to solve them. Consequently parallel AI is no field of research inside DAI [7].

Distributed Problem Solving: Research in the area of DPS is concerned with the question how a problem can be divided into subproblems such that distributed, cooperating agents can act together to solve it by solving the sub-problems. The task of problem solving is collaborative in the sense that sharing information and knowledge is essential to solve the overall problem [8]. DPS-systems are "a kind of top-down designed system, since agents are designed to conform to the requirements specified at the top" [9].

Multi-Agent-Systems: In multi-agent-systems agents are designed first, the solution strategy is defined later - depending on the problem. Agents are therefore pre-existing, autonomous and often heterogeneous [9], working cooperatively towards individual goals that interact. MAS are able to develop their coordination structure by themselves and to rearrange their roles if the context has changed. "A MAS can also be viewed as a bottom-up designed system" [9].

DPS and MAS can be seen as two extremes of the same spectrum [9]. This point is also expressed in the definition of Dezentralized Artificial Intelligence (DzAi). DAI deals "with the cooperative solution of problems by a decentralized group of agents" [5], whereas DzAI is "concerned with the activity of an autonomous agent in a multi agent world" [8]. Following Demazeau and Müller DPS can be stated as the main issue of DAI, MAS as the one for DzAI. It has to be said that the term multi-agent-system has to be used carefully, cause this term is often also used for a collection of agents that do not fulfill the aspects of MAS [7].

2.1 Classification of Systems

Work in DAI can be classified by eight dimensions; first introduced in [10], as shown in Tab. 1:

Dimension	Spectrum of Values
System Model	Individual .... Commitee .... Society
Grain	Fine .... Medium .... Coarse
System Scale	Small .... Medium .... Large
Agent Dynamism	Fixed .... Programmable .... Teachable .... Autodidactic
Agent Autonomy	Controlled .... Interdependent .... Independent
Agent Resources	Restricted .... Ample
Agent Interactions	Simple .... Complex
Result Formation	By Synthesis .... By Decomposition

Table 1. Dimensions for categorizing DAI Systems

The dimensions are [9]:

System Model: Is the system represented as one single agent, consisting of distributed components, or are there multiple agents, forming committees or societies?

Granularity: How detailed can the problem be decomposed?

System Scale: The number of processing units. Is there one large system or is the computation done by many smaller units?

Adaptiveness: Do the units of the system have the ability to learn? Simple units are often programmed and therefore static, whereas others may have the ability to learn.

Control Distribution: How independent are the elements of the system. Is there a master-slave relation or are they autonomous? Are there units that have control over others? To which extend do elements need others to solve their local problems?

Resource Availability: Who has access to which resources and which access restrictions exist? Are there limited resources?

Communication: How complex are the interactions between elements? Is there a simple or standardized communication or do complex interactions exist?

Problem solving: How is problem solving done? Are the problems decomposed (top-down) or are solutions found due to synthesizing results from different elements (bottom-up)?

2.2 Classification of Agents

There exist multiple points of view, what an agent is or what an agent should be. To avoid the definition of a universal agent usable for all kinds of problems, a hierarchy of agents as shown in Fig. 1 seems to be useful [11].

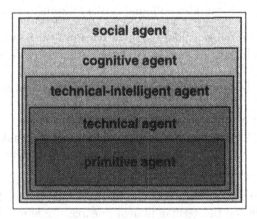

Fig. 1. Agent hierarchy

Primitive agent: Sensor-actor system that reacts on information kept by its sensors in a fixed manner.

Technical agent: Robots and flexible transport-systems, as they are available nowadays. They often include some sort of programmable control.

Technical-intelligent agent: Autonomous systems that are capable of solving a class of defined problems on their own under circumstances that do not have to be known a priori.

Cognitive agent: Agents that are able to learn.

Social agents: The discussion which capabilities are necessary to characterize a social agent is still open. Therefore no definite description could be given.

This hierarchy does not imply, that the capabilities of an included agent must be available for all upper agents. Every agent has features that are not of interest for others - for example an agent responsible for some kind of optical control in manufacturing has the capability to decide quickly if the current piece of work is in a regular condition. This ability is not mandatory for social or even cognitive agents [11].

2.3 Bibliography

Literature concerning Distributed Artificial Intelligence as a new field of research is not as easy to find as for other research directions (for example theoretical computer science, artificial intelligence, . . .), therefore a bibliography of good introductory texts and overview of DAI will be given. The work of Huhns, Bond and Gasser can be seen as the basic literature in the area of DAI (Huhns published 1987 the first book in DAI) [5, 6, 12]. The series "Dezentralized Artificial Intelligence" edited by Yves Demazeau et al. contains the proceedings of the European Workshop on Modeling Autonomous Agents in a Multi-Agent World [1] held every year in a different city in Europe. Therefore these publications give a good overview of the ongoing research in Europe in the field of DAI.

Surveys of research can also be found in the following journals:

(Special Issue on Intelligent and Cooperative Problem Solving) International Journal of Intelligent & Cooperative Information Systems, vol.1, no. 2, June 1992

(Special Issue on Distributed AI) Group Decision and Negotiation, vol.2, no. 3, 1993

(Special Issue on Mathematical and Computational Models of Organizations: Models and Characteristics of Agent Behaviour) International Journal of Intelligent Systems in Accounting, Finance, and Management, vol. 2, no. 4, 1993

An overview of current work in DAI can also be found in [13, 14]. The book of Müller [14] concerns on the german map of research. An annotated bibliography

---

[1] The title of the wokshops differs slightly over the years. For example in 1991: Modeling Autonomous Agents and Multi-Agent Worlds

3 Cooperation and Communication

Cooperation and communication are basic aspects of DAI (among others like representation and reasoning about plans, recognizing and reconciling conflicting intentions) [6]. Bond and Gasser define coordination as "a property of interaction among some set of agents performing some collective activity" [6] whereas cooperation is stated as "a special case of coordination among nonantagonistic agents". In our case we will make use of a definition given by Demazeau and Müller: "Cooperation: in order to perform a personal task, an agent will have to cooperate with others either because it is not able to accomplish it itself (restricted possible solutions), or because others successfully accomplish it more efficiently (e.g. within a shorter interval of time)." [8] The field of communication deals with the specification of protocols, messages, and the kind of information exchange between agents. Different languages have been defined for DAI purposes so far.

3.1 Cooperation

Agents or distributed problemsolvers can coordinate in many different ways - they often have to cooperate to solve complex problems. One question now is why should they cooperate or what are the goals of cooperation? Durfee, Lesser and Corkill [16] state the following goals:

Performance: Working in parallel could improve the performance of the problem solving task. The grade of improvement depends on the structure of the problem to be solved.

Alternatives: Allowing agents to form local solutions without being influenced by others increases the variety of solutions.

Confidence: Agents can verify results of other agents, possibly using own data and problem solving strategies.

Fault tolerance: Assigning important tasks to multiple agents increases the probability of finding a solution, even if some agents fail.

Less effort: Letting agents recognize and avoid useless redundant activities reduces the amount of unnecessary tasks (computations).

Improvement: Permitting agents to exchange predictive information improves the general problem solution.

Less communication: The selection of types of messages to be exchanged reduces the amount of communication.

Balance computation: Agents that are allowed to exchange tasks between computational units can make better use of these resources through balancing the load.

Select agents: The individual agent expertise can be used better if the agents are able to exchange tasks with the purpose that the task is performed by the most capable agent.

Time: Coordinating activity with respect to the time agents are waiting for results from each other.

Cause these goals conflict with each other, agents could not achieve the goals simultaneously [16]. Depending on the task to be performed and the structure of the problem to be solved the concrete form of cooperation has to be determined for each system uniquely.

Cooperation can be done with or without communication. There exist different opinions whether communication is necessary or not - depending on the underlying definition of agents. Following Martial [17] communication is a necessity for coordination in the multi-agent planning area. Genesereth, Ginsberg and Rosenschein claim that "intelligent agents must be able to interact even without the benefit of communication" [18].

Cooperation without communication. Cooperation is no problem if there are common, non-conflicting goals valid for all agents. This is an optimistic point of view. In a world of limited resources conflicts have to be considered. The basic question is therefore how can two or more agents work together (achieve their goals) such that both (all) are able to reach their goals? Two approaches can be seen here: intuitive and the theory of games.

Intuitive: Rosenschein [19] states, that often people choose the same solution even if there are a lot of equivalent ones available. The assumption is that if there are multiple solutions every person chooses the one that seems to differ a little bit from the others. For example [11]: Two persons are sent in different rooms. Both have to split a staple of 100 pieces of paper (banknotes) in two staples. They know that, if they choose the same solution, every one will get a mercedes. Most people split the staple in two staples with 50 pieces each. The 50-50 solution is the only one where the two staples are equal. This seems to be the slightly different one from all other solutions.

Game theory: To solve problems together agents have to know that there are other ones, have to assume what these agents know and how they behave. They have to have common knowledge: knowledge of the payoff-functions and some sort of common knowledge about the behavior of the other agents (behavior in the context of selection of possible actions). Payoff functions represent the evaluation of the result of each course of action an agent can follow, this includes the assumption that every agent knows enough to evaluate the results of his possible actions [5].

Cooperation with communication. Cooperation using communication can also be seen as a way to reduce the amount of data being exchanged between agents. If agents cooperate they do not have to exchange information about their current state, how they solve a problem or doing a computational task.

Communication to provide cooperation is used as well to transmit some kind of meta-level information. For example: If someone agrees with another that he will phone Mr. X, he will not have to tell him, that he takes some coins out of the pocket, throw them into the slot, etc. The term "phone to" includes implicitly all these actions: meta-level information or knowledge. Three kinds of message transfer to support cooperation can be seen: actors, contract-nets and black-boards. The underlying communication has great influence how coordination can be realized in the system.

Actors. Actors have been defined first from Hewitt and Agha [22]. An actor is "a computational agent that carries out its actions in response to processing a communication" [22]. Therefore message-passing is used as the base for concurrent computation. Every actor has its own mail address and they work (carry out their actions) in parallel. The current state of an actor can be described by his internal variables, which can change over time, and their values. A protocol defines to which types of messages the actor reacts and in which manner. Possible actions of an actor are:

- send messages to other actors or to itself
- create actors
- change state or protocol of actor

The behavior of each mail-address (actor) can be described by specifying a local behavior function. The combination of a communication and the target (the mail-address) is called a task. An actor processes each message sent accordingly to his protocol. A message which type is not known by the protocol of the recipient (no method exists in the protocol for this type) will be rejected. To response to a message an actor may send different messages to other actors it knows. Reconfiguring the system of actors dynamically is possible through changing the mail-address of a communication.

Contract-net. The contract-net protocol was first introduced by Reid G. Smith [20]. A contract-net consists of a set of nodes (agents) which negotiate with one another using asynchronous message passing. Contracts exist between two nodes. Each agent can play different roles during the act of problem solving. These roles can be changed dynamically over time. Roles - also mentioned as classes of nodes - are:

Manager: He identifies tasks to be done, decomposes these tasks and distributes the subtask within the net. Monitoring the execution of the tasks and processing their results is also in his responsibility.

Bidder: This is a node that offers to perform a task.

Contractor: A successful Bidder. He is responsible for the execution of the task for which his bid has been accepted. He is allowed to split his ask and award contracts to others.

Communication between agents consists of different types of messages. The messages and the processing of each message are as follows:

Task announcement: A manager announces a task to be processed. The message consists of a description of the task, a bid specification and expiration time.

Bid: Bidders send this message to show their willingness and availability to process the announced task. The message contains a node abstraction showing the relevant capabilities of the node with respect to the announced task. The manager receives these bids and ranks them relatively. A contract is awarded to the bidder that has sent the most satisfactory bid.

Award: Manager sends the award message to the node with the successful bid. This node is now called contractor.

Acknowledgment: The contractor acknowledges the award to the manager. This acknowledgement can be either rejecting or accepting. Every node can bid to several announcements, therefore it is possible, that a bid sent is not longer valid if the award for this bid arrives.

Report: This message is used by a contractor to inform the manager about the current status of task execution (interim report) or the termination of a task (final report). The result of the execution is part of this message.

Termination: A manager makes use of this message type if the contractor should interrupt or terminate task processing prematurely. A contractor receiving this message stops task execution and all outstanding subcontracts.

Availability Announcement: A node that is idle and searching the tasks broadcasts this message giving specifications of tasks and own capabilities.

This protocol, defined through the type and processing of each message, shows the dynamic character of contract-nets. Cause of the direct agent-to-agent communication there is no need for an overall control structure allowing a finer degree of control than possible with traditional mechanisms [20].

Blackboards. The basic idea of blackboards was first published by Newell [21]. "Metaphorically we can think of a set of workers, all looking at the same blackboard: each is able to read everything that is on it, and to judge when he has something worthwhile to add to it. This conception is just that of Selfridges Pandemonium: a set of demons, each independently looking at the total situation and shrieking in proportion to what they see that fits their nature." [21].Blackboards are a special kind of data structure, often separated in different levels or regions. Knowledge Sources (KS), as the independent processes are called, use this data structure as shared memory. KS write messages and partial results of their computations to the blackboard where other KS's are able to read them. The levels allow different representations or different levels of abstraction of a problem. Agents which work on different levels see the corresponding blackboard-level and their neighbor levels. Therefor generated data could be given to upper levels whereas goals can be written to lower levels of the blackboard [7]. Often a control system is used to supervise the blackboard systems, for example to synchronize access or to avoid or recover deadlocks.

The different experts (agents in our case) monitor the blackboard that contains the current state of problem solving. Each agent recognizes if it is able to support the problem solving task, solves the specific subproblem if possible and writes back the result - which can be a solution of the problem or some decomposed subproblems. The following assumptions have to be made:

Independence of expertise: Every agent solves its problem without help of others. It does not even know that there exist other agents. Communication is only done using the blackboard. As a consequence it is easy to add or remove agents from the system.

Different problem solving techniques: Cause each agent behaves independently it could choose its own problem solving technique. Therefor it could use different methods for every subproblem.

Flexible representation: The form of problem representation can be defined different for every application.

3.2 Communication

After the decomposition and distribution of a problem and instantiating different tasks, one has to make considerations how agents interact and communicate to solve their subproblems. Following Gasser [23] DAI designers have to consider:

Unit of interaction: Problems can be decomposed in different levels of granularity. Talking about strategic goals needs other forms of communication than interchanging pure facts, for example machine dependent data or construction details.

Structures and processes of interaction: There exist different modes of interaction as there are marketplace transactions, forum-based discussion or master-slave relationship. Organizational structures influence interaction too.

Protocols and languages: How communication takes place is highly dependent on the protocols and languages used. Protocols describe the semantics and rules of the communication between agents. The language used for communication can have a small amount of message types and small vocabulary -therefor needing a highly structured syntax- or can be more flexible allowing dialogues with rich semantic. Another point we have to mention is the kind of information we want to communicate - the content of a message.

Demazeau and Müller distinguish among three types of information [8]:

Knowledge: Agents interchange knowledge to come to a consistent and complete (with respect to the problem to be solved) description of the current situation. Cause of their different points of view there could exist contradictory descriptions of the environment. In this case a decision has to be made which description matches better.

Possible solution: If agents have to agree to a common solution or plan they have to exchange their solutions. Cause of different problem solving capabilities and strategies there may be no common solution acceptable for all agents. In this case the agents have to exchange knowledge to find other solutions.

Choice and result: If there are multiple possible solutions agents have to negotiate about the "common" solution. Choosing the first avoids this problem, but is sometimes not optimal. Every agent makes his own ranking of solutions, where the high ranking of one agent may be rejected by others that do not accept this solution. In the case where agents offer tasks to others choice has to be communicated too. If there exists a common choice the selected result has to be transmitted.

We also have to think about how communication can take place in multi-agent-systems. Two basic principles have to be considered:

Shared memory: Every agent is able to write the information it wants to share with others to a specific place to which every agent in the system has access to. Blackboards, as mentioned above, make use of this principle. The agents do not know who uses this information and therefore there is no need for them to know the other agents (or their addresses) in the system. To avoid inconsistency some kind of access coordination is needed - like in operating or database systems. This control can be done explicitly by one separate agent, acting like a gate keeper, or implicit through the system using shared memory (some kind of access rule). One problem of blackboard systems is the dependency of the system from one specific node - the blackboard. If the processing unit running the blackboard goes down, the whole system is unable to work. One great advantage can be seen in the anonymous character of the information exchange. No agent needs to know the addresses of the others, simplifying the management of the communication. To remove or to add other agents to those systems is therefor easy.

Message passing: Communication can also take place using message passing. Agents communicate through receiving and sending messages to each other. To do this they have to know either the addresses of the other agent or the address of a port the designated agent is listening to. One port can be used from multiple agents. Contract nets and actor systems make use of this principle. The theory of speech acts [24] deals with the semantic of messages. It distinguishes between sending a message, the intention of the sender and the effects of the sending of a message to the environment. The last point is only visible through the behavior of the receiving agent that can be observed from outside. One problem of systems using message passing for communication is the reconfiguration of these systems (add or remove agents) cause new agents (their addresses) have to be inserted in the list of addresses of each agent which has to communicate with them (or one of them) resulting in lots of update activities. The same has to be done if an agent has been removed. On the other side the system is more fault tolerant. Communication is not dependant on one specific agent.

Clearly these two basic principles can be mixed. One possibility are teams of agents that use shared memory internally and message passing for communication between teams. Information exchange can also be done using mailboxes. Messages can be send to agents and roles. These roles can be played by different agent. A mailbox stores the messages addressed to roles. Agents performing the addressed role could read these messages later [25].

4 The Hierarchical Blackboard-System HBBS

Weiß introduces in his PhD-thesis a hierarchical, distributed blackboard system for concurrent engineering [7]. In the area of concurrent engineering experts of different disciplines often geographically distributed and using different tools have to work together. Standardization and unification of schemes for data and processes would lead to an overwhelming amount of work. Main issues of concurrent engineering are therefore communication and coordination of engineering agents. Two directions can be identified: DAI that deals among other things with the integration of human and artificial (computational) agents, and Computer Supported Cooperative Work (CSCW) which deals with interactions between humans. Table 2 illustrates the embedding of DAI.

Concurrent Engineering		
CAD	DAI	CSCW
Artificial agents		Human agents

Table 2. Embedding of DAI

HBBS is designed as an infrastructure for the development of multi-agent-systems. The infrastructure should simplify the development and implementation of agents [26]. The knowledge and details of communication are separated from the agents and therefore no longer part of the agent definition task. Elements of HBBS are meta-agents that do not have, in contrast to agents that are connected to HBBS, any application knowledge. They only provide coordination knowledge. Each meta-agent consists of a blackboard, an agenda, a scheduler and designated "ambassadors", which themselves are represented as knowledge sources [27].

HBBS basically consists of four layers:

Application layer: This layer contains the whole application knowledge but no knowledge about control. Agents in this layer are all computational units or humans, which communicate with meta-agents of the underlying control layer. From the point of view of the agents they work together via a common virtual working area; communication and HBBS are transparent for them. Information exchange between agents is defined by an interaction language, providing rules and syntax of messages. Agents that do not know the interaction language make use of front- end systems. These translate the internal knowledge representation of the agents to the format of the interaction language.

Control layer: This layer contains the meta-agents. Each knowledge source is defined by a condition part and an event action part. The condition part consists of a trigger that defines the goal the ambassador reacts to, a precondition, which has to hold after the trigger was successful, and a post-condition which holds after performing the designated action. The event part defines the action to take place if the condition part holds. There are two kinds of information exchange between agents and blackboards: the transmission of solutions to other blackboards and the transmission of goals to start some problem solving process on another blackboard. We can also distinguish between local and non-local actions, which effect remote blackboards.

Cooperation layer: This layer consists of mechanisms for the coordination of agents and meta-agents, and the construction of hierarchical multi-blackboard structures. If the agenda contains goals, the one with the highest priority will be selected and the condition parts of all knowledge sources will be tested. If one holds the corresponding action part will be executed. As the result of an action local goals could be generated which will be inserted into the agenda. If the agenda is empty, the meta-agent waits for incoming messages that may contain a goal. To avoid deadlocks meta-agents are allowed to wait for determined messages or events and to store events, which should not be handled in a buffer, where they have to wait until some lock is raised.

Hierarchies of meta-agents can be reached by exchanging agents with meta-agents. Knowledge sources make no distinction between the different kind of agents. Therefore agents can be replaced by whole blackboard- systems easily [7].

Communication layer: The communication layer is responsible for the physical communication. To support naming and location transparency every agent is known only by his logical name. The mapping of logical names to physical addresses is done using a central name database (also called configuration database). Physical communication is based on TCP/IP.

Phillip and Weiß [26] implemented a prototype for distributed MPC-systems. Manufacturing planning is done in units of manufacturing orders. Machine orders are the units of manufacturing control. In their model similar manufacturing machines, those which are able to perform the same machine orders, form machine groups. Planning and control is done using well known, traditional algorithms.

For each machine there exists one machine agent. It gets orders from the higher level machine group blackboard. This blackboard receives orders for the machine group and organizes the cooperative work of the machine agents to perform the tasks. Every machine group blackboard is supported by a scheduling agent that schedules the different orders. The next higher level is build by the machine order blackboards and blackboards responsible for transport, human resource planning and inventory. To have access to these resources the machine order blackboard creates corresponding orders. Manufacturing planning is done by material requirements planning and capacity planning agents. The resulting orders are given to the manufacturing orders blackboard. There the orders are decomposed by another agent and scheduled, resulting in machine orders which are forwarded to machine order agents.

Cooperation and competition take place in this prototype. Machine, transport, and scheduling agents have to cooperate to fulfill their tasks (the right material at the right time on the right place). Material requirement planning and capacity management agents cooperate to formulate valid manufacturing orders and schedules. Competition exists between machine order blackboards cause of limited resources. Every blackboard wants to solve its own problems, leading probably to deadlocks in the manufacturing system. The prototype uses queues to avoid deadlocks.

References

1. Martin, Hans: Auswirkungen auf die Arbeitssituation, in: Geitner, Uwe W. (ed.): CIM-Handbuch, Braunschweig, Vieweg 1991, pp. 645-652
2. Kurbel, Karl: Produktionsplanung und -steuerung, München ..., Oldenburg 1993
3. Lenke, Christian: Zukunftsorientierte Konzepte zur Produktionsplanung und -steuerung auf der Basis moderner Informationstechnologien und flexibler Fertigungsstrukturen, Diplomarbeit, Universität Essen 1994
4. Beckenbach, Niels; van Treeck, Werner: Betriebliche und soziale Auswirkungen, in: Geitner, Uwe W. (ed.): CIM-Handbuch, Braunschweig, Vieweg 1991, pp. 653-661
5. Huhns, Michael N. (ed.): Distributed Artificial Intelligence, Los Altos, Morgan Kaufmann 1987
6. Bond, Alan H.; Gasser, Les (eds.): Readings in Distributed Artificial Intelligence, San Mateo, Morgan Kaufmann 1988
7. Weiß, Michael: HBBS: Ein hierarchisches Blackboard-System für den Verteilten Entwurf, Dissertation, Universität Mannheim 1993
8. Demazeau, Yves; Müller, Jean-Pierre: Decentralized Artificial Intelligence, in: Demazeau, Yves; Müller, Jean-Pierre (eds.): Decentralized A.I.: Proceedings of the First European Workshop on Modelling Autonomous Agents in a Multi-Agent World, Cambridge, England, August 16-18, 1989, Amsterdam, North-Holland 1990, pp. 3-17
9. von Martial, Frank: Coordinating Plans of Autonomous Agents, Berlin ..., Springer 1992
10. Sridharan, N. S.: Workshop on Distributed AI, AI Magazine, Fall 1987, pp. 75-85
11. Fischer, Klaus: Verteiltes und kooperatives Planen in einer flexiblen Fertigungsumgebung, Sankt Augustin, Infix 1993
12. Gasser, Les; Huhn, Michael N.(eds.): Distributed Artificial Intelligence II, Los Altos, Morgan Kaufmann 1989

13. Avouris, Nicholas M.; Gasser, Les (eds.): Distributed Artificial Intelligence: Theory and Praxis, Dordrecht ..., Kluwer Academic 1992, pp. 9-30
14. Müller, Jürgen (ed): Verteilte Künstliche Intelligenz - Methoden und Anwendungen, Mannheim ..., BI-Wissenschaftsverlag 1993
15. Jagannathan, V.; Dodhiawala, Rajendra: Distributed Artificial Intelligence: An Annotated Bibliography, in: [5], pp. 341-390
16. Durfee, Edmund H.; Lesser, Victor R.; Corkill, Daniel D.: Cooperation Through Communication in a Distributed Problem Solving Network, in [5], pp. 29-58
17. von Martial, Frank: Planen in Multi-Agenten Systemen, in: [14], pp. 92-121
18. Genesereth, Michael R.; Ginsberg, Matthew L.; Rosenschein, Jeffrey S.: Cooperation without Communication, in [6], pp. 220-226
19. Rosenschein, Jeffrey S.; Kraus, Sarit: The Role of Representation in Interaction: Discovering Focal Points among Alternative Solutions, in: Steiner, Donald D.; Müller, Jürgen (eds.): MAAMAW'91: Pre-Proceedings of the 3rd European Workshop on "Modeling Autonomous Agents and Multi-Agent Worlds", Kaiserslautern, August 5-7, 1991, pp.
20. Smith, Reid G.: The Contract Net Protocol: High-Level Communication and Control in a Distributed Problem Solver, in [6], pp. 357-366
21. Newell, A.:Some Problems of Basic Organization in Problem Solving Programs, in: Yovits, Jacobi, Goldstein (eds.): Proc. Conference on Self Organizing Systems, Washington, Spartan Books 1962, pp. 393-423
22. Agha, Gul; Hewitt, Carl: Concurrent Programming Using Actors: Exploiting Large-Scale Parallelism, in: [6], pp. 398-407
23. Gasser, Les: An Overview of DAI, in: [13], pp. 9-30
    in: Müller, Jürgen: Beiträge zum Gründungsworkshop der Fachgruppe verteilte Künstliche Intelligenz, Saarbrücken, 1993, DFKI Document D-93-06, pp. 77-89
24. Austin, J. L.: How to do things with words, Oxford, University Press 1962
25. Tomalla, Alf: Verteilte Künstliche Intelligenz in der Produktion, Diplomarbeit, Universität Essen 1994
26. Philipp, Mathias; Weiß, Michael: Ein hierarchischer Blackboard-Ansatz für Verteilte PPS-Systeme, in: Müller, Jürgen: Beiträge zum Gründungsworkshop der Fachgruppe verteilte Künstliche Intelligenz, Saarbrücken, 1993, DFKI Document D-93-06, pp. 77-89
27. Weiß, Michael: HBBS: Object-Oriented Design of a Distributed Blackboard Kernel, in: Engineering SEKE-93, San Francisco, IEEE Press, 1993, pp. 285-287

A prominent example of the stochastic search processes is the Monte Carlo method: It generates at random a number of trial solutions and uses the best of the solutions generated in this way as a solution of approximation for the problem. At the beginning of this search process, the Monte Carlo method shows very rapid improvements on the solutions as it searches large areas of the search space by random sampling and so rapidly finds areas of the search space which contain good solutions to the given problem. Since each attempt of a solution is independent of the preceding ones, the Monte Carlo method cannot make use of the solutions already examined. Therefore, shortly after the beginning of the process hardly any more significant improvements will be achieved.

In contrast to the Monte Carlo method, the neighborhood search [2] benefits from those points of the search space inspected before in the course of the process. Starting out from a trial solution for the given problem, a number of adjacent solutions is generated by partial modifications of the initial solution, i. e. solutions which differ only slightly from the initial solution. The best of the solutions in the neighborhood will become the new initial solution for the search, and the process will start again. A drawback of the method is that, by generating new trial solutions derived from the initial one, only a rather small part of the search space will be examined. Therefore, most of the time the search delivers only local optima depending on the initial solution.

A repeated start of the neighborhood search from different points of the search space as well as modifications such as the tabu search or the "sintflut" algorithm [11] shall prevent that the search is stuck in local optima. The most aggravating drawback of this and other enumerative procedures is, however, that each of the trial solutions can only be examined as a whole in respect of their suitability. In the case of scheduling problems, the search space is, however, so big that during the periods of time relevant to the practical application only a very small part of the search space can be examined.

A category of procedures which combine elements of the neighborhood search as well as of the Monte Carlo method, are the genetic algorithms (GA). GA's can examine large parts of the search space very efficiently. Since GA's also belong to the category of universal search methods, they have already successfully and often been used for the solution of most varied problems.

In the course of this treatise, different approaches to the solution of selected scheduling problems by means of genetic algorithms will be described.

2 Principle of Genetic Algorithms

The development of the GA's is based on Holland's treatises on artificial evolution [21]. Starting out from the thesis that the biological evolution is a very efficient optimization method the efficiency of which is also demonstrated by the development of the human being, Holland tried to formalize the principles of evolution and to employ them for the optimization of artificial systems.

Production Scheduling
and
Genetic Algorithms

Michael Neubauer

Wirtschaftsinformatik der Produktionsunternehmen
Universität Gesamthochschule Essen

Abstract. This treatise deals with the applicability of genetic algorithms to the area of production scheduling. To begin with, an introduction to the principles of genetic algorithms is given. After having outlined a standard genetic algorithm, first approaches to the traveling salesman problem are explained. On this basis, a survey on several approaches to different production scheduling problems is given.

1 Introduction

In the field of production planning, scheduling is intended to allocate the available resources to the production orders to be processed. Often not only the mere establishment of a valid schedule is aimed at, but the creation of a schedule which is optimal for the given problem. The quality of schedules can be measured with the aid of different performance functions. To this effect, mostly targets of cost or time are examined. This treatise deals with differently structured scheduling problems which all have in common that they belong to the category of NP complete problems [12]. Therefore, due to the complexity of time and memory exponentially increasing with the problem size, conventional optimization methods such as Branch & Bound and Dynamic Programming [2] can accurately solve only very small scheduling problems. Methods of approximation with a practicable complexity of time and memory by means of which bigger scheduling problems can be solved, cannot guarantee that an optimum solution is found. In practice, this is often not necessary. When scheduling, it will generally suffice to find a near optimum solution for the given problem, since the economic goals being the basis of the optimization efforts cannot be accurately outlined by means of the performance functions used.

Methods of approximation include priority rules which can easily be applied and therefore belong to those methods most frequently used in practice. Each time a resource is set free, they allocate an ordinal to the waiting orders, namely the priority. The order with the highest priority will then be allocated the free resource. The literature knows a multitude of different priority rules for simple scheduling problems [27]. The solution of complex scheduling problems for which only insufficient heuristics can be formulated, is often approached by stochastic search processes.

Rather than simulating analytic reasoning processes, the so created optimization method is, on the contrary, based on the successive application of the biological principles of inheritance and natural selection.

Genetic algorithms work on a population. This population consists of a number of organisms, also called individuals. The organisms live in an environment and compete for limited resources. The better an organism is adapted to its environment, the higher is the probability of its standing up against its competitors in the battle for the limited resources and of its living long enough to transmit its genotype to offsprings. Offsprings are produced by the mating of two parent organisms. During the mating, the offspring's genotype is produced by a recombination of the parents' genotype. The mating of two parents who have been adapted above average to the environment, involves the hope that the "good" qualities of both parents will be passed on to the offspring. Thus there is a great probability that such an offspring will even be better adapted to the environment than its parents and that it can again transfer its ameliorated genotype to offsprings. Since the resources in the environment are limited, these offsprings will supersede the offsprings of those organisms from the parent population which are less adapted to their environment. In this way a new population with new organisms will be created. These successive populations are called generations. In this manner, a collective learning process will take place from generation to generation with the result that the organisms will adapt themselves ever better to their environment.

After genetic algortihms were presented as "reproductive plans" for the first time in [21], they have been subject of numerous examinations. From these resulted not only analyses, but also alterations and extensions with the goal of increasing the performance of the GA's or of adapting them to certain problems. A critical assessment of the most important extensions would go beyond the scope of this work. We confine ourselves to the description of genetic algorithms as defined by Goldberg [14]. Most of the treatises on GA's are based on the definition employed by Goldberg. This kind of GA is also called a canonical or standard GA. The following paragraph briefly describes this standard GA and then looks into the question in which way GA's can be employed for production scheduling.

3 Standard Genetic Algorithm

In this section the most important parts of the standard GA are explained. These are the underlying representation, the evaluation function, the genetic operators, and the termination criterion. First a rough characterization of the standard GA's functioning is given:

The standard GA is initialized by the creation of an initial population. This initial population is usually generated at random. After initialization an interative process of evaluation, selection, and recombination is started. This process will continue until a certain termination criterion is satisfied.

The following pseudo-code outlines this process in more formal way:

```
procedure GA;
begin
t := 0;
initialize P(t);
while termination criterion = false do
        evaluate P(t) ;
        t:= t + 1 ;
        select P(t) from P(t-1);
        apply crossover & mutation to P(t);
        od;
end;
```

Fig. 1. Pseudocode of a genetic algorithm

3.1 Representation

Each organism represents a trial solution to the optimization problem. The trial solutions are represented by a string of a fixed length over a finite alphabet. Analogous to nature, this string is also called chromosome or genotype while the represented trial solution is called phenotype. In each position of the string a gene is situated the values of which are called allele. The principle of minimal alphabets [14] means that that coding with the smallest cardinality is selected which is derived in a natural way from the problem, because minimal alphabets lead to a more efficient search than bigger alphabets. After the development of genetic algorithms, these were at first nearly exclusively employed for the optimization of functions of the form $\mathbb{R}^n \rightarrow \mathbb{R}$. Therefore most of the studies utilize a binary or Gray coding over the alphabet $\{0, 1\}$. To give an example, the integer values 112 and 93 are represented in binary coding by the following two genotypes:

$$112 \equiv \boxed{1|1|1|0|0|0|0}$$

$$93 \equiv \boxed{1|0|1|1|1|0|1}$$

Fig. 2. Binary coded genotypes

3.2 Evaluation Function

The environment to which the organisms must adapt themselves, is determined by the objective of the optimization problem. The degree of the organism's adaptation to the environment is given by the objective's value of the trial solution which it represents, namely the phenotype, and is called fitness: Assuming for instance that the function $f(x) = x^2$ shall be maximized at the interval of $[0..127]$, the objective's value for both genotypes will be determined as follows:

$1|1|1|0|0|0|0 \equiv 112$ and $f(112) = 12544$

$1|0|1|1|1|0|1 \equiv 93$ and $f(93) = 8649$

Fig. 3. Evaluation of genotypes

3.3 Genetic Operators

In order to mimic the process of natural selection and inheritance on the population of genotypes, there basically exist three genetic operators, reproduction, crossover, and mutation.

Reproduction. Starting out from an initial population generated at random, the reproduction operator generates the following population by random selecting the individuals of the new population from the individuals of the old one. The number of copies of an individual being found in the next population, is proportional to the ratio of its fitness in relation to the average fitness of the old population.

Crossover. The crossover operator is the primary operator in the GA. Its task is to generate two offsprings out of two parents. This happens by random selecting a position in the genotype and by interchanging all allele from this position onward to the end of the genotype with the corresponding allele of the other genotype. As an example the fifth position has been chosen as crossover point:

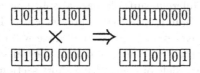

Fig. 4. Standard crossover operator

Mutation. After applying the crossover, the allele of the offspring are deliberately modified at a probability within the order of 0.001. Therefore, the mutation operator is a background operator. It is intended for introducing allele not occurring in the gene pool of the population, investigating new areas of the search space. A too frequent change of an allele by the mutation operator would degrade the GA to a mere random search. The example shows the mutation of the sixth allele:

Fig. 5. Mutation operator

3.4 Termination Criterion

The GA iteratively generates new generations, using the aforementioned genetic operators until a termination criterion is satisfied. Reaching a certain quality level for the trial solutions or a certain convergence of the population, the stagnation of the search process or exceeding a given time limit can constitute such a termination criterion. Often, also the number of generated trial solutions is employed as criterion.

4 Combinatorial Optimization Using Genetic Algorithms

In order to design a genetic algorithm for solving scheduling problems, four steps are required. First, an adequate coding of the trial solution in genotypes must be chosen. This representation should be derived as naturally as possible from the problem posed. In order to be able to evaluate the genotypes in respect of the objective, the second step must define a decoding function which again generates the phenotype, i. e. a trial solution from a genotype. In the third step the genetic operators crossover and mutation are designed which work on the genotypes. The fourth and last step is the configuration of the GA in which the parameters are defined determining the course of the GA.

4.1 Representation

The GA functions in that way that individuals which consist of outstandingly good partial solutions, multiply at a frequency above average whereby the advantageous heredity is spread in the population. By means of the crossover it is hoped to combine different suitable partial solutions of different individuals in order to obtain an individual of even greater capability. Since through the standard crossover, longer sequences of good allele will be destroyed with great probability, for the function of the GA mainly partial solutions are important which are coded by the allele of consecutive short sections of the genotype. According to the building block hypothesis [14], these sections are called building blocks.

When selecting a representation the difficulty is to find one which permits the taking shape of building blocks and simultaneously enables the crossover operator to recombine the parents' building blocks to offsprings' genotypes so that these again represent a legal schedule and that the information contained in the building blocks is least possible distorted. In the following we shall see that this task is in no way trivial.

With the exception of a first feasibility study by Davis [9], the development of approaches to solve scheduling problems is an integral part of the approaches to a solution of the traveling salesman problem (TSP). Both problems are NP-complete, and sequencing is of paramount importance. Contrary to the TSP which, though NP- complete, is of a very simple structure, the scheduling problem is not only computationally hard, but in addition very complex and has many facets. Encouraged by the results achieved with regard to the travelling salesman problem there have been many attempts to apply the methods and techniques for solving the TSP to scheduling problems. For the sake of a better understanding of these approaches, first of all different approaches to the solution of the TSP will be described.

4.2 Traveling Salesman Problem

The traveling salesman's problem is to find the shortest tour through a number of cities to that effect that all cities are visited exactly once and that the traveling salesman returns to his starting point at the end of the trip. The solution space for the TSP consists of all possible permutations of the cities. Since there exists no natural binary coding for this problem [35], in the early days of the GA's application to TSP's experiments were made with different representations and special genetic operators defined on the specific representations. As the outcome the ordinal, adjacency and path representation were achieved which also constitute the basis for different GA approaches in respect of scheduling. Unfortunately, there exist no comparing studies on these representations so that a discussion of the different approaches must be dispensed within this place.

Ordinal Representation. In the event of an ordinal representation [17] a genotype consists of a vector of the length n, the i'th component of which consists of an integer of between 1 and $(n - i + 1)$. To set up the tour described by the vector, first an ordered list of the city names, the so-called "Free list" as well as an empty "Tour list" will be produced. Then the genotype's vector components will be used one after the other for indexing the Free List. The respective city will be deleted from the Free List and added to the end of the Tour List.

Genotype	Tour list	Free list	Genotype	Tour list	Free list
2 1 3 2 1 ↑	()	(a,b,c,d,e)	2 1 3 2 1 ↑	(b,a,e)	(c,d)
2 1 3 2 1 ↑	(b)	(a,c,d,e)	2 1 3 2 1 ↑	(b,a,e,d)	(c)
2 1 3 2 1 ↑	(b,a)	(c,d,e)	2 1 3 2 1 ↑	(b,a,e,d,c)	()

Fig. 6. Decoding a tour in ordinal representation

The ordinal representation permits the use of the already described standard crossover operator. Each genotype thereby generated represents a valid tour. Inspite of that, a GA with ordinal representation did not achieve better results in experiments [17] than the standard Monte-Carlo method. The main reason is the effect of the crossover. Whereas the subtour which is defined by that part of the genotype which is situated before the crossover point, remains unaltered, the subtour of the phenotype which is defined by the back of the genotype, is mixed up at random.

Adjacency Representation. Another approach to the representation of tours in the TSP is the adjacency representation presented by Grefenstette et al.. A tour is represented by a permutation of the indices of an ordered list of the city names. In this case j stands exactly in the ith position, if according to the represented tour the next city to be visited from the ith city is the jth city.

Cities: (a,b,c,d,e)

Genotype: | 5 | 1 | 2 | 3 | 4 |
             ↑ ↑ ↑ ↑ ↑
             1 2 3 4 5

Tour: (a → e → d → c → b)

Fig. 7. Decoding a tour in adjacency representation

Each tour in the adjacency representation is outlined irredundantly by only one genotype. That means that in case of n possible permutations there exist $(n - 1)$ invalid representations which are composed of several cyclic subtours. There is very little probability that the standard crossover of two valid genotypes again generates a valid offspring. For this reason Grefenstette et al. developed three new operators, the alternating edges, the subtour chunk and the heuristic crossover which are described here in short.

Alternating edges crossover. The alternating edges crossover (AEX) builds up the genotype of the offspring by selecting an arbitrary edge out of one of the parents and integrating it into the offspring. Starting out from this edge, the AEX completes the tour by taking over the adjacent edges alternately from both parents. If one of the so selected edges results in a cycle in the offspring, one edge will at random be selected among the remaining edges which does not result in a cycle. With regard to the TSP, the AEX supplies rather poor results, since it cuts good subtours which cover several cities, into pieces and thereby prevents the aggregation of elementary building blocks to bigger units. The objective of developing the subtour chunking crossover was to eliminate this deficiency.

Subtour chunking crossover. The subtour chunking crossover (STC) alternately selects a subtour of any length out of both parents and integrates it into the offspring. If one of the edges so selected results in a cycle in the offspring, the same as with the AEX one edge will at random be selected among the remaining ones which does not result in a cycle. The performance observed was better than in case of the AEX, but still not competitive with conventional heuristic methods. Continued efforts toward an increase of performance led to the development of the heuristic crossover.

Heuristic crossover. The heuristic crossover (HX) begins with the random selection of a city and successively completes the tour by that edge of one of the parents which represents the shorter connection to the next city in the parent genotype. If the selected edge produced a cycle, an edge selected at random the same as with AEX and STC will be integrated into the tour. In experiments [17] the HX proves its clear superiority to the AEX and STC. This is not surprising, because, contrary to all other methods described in this treatise, the heuristic crossover makes use of information over the length of partial distances. All other GA methods only utilize the overall length of a tour. If the length of partial distances is unknown with a TSP, it will also be called a "blind TSP".

When deciding which edge completes the subtour, the HX uses the "nearest neighbour" heuristic. The observation that by means of HX the GA very quickly achieves good tours, though not reaching the optimum because some edges cross each other, suggests that the HX does not converge against the length of the tour, but against the suboptimal nearest neighbor characteristic. In experiments [33] Suh and Gucht compensated this behaviour by modifying the HX into a modified heuristic crossover (MHX) on the one hand, and by integrating a local optimizer into the GA on the other hand. The HX is modified in that way that with impending formation of cycles when the shorter edge has been selected, the (longer) edge of the other parent will be selected. Only if this, too, led to a cycle, a random edge would be selected for completion of the tour. The local optimizer is based on the 2-opt heuristic by Kerningham and Lin [22]; it selects two arbitrary edges and exchanges the final points if thereby the sum of the edge lengths will be reduced. The 2-opt heuristic is applied several times to each offspring. The results obtained were significantly better than those with the unmodified HX [33]. Unfortunately, no comparisons with an iterated nearest neighbor heuristic in combination with the 2-opt heuristic have been made, in order to prove the advantage of the GA approach as compared with the purely heuristic approach.

Path Representation. The representation which is surely the most natural and most frequently used, is the path representation. A circular tour is represented as a permutation of the city names.

Genotype: [a|e|d|c|b]

Tour: $(a \to e \to d \to c \to b)$

Fig. 8. Decoding a tour in path representation

Partially mapped crossover. The partially mapped crossover[1] (PMX) belongs to the first operators developed for the crossover of permutations; it was presented by Goldberg and Lingle [15] in 1985.

Starting out from the genotypes A and B, two crossover points are fixed at random. The section of the genotypes which is situated between the crossover

---

[1] Later on Goldberg used the name of "partially matched crossover".

points, is designated as mapping section and interchanged between the two parents. Duplicates produced by the exchange are eliminated in that way that all exchanges of cities within the mapping section are also made outside of the mapping section.

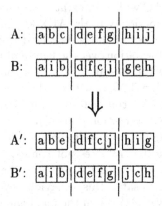

Fig. 9. Partially mapped crossover operator

Order crossover. The order crossover (OX) applied by Oliver et al. in [26] is a variant of the modified crossover (MX) developed by Davis (see [9]).

The order crossover at random determines two crossover points and copies the section of the parents which is situated between the crossover points, into the offsprings. Thereafter all other elements, beginning with the first position behind the crossover section, are taken over in that order in which they occur in the other parent genotype respectively. When reaching the end of the genotype, it continues from the beginning of the genotype to the beginning of the crossover section.

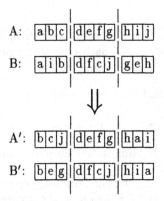

Fig. 10. Order crossover operator

In case of the modified crossover according to Davis the first crossover point is always situated in the first position of the genotype.

Cycle crossover. The development of the cycle crossover (CX) [26] is based on the concept that the position of each city in the offspring genotype is taken from one of the parent genotypes. An offspring is built by copying the first city of one parent. This is the city "a" in the given example. Since each city in the offspring shall be situated in the same position as in one of the parents, the next city to be copied is city "h"; it is moved to position eight, as the first position is already occupied by city "a". After determining "h", however, also the cities "j", "e", "d" and "b" are successively fixed. The cycle is closed, and the remaining cities are taken from that parent out of which the first city was selected. In this manner the cities "c", "f", "g" and "i" will be positioned.

A: | a | b | c | d | e | f | g | h | i | j |

B: | h | a | i | b | d | f | c | j | g | e |

$$\Downarrow$$

A': | a | b | i | d | e | f | c | h | g | j |

B': | h | a | c | b | d | f | g | j | i | e |

Fig. 11. Cycle crossover operator

Edge recombination crossover. While the approaches presented hitherto concentrated on the recombination of the cities, Whitley et al. in [35] used a completely different approach. The edge recombination crossover (ER) developed by them focuses on the edges which connect the cities mutually. To build up an offspring, first an edge table is constructed which contains the direct neighbors of each city in both parents. Thereafter the first city of one of the parents will be fixed as starting position of the new tour. Then one of the neighboring cities will be selected as next city from the edge table. In order that no city is isolated, the city with the lowest number of entries is preferred while selecting. If several cities are concerned, a random selection will be made among these. The selected city will be deleted from the edge table and the selection of the next neighboring city will continue until the tour is completed.

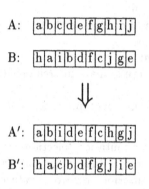

A: | a | b | c | d | e | Edge table:

$$\Longrightarrow$$

B: | b | d | e | c | a |

AB': | b | d | e | c | a | $\Longleftarrow$

city	has links to
a	b, c, e
b	a, c, d
c	a, b, d, e
d	b, c, e
e	b, c, d

Fig. 12. Edge recombination operator

Enhanced edge recombination crossover. The ER was further improved by Stark-weather et al. to the enhanced edge recombination crossover (EER) [31]. The EER supports the inheritance of common subtours by marking edges which occur in both parents with a minus in the edge table. When designing the offspring, the marked edges of the edge table will be preferred when selecting the cities.

Mutation. In contrast to the definition of a suitable crossover operator, it is pretty easy to determine mutation operators for the TSP. Syswerda investigated the effect of different operators on the traveling salesman problem [34].

- Position-based mutation:
  Two cities are selected at random, and the second city will be inserted before the first in the genotype.
- Order-based mutation:
  Two cities selected at random are interchanged.
- Scramble mutation:
  A partial distance of the genotype is selected, and the order of the cities arbitrarily altered.

A comparison based on a mutation-selection- strategy[2] showed better results for all three mutation operators than a random sampling. While the position-based and the scramble-mutation operator produced similar results, the order-based mutation operator proved to be superior to the others.

4.3 Comparison of the Operators.

Oliver et.al. in [26] made a first comparison of different crossover operators applied to the traveling salesman problem and came to the result that the OX is superior to the PMX and this, in turn to the CX. These results were confirmed by Starkweather et al. in [31]. Their experiments showed better results for the EER than for the OX, while this obtained better results than the PMX.

The results observed are easily explained. The CX, which receives the absolute position of a city in the genotype, attains so bad results because the absolute position of a city does not contain any information on the represented tour. The OX takes into account the relative order of the cities between each other, i. e. the information which city will be visited prior to an other, and thereby understandably attains definitely better results. The PMX tries to obtain absolute positions as well as relative positions in the crossover. Due to this hermaphroditic quality, it is the most frequently used operator, though or just because, as a consequence thereof, it supplies average results. The ER operators, when designing a tour, employ the information on the direct neighbors of a city in both parents. Thus, they make use of the information contained in the genotype to a much greater extent than the OX. In addition, the EER exploits information on common subtours and is thereby the most successful operator for the blind TSP up to the present.

---

[2] See [28] and [30].

5 Scheduling Using Genetic Algorithms

5.1 Scheduling Problem Definition

Contrary to the TSP which has a clearly defined problem constellation, production scheduling is a generic concept which covers a multitude of different problems. Abstracted from the actual production environment, the literature deals with different scheduling models which, by means of clearly defined model assumptions, stake out the basic conditions of the examined scheduling problems. As to the complete classification of scheduling problems, we refer to the description in [5].

Within the scope of applying GA's to production scheduling, the emphasis of the investigations lies on the three scheduling models briefly explained in the following: Single Machine Scheduling (SMS), Flow Shop Scheduling (FSS) and Job Shop Scheduling (JSS)[3].

If not explicitly mentioned, the studied three models are based on the following additional properties which in part essentially reduce the complexity of the scheduling problems:

- The viewed problem is static, i. e. all jobs are known at the beginning of the planning period, and no new jobs will be added during the processing.
- The processing times of the operations of a job are known; possible set-up times are included in the processing times.
- All jobs are ready for processing by the time 0.
- Each machine can simultaneously execute only one operation, and each job can only be processed by one machine at the same time.
- The model is non-preemptive, i. e. the processing of operations on machines cannot be interrupted.
- If a job can be processed by a machine, it will be processed without delay, i. e. there exist no inserted idle times.

The different constellations are described in more detail in the following. The quality of a schedule strongly depends on the actual production situation. Many criteria for evaluating a schedule can hardly be quantified. Objectives that can be assessed at justifiable expense therefore always represent merely an approximation of quality criteria. The approaches regarding scheduling make use of a multitude of most varying objectives. The studies described in the following use the following elementary objectives:

- The makespan (C_{max}) is that time at which the complete schedule has been fully processed; it is also called the length of the schedule.
- The lateness (L) is calculated by the completion time's deviation from the due date and summed up over all jobs. In this way earlinesses and tardinesses might neutralize one another.

---

[3] An introduction to these scheduling models is also given by Gerd Finke in the article "Planning and Scheduling".

- The earliness (E) is calculated as the sum of the absolute value of the earlinesses of all jobs.
- The tardiness (T) is calculated as the sum of all tardinesses of all jobs.
- The utilization (U) measures the utilization of the resources during the time required for the execution of the schedule.
- Set-up costs, idle costs and processing costs measure the costs involved in the execution of the schedule.

Single Machine Scheduling Problem. The single machine scheduling problem (SMSP) designates the simplest scheduling problem. A number of jobs is given. There is one machine available for the processing. The problem is to find a sequence of processing which optimizes the respective objective.

Flow Shop Scheduling Problem. The flow shop scheduling problem (FSSP) deals with the case in which not only one, but several machines exist and in which each job is composed of several operations each of which must be executed on another machine. The order of operations, is the same for all jobs.

Job Shop Scheduling Problem. The job shop scheduling problem (JSSP) is an expansion of the FSSP for which the order of passing through the machines can be different for each job. Here even exist several alternatives for the sequence of processing by the different machines in one job. Alternative process plans add further complikation to the JSSP.

5.2   Single Machine Scheduling Using GA's.

The first attempt to apply approaches of a solution of the TSP to the single machine scheduling, was made by Liepins et al. [20]. Schedules are represented by a permutation of the job identifications which is interpreted as priority list by the evaluation function. The jobs are processed on the machine in that order in which they appear in the priority list. Though the representation is similar to that of the TSP's path representation, its semantics are different. While in case of the TSP, the city from which to start is not determined and, for this reason there exist n representations of a tour, the scheduling problem is acyclic and asymmetric. Each representation defines another schedule.

Liepins et al. choose the minimum lateness (L) as objective (see [2]). The crossover operators used are the PMX (see [15]) and two greedy crossovers which, to a large extent, correspond to the MHX (see [33]). The MHX is based on the nearest neighbor heuristic which is based on the euclidian distance of each two cities. With respect to scheduling, however, there is no euclidian distance defined. Instead of that, the greedy crossovers use the processing time or the slack time of the single jobs. Therefore, the schedules resulting from the application of the greedy crossover tend to schedules as they are generated by the priority rule of the shortest processing time (SPT) or least slack time (LST).

The result proves the greedy crossover with SPT to be superior to the PMX and the latter to be superior to the greedy crossover with LST. Such results do not surprise since the SPT rule for the problem used supplies the optimum solution [2]. Rough approximations such as the LST rule are rather disadvantageous, since they lead the search into a wrong direction so that the result will even be inferior to the PMX with only average optimization. But if the optimum heuristic is known, the search with a GA is, however, superfluous. Therefore, the greedy crossovers used are hardly suitable for practical problems. The described approach is of hardly any practical use, since, by means of the priority rule SPT, the lateness criterion can be optimized much faster. The merits of the studies of Lipiens et al. are rather to be seen in the fact that it points out a feasible way how to use GA's for scheduling and thereby to have provided the basis for the approaches to the FSS.

5.3   Flow Shop Scheduling Using GA's.

Most of the approaches to the FSS are based on the manufacturing of printed circuit boards as practical application. Inspite of this common basis and the uniform representation of schedules as priority lists, the obtained results cannot be compared to each other, since the approaches differ by the crossover operators and the objectives used. With equal representation, the crossover operators' efficiency very much depends on the objective applied because different sections of the information coded in the genotype will be relevant. We should therefore refrain from a comparing evaluation of the approaches.

One of the first approaches originates with Cleveland et al. [8]. The total flow shop is considered as a complex single machine for which only the sequence in which the jobs are loaded into the first machine is scheduled. The jobs' sequence in the first machine is transferred to all other machines. Due to this assumption, the FSSP is reduced to the already described SMSP. Schedules are represented by priority lists. $E + T^2$ (see [2]) is selected as the objective to be optimized.

The crossover operators used are different variants of the subtour chunking crossover (see [17]) and a modified PMX (see [15]). The Scramble operator serves for the mutation. With the performed experiments, the weighted subtour chunking operator (WSTC) obtained the best results. The WSTC is based on the chunking operators described in [17] and [16]. A chunk of jobs is alternately selected out of the parents and reduced until no duplicates will occur in the offspring. When placing the chunk in the offspring, the WSTC takes into account information on the due dates of the jobs. The objective selected by Cleveland is $E + T^2$; for its optimization in general inserted idle times will be necessary. These have not yet been provided for in the scheduling model used. Since the selected approach comprises sequence-dependent set-up times, the GA will tend to prevent an earliness of the jobs by unnecessary set-ups.

The approach of Whitley et al. [35] published at the same time views an enlarged problem in which each work cell consists of two parallel machines. The target of the optimization is to minimize the costs entailed by the schedules, namely the sum of the processing, sequence- dependent set-up and idle times. The ER serves as crossover operator. Whitley et al. pursue the original approach in [36], [37], [31] and [32] with slight variations of the problem constellation and of the GA to be used. The emphasis of their studies, however, is on two different types of evaluation. A first approach called FIFO in which a FIFO queue is formed before each work cell. As soon as one of the parallel machines of one work cell has processed a job, the first job of the queue will be selected for processing. In a second approach called HYBRID, each work cell has its own look ahead scheduler which heuristically tries to minimize set-up and idle times. The experiments carried out, however, proved the hybrid approach to be inferior to the FIFO approach, since, despite a considerably higher expense of computation, it did not obtain better results than the FIFO approach.

The approaches by Biegel and Davern [4] as well as by Stöppler and Bierwirth aim at minimizing the makespan of the generated schedule. To this effect, Biegel and Davern use the PMX (compare (GL85)), Stöppler and Bierwirth compare PMX and OX (compare (OSH87)) whereat the OX attains slightly better results than the PMX.

In [4] Lee et al. present an approach to the problem of simultaneous lot sizing and scheduling. The problem constellation includes sequence-dependent set-up times as well as limited buffers between work cells. The objective is to minimize the makespan of the resulting schedule. The principle of the priority list is used as representation. When starting the optimization, all lots are partitioned into small partial lots of equal size. An identification is assigned to each partial lot. Now the genotype consists of a permutation of the partial lots.

After the iteration of a number of generations by the GA, the best genotype of the population will be selected and condensed. Condensing takes place by the combination to one lot of neighboring partial lots with the same job type. The GA will be started again. The initial population consists of the condensed genotype of the previous pass as well as of a number of genotypes produced at random. By the combination of partial lots, the genotypes will become shorter and shorter with each run of the GA. The process ends when there are no more combinations of lots possible.

5.4 Job Shop Scheduling Using GA's.

The FSS approaches showed that by the transfer of the approaches to the TSP simple scheduling problems can be solved. With the JSS this simple transfer is no more possible. Nevertheless, there exists a number of approaches to the JSS with GA's. The basic two development trends could be described as hybrid sequencing approaches and evolution programs.

Hybrid Sequencing Approaches. These approaches use conventional scheduling methods which are controlled by job sequences.

Nakano and Yamada [24] describe a method for optimizing the makespan in which the schedule is coded in bit strings. Each bit describes a precedence relation between two jobs in respect of a machine. By the use of the standard crossover and mutation, invalid representations can be created. A so-called forcing mechanism replaces invalid genotypes by valid ones with a minimum hamming distance. Thereafter, a simple deterministic scheduler generates the schedule from the binary representation .

Bruns [6] describes the coupling of a knowledge-based scheduling system to a GA in that manner that the GA works on permutations of job sequences which, in the evaluation step, will then form the input for the scheduler. The scheduler works job-based, i. e. it successively schedules all operations of a job before passing to the next job of the sequence. Therefore different schedules will result as a function of the job sequence. In view of the GA, this procedure is not unproblematical, as a good scheduler will have the tendendy to generate good schedules also from less qualified job sequences. The evaluation of the generated schedule, however, serves the GA as feedback with regard to the quality of the job sequence. That means that important information needed for the selection of good job sequences will be withheld from the GA.

The studies by Dorndorf and Pesch [10] in respect of optimizing the makespan are based on the algorithm by Giffler and Thompson (see [13]). For a given problem, the Giffler & Thompson algorithm can generate all active schedules. Active schedules are those without inserted idle times. When building a schedule, the algorithm must repeatedly make a selection among the number of not yet scheduled operations. This selection is controlled by a series of priority rules; the assignment of the priority rule which will be applied for which decision, is controlled by a genetic algorithm. A further approach by Dorndorf and Pesch consists of the integration of a GA into a shifting bottleneck scheduler.

Evolution Programs. The concept of evolution programs (EP) used by Michaelewicz [23] describes a significant enlargement of the standard GA. Instead of strings, EP's work on complex data structures. On the data structures used, crossover and mutation operators specific to the problem are defined.

The first approach to scheduling in this direction originates from Davis [9]. Davis composes the genotype of a number of partial genotypes. Each partial genotype is allocated to a machine and contains a number of priority lists. Each priority list is made up of a number of jobs and the commands "wait" and "idle" as well as of the indication of the time at which it becomes valid. The crossover operator interchanges partial genotypes between parent genotypes, and the mutation operator scrambles individual priority lists. A run-idle operator inserts idle times in the schedule. As objective, Davis considers the sum of the processing costs, idle costs and set-up costs.

In [1] Bagchi et al. compare three different representation studies. The first consists of a series of job identifications only, while the second additionally contains the process plan which determines the machines required for processing, and the third, in addition, contains an allocation of the operations to certain machines for each job. The GA selects among two crossover operators. The first crossover operator uses the PMX, in order to recombine the sequence of the jobs.The other crossover interchanges between two parent genotypes the information on the process plan used as well as the allocation of the machines. A scheduler generates the complete schedule out of the genotypes, while heuristically completing the missing information on process plans or selected machines with regard to the first two representations. Results of the experiments by Bagchi et al. prove that the quality of the schedules produced by the GA which in this approach is measured by the overall machine utilization, increases proportional to the degree at which the representation contains problem-related knowledge.

Bruns [7] pursues the approach of Bagchi et al. by additionally including in the representation the beginning and end of the processing times for each operation. As objective, Bruns uses the squared lateness (L^2). The crossover generates new genotypes by copying a series of non- overdue jobs out of one parent and the remaining jobs out of the other parent. The mutation operator modifies the process plan for individual jobs and the allocated machine for individual operations, or it shifts the beginning of the processing time to the earliest possible time.

The approach by Nakano and Yamada [25] to optimize the makespan works on a population of active schedules. The crossover designs schedules by using the already mentioned Giffler & Thompson algorithm. With each option out of a number of operations, the crossover makes either a random decision (mutation) or, with the same probability, selects one of the two parent schedules and that operation which has also been selected by the parent schedule.

6 Conclusion

As already explained, the GA approaches to the TSP are the basis of the GA approaches to production scheduling. Chronologically, the approaches first concentrate on the simplest scheduling model, namely the single machine scheduling. Then approaches to the flow shop scheduling and the job shop scheduling have succesively been developed. The field of study of production scheduling by means of genetic algorithms is, however, so young that there exist neither sound findings nor a basic theory in this field. The approaches outlined hereinbefore can hardly be compared with each other, so that they can only be considered isolated. Without further comparative studies, they are hardly suitable as a basis for more intensive examinations. Even though most of the authors describe their results to be encouraging, according to the present state of research, no definite statements can be made in respect of the suitability of GA's for production scheduling.

References

1. Sugato Bagchi, Serda Uckun, Yutaka Miyabe, and Kazuhiko Kawamura. Exploring problem-specific recombination operators for job shop scheduling. In Belew and Booker [3], pages 10–17.

2. Kenneth R. Baker. *Introduction to Sequencing and Scheduling*. John Wiley & Sons, 1974.

3. Richard K. Belew and Lashon B. Booker, editors. *Proceedings of the Third International Conference on Genetic Algorithms*. University of California, San Diego, Morgan Kaufmann, July 13-16 1991.

4. John E. Biegel and James J. Davern. Genetic algorithms and job shop scheduling. *Computer Industrial Engineering*, 19:81–91, 1-4 1990.

5. J. Blazewicz, K. Ecker, G. Schmidt, and J. Weglarz. *Scheduling in Computer and Manufacturing Systems*. Springer-Verlag, 1993.

6. Ralf Bruns. Incorporation of a knowledge-based scheduling system into a genetic algorithm. In W. Görke, H. Rininsland, and M. Syrbe, editors, *Information als Produktionsfaktor*, pages 547–553. Springer Verlag, 1992.

7. Ralf Bruns. Direct chromosome representation and advanced genetic operators. In Stephanie Forrest, editor, *Proceedings of the fifth International Conference on Genetic Algorithms*, pages 352–359. Morgan Kaufmann, 17-21 July 1993.

8. Gary A. Cleveland and Stephen F. Smith. Using genetic algorithms to schedule flow shop releases. In Schaffer [29], pages 160–169.

9. Lawrence Davis. Job shop scheduling with genetic algorithms. In Grefenstette [18], pages 136–140.

10. Ulrich Dorndorf and Erwin Pesch. Evolution based learning in a job shop environment. *Computers & Operations Research*, 1992.

11. Gunter Dueck, Tobias Scheuer, and Hans-Martin Wallmeier. Toleranzschwelle und Sintflut: neue Ideen zur Optimierung. *Spektrum der Wissenschaft*, pages 42–51, 1993.

12. M. R. Gary and D. S. Johnson. *Computers and Intractability: a Guide to the Theory of NP-Completeness*. Freeman, 1979.

13. B. Giffler and G. L. Thompson. Algorithms for solving production-schedule problems. *Operations Research*, 8:487–503, 1960.

14. David E. Goldberg. *Genetic Algorithms in Search, Optimization, and Machine Learning*. Addison-Wesley, 1989.

15. David E. Goldberg and Robert Lingle. Alleles, loci, and the traveling salesman problem. In Grefenstette [18], pages 154–159.

16. J.J. Grefenstette. Incorporating problem specific knowledge into genetic algorithms. 1987.

17. John Grefenstette, Rajeev Gopal, Brian Rosmaita, and Dirk Van Gucht. Genetic algorithms for the traveling salesman problem. In Grefenstette [18], pages 160–168.

18. John J. Grefenstette, editor. *Proceedings of the First International Conference on Genetic Algorithms and Their Applications*. Carnegie Mellon University, Pittsburgh, Lawrence Erlbaum Associates, July 24-26 1985.

19. John J. Grefenstette, editor. *Proceedings of the Second International Conference on Genetic Algorithms and Their Applications*. Massachusetts Institute of Technology, Cambridge, MA, Lawrence Erlbaum Associates, 28-31 July 1987.

20. M. R. Hilliard, G. E. Liepins, Mark Palmer, and Michael Morrow. Greedy genetics. In Grefenstette [19], pages 90–99.

21. John H. Holland. *Adaption in Natural and Artificial Systems*. University of Michigan Press, 1975.

22. S. Lin and B. W. Kerningham. An effictive heuristic algorithm for the traveling salesman problem. *Operations Research*, pages 498–516, 1972.

23. Zbigniew Michalewicz. A hierarchy of evolution programs: An experimental study. *Evolutionary Computation*, 1(1):51–76, 1993.

24. Ryohei Nakano and Takeshi Yamada. Conventional genetic algorithm for job shop problems. In Belew and Booker [3], pages 474–479.

25. Ryohei Nakano and Takeshi Yamada. *Parallel Problem Solving from Nature*, chapter A Genetic Algorithm Applicable to Large-Scale Job-Shop Problems, pages 281–291. Elsevier Science Publishers B.V., 1992.

26. I. M. Oliver, D. J. Smith, and J. R. C. Holland. A study of permutation crossover operators on the traveling salesman problem. In Grefenstette [19], pages 224–230.
27. S. S. Panwalkar and W. Iskander. A survey of scheduling rules. *Operations Research*, pages 45–61, 1977.
28. Ingo Rechenberg. *Evolutionsstrategie: Optimierung technischer Systeme nach Prinzipien der biologischen Evolution.* Frommann-Holzboog Verlag, 1973.
29. J. David Schaffer, editor. *Proceedings of the Third International Conference on Genetic Algorithms.* George Mason University, Arlington, VA, Morgan Kaufmann, June 4-7 1989.
30. Hans-Paul Schwefel. *Numerical Optimization for Computer Models.* Wiley, 1981.
31. T. Starkweather, S. McDaniel, K. Mathias, D. Whitley, and C. Whitley. A comparision of genetic sequencing operators. In Belew and Booker [3], pages 69–76.
32. T. Starkweather, D. Whitley, K. Mathias, and S. McDaniel. Sequence scheduling with genetics algorithms. In Günter Fandel, Thomas Gulledge, and Albert Jones, editors, *New Directions for Operations Research in Manufacturing. Proceedings of a Joint US/German Conference*, pages 129–148. Operations Research Society of America (ORSA), Deutsche Gesellschaft für Operations Research (DGOR), Springer Berlin, 1992.
33. Jung Y. Suh and Dirk Van Gucht. Incorporating heuristic information into genetic search. In Grefenstette [19], pages 100–107.
34. Gilbert Syswerda. Schedule optimization using genetic algorithms. In *Handbook of Genetic Algorithms*, pages 332–349. Van Nostrand Reinhold, 1991.
35. D. Whitley, T. Starkweather, and D. Fuquay. Scheduling problems and the traveling salesmen: the genetic edge recombination operator. In Schaffer [29], pages 133–140.
36. D. Whitley, T. Starkweather, and D. Shaner. Using simulations with genetic algorithms for optimizing schedules. In B. Svrcek and J. McRae, editors, *Proceedings of the 1990 Computer Simulation Conference*, pages 288–293. SCS, San Diego, CA, 16-18 July 1990.
37. Darrel Whitley, Timothy Starkweather, and Daniel Shaner. The traveling salesman and sequence scheduling: Quality solutions using genetic edge recombination. In Lawrence Davis, editor, *Handbook of Genetic Algorithms*, chapter 22, pages 350–372. Van Nostrand Reinhold, New York, 1991.

Implementation of Systems
with Declarative Constraints

Julius Csonto

Department of Cybernetics and AI, Technical University of Kosice,
Letna 9, 041 20 Kosice, Slovakia
e-mail: csonto@ccsun.tuke.sk

Abstract. The methodology of constraint logic programming (CLP), combining the declarative aspect of standard logic programming languages with efficient constraint solving techniques is presented. An experimental scheduling system, based on modified Allen's temporal relationships and implemented in the ECL^iPS^e programming environment is described. Formulations of scheduling problems and results are presented.

1 Introduction

Class of problems defined by a constraint system can be called design synthesis problem: given a set of design rules, the problem is to find one or more legal designs which conform to the rules. Put another way, given a set of constraints (to each constraint is assign a finite or infinite domain), the problem is to find an assignment of values to variables that is consistent with the constraints. From this point of view we can speak about *constraint satisfaction problem (CSP)*.

2 Constraint Logic Programming

For the solutions of these tasks several mechanisms with different descriptive (which types of constraints it enables to represent) and procedural (what type of reasoning - constraint "spreading" - can be implemented) power can be used. We want to concentrate on the use of *constraint logic programming (CLP)* [10, 15, 16].

CLP is a generalization of classic logic programming, it is a new programming paradigm at which unification is replaced by the investigation of *solvability* of a constraint system. There are several CLP implementations depending on the domain definition: CLP($\Re$) [9] for real numbers, Trilogy [2] for integers, CLP(Σ^*) [17] for regular sets, PROLOG III [6], for rational numbers and logic values, CAL [1] for complex numbers.

One of the most powerful CLP system is ECL^iPS^e [5] which using *forward checking* and *consistency checking* enables to reduce the search space substantially. Programming systems CHIP [7] and CONDOR [14] use analogical strategy.

Some specific applications which can be defined by a constraint system:

- planning of activities with temporal constraints [11]
- planning of operations with limited sources [14]
- layout and placement tasks [8]
- inference processes in expert systems with uncertainty handling [13].

3 Scheduling

The problem of representing temporal knowledge and temporal reasoning arises in a wide range of applications. It plays an important role in preparing schedules in operation management domains, e.g. factory management, airline gate scheduling, flight operation control. Constraint-solving techniques are very useful in the areas mentioned above because they allow to describe the problem in a simple declarative way.

Scheduling of projects is traditionally solved through the application of heuristics embedded in procedural languages. Specialised systems of this type have been developed [12], nevertheless general purpose *constraint-solving systems* may be used for this class of problems [16].

The general principle of solving scheduling problems using CLP systems is as follows:

- a domain-variable *Beg* representing the starting time is associated to each task (the corresponding domain can be defined as the interval $0..MaxD$, where *MaxD* is the summation of all task durations)
- the constraints (representing the time relationships) are imposed on these variables
- the constraint propagation process reduces the domains
- a labeling procedure assigns values to domain-variables; since every instantiation wakes all constraints associated to the variable, and changes are propagated to the other variables, the search space is usually quickly reduced and either an early failure occurs or the domains of other variables are reduced or directly instantiated.

4 A Simple Scheduler

Let us consider an example scheduling problem: There are five jobs A, B, C, D, E (the corresponding durations are $4, 8, 9, 2, 12$) and the following precedence constraints are required: *A precede B, A precede D, B precede C, D precede E.*

We can write a simple scheduler using the *formulate constraints & generate* method (which is much more efficient than the usual *generate & test* method):

```
schedule0(L,Fin) :-
            constraints(L,Fin),
            generate(L).
constraints(L,Fin):-
            L = [A,B,C,D,E,Fin],
            L::[0..100],
            A+4    #<= B,
            A+4    #<= D,
            B+8    #<= C,
            D+2    #<= E,
            C+9    #<= Fin,
            E+12   #<= Fin,
            min_max(indomain(Fin),Fin).
generate([]).
generate([H|T]):-
            indomain(H),
            generate(T).
```

The subgoal $L :: 0..100$ declares all the variables in the list L as domain-variables with a domain $0..100$ (interval $<0,100>$), the built-in operator $\# > =$ defines an inequality constraint between two linear terms (linear integer combinations of domain-variables) and the built-in predicate $\text{min\_max}(G, C)$ finds the solution of the goal G that minimizes the cost C. The domain-variable *Fin* corresponds to a dummy job representing the actual project completion time.

There are ten solutions L of our simple example scheduling problem, formulated as the question schedule0(L, Fin) (the value of *Fin* is *21*).

In the case of real-life problems, the number of satisfactory solutions is enormous and it is difficult to present them in a well arranged form. The method of forward checking provides an efficient solution of this problem: the execution of the constraints subgoal imposes the constraints (representing the time relationships) on variables (representing the start-times) and the constraint propagation process reduces the domains of these variables or directly instantiates them to a single constant (when the job is on the critical path). So the question constraints(L, Fin) is answered as follows:

```
L = [0,4,12,D::[4..7],E::[6..9]]
Fin = 21
Delayed goals: E - D - 2 >= 0
```

The jobs *A, B, C* are on the *critical path* and there are *4*4=16* combinations of possible start-times for jobs *D, E*. The delayed goal *D+2 <= E* reduces the number of solutions to *10* (the meaning of this additional condition is that task *E* has to start at least 2 time-units later than task *D*.

The currently used methods yield only the "first" solution, corresponding to the minimum values in these domains. Our system avoids the last (usually the most time-consuming) step of the scheduling process - namely the labeling - and yields only the acceptable start-time intervals for each job completed with a few additional conditions, which must be satisfied (there are usually not too many of these additional conditions). This method enables us to represent all solutions in a compact and readable form.

There exists a condensed and transparent mode of representing the start-time domains in the form of a bar-chart (see Fig. 1). Each horizontal bar indicates the duration of a job: hollow rectangles represent jobs with fixed start-times (i.e. jobs in the critical path), other rectangles are divided into upper and lower parts and shaded such that the upper bar represents the constrained interval for the start-time of the job, and the lower one represents the constrained interval for its end-time.

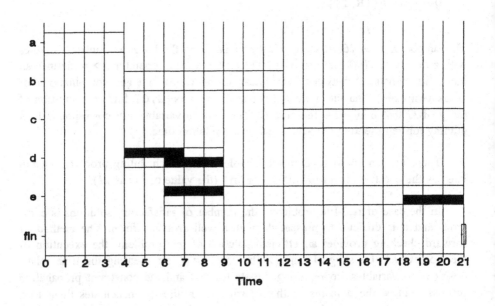

Fig. 1 Bar chart of the example scheduling problem

The above form of declaring the precedence constraints is not the most suitable one:

- the durations values must be written at each occurrence of a job (see job *A*)
- other time relations (we will discuss them in section 6) may occur in the project network and it is inconvenient to write the corresponding constraints in the explicit form

A better way is to represent the durations and the precedence constraints in two separate data structures. Perhaps a list of lists in arguments of durations and precedences relations:

```
durations([   [a,4], [b,8],   [c,9],
              [d,2], [e,12],  [fin,0]  ]).
precedences([ [a,b], [a,d],   [b,c],
              [d,e], [c,fin], [e,fin]  ]).
```

The core of a more sophisticated scheduler is the following predicate:

```
schedule(LVars) :-
      durations(LDurs),                  % 1
      genvars(LDurs,LVars,Fin),          % 2
      precedences(LPrec),                % 3
      gencsrts(LPrec,LVars),             % 4
      min_max(indomain(Fin),Fin).        % 5
```

1. returns the list of elements *[N, D]*, where *N* are the names and *D* the durations of jobs

2. returns a list *LVars* of elements *[N, D, V]* (where *V* are domain variables representing the start-time of the job *N*) and returns the domain-variable *Fin* corresponding to the dummy job *fin* representing the actual project completion time

3. returns the list *LPrec* of precedences

4. imposes the constraints on the domain-variables *V*

5. finds the solution of goal indomain(Fin) that minimizes the cost - the value of *Fin* (in our example a solution with the cost *21* was found).

The answer to the question schedule(L) is:

```
Found a solution with cost 21
L = [   [a,4,0],[b,8,4],[c,9,12],
        [d,2,[4..7]],[e,12,[6..9]],[fin,0,21]]
Delayed goals:  E - D - 2 >= 0
```

5 A More Complex Example

Let us consider as the next example of scheduling problem a project of building a house. The jobs and their durations are represented in the following form:

```
durations([ [excavate,                    2],
            [foundation,                   4],
            [rough_wall,                  10],
            [roof,                         6],
            [rough_exterior_plumbing,      4],
            [rough_interior_plumbing,      5],
            [wall_board,                   8],
            [flooring,                     4],
            [interior_painting,            5],
            [interior_fixtures,            6],
            [exterior_siding,              7],
            [exterior_painting,            9],
            [exterior_fixtures,            2],
            [rough_electric_work,          3],
            [finish,                       0]    ]).
```

The required temporal relations are as follows:

```
precedences([
    [excavate,                   foundation],
    [foundation,                 rough_wall],
    [rough_wall,                 rough_exterior_plumbing],
    [rough_wall,                 roof],
    [roof,                       exterior_siding],
    [rough_wall,                 rough_electric_work],
    [exterior_siding,            exterior_painting],
    [exterior_painting,          exterior_fixtures],
    [rough_exterior_plumbing,    rough_interior_plumbing],
    [rough_interior_plumbing,    wall_board],
    [wall_board,                 flooring],
    [wall_board,                 interior_painting],
    [interior_painting,          interior_fixtures],
    [rough_electric_work,        wall_board],
    [interior_fixtures,          finish],
    [exterior_fixtures,          finish],
    [flooring,                   finish]    ] ).
```

The system finds the solution with cost 44:

```
excavate                    0
foundation                  2
rough_wall                  6
roof                        [16..20]
rough_exterior_plumbing     16
rough_interior_plumbing     20
wall_board                  25
flooring                    [33..40]
interior_painting           33
interior_fixtures           38
exterior_siding             [22..26]
exterior_painting           [29..33]
exterior_fixtures           [38..42]
rough_electric_work         [16..22]
finish                      44
```

The schedule is reproduced in the form of a barchart for ease of interpretation (see Fig. 2). The following additional constraints must be satisfied:

exterior_fixtures-exterior_painting-9 > =0
exterior_painting-exterior_siding-7 > =0
exterior_siding-roof-6 > =0

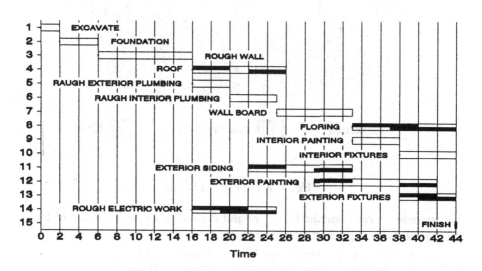

Fig. 2 Task scheduling for building a house

6 Resource-limited scheduling

The problem is to find a schedule that minimizes the time to build a two-segment bridge (Fig.1), what is a simplified version of the example taken from Bartusch's Ph.D. thesis on scheduling problems [3]. In addition to the usual *precedence constraints* (the task a must start after the completion of the task b) there are *disjunctive constraints* due to resources (for instance excavator, crane, a squad of brick-layers), which must be shared by several tasks. A disjunctive constraint states that two tasks a and b cannot be executed simultaneously, i.e. the task a must precede the task b or the task b must precede the task a. Such constraints completely change the nature of the problem, and no efficient polynomial algorithm can be exhibited for solving all scheduling problems involving disjunctive constraints. This problem is discussed in details in [15]. The durations of all tasks in the project are stored as a list of lists in the argument of the predicate **tasks**:

```
tasks([  [excavation1,4],  [foundation1,8],  [masonry1,16],
         [excavationP,2],  [foundationP,5],  [masonryP,10],
         [excavation2,3],  [foundation2,6],  [masonry2,12],
         [bearer1,   12],  [bearer2,   12],
         [filling1,  15],  [filling2,  10],  [finish, 0]]).
```

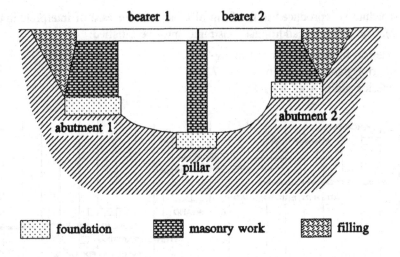

Fig. 3 A two-segment bridge

The *precedence relationships* for the project network are stored as a list of lists in the argument of the predicate **precedence**:

```
precedences([  [excavation1,  foundation1],
               [excavationP,  foundationP],
               [excavation2,  foundation2],
               [foundation1,  masonry1],
               [foundationP,  masonryP],
               [foundation2,  masonry2],
               [masonry1,     bearer1],
               [masonryP,     bearer1],
               [masonryP,     bearer2],
               [masonry2,     bearer2],
               [bearer1,      filling1],
               [bearer2,      filling2],
               [filling1,     finish],
               [filling2,     finish]              ]).
```

Lastly, the *resource requirements* for all tasks in the project are represented as a list of lists in the argument of the predicate **resources**:

```
resources([
  [excavator,      [excavation1,excavationP,excavation2]],
  [concrete_mixer, [foundation1,foundationP,foundation2]],
  [bricklayers,    [masonry1,    masonryP,    masonry2    ]],
  [crane,          [bearer1,     bearer2                  ]],
  [caterpillar,    [filling1,    filling2                 ]]]).
```

The *scheduler* for the bridge construction is defined by the following predicates:

```
bridge :-    setup(K, Finish),
             min_max(solve(K, Finish), [Finish]).
setup(K, Finish) :-
             tasks(L),
             gen_vars(L, K),
             member([finish, _, Finish], K),
             precedence(M),
             gen_precedence(M, K).
solve(K, Finish) :-
             resources(R),
             gen_disj(R, K),
             indomain(Finish),
             show_result(K).
```

The auxiliary predicate **gen_vars** generates a list K of elements of the form *[Task,Dur,Beg]*, where *Task* and *Dur* represent the name and the duration of a task and *Beg* is a domain-variable representing the start-time of this task. *Finish* is the start-time of a dummy task *finish*, representing the project completion. The next auxiliary predicate **gen_precedence** imposes the precedence constraints on the domain-variable *Beg* for each task. The third predicate **gen_disj** imposes the disjunctive constraints on these variables. The built-in predicate indomain instantiates the domain-variable *Finish* to the first element of its domain. Than the predicate show_result displays the values/domains of starting-time for each task (together with additional conditions, which have be to fulfilled) and draws a bar-chart of the project schedule. Lastly, the built-in predicate min_max yields the solution of solve(K,Finish) that minimizes the value of *Finish*.

A minimised project schedule with the cost (completion time) 70 time-units was computed by ECL^{i}PS^e within 2 secs on a SUN Sparc Station IPX. The corresponding start-time values/domains are as follows:

```
excavation1    [ 2 ..  5]
foundation1    [ 7 ..  9]
masonry1           17
excavationP         0
foundationP         2
masonryP            7
excavation2    [ 6 .. 27]
foundation2    [15 .. 30]
masonry2       [33 .. 36]
bearer1            33
bearer2        [45 .. 48]
filling1           45
filling2           60
finish             70
```

The real start-times have to satisfy some additional conditions:

```
foundation1 >= excavation1 +  4
bearer2     >= masonry2    + 12
masonry2    >= foundation2 +  6
foundation2 >= foundation1 +  8
foundation2 >= excavation2 +  3
excavation2 >= excavation1 +  4
```

The problem has 67 584 (4*3*22*16*4*4) solutions satisfying the above mentioned starting-time domains. The methods presented in [4, 14] yield only the "first" solution, corresponding to the minimum values in these domains. There exists a condensed and transparent mode of representing such a large amount of data (see Fig. 1). Each horizontal bar indicates the duration of a task: hollow rectangles represent tasks with fixed start-times (i.e. tasks in the critical path), other rectangles are divided into upper and lower parts and shaded such that the upper bar represents the constrained interval for the start-time of the task, and the lower one represents the constrained interval for its end-time.

The additional conditions reduce the number of solutions to 14 995.

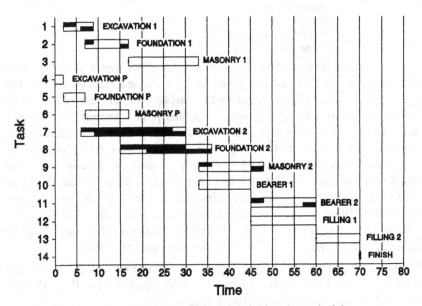

Fig. 4 Bar-chart of the minimised project schedule

7 Conclusions

ECL^iPS° combines the declarative aspect of standard logic programming languages with efficient constraint solving techniques allowing to prune the search-space in an a priori way (i.e. before detecting a failure). This is made possible by the removal of combinations of values which cannot co-exist in a solution. The application of the constraint & generate paradigm in ECL^iPS°, allows a programmer to avoid the use of the passive and inefficient generate & test paradigm. User defined scheduling heuristics also find a natural expression as rules in a Prolog-like formalism. These heuristics coupled with ECL^iPS°'s built-in constraint manipulation techniques, provide an efficient solution mechanism.

References

1. Aiba, A., K. Sakai, Y. Sato, D.J. Hawley and R. Hasegawa: Constraint Logic Programming CAL. In.: *Proc. of the International Conference on Fifth Generation Computer Systems*, pp. 263-276. ICOT, Tokyo.
2. Andrews, J.: *Trilogy (users manual)*. Complete Logic Systems Inc., North Vancouver.
3. Bartusch, M.: *Optimierung von Netzplanen mit Anordnungsbeziehungen bei knappen Betriebsmitteln. Ph.D. Thesis.* Technische Hochschule Aachen 1983.
4. *Bridge - a CHIP Demonstration Program.* München, ECRC 1989. 5 p.
5. Brisset, P., Y. Frühwirth, P. Lim, M. Meier, T. Le Provost, J. Schimpf and M. Wallace:*ECL¹PSᵉ 3.4 (Extensions User Manual)*. ECRC, München.
6. Colmeraurer, A.: An Introduction to Prolog III. *Com. ACM*, **33**, 69-90.
7. Dincbas, M., P. Van Hentenryck, H. Simonis, A. Aggoun, T. Graf and F. Berthier: The Constraint Logic Programming Language CHIP. In.: *Proc. of the International Conference on Fifth Generation Computer Systems*, pp.693-702. ICOT, Tokyo.
8. Dincbas, M., H. Simonis and P. Van Hentenryck: Solving a Cutting-Stock Problem in Constraint Logic Programming. In: *Proc. of the 5th international conference on Logic Programming*, pp. 42-58. MIT Press, Cambridge.
9. Jaffar, J. and S. Michaylov: Methodology and Implementation of a CLP system. In.: *Proc. of the 4th Logic Programming Conference*, pp. 196-218. MIT Press, Cambridge.
10. Kriwaczek, F.: An Introduction to Constraint Logic Programming. In.: *Advanced Topics in Artificial Intelligence*, pp.82-94. Springer-Verlag, Berlin.
11. Leng, Y.W. and L.F. Pau: Temporal reasoning in Blood Gas Diagnosis. *Expert Systems*, **8**, 159-170.
12. Meng, A.Ch. and M. Sullivan: LOGOS, a Constraint-Directed Reasoning Shell for Operations Management. *IEEE Expert*, **6**, pp. 20-28.
13. Sabol, T.: The Use of Multiple-Valued Logic in Expert System Knowledge Basis (in Slovak). In: *Proc. of ASRTP'92 Conference*, pp. 124-131. Technical University, Kosice.
14. Tay, D.: COPS - A Constraint Programming Approach to Resource-Limited Project Scheduling. *The Computer Journal*, **35**, A237-A249.
15. Van Hentenryck, P.: *Constraint Satisfaction in Logic Programming*. MIT Press, Cambridge.
16. Van Hentenryck, P.: Constraint Logic Programming. *The Knowledge Engineering Review*, **6**, 151-194.
17. Walinsky, C.: CLP(Σ*) - Constraint Logic Programming with Regular Sets. In.: *Proc. of the 6th International Conference on Logic Programming.* pp. 181-198. MIT Press, Cambridge.

This work was supported by MES grant 1686/94.

Qualitative Reasoning and CIM

Ivo Marvan, Olga Štěpánková

Czech Technical University of Prague, Faculty of Electrotechnical Engineering,
Department of Control Engineering. Czech Republic.

Abstract: Qualitative reasoning has been suggested and studied in artificial intelligence as a tool for knowledge representation, mental simulation and simulation of behavior of physical systems. This paper is an attempt to formalize various aspects of human thought covering common sense reasoning about physical reality. We try to point to those CIM tasks, which are of qualitative nature. This review helps us to identify those existing qualitative techniques, which are ready to be used in CIM. This paper introduces readers to some techniques of qualitative simulation on examples.

1 Introduction

CIM attempts to integrate all phases of production in a complex computer system including design, marketing, production planning, process control as well as economic agendas. Domain specific knowledge crucially influences decisions and strategies in each of these fields. This has to be reflected in the corresponding computer systems. Such an ambitious goal is impossible to be achieved without utilisation of recent results of computer science and AI, which tries to develop methods allowing computers to handle tasks usually solved by humans. Knowledge representation and reasoning is obviously one of the central topics of AI. Qualitative reasoning has occurred in eighties as an attempt to formalise those aspects of human thought covering common sense reasoning about physical reality.

We shall review briefly principles of qualitative reasoning in order to identify those CIM tasks which can potentially benefit from this approach. On the other hand, we shall try to summarise those methods of qualitative reasoning, which are ready to be applied in CIM.

2 What is qualitative reasoning?

In everyday life, people do not use numerical calculations when considering the problems of physical world surrounding them. That is why the corresponding strategies are referred to as qualitative ones in contrast to the quantitative ones of classical physics. Our treatment of the problem will be limited to the description of some available formalisms and corresponding strategies. In a broader sense, the term qualitative reasoning refers to reflections on behaviour of physical objects covered in human experience of everyday-life by common sense. Another term, namely "naive physics", seems to be a more adequate name for this broader topic.

When observing human common sense reasoning we can identify some typical patterns:

First of all the reasoning proceeds on the level of natural language, without explicit utilisation of numbers. The arguments work with a limited amount of unprecisely defined values characterising the considered quantities ("small", "large", "heavy"). The actual meaning of these qualitative notions is not fixed, but it is determined by the considered context. The intended meaning of "small" is different in astronomy and in architecture.

Both phenomena (utilisation of qualitative terms and their flexible interpretation) reflect the important property of human problem solving strategy, namely abstraction and choice of its appropriate level. Suppose the attributes of a physical system are characterised by its qualitative values (e.g. $x1$ is small, $x2$ is large). Even in very precise context, this description covers a number of different qualitative states characterised by specific precise values. That is why the qualitative description offers an abstraction which is difficult to achieve on the level of the original quantitative information.

The crucial goal of qualitative reasoning is search for such formal methods based on the discussed type of abstract description, which provide means for expression of causal relations and which can lead to derivation of the new facts. They should not lack the ability to move on the abstraction scale in the search for necessary compromise between :

- the precise specification (the complexity of which makes the problem computationally intractable) and
- unprecise descriptions (inefficient due to its low predictability).

3 Which CIM tasks are of qualitative nature?

The main motivation for development of qualitative reasoning was to compete with human's ability to cover broad scale of different tasks through common sense including reasoning about complex technical systems from the first-principles [6]. Why this is necessary in CIM? Let us observe and identify problems which people have to face in the domain of perspective application of CIM systems.

In the preparatory phases of the product life cycle, there appear many tasks demanding thorough analysis of the actual situation and corresponding decision-making based on available information, which is often far from precise (strategic planning, market trends, customer's behaviour, etc.). As a rule, these very complex systems (often of macroeconomics nature) defy the description by a classical apparatus of real mathematics with exact parameters. All over it, people do analyse these situations and base their strategic decisions on them. Here is surely one of the places in CIM where qualitative reasoning can play an important role.

Know-how of many construction tasks is based on an appropriate choice and combination of already known prepared elements. This is a common feature of a construction in different real life domains ranging from mechanical or electrical engineering up to software construction and process scheduling. In the search and verification of possible combinations "the brute force" of a computer significantly exceeds that of a human - computer can help to find the solution. But often the considered tasks are far too complex or even intractable (NP-complete problems or even problems lacking algorithmic solution). In such cases, it is necessary to transform the original task into a new one - this transition can be achieved e.g. through the utilisation of simpler description language, shifting the task to a more abstract (but less precise) level.

The efficiency of the mass production is determined by many factors, an important one is a perfect on-line quality control and fault-less function of used machinery. Its regular preventive diagnostics ensures continuity of production without unexpected interruptions - these should be avoided, because re-start of a production line can be very costly. Support of diagnostic processes was one of the practical motivation sources for the study of qualitative reasoning. It has met these hopes in medical diagnostics, already. The system KARDIO [1] uses a qualitative model of human heart in order to depict explicitly relations between combinations of the elementary physiological dysfunction in the components of the heart system and the symptoms observable on ECG of human patient. This method makes it possible to obtain diagnostic rules nearly automatically from the qualitative model through qualitative simulation. This approach allows to overcome the knowledge acquisition bottle-neck of expert systems, namely with respect to the search for rules binding multiple faults of the system with corresponding observable effects. Moreover, it can bring new knowledge - this was proved by the KARDIO system, which has succeeded to find a multiple-cause rule explaining observed behaviour of a specific type. The resulting rule uncovered its causes, thus it made it possible to suggest an original method of its treatment. Qualitative reasoning in diagnostics is extensively studied [18], [6].

The core of production itself are complex physical systems, which work up, shape and change the intermediate products. They are governed by different laws determined by its constitution (rotating machinery, melting-furnaces, chemical plants, mechanical parts of robots, transporters, etc.). But all of them have to be controlled. The theory of control has developed a number of methods for this purpose. But their ability is limited. Troubles arise when there is not available a proper description of the system, or if this description does not meet the requirements for solvability by classical methods - this is the case of strongly non-linear systems, systems with variable parameters or those under significant external influence.

Qualitative analysis of the controlled system could support the choice of the appropriate control. There were reported first experiments on this topic [2], [8]. Fuzzy controllers seem to offer another promising paradigm in this domain. Both fuzzy and qualitative approach study systems in "unprecise" terms, on a level of abstraction closer to human language. Those the relation between both these approaches remains to be clarified, it seems that they can successfully co-operate. In [19], [8], there is documented that fuzzy notions can fruitfully enhance the expressive power of qualitative reasoning. This topic will be treated in more detail in 4.4.

Even highly automated systems rely on decisions of human operators. Their responsible role is facilitated by complex computer systems providing supporting information. Not to overwhelm operator with mass of meaningless data, information has to be offered in appropriate amount and in relevant context. If necessary, explanation of the actual situation should be provided by a knowledge system. Qualitative reasoning can significantly help the user to understand what is going on in the controlled system when providing a qualitative simulation of the process. Qualitative subsystems representing the deep knowledge about the controlled system and allowing to switch between various abstraction level will certainly find their place in communication systems for human operators [23].

Obviously, methods of qualitative reasoning are ready to solve specific types of problems only. They can offer rough prediction of system's behaviour or support the appropriate choice of abstraction to be used in more precise simulation. Qualitative reasoning can be useful in close co-operation with classical quantitative methods, complementing them whenever these prove to be deficient due to e.g. lack of information. The possibility of mutual "communication" between both approaches working on different levels of abstraction is an interesting and challenging research topic. Here is probably hidden the key to practical applications of qualitative reasoning.

4 Review of developed methods of qualitative reasoning

4.1 When to apply a qualitative method

The CIM system constructor has an impressive choice of many computer systems developed to support singular subtasks of production. They are based on well formalised methods and utilise often precise quantitative computations. The difficult problem is their integration into one concise system. But humans are able to handle this task. Is it their intuition or a chance what leads them to the accepted decisions? Is there a space for "common sense"? If yes, do actual methods of qualitative reasoning offer the means necessary for CIM tasks?

Let us identify those circumstances, which can force the CIM system to apply a qualitative argument:

a) The most typical domain for QR represent situations when there is not available the quantitative description or model of the considered task. This deficiency can be either temporary (the relevant data have not arrived yet), wilful (the data acquisition is too costly) or principal (there is no deep exact knowledge about the system). In the last case, the qualitative reasoning seems to be the only step towards any solution.

b) Travelling along the abstraction scale can be fruitful namely when communicating with human user/operator. It is necessary to find a compromise in the choice of the abstraction level so that all relevant information is provided, but manager/operator does not become lost in the jungle of great amount of very precise (but "meaningless") data. Here the precision reduction can become the desired goal and condition for communication. Similar situation can be encountered in CAT (Computer Aided Teaching), too. The computer system works with a precise quantitative model, but the explanation offered to the human student has to meet the student's model and use natural language at best.

c) Another place where humans apply "common sense" reasoning is characterised by situations when the "exact" methods fail though there is available a correct qualitative model of the studied system. This can happen e.g. if mathematics is lacking the algorithms leading to exact solutions (they do not exist for some profound reason) or if the resources necessary for search of solution are not adequate. This can happen e.g. if they exceed the required time limit (real-time systems). The system's description can be simplified by a shift to a more abstract level allowing to generate some (less exact) solution.

The search for solution can be divided into two steps:

First, the problem has to be described by an appropriate formal qualitative language. This language must correspond to the chosen level of abstraction, it must be able to express the laws governing the system's behaviour. The resulting theory has to be provided by a decision/solution algorithm which allows to answer typical question by formal manipulation of the problem description. The search for appropriate representation language/system for qualitative reasoning seems to be the key issue in its further development and real-life application.

Naturally, the second step is problem solution itself utilising methods which are an organic part of the chosen qualitative system. They are based on classical AI techniques (state space search, theorem proving).

Qualitative simulation (QS) is a crucial part of qualitative reasoning. Its goal is to derive the dynamic behaviour description of the considered system from its (static) schema characterising mutual binding of elementary components extended by the list of relevant physical laws (responsible for its dynamic behaviour). Let us briefly mention some methods of QS.

4.2 QSIM and its descendants

Basic vocabulary of qualitative reasoning is close to expressions used in natural language communication - information about the values of the considered quantities is expressed in a finite domain of discrete values referred to as *landmark* or *distinguished values*. Distinguished values mark the borders of characteristic behaviour of the observed system - it can be the boiling temperature of considered liquid or the capacity of the used container. Qualitative domain is often simplified as much as possible - then it is limited to 3 values denoting the sign of the considered quantity (*positive, negative, zero*). On the other hand, qualitative description usually refers to local dynamic properties of the considered variable, too. These properties are represented by the development trend of the variable given in terms of the sign of time derivative of the variable at the considered moment. Qualitative value of an attribute is thus represented by a pair - its first member carries information about the placement among the considered landmarks, while the second member describes its development trend. *To simplify the notation we introduce the function first the purpose of which will be to give the first member of the pair representing the qualitative value of the variable.*

Principal simplification of the domain leads to a specific choice of basic functions/relations for expressing the relations among considered qualitative variables. Different authors build on different sets of these primitives.

Time is submitted to discretization in qualitative systems, too. There are distinguished only such time instances (time-points) in which there appears a phenomenon which can be recognised in terms of the chosen vocabulary, e.g. one of the following conditions is met

- one of variables reaches its landmark value,
- the development trend of one of the variables has changed.

The *state* of the observed system is characterised by qualitative values of all used variables at a specified instant. Of course, there can be traced no difference between 2 instants in an interval between 2 neighbouring distinguished time points. That is why there are used 2 types of time instant specification - a *time-point* and a *time-interval*. This distinction appears in states, too. The *history* of the system is expressed as a sequence of time-point and time-interval states which alternate regularly.

The core of each simulation system consists of the rules which allow to perform prediction of future states of the studied system. The input of a simulation algorithm has two parts: the description of the studied system and data on the starting state, both expressed in terms of simulation language. The output of simulation can be a graph the nodes of which depict possible future states of the considered system and the edges correspond to transition between them.

Let us briefly summarise a theoretically well founded simulation system QSIM for qualitative domain developed and implemented by B. Kuipers [11]. This system is intended to reason about such physical quantities which can be visualised as continuous and smooth functions. It works with a very restricted set of relations, the aim of which is to specify mutual constraints among the values of the considered quantities, which hold throughout the life-cycle of the considered object. These constraints are implied e.g. by construction of the object. One of such constraints is the relation M^+ { M^-} - a qualitative abstraction of the proportional relation between its arguments:

"$M^+(X,Y)$ { $M^-(X,Y)$}, holds iff the development trends of both variables X,Y are the same {are opposite}". The set of relations used by QSIM is explicitly introduced in the following table:

relation	meaning
$M^+(x,y)$	x and y are "qualitatively dir. proportional"
$M^+_0(x,y)$	- " - and x=0 <=> y=0
$M^-(x,y)$	x and y are "qualitatively reciprocal"
$M^-_0(x,y)$	- " - and x=0 <=> y=0
add(x,y,z)	z = x + y
mult(x,y,z)	z = x * y
deriv(x,y)	y is the time derivation of x
const(x)	x is constant

Tab.1.

The relations M^+ (M^+_0) and M^- (M^-_0) are denoted mplus (mplus0) and mminus (mminus0) in the rest of the paper.

The development rules of QSIM are derived from knowledge of behaviour of smooth functions as it is characterised by classical mathematical analysis.

Our system QUASIMODO [17] is based on QSIM-like ideas, but it allows to reason about discontinuous changes, too. This part is inspired by [26] - discontinuous changes can appear in distinguished time points under well specified conditions. These changes are characterised by the transition rules together with the constraints of the system. This can be done in a very natural way due to the fact that QUASIMODO is implemented in Prolog. Let us illustrate its function and the results it is ready to produce on a simple example.

4.2.1 Example 1.: Qualitative simulation of a bell.

<u>Definition</u>: Simulate behaviour of the electromagnetic bell.

<u>Analysis</u>: The electromagnetic bell (buzzer) is a device consisting (among others) of a mobile hammer with a contact, electromagnet, metal vessel (gong) and a spring (see Fig. 1). Electric current can flow through the electromagnet only provided that the hammer is at its leftmost position (thus it connects the circuit with the electromagnet by its contact). Otherwise, the circuit is disconnected. The permanent force of the spring pulls the hammer in the direction to the contact (in our case to the left). On the other hand, if a current flows through the electromagnet, the force of the electromagnetic field attracts the hammer to itself (this force is bigger than that of the spring and it acts in the opposite direction, i.e. to the right).

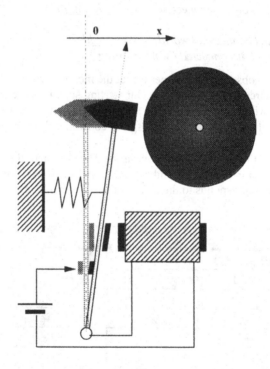

Fig 1.

<u>Remark:</u> To get a model as simple as possible, let us concentrate on the changes in the system caused by connection/disconnection of the circuit, while neglecting problems occurring at the moment when the hammer clides the gong. In other words, we the bang is treated as an elastic collision, any discontinuous change is neglected.

Let us observe the position of the hammer x and its velocity v provided the fixed contact is at x = 0 and the hammer can move towards the positive values of x.

We formulate the constraints on the observed variables separately for the case of the circuit disconnected (the name of the corresponding world is spring):

$deriv(x,v)$, $deriv(v,a1)$, $x{\geq}0$, $a1 = const. < 0$,

and partly for the case of the circuit connected (the name of the corresponding world is el-mag-field):

$deriv(x,v)$, $deriv(v,a2)$, $x=0$, $a2 = const. > 0$.

For the transition from one such "world" to the other one we define rules:

1. x=0, v<0, contact disconnected / world=spring   =>
   x:=x, v:=-v, contact connected / world= el-mag-field

2. x=0, contact connected / world= el-mag-field   =>
   x:=x, contact disconnected / world=spring.

Let us start to observe the hammer e.g. at the moment when it is receding from the contact and the circuit is disconnected. Our starting state is in the world "spring", it is a time-interval with first(x)>0, first(v)>0. Full qualitative values of all the other variables can be determined so that the "spring constraints" are met.

The result of the simulation is given in the Tab. 2, where the qualitative value is represented by a pair (both its members are separated by an operator "/") the first member of which corresponds to the actual value, the second one to the trend.

number	time	x	v	world
0	t(0) .. t(1)	+/+	+/-	spring
1	t(1)	+/0	0/-	spring
2	t(1) .. t(2)	+/-	-/-	spring
3	t(2)	0/-	-/-	spring
4	t(2)	0/+	+/+	el-mag-field
5	t(2) .. t(3)	+/-	+/+	el-mag-field

Tab. 2.

The obtained graph of transitions has the form depicted at the figure Fig 2.

This graph can be interpreted like this: We observe the system from the moment when the hammer is receding from the contact, the electrical circuit is disconnected and the spring is acting against the direction of the motion of the hammer (state 0). The motion of the hammer will stop eventually (1) - the bell gives a stroke, and the hammer will start moving back to the contact (2) until it touches it and connects the circuit (3, 4). The force of the electromagnetic field becomes the dominant force affecting the hammer now, it pulls the hammer away of the contact and the hammer starts moving in its direction (5). At this moment, disconnection of the contact occurs and the whole process will be repeated again (0) and again. The bell rings.

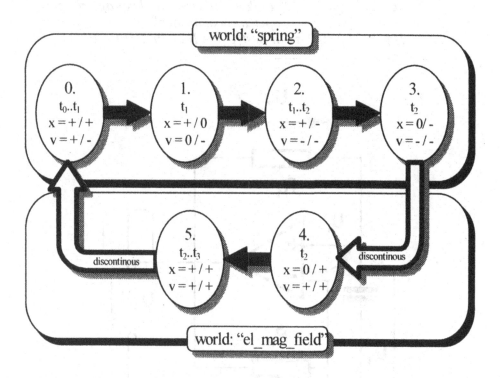

Fig. 2.

4.2.2 Example 2.: Qualitative simulation of an astable trigger

Let us consider an electronic circuit specified at the Fig. 3. It is an astable trigger using an operation amplifier. Input resistance of an ideal amplifier is infinitely large, that is why the current flow into the input clips (denoted by +, - in the schema on the Fig. 3.) is zero. Output voltage U_o is constant under non-zero input voltage U_i, moreover U_o has the same polarity as U_i (this is due to properties of operation amplifier in the saturation region). The basic facts about this circuit are expressed by the following set of constraints:

(1) $U_1 + U_i = U_4$ *"Kirchhoff's law for voltage"*
(2) $U_1 + U_2 = U_o$ *"Kirchhoff's law for voltage"*
(3) $U_3 + U_4 = U_o$ *"Kirchhoff's law for voltage"*
(4) $mplus0(I_1, U_2)$ *"Ohm's law for resistor R_1"*
(5) $mplus0(I_2, U_3)$ *"Ohm's law for resistor R_2"*
(6) $mplus0(I_2, U_4)$ *"Ohm's law for the resistor"*
(7) $deriv(U_1, I_2)$ *"condenser current and voltage relation"*
(8) $const(U_o)$ *"operation amplifier"*
If $first(U_o) = H$ **then** $first(U_i) = H$ **or** $first(U_i) = 0.$

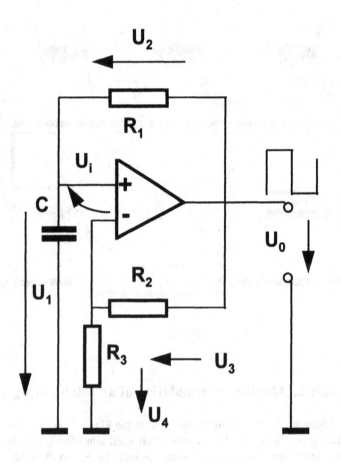

Fig. 3

The constraints (1)-(8) describe continuous change in the circuit. But there is one more rule which has to be added in order to specify possible discontinuous change in the trigger:

(9) *If at the time instant t_N the voltage U_i has the qualitative value 0/D*
     and the voltage U_1 has the qualitative value X/_,
     then in the neighbouring time interval $t_N ..t_{N+1}$
     the value of U_o is D/0 and U_1 remains X/_.

(The symbol "_" represents any value.)

"When the input voltage is in the neighbourhood of 0, the operational amplifier transforms the input voltage to a constant output value (of the same polarity as the trend of the input voltage). The condenser voltage is changing in a continuous way."

Now, we are ready to reason about behaviour of the circuit. Let us start from the initial state denoted by t_0, characterised by the evaluation:

(10) t_0: $U_o = +/0$, $U_1 = -/+$.

"the voltage U_0 on the output of the amplifier is positive (and constant), while the condenser voltage U_1 is negative and it grows".

The system QUASIMODO [17] generates a graph of all future states the considered system can reach. The table Tab. 3. summarises the values of all relevant variables at all relevant states.

state	time	U_i	U_o	U_1	U_2	U_3	U_4	I_1	I_2
0	$t_0..t_1$	+ / -	+ / 0	- / +	+ / -	+ / 0	+ / 0	+ / -	+ / 0
1	t_1	+ / -	+ / 0	0 / +	+ / -	+ / 0	+ / 0	+ / -	+ / 0
2	$t_1..t_2$	+ / -	+ / 0	+ / +	+ / -	+ / 0	+ / 0	+ / -	+ / 0
5	t_2	0 / -	+ / 0	+ / +	+ / -	+ / 0	+ / 0	+ / -	+ / 0
6	$t_2..t_3$	- / +	- / 0	+ / -	- / +	- / 0	- / 0	- / +	- / 0
7	t_3	- / +	- / 0	0 / -	- / +	- / 0	- / 0	- / +	- / 0
8	$t_1..t_2$	- / +	- / 0	- / -	- / +	- / 0	- / 0	- / +	- / 0
11	t_4	0 / +	- / 0	- / -	- / +	- / 0	- / 0	- / +	- / 0

Tab. 3.

The relation of neighbourhood between these states is depicted at Fig. 4. - the double-line edges connecting the states 5-6 and 11-0 correspond to discontinuous change. One interpretation of this graph is straightforward - it describes an infinite loop of discontinuous changes of U_0 between fixed positive and fixed negative value. (The states denoted by 3,4, 9 and 10 are spurious ones [23].)

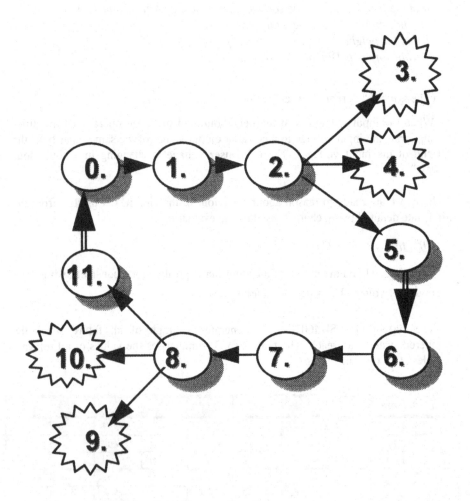

Fig. 4.

The qualitative model of the astable trigger carries a lot of information about behaviour of the circuit. We shall show, that there can be based on this model a "discussion" with the human user, which can test (and improve) his understanding of the system's function. Qualitative reasoning can play a similar role in the explanation modules of KB systems. Let us suggest a fictive dialogue between the student and a MSP system ("systems for training of Mental Simulation Processes"). The basic goal of the intercourse between the human user and the computer is to lead the user towards correct specification of the rules governing behaviour of the system:

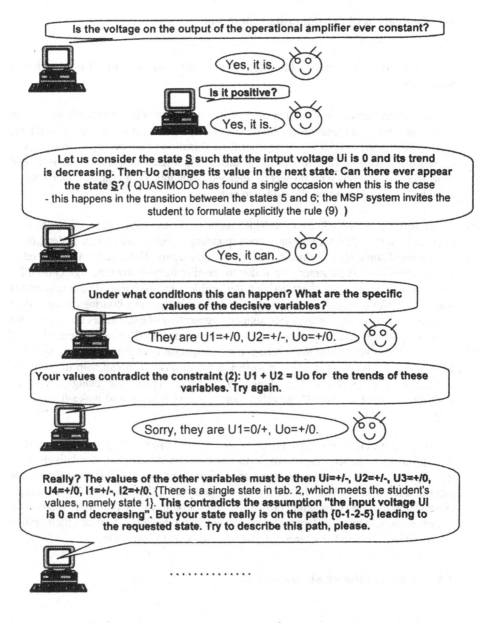

We have just seen that using qualitative reasoning/simulation there can be animated a lively discussion on the considered subject.

4.3 Evaluation of the described methods

What are the main advantages and draw-backs of the qualitative simulation?

It has been proven in a number of experiments that qualitative systems are able to solve problems with vague descriptions. They often offer several variants of solutions, each of them corresponding to a significant combination of unspecified parameters of the modelled system. This possibility to identify significant values of parameters and their decisive relations with respect to intended behaviour is of crucial importance, it provides a "global view" on the modelled task.

A serious draw-back of QSIM-like methods is too rich space of envisioned (predicted) states. Besides the states corresponding to behaviour of real system, there are predicted some states, which have no real counterpart. These states are referred to as spurious states. Their generation is due to the simplicity of the language chosen for expression of constraints characterising the problem environment. These constraints restrict the relations between the qualitative values in the neighbouring or equivalent states only. This limited information supports generation of extra states but it does not allow to give reasons why these states should be doubt. Obviously, generation of spurious states undermines the potential user's intention to utilise qualitative systems. That is why lot of effort is devoted to search for means which can help to eliminate them (e.g. [4], [Kuipers 93], [Fouch 92], [24]). There are being developed sophisticated methods which can refine qualitative simulation and make it closer to real-life applications.

In practical problems, there are often encountered problems having only a part of their problem statement vague, while the rest of the description is relatively well specified. All over it, the upper mentioned algorithms are not able to utilise more precise knowledge - this deficiency in integration is their serious disadvantage. Observation of human reasoning proves that the strength and usefulness of abstraction can excel namely when different abstraction levels can be considered and their results combined. This is another interesting and active research topic in QS.

4.4 Fuzzy qualitative simulation

Very promising idea how to integrate fuzzy and qualitative reasoning is presented in [19]. The resulting system eliminates (at least partially) the upper mentioned problems of qualitative reasoning. Qualitative values of considered attributes are expressed there using fuzzy numbers with trapezoidal membership function. The same features are observed in their system as in QSIM, namely value and trend (first order time derivation). There is given a finite list of possible values for each qualitative variable. This is close to "distinguished values" of systems from QSIM family. In this special case the possible value is expressed in the considered fuzzy language - each fuzzy number is represented by a quadruple of real numbers pointing to the vertices of the trapezoidal membership function.

The modelled system's behaviour is characterised by a set of constraints to be meet in every state. Again, these constraints substitute algebraic or differential equations used in classical approaches to physics or system's behaviour prediction. In the following paragraph, we are going to explain how constraints are handled in the fuzzy domain.

If trapezoidal membership functions are considered, there are well known the formulas characterising behaviour of fuzzy numbers when different arithmetical operations are applied (addition, multiplication and their reverse operations). They ensure that the result of an arithmetical term has a well defined value in the fuzzy domain - it is a fuzzy number with trapezoidal membership function again. The originality of the described approach [19] to qualitative reasoning lays in the following step. The result of an arithmetical operation is not just the fuzzy number resulting from the mechanical application of formulas for arithmetical operations but one of the fuzzy numbers identified among the possible values of the considered variable. More precisely, it is the possible value closest to the fuzzy value obtained as an evaluation of the considered arithmetical term. The distance between the fuzzy numbers is specified by a chosen metrics. This approach maintains the qualitative nature of reasoning with small (finite) amount of values, what decreases the complexity of the treated problem. Besides the simple classical arithmetical operations, the user can define his/her own functions. This is done through fuzzy rules or through tabulation of the function's values on the given list of fuzzy numbers. In this way, some non-linear bindings can be described naturally. This approach can be considered as an extension of the QSIM set of constraints, namely the relations mplus, mminus, etc. These constraints are used for elimination/filtering out some states characterised by evaluations obtained as a result of application of the transition rules for variable development.

The transition rules for variable development follow the philosophy of QSIM - they constrain the qualitative value of a variable in the next future state w.r.t. the actual qualitative value under assumption of continuous development. Another difference of Shen's system is due to more precise treatment of time. This is the result of utilisation of fuzzy numbers for characterisation of the variable's trend. It can be taken for an estimate of the value of first-order time-derivation. This estimate is used for finding another estimate, namely that of persistence time of each time interval. This is obtained from the quotient between the possible fuzzy values of the variable at both ends of the interval and the variable's trend given by the fuzzy number. Because both values are fuzzy numbers, the authors suggest a procedure how to interpret this quotient as an interval. In a similar way, there is introduced time necessary for the change of state (arrival time).

This novel understanding of time makes it possible to extend the filtering methods for elimination of spurious states:

Besides constraint filtering used in classical QSIM there can be expressed and exercised control demanding similar time scale for changes of values of considered variables - temporal filtering.

The Shen's approach utilises further methods of global filtering, too. The global filtering is based on some heuristic information or it reflects other special knowledge about the considered system. One of the heuristics applied is non-intersection constraint [14], [21].

The described method overcomes some draw-backs of classical QSIM by providing new means for elimination of spurious states. In contrast to QSIM it offers approximate information about the time needed for the observed changes of the system as well as about the actual values of individual considered variables.

5 Conclusions

In the Chapter 3., we have brought up some arguments pointing to the fact that a number of problems related to CIM are of qualitative nature. Within last 10 years, there has been developed a number of algorithmic methods which allow to treat qualitative information formally. Further research in this domain is rather active. It can be hoped, that the new results will allow to scale up the application domain of qualitative reasoning to real-life problems encountered within CIM tasks.

References

1. Bratko I., Mozetič I., Lavrač N., KARDIO: a Study in Deep and Qualitative Knowledge for Expert Systems. MIT Press. 1989.

2. Bratko I., Qualitative Modelling: Learning and Control, Conf. Proc. Artificial Intelligence Applications, Prague, June 1991.

3. Bylander T., A Critique of Qualitative Simulation from a Consolidation Viewpoint, IEEE Transactions on Systems, Man, and Cybernetics, Vol. 18, No. 2, March/April 1988, pp. 252-263.

4. Cem Say, A. C., Kuru, S. Improved Filtering for QSIM Algorithm, IEEE Transactions on Pattern Analysis and Machine Intelligence, Vol. 15, No. 9. September 1993.

5. de Kleer,J. and Brown,J.S. A Qualitative Physics Based on Confluence's, Artificial Intelligence 24 (1984), 7-83

6. de Kleer, J., Brian, C., W., Diagnosing Multiple Faults, Artificial Intelligence 32 (1987), 97-130.

7. de Kleer,J. and Weld, D.,S., Readings in Qualitative Reasoning about Physical Systems, Morgan Kaufmann Publishers, Inc., Sant Matheho 1990.

8. Feulloy, L., Qualitative Control and Fuzzy Control: Towards a Writing Methodology, AICOM Vol. 6. Nrs. 3/4 Sept/Dec. 1993.

9. Fouché P., Kuipers B. J., Reasoning About Energy in Qualitative Simulation, IEEE Transactions on Systems, Man, and Cybernetics, Vol. 22, No. 1, January/February 1992.

10. Klein,G.A. and Calderwood,R. Decision Models: Some Lessons From the Field, IEEE Transactions on Systems, Man, and Cybernetics, vol. 21, no.5, September/October 1991, 1018-1026

11. Kuipers,B. Qualitative Simulation, Artificial Intelligence 29 (1986), 289-338

12. Kuipers B. J., Chiu, Ch., Molle, D. T. D., Thtoop D. R., Higher-order derivative constraints in qualitative simulation, Artificial Intelligence 51 (1991) pp. 343-379, Elsevier Science Publishers B.V.

13. Kuipers B. J., Qualitative Reasoning: Modelling and Simulation with Incomplete Knowledge, Automatica, Vol. 25. No 4, pp. 571-585.

14. Lee, W. W., Kuipers B. J., Non-intersection of trajectories in qualitative phase space: A global constraint for qualitative simulation, Proc. Seventh Nat. Conf. Artificial Intell., pp. 286-290, 1988.

15. Leitch, R., Stefanini, A., Task depend tools for intelligent automation, Artificial Intelligence for Engineering, 1989, Vol. 4. No. 3. pp 126-143.

16. Leitch, R., Stefanini, A., On extending the quantity space in qualitative reasoning, Artificial Intelligence for Engineering, 1992, No. 7. pp 167-173.

17. Marvan,I. and Štěpánková,O. QUASIMODO - System for Qualitative Simulation and Modelling, Research Report TR-PRG-IEDS-7/93 FAW Linz

18. Reiter, R., A Theory of Diagnosis from First Principles, Artificial Intelligence 32 (1987) 57-95.

19. Shen, Q., Leitch. R., Fuzzy Qualitative Simulation, IEEE Transactions on Systems, Man, and Cybernetics, Vol. 23, No. 4, July/August 1993.

20. Simmons, R.G. The roles of association and causal reasoning in problem solving, Artificial Intelligence 53 (1992), 159-207

21. Struss P., Global filters for qualitative behaviours, Proc. Seventh Nat. Conf. Artificial Intell., pp. 275-279, 1988.

22. Štěpánková, O., An Introduction to Qualitative Reasoning, Advanced Topics in Artificial Intelligence (Mařík, Štěpánková, Trappl eds.) Springer Verlag, 1992, LN in AI 917, 404-418.

23. Štěpánková, O., Marvan, I., Qualitative Simulation in knowledge Systems, Proceedings of the International Conference on Computer Aided Engineering Education (CAEE'93), Bucharest, Romania, September 22-24, 1993.

24. Štěpánková, O., Marvan, I., Time Independent Global Constraints and QSIM, Proceedings of the Conference on Modelling an Simulation 1994 (ESM'94) eds. Guasch and Huber, Barcelona, June 1-3, 1994.

25. Weld, D. S., Exaggeration, Artificial Intelligence 43 (1990) pp. 311-368, Elsevier Science Publishers B.V. (North Holland)

26. Yoshida,H., Izumi,H., Harita,R.: An Expert System Based on Qualitative Knowledge, Fujitsu Sei. Tech. Journal, 24, 2, 1988.

Comparison and analysis of selected CIM-laboratory concepts and their importance for the improvement of staff training

Peter Starzacher, Gerald Quirchmayr, Roland Wagner

Research Institute for Applied Knowledge Processing (FAW).
University of Linz, Austria

Abstract. Changing market influences force more and more companies to use new promising strategies such as Computer Integrated Manufacturing. In this connection the integration of technology, organisation and personnel comes in the centre of interest. The increasing importance of personnel in the field of CIM leads to an intensified attention concerning the training of employees, students and pupils.
In the first part of this article the importance of CIM and the need for new qualifications is addressed. Subsequently a systematization of CIM-laboratories from a technical point of view is proposed. The contribution closes with a description of selected CIM-laboratories and final conclusions.

Key Words: CIM, qualification, CIM-laboratories, simulation, model factories

1 Importance of CIM-training for the companies

1.1 Changing influences on production companies

The transformation of the market from a selling market into a buyers' market leads to major changes in a substantial number of companies. These changes can be devided into the following two groups:

- external changes = changes of the market
- internal changes = in-house (internal) changes

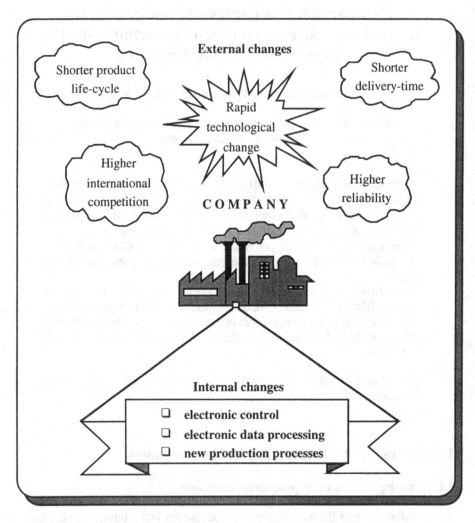

Fig. 1. Changing influences on the production companies

As a result of the higher international competition and the rapid technological change the product life cycle is getting shorter and shorter. Customers also demand a shortening of delivery time and higher reliability.

On the other hand, the most significant changes inside the company are the intensified use of electronic control mechanisms for machine tools or whole production systems, the increasing use of electronic data processing, the expansion of known production processes and the development of new ones.

For the reasons of these heterogeneous influences the companies are forced to produce goods of a high technical standard in small quantities and in numerous variants. One strategy to meet these requirements is to implement a CIM-system in the company. CIM-systems support the following goals:

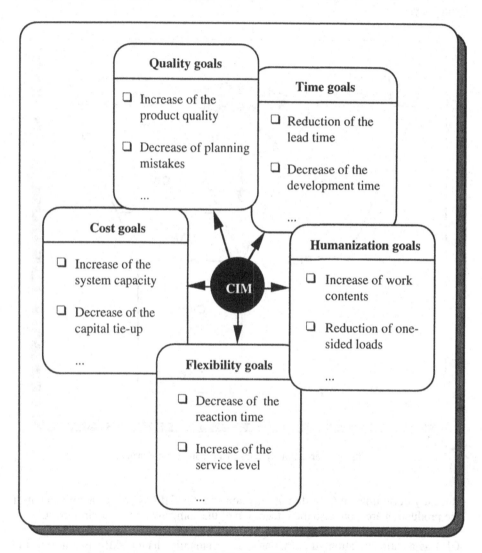

Fig. 2. Goals of CIM

If it is possible to reach so many goals with one strategy, the question, what CIM really means, arises almost automatically.

1.2 The focus of CIM

August-Wilhelm Scheer [4] characterizes Computer Integrated Manufacturing as the integrated information processing for operational and technical tasks of a production company.

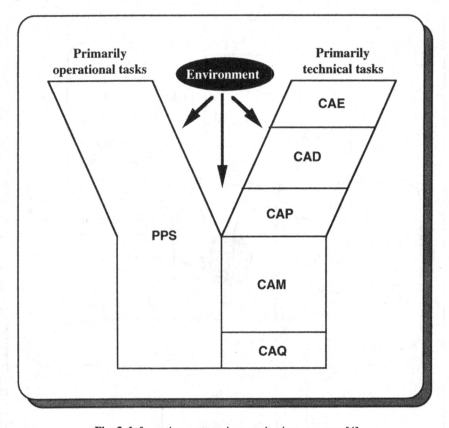

Fig. 3. Information systems in a production company [4]

Recently definitions of CIM do not only consider the CIM-system as a component of the production area, but also the integration of the company into its environment.

CIM as a rationalization strategy has seen a semantic change. CIM is not only the computer aided integration of different operational function areas, as it is often done in companies, but is often equated with a change of structure of the organisation. Therefore planning a CIM-system does not only mean to solve technical problems, but also the integration of technology, organisation and personnel.

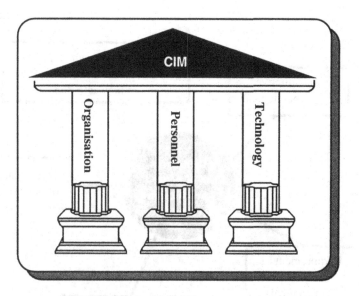

Fig. 4. Triad technology-organisation-personnel

Unfortunately, companies do hardly pay attention to this integral co-operation between technical, organisational, and personal aspects. Technically oriented rationalization perspectives and not the staff members are in the centre of the planning. The secret of successful and integral planning especially lies in the co-ordination of technology and personnel.

A new production technology in particular can have the following effects on the personnel and organisation.

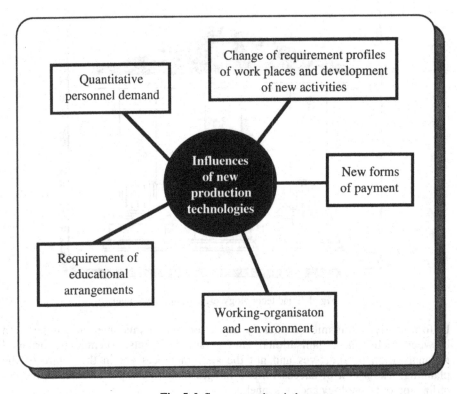

Fig. 5. Influences on the triad

Whether it is possible to plan and realize the triad technology, organisation, and personnel in an integrated way depends on the qualification of all participants. The qualification of staff members does not only gain in significance because of the new production technologies, but also by the use of new forms of organisation in the company, which are required by the new technology.

The centralized and taylorized organisation with a consequent division of labour is going to be replaced by decentralized and qualified production work as a consequence of the changing market requirements and the new technologies. Qualified production work is characterized by a small functional and professional division of labour. A production island represents a typical concept of a qualified work structure. Decentralized solutions in contrast to technically oriented concepts take the limitations, respectively the limited efficiency, of technical solutions into consideration. Only technical solutions in connection with qualified staff members will lead to the desired effects such as an increase in productivity and flexibility.

1.3 The synchronization of CIM-planning/realization and training

Because qualified production work demands the involvement of the employees, it is very important to focus on motivation and acceptance of the CIM-implementation. Timely information and training of all participants do increase the chances of success.

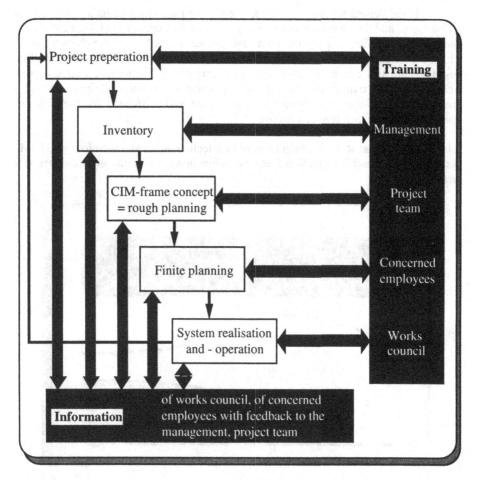

Fig. 6. Planning and realization steps

We don't want to go into the single planning steps in detail, but emphasize the fact that permanent education and training of employees must start as early as the first planning step. Due to a timely participation of staff members in the CIM-planning process it is possible to increase their motivation and acceptance and to benefit from their experience and theoretical knowledge to organise the system effectively and to make the system's installation easier [2].

1.4 The need for new qualifications

After a survey about the demand of the increasing employee qualification the question arises which qualification requirements are relevant when a company decides to implement a CIM-system.

As mentioned before, the use of more and more complex information and production technologies and the installation of new organisational structures lead to an increasing demand for qualified employees. To a high degree this is true of the characteristics which enable staff members to independent and technical acting and to an intensified co-operation and communication between employees. Expensive and complex technical systems require fast diagnosis competence in order to find the reason for a breakdown, to eliminate it quickly and to ensure a departmental overlapping co-operation between different employees.

It is remarkable that at the establishment of new technologies, and therefore also CIM-systems, the demand for functional and extra-functional qualifications are dependent on time [1].

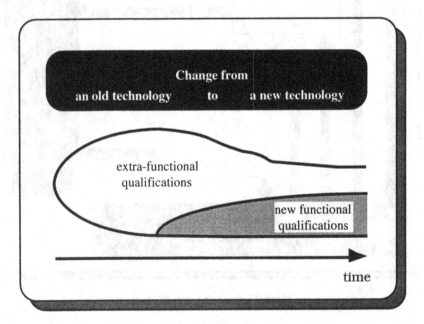

Fig. 7. Qualification demands

Fig. 7. shows that the importance of functional qualification increases with the use of the new technology. However, extra-functional qualifications or key qualifications are more important during the planning and installation phase, because employees must be prepared for new requirements such as independent and technical acting. They enable employees to really use their specialized knowledge.

Therefore these qualifications are especially important for a CIM-implementation. They are responsible for employee's acquisition of that acting competence, which is closely connected to the expansion of the decision and acting margin. As a consequence the personnel must obtain a general view of the existing production connection and of the new reorganisation by training arrangements.

A complete acting competence, which is the basic requirement for the professional acting ability, consists of a specialized, methodological, social and learning competence [1].

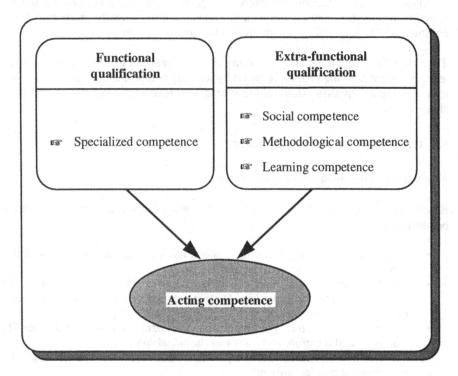

Fig. 8. Acting competence

The specialized competence can primarily be assigned to the functional qualifications and the social, methodological and learning competence to the extra-functional qualifications.

Qualifications about profession, equipment and data processing are integrated in the specialized competence. Therefore these qualifications enable employees to handle machines and equipments in an efficient and secure way. The teaching of knowledge, which is not limited to tasks of one department, gains in significance. Here the close connection to other competencies like the methodological competence can also be seen.

The ability to think in an abstract, systematical and logical way, as well as the planning, decision and problem solving talent are integrated in the methodological competence. A well trained methodological competence is especially required for the systematical search, identification and elimination of faults.

Social competence is the ability to notice ones necessity to work in a co-operative way, to also articulate complex facts and so on. Therefore it is the basis of a departmentally internal and departmentally overlapping co-operation in a company. This departmentally overlapping co-operation grows on the basis of the increasing complexity of the production and information technology. Social or personal abilities like the co-operation and communication talents are not directly teachable. These abilities must be acquired by sensibilization in a group learning process.

The ability to get qualified and the willingness to understand the professional activity as a permanent learning process is meant by learning competence. On the basis of the rapid changing qualification requirements the learning competence gains in significance.

1.5 Training actions and consequences

New demands for the qualification actions

Based on these facts, listed before, the following demands for qualification actions can be identified:

- The structuring of the learning methods has to occur in such a way that it stimulates the motivation of the learner.
- The learning contents must not only comprise the understanding of the functions but also the procedure. By that means the ability to think in a more flexible way, to analyse and to solve problems can be improved.
- The use of specific forms of organisation for learning purposes shall intensify teamwork and therefore also increases the social competence.
- The working material should help to teach both specialized knowledge and strategies for problem solving.

Key requirement for the correct and specific structure of qualification actions is a thorough knowledge of qualification demands, which can be identified centralized on the basis of organisational data or decentralized on different internal planning positions.

Formation of training groups

In order to be able to offer employees an overall qualification, they should be comprehended into homogeneous training groups. In this manner both an individual adaptation of the learning depth respectively of the training unit to the group and the training at the right time will be enabled.

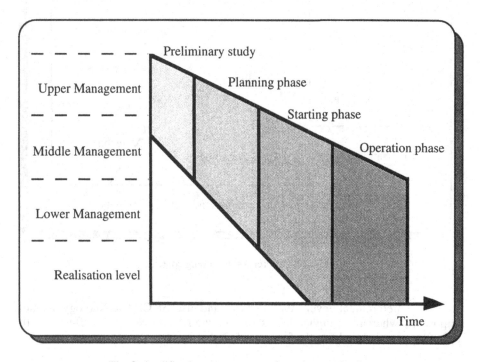

Fig. 9. Qualification arrangements dependent on the time

As it can be seen in Fig. 9 the qualification of the personnel should occur from top to bottom. During the first stages, that means in the preliminary study and in the planning phase, the upper hierarchical levels should be trained primarily, whereas during the starting and operating phases especially the lower hierarchical levels should be educated.

However, there are not only differences in the timing of the training demands of the single training groups but also in the training and learning contents.

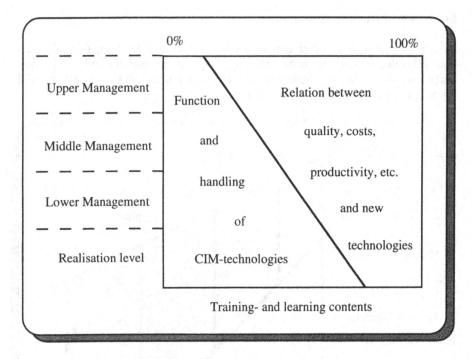

Fig. 10. Differences in training contents

At lower hierarchical levels the function and use of CIM-technology is most important, whereas at higher levels the individual and economic effects of the technologies are placed into the foreground [3].

Extensive training program

A training program can be called extensive if it offers professional knowledge, the basis of data processing, system-specialized knowledge, departmental overlapping knowledge and social qualifications.

A modular structure of the single training units seems to be optimal for the realization of such a comprehensive training program and the individual adaptation to the different training groups. Thus the timely and technical co-ordination of training priorities on single learning and qualification goals can be granted.

For a well designed and balanced training program it is important that there also exist the possibility to gain practical experience with the new technology besides theoretical training. Neither an exclusive practical training (corresponds to learning by doing) nor an exclusive theoretical training can be seen as adequate for an extensive training [1].

Motivation for and during the training

The motivation of employees for and during the training is a further requirement for the successful realization of a training program [1].

☞ Motivation during training and continued education
It is for instance possible to increase the motivation of learners or to avoid a passive behaviour of participants by the use of specific learning instruments. Enclosed audio-visual supported lecture methods like videos, transparencies and especially simulation programs and CIM laboratories are particularly appropriate.

☞ Motivation for training and continued education
Because employee training only makes sense, if employees are prepared for it, it is not enough to motivate staff members with a well-organised course structure. Rather it is important to make the way to a higher learning competence palatable with a financial stimulus, with an exemption for training, and so on.

Training environment

First of all the room where the training takes place should correspond to the above mentioned requirements. Second, it is profitable for a company that it falls back on both external and internal training actions, especially if they want to use a CIM-laboratory for training purposes. By that means it is possible to avoid high costs of an exclusively external training and the danger of a "company blindness" by using an exclusively internal training and to benefit from the advantages of internal and external training (like an exchange of experiences, the use of different learning facilities, training on the job, the consideration of specific necessities,...).

2. Systematization of CIM-laboratories from a technical point of view

Because organisational and technical innovations need highly qualified employees and because there are only few people on the labour market who correspond to these requirements, companies have to care for an individual internal training or they must claim the assistance of external training centres. On the other hand such training centres like universities and technical colleges should try to decrease this lack of qualified people on the labour market.

Traditional professional training and internal training contain a premature specialization, a poor consideration of tactical activities and a concentration on manual work like the item machining. Furthermore, such trainings do not emphasize enough methodical thinking and planning. As a consequence of these wrong training priorities and because of starting the training too late, employees are overcharged. This fact in connection with a modest problem solving ability and a limited social competence of staff members can lead to serious problems for a company. The desired increase in productivity would for instance not be possible due to reduced capacity and higher idle times [1].

On the other hand the education at universities is concentrating exclusively on teaching theoretical knowledge by frontal instruction. Therefore the collection of practical experience and the transformation of theoretical knowledge is not possible for students.

In order to overcome these deficiencies the simulation on a physical model has proven to be a useful learning material to decrease training and education deficits. In this way action and production oriented transformation of technical education is possible. And by that approach a systematic instruction of key qualifications could be realized.

Simulation models, which can be used to reproduce CIM-Systems, can be subdivided into two categories [5]:

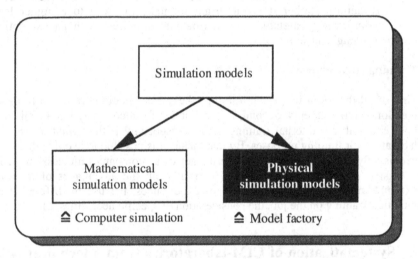

Fig. 11. Systematization of simulation models

☞ Mathematical simulation models = computer simulation programs

These are programs which mainly represent parts respectively special machines of a CIM-factory and which run off interactively with the user of the program. It is possible to simulate different operating runs by changing the parameters.

Up to now computer simulations belong to the conventional concepts of a technological information and communication training, which can be designated as an action oriented instruction. Therefore computer simulations make the instruction of technical facts feasible, but demand an abstract intellectual power and do because of their model character not provide a production oriented training. This is the reason why they cannot go beyond the level of the symbolic representation. Accordingly, giving the feeling of a real production as a product and production oriented activity, which also stimulates the practical knowledge, is not possible.

☞ Physical simulation models = model factories

They are characterized by an extensive objective correspondence. In order to limit the modelling effort, simplifications during the development of the model are a prerequisite. Like computer simulations, this kind of model is often used for planning and checking systems if the direct examination of the real System is too expensive or too dangerous for training purposes. Recently more and more miniature or conventional industrial machines and robots, which are integrated by information and communication technologies, are used for these CIM-teaching purposes.

Model factories can be divided into two categories:

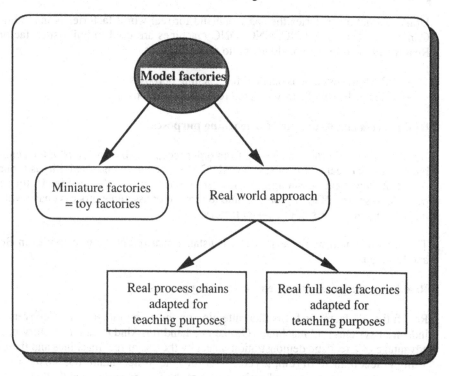

Fig. 12. Systematization of model factories

2.1 Miniature (toy) factories

Miniature or toy factories are CIM-factories on a small scale. These factories are made of a construction set from Festo, Lego or similar firms. On the basis of the use of miniature machines the limited space consumption is one of the fundamental characteristics. This limited space consumption does also result in an excellent clearness over the system and represents for that reason the essential difference to the real world systems. Another characteristic of these factories is that they focus on the integration of as many CIM-components as possible. On the other hand employees often find it difficult to accept these miniature factories because they view them as a toy.

This kind of model factory is for instance realized at the universities in Lyngby, Ulm and Saarbrücken.

2.2 Real world approach

Unlike miniature factories the spare demand for real world factories is much higher. Conventional industrial NC, CNC, DNC machines are used to build such factories. Real world factories can be divided into two categories:

- Real process chains adapted for teaching purposes
- Real full scale factory adapted for teaching purposes

Real process chains adapted for teaching purposes

Not the integrated order processing from order reception to dispatch is in the centre of interest, but the realization of single CAD/CAM process chains. Flexibility is probably the most important advantage of these solutions. With such a model factory it is possible to solve different problems and to remove or assemble process chains without influencing other parts of the model factory.

This kind of a real world factory is for instance realized at the universities in Berlin and Erlangen.

Real full scale factory adapted for teaching purposes

Real full scale factories focus their attention on the entire integration of conventional industrial systems and machines. On the one hand this kind of model factory has the advantage of a realistic demonstration object by the use of real machines and therefore a high acceptance of employees. But on the other hand the costs for the implementation, operation and maintenance of the factory are extremely high and as a consequence these factories can hardly be financed.

This kind of a real world factory is for instance realized at the universities in Bochum and Karlsruhe.

3. Description of selected CIM-laboratories

To get a better overview over the systematization we will give an example for each category of a model factory. All these factories have been built in the course of a CIM technology transfer project. The goal of this kind of project was to train small and medium-sized companies in planning and realization of computer integrated production systems by seminars and information arrangements.

First of all we will start with the description of the toy-factory in Saarbrücken.

3.1 Saarbrücken

Two CIM-miniature factories have been realized at the CIM technology transfer centre in Saarbrücken, which are used for demonstration purposes. These toy factories are based on the Y-CIM-model of Scheer. This training centre has especially emphasized the representation of as many CIM-components as possible. Therefore, the entire integration of different electronic data processing systems, which are used in industry, was in the centre of interest. With this model factory the visitors should not only improve their theoretical knowledge, but they should also get an idea of how a CIM oriented company could look like in reality.

In the course of our article we want to give a short description of the button production in Saarbrücken.

Button production

The institute in Saarbrücken has established a computer integrated production cell for a customer individual and fully automated production of pin buttons in co-operation with other firms. The production can be designated as customer individual because the customers are allowed to choose different colours of buttons and an individual imprint. Characteristic for the whole factory is the integrated order processing from order reception to dispatch.

The CIM-factory consist of the following CIM-components:

- Production planning and control and a leitstand
- CAD
- CAM with the following processing stations:
  - store/commissioning
  - assembly
  - lettering
  - quality control (which belongs to CAQ)
  - dispatch
  - conveyor belt
- CAQ

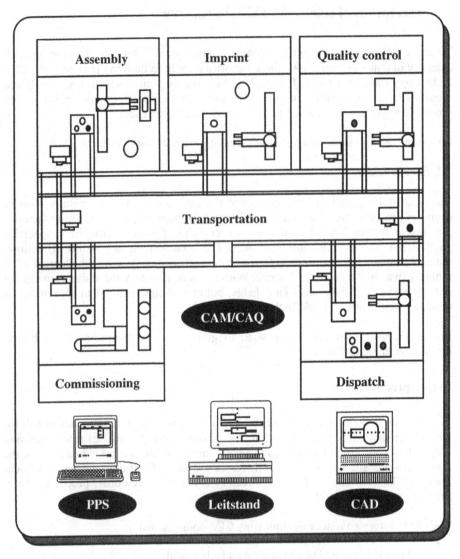

Fig. 13. CIM-components of the toy factory

Throughput of orders

The order of the customer is registered in the production planning and control system by the input of the desired colour and imprint. After the planning of the order it is transferred to the leitstand, which is responsible for the finite scheduling. These finite scheduled operating sequences are transmitted to the process control system after the order's release.

Then the order is assigned to a pallet by the control of the conveyor belt. Afterwards these order data and pallet numbers are transmitted to the first processing station, the commissioning.

If the pallet, which is assigned to the order, is approaching the commissioning station, the pallet number will be decoded, identified as the right one and transferred to the commissioning station by a cross connection.

After the machining in the commissioning cell the pallet is transported to the next processing stations by the conveyor belt. The machining of the button ends with the deposition of the buttons in one of three previously specified final storages in the dispatch station.

The order processing on the process level ends with a completion note or with a production break note to the process control system, which transmits this note to the production planning and control system [8].

The use of the toy factory for training purposes

The miniature factory in Saarbrücken is used both for training and education of staff members and for instruction of students. In the course of the training of employees the model factory serves for demonstration purposes. By means of this factory it is possible to show how a CIM-solution can look like in rough outlines. For the education of students it is also a valuable demonstration object and is used to deliver practical program studies and theses.

3.2 Karlsruhe

The integration of different computer and software systems can be described as the main priority of the technology transfer in Karlsruhe, because the interfaces between different areas of the computer integrated production represent a major problem.

The model factory is subdivided into three fields, which deal with different priority themes. In the process of this article we want to concentrate on the flexible production and assembly system for the production of grabs of industrial robots for the Bosch company. Therefore, the product, which is produced in this factory, is used and sold in the real market.

Configuration of the flexible production and assembly system

Essentially the configuration of the flexible production and assembly system contains of three domains:

- Production:
  The production system is composed of a processing centre for drilling and milling, a turning cell, a pallet washing-machine and a control measuring machine for quality control.

- Transportation:
  The single parts, which are produced in this way, are transported to the assembly line by an automatically controlled ground conveyor.

- Assembly:
  At the assembly line the single parts are commissionized on a work holder by means of an industrial robot. These single parts are assembled to the desired final product on six assembly stations, which are connected by a conveyor belt system. Subsequently the final product is checked whether it meets the quality standards or not.

Because in this laboratory special attention is given to the integration of different CIM-areas we will describe these connections.

The integration of primarily operational planning functions

Production within the flexible production and assembly system is planned and controlled on the basis of a hierarchically structured concept.

The long and medium ranged planning of the production is realized by means of a production planning and control system. This production planning and control system proceeds from resultings of the construction and operations planning, and from specific order data, which are provided by the sales department. On the basis of these releases the production orders are supplied by the production planning and control system and transmitted to the shop floor control by file transfer.

The finite scheduling of orders is realized in the shop floor control. The production orders, which are scheduled in detail now, are passed on to the different work places by the production data capturing system.

This production data capturing system can be seen as the link between control level and production realization or operational level, because both single control data and feedback data are transmitted by the production data capturing system.

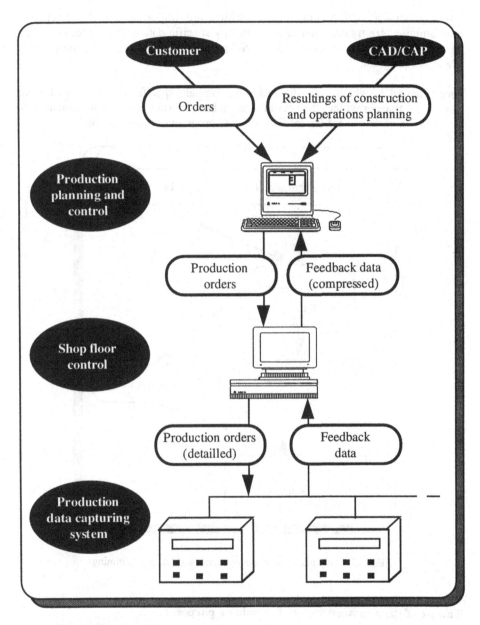

Fig. 14. The hierarchical structured concept of the planning and control system

The integration of primarily technical functions

From a technical point of view the production of robot grabs starts with the construction of a volume model by a 3 dimensions CAD system. Afterwards this data is used by a 2 dimensions CAD system, where the dimensioning is done.

The geometrical and technical data, which are important for the later NC-programming, are transformed into a company specific data exchange format. Then the transformed data is transferred to the NC-programming system, which supplies the whole NC-program.

Furthermore data, which is transformed into a company specific data exchange format, can also be transferred into the neutral data exchange format IGES, which enables the transmission of data to a different CAD- or NC-programming system [6].

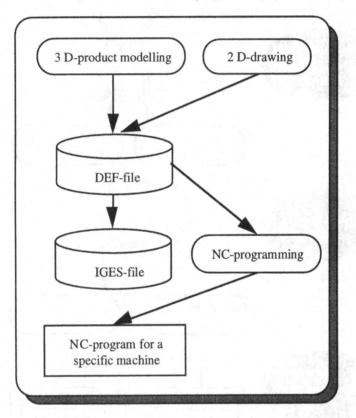

Fig. 15. Integrated product development and NC-programming

The use of the real word factory for training purposes

The CIM-laboratory in Karlsruhe has been established for training and education purposes of employees, engineers and students. All in all this factory is used in quite the same way as the toy factory in Saarbrücken. That means, for the training of employees it serves mainly as a demonstration object of an exemplary CIM-solution with a high level of integration. In the course of the education of students the real world factory is used for laboratory practices and theses.

3.3 Erlangen

Within the scope of a CIM co-operation project a real world factory was built at the university in Erlangen-Nürnberg and is permanently expanded. This laboratory is concentrating primarily on the development of new solutions with industrial robots for small and medium sized companies and is also representing the interface between university and companies.

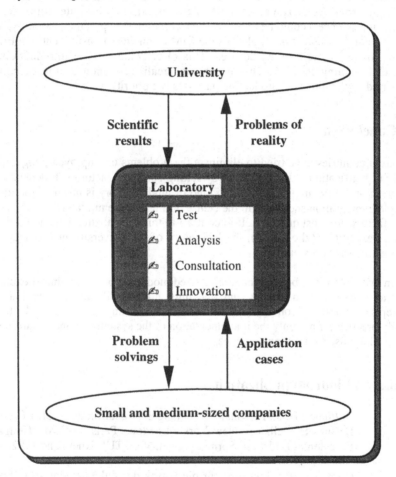

Fig. 16. The laboratory as the interface between university and companies

A number of machine systems for prefabrication and assembly represent the technical basis of the CIM-factory. In order to demonstrate the close relationship of the model factory to practice, conventional industrial machines of medium-sized companies are used. These single partial systems are connected by an automatically controlled ground conveyor.

In the centre of interest is the development of single CAD/CAM process chains and not the entire integration of all machine groups [7].

The use of the real process chains for training purposes

Unlike other universities the university of Erlangen applies its real world factory primarily as a research object for the education of the students and secondarily for the training of employees. The co-operation between different institutes (other technical and natural sciences) could be improved by a common set-up of the laboratory. Within the scope of the training of employees the CIM technology transfer centre in Erlangen offers post-graduate training courses. This offer contains both introductional and extended training in CIM. That is why the realistic demonstration of computer supported applications for production purposes is a priority.

4. Conclusion

Different countries have tried to eliminate the problems of employee-, engineer- and student-qualification in the CIM area by setting up model factories. It depends on the intention of the training centre, which kind of model factory is taken. If the idea of a complete integration should be in the centre of interest, the miniature and the real full scale factory are most qualified. If in contrast to this special attention should be given to the flexibility of the system, the laboratory of the university in Erlangen would probably be the best solution.

But in principle it can be said that the realized model factories are almost exclusively used as demonstration objects and that until now significant pedagogical frame concepts are missing. From our point of view an establishment of such model factories is only practical, if not only the implementation of the system is planned carefully, but also the later use for training purposes.

Books and journal publications

[1]   Bullinger, H.-J.: Personalentwicklung und -qualifikation, CIM-Fachmann, Berlin - Heidelberg - New York - London - Paris - Tokyo - Hongkong - Barcelona - Budapest: Springer, Köln: Verl. TÜV Rheinland 1992.

[2]   Cronjäger, L.: Bausteine für die Fabrik der Zukunft: eine Einführung in die rechnerintegrierte Produktion (CIM), CIM-Fachmann, Berlin - Heidelberg - New York: Springer, Köln: Verl. TÜV Rheinland 1990.

[3]   Scalpone, E.W.: Education Process is vital to realisation of CIM-Benefits, Handling of bit falls, In: Industrial Engineering, Norcross (GA) 16 (1984) 10, p. 110-116.

[4]   Scheer, A.-W.: CIM, der computergesteuerte Industriebetrieb, zweite, durchgesehene Auflage, Berlin - Heidelberg - New York - London - Paris - Tokyo: Springer 1987.

[5]   Wiendahl, H.-P.: Analyse und Neuordnung der Fabrik, CIM-Fachmann, Berlin - Heidelberg - New York: Springer, Köln: Verl. TÜV Rheinland 1991.

Research laboratory publications

[6]   CIM-TT Karlsruhe: Möglichkeiten und Ressourcen am CIM-Technologietransferzentrum Karlsruhe, Informationsbroschüre.

[7]   FAPS: Der Lehrstuhl für Fertigungsautomatisierung und Produktionssystematik stellt sich vor, Informationsbroschüre.

[8]   Institut für Wirtschaftsinformatik an der Universität des Saarlandes (IWI), Informationsbroschüre.

The Use of the Active Database SIMON for a Short Term Production Planning System in the Automotive Industry

A.M.Tjoa [1,2] , R.R.Wagner [1] , Ch.Gierlinger [1] , Ch.Kaiser [1] , C.M.Lechner [1] ,
J.Laimer [1]

1 Research Institute for Applied Knowledge Processing (FAW-Linz)
University of Linz, Austria
rwagner@faw.uni-linz.ac.at

2 Department of Information Engineering
University of Vienna
tjoa@ifs.univie.ac.at

Abstract. The goal of this paper is to describe a short term production planning system based on the behaviour integrated entity relationship approach (BIER). The BIER model is characterized by its ability to support both the static and dynamic aspects of a workflow model. The implementation described in this paper has been used in an application for short term production planning in the automotive sector.
Keywords: active data bases, workflow system, production planning systems, dynamic models

1 Introduction

Although a lot of research is being done in the field of active databases and a few commercial relational DBMS already provide support for some active database capabilities (e.g. Ingres, Interbase, Oracle, and Sybase), there are at present no design tools available that can take full advantage of these new capacities which are very useful for applications in computer integrated manufacturing in general and production planning systems in particular [12].
We propose the modelling of the production planning system based on the behaviour integrated entity relationship approach as introduced by the authors in [3]. In this model the active database behaviour is modelled by extending the entity-relationship with temporal and data features which allow the representation of behaviour by high-level Petri nets.

The active database concept based on this model was used and implemented for operating a short term production planning module of an integrated manufacturing system in the automotive industry.

It should be stressed, however that our system SIMON (System zur integrierten Manipulation mit Objektorientierten Normen - System for Integrated Manipulation with Object-oriented Norms) is generic and not restricted to applications in the automotive sector. The system was created for the automotive industry between 1991 and 1993. Yet, due to its fundamental characteristics, SIMON may also be used for general integrated applications which can be served by an active database.

The ultimate goal of the system is to build a fully object-oriented solution based on an object oriented database. However, the system which was developed at the Research Institute for Applied Knowledge Processing (FAW) uses as underlying database the ORACLE DBMS.

Actual experience with this system has been gathered in the field of short term production planning and control in the automotive industry.

The following section will review and characterize some of the functionality and of the tools used for short term production planning and control. Next, we will define very briefly the behaviour model which is used for the active database approach. The implementation of the active database language and its CASE-tool and graphical user interface will be described in section 4 and 5.

In section 6 the modelling of the short term production planning system based on the behaviour integrated database model will be presented.

Section 7 and 8 will illustrate experiences made by building two prototypes for the same short term production planning system at the research lab at FAW, i.e. one for the ORACLE-Database with an underlying conventional conceptual model (relational model without behaviour) and one for the object-oriented system DAMOKLES, which will be presented along with lessons learned.

The comparison between the three approaches will finally be explored in some detail in section 9.

The whole system, i.e. STPPS and the underlying active database tool BIER-CASE which was constructed especially for STPPS despite its generality), was designed within two and a half years from 1991 to 1993. Four research assistants of FAW were working on this project at full time basis. As a whole (i.e. with considerations about the management and meeting hours) 12 person-years were invested into this project which was finished in August 1993.

2 Short term production planning systems (STPPS)

2.1 Motivation

The requirements for flexibility and production (and assembling) have been steadily increasing over the last years due to the tough competition conditions.

Small/middle enterprises but also economically and organizationally "self-sufficient" departments of huge companies are tending towards the production of small sets, variants and individually customized articles. Despite this requirement for type variety, market conditions call for 'just-in-time' delivery. Short term production planning systems with very flexible properties are therefore a major challenge to modern computer integrated manufacturing systems.

The so called "Leitstand"-Sytem which has been successfully developed in Germany over the last few years is a typical showpiece of short term production planning systems.

The system described in this paper, which was built for one of the largest automotive companies of Germany, is intended to be a further development of the conventional Leitstand system.

To achieve the short and medium term planning goal, an enforcement tool for planning and environment is a necessary prerequisite.

Nevertheless, the decision maker (production planning manager) is not released from her/his tasks - the system "solely" supports him/her in his/her decision making process in an optimum way. It is obvious that this system provides a solid basis for cutting unnecessary administrative personal overheads in the production planning.

The experiences with SIMON based on production planning systems in two different branches (automotive, VLSI-production) meet these expectations.

The graphical representation of the production allows an optimum holistic view of the situation of production and user-friendly interference in short term planning by the manager. Analyses of changes in the field of short term production planning can be delivered in a convincing way. However, these simulation features would go far beyond the scope of this paper.

2.2 The structure and functionality of STPPS

The STPPS is designed as an independent module of a PPS. However, one of the main requirements of STPPS is its autarchy. A STPPS should have the ability of existing/running independently without a "superordinated" PPS. This is especially important for small/medium companies or departments. If such a PPS exists, then the task of managing the common main data and order data is obviously delegated to the PPS.

In practice one problem might still occur in this case, namely that the databases of the STPPS and the PPS might be different/heterogeneous, thus causing consistency problems. In our (FAW's) STPPS this problem is solved in the following way: The common data (which are also used in the PPS or other modules of a computer integrated manufacturing system) are managed by an autonomous module of STPPS which can also be inserted into a PPS. This module (called Edtools) has the function of a "data distributor" in case of a fully integrated manufacturing system and must at least simulate the most primitive functions of production planning system (PPS) in case such a system does not exist.

With regard to order management this module handles the following production order data:

      Part identifier
      Workplan, Working schedule
      Priority (normal order, special order ranking),
      Earliest beginning of production,

Latest beginning of production,
Earliest termination of production,
Latest termination of production,

This module (Edtools) should also be considered as an interface between the production planning system and the STPPS. Production orders are sent from the PPS to this interface. From this moment on the STPPS takes over control and fulfills the following tasks (see Fig. 1):

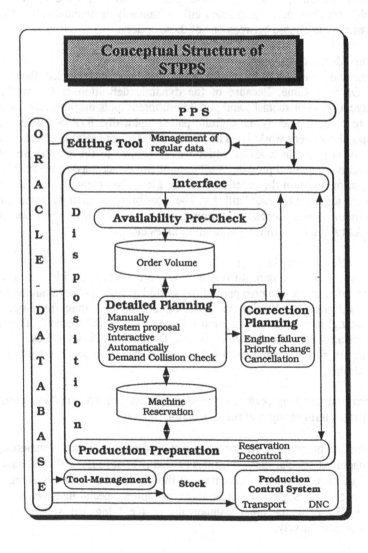

Fig. 1

- Availability pre-check: This task proves successful if all needed materials and tools for a production order are available within the potential production-time-interval (i.e. from the earliest production beginning to the latest production termination). Beyond this proof a plausibility/consistency check of the common main data is performed in this phase.

- Insertion of an order into the Order Volume (Relation): After a positive availability pre-check the order will be inserted into the Order Volume.

- Detailed Planning: The data in the order volume form the basis of this step. Basically this task can be performed either manually or automatically. The user has the possibility to choose between one of these options.

a) manual detailed planning:
   In this case the user (e.g. the production manager) has to select the orders out of the Order Volume. Because of the dynamic definition of the workplans, i.e. different paths of subtasks and possibly different tools/machines for every subtask, the user has to specify the definite path graphically. STPPS delivers a standard proposal for every workplan. The standard system proposal is based on a heuristic method (e.g. in our case one which was developed by the automotive industry). If the heuristic method provides a solution, it can be accepted or rejected by the user. In case of rejection the order is available for further manipulation from the Order Values again. The user still has the possibility to change particular subtask selections of the proposed solution. The system will then carry out plausibility checks on transport times, preparation times etc.

b) automatic detailed planning
   If this option is chosen, all orders of the order volume will be considered. For every order one alternative path of the workplan will be chosen automatically. The path selection criterion is based on a global optimum/semi-optimum (i.e. balancing) of the machine capacities (branch and bound method). The last step of the detailed planning module is the Demand Collision check. This step is necessary for the management and use of special tools (e.g. instruments like pallets etc.)

- Correction planning performs the adaptation of the machine/tools assignment in case of short term changes of the production programme.

- Production preparation constitutes the step where all necessary material, standard tools/machines, special tools/machines are being kept in the different stacks and the control of every subtask is executed by the production control system, under the condition that all needed materials, tools, machines, special tools etc. are confirmed to be available by the other components of the integrated system and by their interfaces, respectively.

The external components (modules) necessary for the STPPS data exchange are:

- the production planning system
- the production control system
- STOCK management
- Tool-Management.

Finally it should be stressed that many qualities of STPPS may only be appreciated by means of a very detailed description of the functionality of STPPS. Yet, this is beyond the scope of this paper whose primary goal is the description of the role of the active industrial database SIMON for STPPS.

Moreover, to give the reader a brief survey of the functionality of STPPS, we will list some features of STPPS in the following:

- Calendar of the plant (holidays etc.)
- Machine defects plus forecasts
- Maintenance duration, periods plus forecasts
- Calculation of the minimum duration between two subtasks of an order.
- For every subtask different influences are considered, e.g. special tools (time of usage of special tools, preparation time for special tools, termination time for special tools, preparation time for reusage of standard tools etc.)
- Necessary continuity of a subtask (e.g. continuous without interrupts should be written within one shift etc.)
- Splitting of subtasks
- Interactive shift of subtasks (on the same machine, on alternative machines) with all its consequences.

3 Synopsis of the BIER Design Method

The BIER-(Behaviour Integrated Entity Relationship) approach was first introduced in [3, 4]. Further developments of this approach are described in [9, 6, 13] and are used in our present approach. The reader who is interested in further details of this approach is referred to the description mentioned above.

BIER is characterized by the following concepts:

A) BIER is based on the entity relationship model with the following extensions to its static properties:

A1) Introduction of the surrogate concept, i.e. every object in BIER is identified by a surrogate which is unique within the whole system. The surrogate-key belongs to its own domain of surrogates. Surrogates are hidden to the users and can only be inserted or deleted but never updated.

A2) Introduction of two different 'time concepts'. The first time concept is introduced as an absolute-time' concept, i.e. a time stamp of the system. The second time concept is introduced as a "state-time' which characterizes the different states of a defined object type. Due to the introduction of the absolute-time every object state is identified by its surrogate and its time stamp. The BIER approach allows the realization of a non- forgetting mode as known in historical databases.

A3) Distinction between 'regular' and 'weak' entities. Entities which are subordinated to other (regular) entities in the sense that they only exist as long as the superordinate entities exist are handled as weak entities.

A4) Generalization concept. The generalization concept is used to introduce roles and states. For every object type the extended concept of the 'timed object' is further refined by specialization within a generalization hierarchy which allows to define the states of an object type. These states object types build the basis for the definition of the dynamic aspects of BIER.

B) The dynamic part of BIER is given by the following concepts which are based on a decomposition approach and Petri-net concepts.

B1) Every object-type is defined by an object-type life cycle. The different states of an object-type are defined by the generalization described above. To allow the transition of pre-states of an object-type to its post-states, so called elementary activities are introduced which enable the firing mechanism. The firing of an elementary activity causes the transition of the surrogates ('tokens') from the pre-states object-types to the post-state object-type.

B2) The whole life cycle of an object-type is described by an elementary process. Every insertion or creation of an entity and all state transitions of the objects which can be modelled prospectively or a-priorily by the designer are the results of some activity which is defined in the dynamic part of BIER, i.e. the tight computing of the dynamic and static part allows us to model all standard state transitions in BIER. An elementary process describes a set of elementary activities belonging to the same object-type. Furthermore, every elementary process possesses a distinguished begin activity and at least one distinguished end activity.

B3) Complex activities. Complex activities are activities where more than one object is involved. Important complex activities are the creation of relationships (R-1-activities), the creation of existence dependent ('weak') entities (E-activities) and the group-by activities (G-activities) which allow a state transition of the superordinate entity after grouping all dependent entities.

C) The integration of the static and dynamic parts of BIER is the most essential concept of this active data model. Every insertion or creation of an entity and all state transitions of the objects which can be modelled prospectively or a-priorily by the designer are the results of some activity which is defined in the dynamic part of BIER, i.e. the tight coupling of the dynamic and static parts allows us to model all standard state transitions in BIER.

4 The CASE-Tool of the STPPS and the active database language.

4.1 The CASE-Tool

The BIER-CASE-Tool consists of the following components which are illustrated in Fig. 2: the graphic editor, the active database language BIEL-SQL, the BIER/C-interface, the scheduler of BIER.

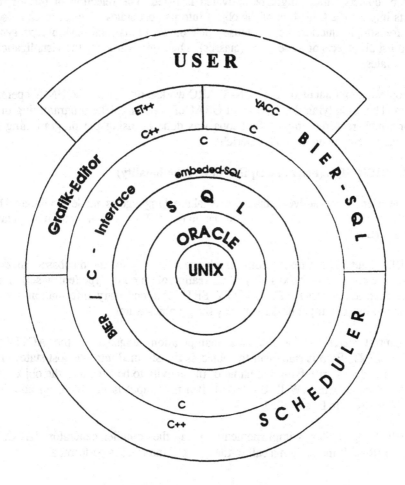

Fig. 2

4.1.1 The graphics editor

The primary aim of the editor is the creation of a BIER diagram. All BIER-diagram symbols are situated on a vertical edge on the left edge in the edge of the screen. Every symbol can be selected by a mouse and placed on the screen. A dialogue window allows the specification of all the information necessary for the creation of the graphically described symbol. The creation of an entity type would force the user to input its name, a short description and the attributes of the entity type.

In this respect the graphics editor functions as a tool for the conceptual (active) database design. The (active) database administrator need not make use of any other design tools except this editor. Another feature of the editor provides the possibility for an authorized user to activate an activity of the BIER-diagram. By a mouse click a list of objects which might be activated is listed. The selection of one of these objects implies the transition of the object into its post-states. The editor can also be used for simple queries, i.e. a mouse click on an entity/relationship type symbol would list all objects of these types (classes). The same is true for the visualization of object-states.

The tool was implemented on a SUN-SPARC workstation under the UNIX operating system. The underlying database is an ORACLE database. The programming of the editor which requires graphic functions was done by using the programming tool ET++ as developed by the ETH Zuerich.

4.1.2 BIER-SQL (brief description of its functionality)

For the user of the active database a database language is made available which usually consists of a data definition language (DDL) and a data manipulation language (DML).

The DDL contains the instructions which are necessary for the database-realization of the different BIER-symbols [10]. The creation of an entity type (entity set) is done by the instruction "CREATE ENTITY SET". All necessary information for this instruction must be input or delivered by the graphics editor.

The central statement for the data manipulation language is the ACTIVATE STATEMENT. By this statement the object is transformed into the post states of the chosen activity by specifying the name of the activity to be fired and the object to be selected for the firing (WHERE-clause). If a user input is provided, an additional INPUT-clause is to be specified.

The BIER-SQL editor is implemented by using the compiler generator YACC. For every BIER-SQL instruction a call to the BIER/C interface is performed.

4.1.3 The BIER/C-interface

The BIER/C-interface consists of a set of routines/procedures. Every routine is assigned to an action which is specified by the user either with the help of the graphics editor or by a BIER-SQL-statement. The creation of a rectangle by the graphics editor implies a call to BIER/C-routine which is responsible for the creation of the entity type.

Each action of a user on the BIER-diagram should imply an action in the underlying database (ORACLE). The BIER/C-interface, which is obviously implemented in C, uses the ORACLE interface for the C-language. This means that within the code of a BIER/C-routine embedded SQL statements are used.

4.1.4 The BIER-schedule

One of the advantages of a dynamic data model is due to the fact that any activity can be fired automatically if the preconditions are fulfilled. The scheduler is an event-driven scheduler. Details of the scheduler are described in [10].

The scheduler is implemented in C++. Data which are needed for the firing of an activity must be extracted from the ORACLE database. This is again performed by the BIER/C-interface. The automatic firing of activities is carried out by the scheduler by a call to the BIER/C-interface. Once an activity is detected which is ready for firing, an "ACTIVATE"-routine is called on.

4.2 The active database language BIER-SQL

4.2.1 Data Definition Language

The first instruction which allows the designer to construct a BIER-diagram is the SET DIAGRAM TO instruction which assigns a diagram name to a BIER-diagram to be created. (SET DIAGRAM TO diagram_name)

After the input of this statement all further specifications of the DDL refer to the diagram which has been assigned. For the creation of the different symbols (parts) of the BIER-diagram different "CREATE"-statements are used. The following subsections (4.2.1.1 - 4.2.2.5) describe the most important statements for the creation of the extended ER-diagram.

4.2.1.1 Creation of an entity type

```
CREATE ENTITY SET        symbol-name
TYPE                     entity-set-type
[DESCRIPTION]            description-string
[                        (list-of-attributes)];
```

symbol-name: Name of the entity-type
entity-set-type: "REGULAR"/"WEAK"
description-string: Short description
list-of-attributes: Attributes of the entity type following the SQL-syntax.

4.2.1.2 Creation of a GROUP

```
CREATE GROUP        symbol-name
[DESCRIPTION        description-string];
```

symbol-name: Name of the state group which is used to structure the different states
of an entity type.

4.2.1.3 Creation of states

```
CREATE STATE        symbol-name
[DESCRIPTION        description-string]
[                   (list-of-attributes)];
```

symbol-name: Name of a state which belongs to a state group
list-of-attributes: Attributes of a particular state (analogue to CREATE ENTITY
SET)

4.2.1.4 Creation of "links"

"Links" between two BIER-symbols are created by specifying the names of the
symbols to be listed in the BIER-diagram.

```
CREATE LINK FROM symbol-name TO symbol-name
[direction]
[ROLE               role-definition]
[DESCRIPTION        description-string]
```

direction:"SINGLE"/"DOUBLE"

("SINGLE" means a single arrow, i.e. the object is annihilated from the pre-state
after the firing of an activity; "DOUBLE" represents a two-directed arc representing
the fact that the object will be annihilated from the pre-state after firing)

4.2.1.5 Creation of relationships

Two different relationships are created in the extended entity relationship diagram,
namely "RELATIONSHIP", representing the usual aggregation of entity-types and
"EXISTENCE" representing the relationship between existence-dependent entity
types (parent-child relationship)

```
CREATE DEPENDENCY        symbol-name
TYPE                     dependency-type
[DESCRIPTION             description-string];
```

dependency-type:"EXISTENCE"/"RELATIONSHIP"

The 'existence dependent relationship' and the 'relationship' itself, respectively, are finally specified by using the CREATE LINK statement.
Example: To define a relationship type C between the entity types A and B the following statements are necessary:

```
CREATE        DEPENDENCY C
TYPE          RELATIONSHIP
CREATE        LINK FROM A TO C
CREATE        LINK FROM B TO C
```

(Note: The links represent the lines in a BIER-diagram whereas the other statements create different kinds of boxes. However, these statements are not used by the end-user working with the SIMON'S BIER graphics user interface)

4.2.1.6 Creation of activities

In the previous subsections (4.2.1.1 - 4.2.1.5) the most important commands for the creation of an extended entity relationship diagram in SIMON, namely the static part of BIER, were described. The CREATE ACTIVITY statement allows the designer to define the dynamic aspects of BIER.

```
CREATE ACTIVITY      activity-name
TYPE                 activity-type
[RELATION TO         symbol-name]
ACTIVATION           activation-definition
[WHEN                condition]
[DESCRIPTION          description-string]
```

```
activity-type:     BEGIN
                   |END
                   |ELEMENTAR
                   |EXISTENCE
                   |GROUP_BY
                   |R-1
activation-definition: "MANUAL"/"AUTOMATIC"
```

The activity-types have the following meanings:

a) BEGIN: this activity represents the creation (insertion) of an entity into an entity-type

b) END: this activity represents the last activity of an object within the object`s life cycle

c) ELEMENTAR: an elementary activity fires an object from its pre-states to its post-states (it can be considered as a Petri net transition)

d) EXISTENCE creates existence dependent objects from a parent-entity

e) GROUP_BY activates an operation on a parent-entity after grouping all existence dependent entities

f) R-1 creates a relationship object as an aggregation of the entities linked by these relationship-types. In general we need the WHERE-clause to specify which relationship should be built

4.2.2 The Data Manipulation

The most important manipulation for the user is the ACTIVATE statement. Note that in an active database environment like SIMON which is based on the BIER-model every manipulation (insert, state-update etc) should be performed by an activity defined in the BIER-diagram. In the routine-case many of the manipulations will be carried out "automatically" by the scheduler. To manipulate data manually, the ACTIVATE-statement is defined as follows:

```
ACTIVATE          activity-name
WHERE             condition
INPUT             (manual-input[,manual-input]...)
```

condition: defines the selection of the object which should be activated (transformation to a post-state)

manual-input: [entity-set-name.] attribute-name=value-
                 |[entity-set-name.] attribute-name=value- list

To support queries in BIER-SQL, a statement which is similar to the SELECT statement in SQL is used:

SHOW columns IN STATE symbol-name WHERE condition

columns: List of attribute-names to be selected, * means all attributes
symbol-name: name of the states
condition: boolean expression

4.2.3 General comments

The EXECUTE-command allows the BIER-SQL user who does not use the graphics user interface to edit DDl statements:

EXECUTE file-name

Every BIER file-name in SIMON has the suffix bier. The EXECUTE command allows to execute a set of BIER-statements to construct a BIER-diagram. Especially in the design phase one can edit a *.bier-file and with EXECUTE a new BIER-diagram will be generated.

The DROP DATA FROM statement allows a user to delete all user-data (i.e. instances) of a BIER-diagram in SIMON. Of course the BIER-diagram is a schema without instance data being preserved.

DROP DATA FROM diagram-name;

The EXIT-command will close a BIER-session

Of course BIER_SQL possesses many other statements which cannot be described within the scope of this paper. The complete grammar of BIER-SQL is given in [10].

5 The graphical user interface of SIMON

In SIMON a graphical user interface is built for the BIER active database model. The interface is programmed in C++ using the ET++ library [7, 5]. The editor is not integrated into the underlying database and thus only constitutes a graphical tool which can be used to design a diagram without correction test. Therefore the linkage to the database is very important. This is performed by the BCI (BIER C-Interface). A symbol which is created by the editor is also created in the database after the OK-button has been 'pressed'. The only exception to this direct creation is the creation of 'actual state entity types' which are not described in this paper.

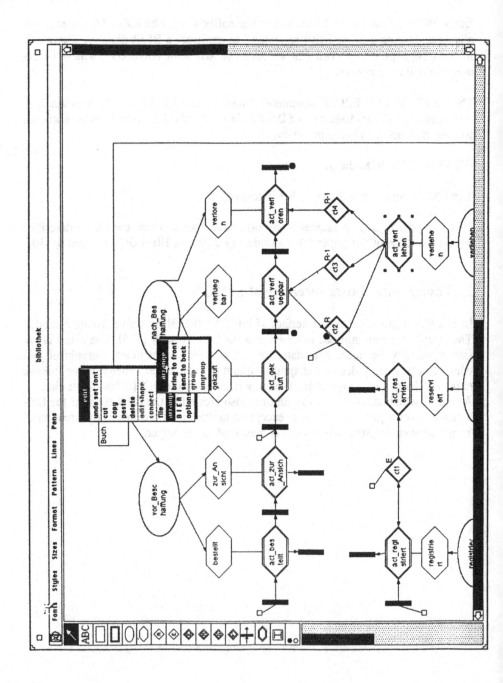

Fig. 3

On the left vertical edge of the frame of the graphical interface beginning with the first rectangle symbol the following BIER-symbols are represented:

(down from the top): Regular entity set (rectangle), weak entity set, group of states, state-entity-type, relationship type (R), existence dependency (E), followed by the activities, creation of existent dependent entities (E), creation of relationships (R), group by activity (G), part-of-relationship (R-1), elementary activity, actual state entity type, input (I), automatic or manual input (erase button).

The different BIER-symbols can be connected by clicking 'connect' in the pop-up menu shown in Fig.X. Connect can only be executed if two BIER symbols are marked in sequel (i.e. click of the left mouse button and simultaneous pressing of the shift button).
In Fig. 4 which appears after the connect activation we can specify the details of the connections. Fig. 5 shows the situation described above.
For more detailed information about the SIMON-user interface, [11] gives a very precise description. The special methods and objects of the graphical user interface with its C++ and ET++ specific features are described in [11].

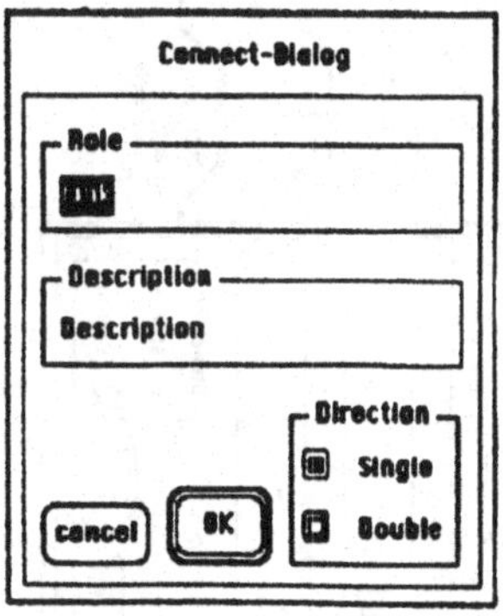

Fig. 4

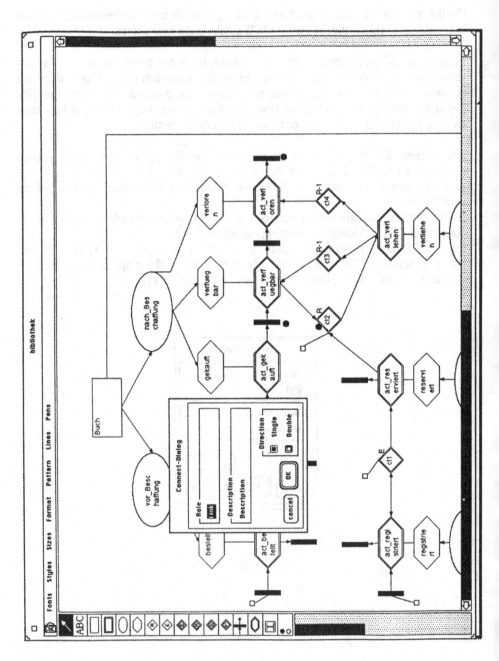

Fig. 5

6 The modelling of STPPS with BIER in the active database system SIMON

In this section we will try to give a description of the modelling of STPPS in SIMON. Of course it is not possible to describe within this paper the whole functionality of STPPS which consists of 27 entity sets .

Therefore we will illustrate this by the example of the 'special tools scheduling' WBEL. In Fig. 6 the part of STPPS which on 'special tools scheduling' is situated within the marked circle. It is only a tiny part of the whole STPPS. However, most of the important features of active data modelling can be shown in this example. [8] gives the BIER-SQL definition of this part.
(Note: The authors decided to present the original diagram and BIER-SQL statements and their comments in German which was used for the automotive application of STPPS)

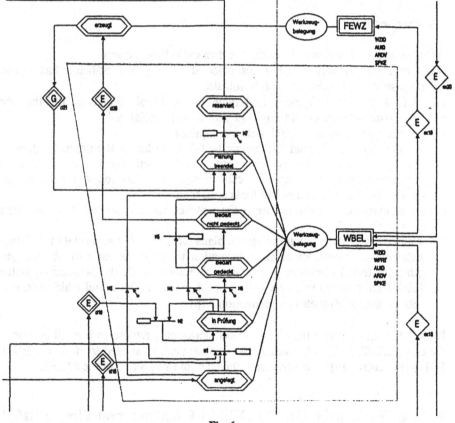

Fig. 6

The entity set Wbel consists of the attributes WZID, WFRT, AUID, ARDV, SPKE, representing:

WZID   special tool identifier
WFRT   corresponding tool
AUID   order identifier
ARDV   status of previous working step
SPKE   identifier of the split

The states marked 1, 2, 3, 4, 5 and 6 have the following meaning:
state #1: start state
state #2: check of the special tool demand performed
state #3: demand can be fulfilled
state #4: demand can not be fulfilled
state #5: planning has been performed
state #6: reservation of the special tool identifier

The activities are:

t41: elementary activity which checks the demand collision check
t45: elementary activity which takes place if the demand collision has proved successful, i.e. the demand can be fulfilled.
t42: elementary activity trying to find another special tool which can fulfil the task. The existent dependent activity ct19 delivers the special tools.
t43: trivial case where no special tools are required
t44: the demand for a special tool cannot be fulfilled (despite the iterations done by t42). [Remark ct20: The complex activity ct20 is activated as a consequence of t44 because state #4 is reached. ct20 creates temporary (existence dependent) entities for the lacking demand for special tools.]
t46: elementary activity which inserts the matched special tool into the short term plan
t47: elementary activity which creates the plans for the use of the special tools. These lists constitute complementary information to the plans for normal tools and other (material) resources. These plans therefore signify the production control [Link from state #6 to ct22 which is not depicted in Fig. 6 and which performs reservations of all tools of a production step]

It should be stressed that the part described above only represents a small portion of the entire STPPS. The interested reader is once more motivated to study the BIER-SQL statements of [8] which represent the DDL of this portion of the STPPS.

7   The STPPS-prototype for ORACLE without modelling in BIER

To allow a comparison of the advantages of the SIMON-based active database approach with the behaviour integrated entity relationship model with `pure'

657

relational modelling and conventional C programming, a prototype and test system was implemented for this purpose. The interested reader is referred to [8] where this prototype and test system is described in great detail. To allow a fair comparison, STPPS was designed as an extended ER-diagram (see Fig. 7 - the part described in section 5 is marked within the circle).

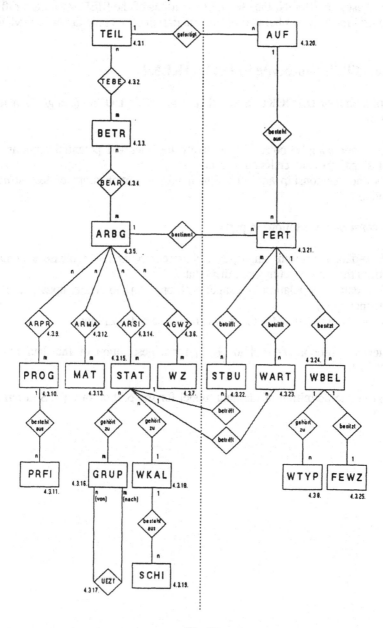

Fig. 7

Twenty-five relations were specified for this prototype and the testbed, respectively, defining the entire data needed for STPPS. In fact, all relations used in this prototype also represent entity types of the BIER model. Of course there is additional information needed for the SIMON-definition of the active database for the existence dependencies and the specification of elementary and complex activities. Due to the surrogates used in SIMON, the ORACLE relations of the BIER-model are different. Another difference occurs because of the integration of the time factor in SIMON.

8 The STPPS prototype in DAMOKLES

The data model of DAMOKLES [1, 2], the so called DODM (design data model) consists of:

- structured and complex objects, respectively, together with potential versions.
- relationships between versioned objects
- objects and relationship attributes which might contain more or less structured information

DODM consists of the following parts:

a) DDL (=data definition language) which represents the database schema and describes the data structures and their states
b) DML (=data manipulation language) with operators to create, update and delete the defined types
c) Consistency constraints for the states and the state-transitions

The interested user is referred to [1, 2] for a description of the data model of DAMOKLES.

Fig. 8 gives the definition of the data model for the portion of STPPS described in section 5.

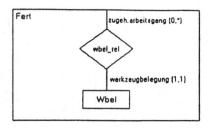

OBJECT TYPE Wbel
    ATTRIBUTES
        wzid   : STRING
        wfrt   : STRING
        auid   : STRING
        ardv   : STRING
        spke   : STRING
        wbbg   : STRING
        wbben   : STRING
END Wbel

RELSHIP TYPE wbel_rel
    RELATES
        werkzeugbelegung   : Wbel
        zugeh. arbeitsgang   : Fert
END wbel_rel

OBJECT TYPE Fert
    ATTRIBUTES
        auid   : STRING
        ardv   : STRING
        spke   : STRING
        fer1   : STRING
        fer2   : STRING
    STRUCTURE IS Wbel, wbel_rel
END Fert

Fig. 8

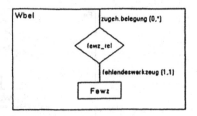

OBJECT TYPE Fewz
    ATTRIBUTES
        wzid   : STRING
        auid   : STRING
        ardv   : STRING
        spke   : STRING
END Fewz

RELSHIP TYPE fewz_rel
    RELATES
        fehlendeswerkzeug : Fewz
        zugeh. belegung   : Wbel
END fewz_rel

OBJECT TYPE Wbel
    ATTRIBUTES
        wzid   : STRING
        wfrt   : STRING
        auid   : STRING
        ardv   : STRING
        spke   : STRING
        wbbg   : STRING
        wbben   : STRING
    STRUCTURE IS Fewz, fewz_rel
END Wbel

Fig. 8 (continued)

9 Comparison of the three approaches

The relational database 'ORACLE' with its purely relational conceptual model, the active database SIMON with its BIER-model and its realization in ORACLE and the DAMOKLES implementation of STPPS for the automotive industry are compared in this section. The following criteria are used for the comparison:

1) The static conceptual data model (based on the entity relationship approach) which is used in all three implementations.

2) The dynamic data model which allows the conceptualization of updates in a generic way.

3) The integration of static and dynamic aspects of the implementation.

4) The non forgetting mode aspect which permits to reconstruct the life cycle of an object (i.e. historical database)

5) The versioning concept which makes it possible to implement different versions of an object life cycle and to switch between different versions (important for the simulation of planning alternatives)

6) Information hiding. Encapsulation of information provides the filtering of relevant information for the user who does not need all the information ballast. This is especially important in computer integrated manufacturing.

7) Query facility. A tailored query system for the conceptual model which integrates static and dynamic elements is very useful

The following table intends to give a survey of the comparison based on the user experiments.

Criteria	purely-relational	SIMON-BIER	DAMOKLES
1)	+	+	+
2)	--	++	--
3)	--	++	--
4)	-	++	--
5)	--	-	++
6)	+	+	++
7)	++	++	++

Legend: --   not suitable
           -       rather suitable
           +   suitable
          ++   very suitable

BIER is the only approach among the three implementations which allows us to define a dynamic schema of an STPPS in a generic way. It is also the only system which delivers a non forgetting mode for objects. DAMOKLES (as a representation of OODBMS) is the system which supplies us with a versioning concept that is very important for the simulation and comparison of different planning alternatives. In ORACLE the versioning concept was not realized. DAMOKLES provides the possibility of encapsulation and information hiding, respectively, through its concept of 'long fields'. For the query system, BIER is slightly more advanced because of the BIER-SQL facilities. In DAMOKLES we can use the 'cursor' concept which is also available in ORACLE.

To draw a resume of the experiment we are convinced that the BIER-approach is most suitable for computer integrated manufacturing purposes because of its integral approach of binding together static and dynamic components. A drawback is its lack of versioning services which should be integrated in SIMON-BIER in future as a result of this comparison. On the other hand, BIER's non forgetting mode allows a retrospective analysis of planning, i.e. statistical analyses of frequently repeated planning alternatives, machine defects etc.

For the future a further experiment with a system based on the object-oriented Behaviour Diagram (OBD) [9] will be compared with the BIER approach. OBD is a variant of BIER based on a purely object-oriented paradigm.

Acknowledgement. This work was supported by the Research Institute for Applied Knowledge Processing and by the ESPRIT-Project KBL of the European Commission (DG XIII).

References

1. K.Abramowicz et al.: DAMOKLES-Database Support for Software Engineering Environments. Technical Report, FZI, University of Karlsruhe, 1987

2. K.Abramowicz et al.: DAMOKLES-Database Mangement System for Design Applications. Reference Manual Release 2.0, March 88, FZI, University of Karlsruhe

3. J.Eder, G.Kappel, A M.Tjoa, R.R.Wagner: BIER - The Behaviour Integrated Entity Relationship Approach. Proc. 5th Int.Conf. on the Entity Relationship Approach, 1986

4. J.Eder,G.Kappel,A M.Tjoa,R.R.Wagner: A Behaviour Design Methodology for Information Systems. Proc. 6th IEEE Int. Conf. on Computers and Communication, 1987

5. E.Gamma: Design and Implementation of ET++. A Scannless Object Oriented Application Framework. In: A. Weinand: Structured Programming, Springer, 1989

6. Ch.Gierlinger, A M. Tjoa, R.R. Wagner: A Methodology for Computer Modelling of Information Systems Based on Extended Behaviour Entity Relationship Model. Proc. EURO-CAST Conf., Lecture Notes in Computer Science, Springer Verlag,1991

7. E.Gamma, A. Weinand, R.Marty: ET** An Object Oriented Application Framework. Internal report University of Zürich, 1988

8. Ch.Kaiser: Implementation of BIER for Production System. Internal Report, University of Linz, 1991

9. G. Kappel, M. Schrefl: Object Behaviour Diagrams. Proceedings of the 7th International Conference on Data Engineering, IEEE Computer Society Press, 1991

10. J. Laimer: Implementation of a database language for the BIER model. Thesis, University of Linz, 1991

11. C.M. Lechner: Implementation of a graphical user interface for the behaviour integrated entity relationship model. Internal Report, The Research Institute for Applied Knowledge Processing (FAW-Linz), University of Linz, 1992

12. S.B.Navathe, A.K. Tanaka, C.Chakravarthy: Active Database Modelling and Design Tools: Issue, Approach, and Architecture. Bulletin of the Technical Committee on Data Engineering, IEEE, Vol.15, 1992

13. T.Urpi, A. Olive: Events and Events rules in Active Databases. Proc. of the 18th VLDB Conf., 1992

Author Index

Springer-Verlag
and the Environment

We at Springer-Verlag firmly believe that an international science publisher has a special obligation to the environment, and our corporate policies consistently reflect this conviction.

We also expect our business partners – paper mills, printers, packaging manufacturers, etc. – to commit themselves to using environmentally friendly materials and production processes.

The paper in this book is made from low- or no-chlorine pulp and is acid free, in conformance with international standards for paper permanency.

Lecture Notes in Computer Science

For information about Vols. 1–903

please contact your bookseller or Springer-Verlag